CLASSICAL NOVAE

Second Edition

Since the first edition of this book was published, knowledge regarding the nova phenomenon has grown significantly. This is due to the advent of new observational facilities, both on the ground and in space, and considerable advances in theoretical work. This second edition has been fully updated and revised and contains new contributions which comprehensively cover the important developments in this field, and reflect on interesting new insights into the outbursts of classical novae.

The book begins with an historical perspective and an overview of nova properties. It then examines in detail thermonuclear processes, the evolution of nova systems, nova atmospheres and winds, abundance studies, the evolution of dust and molecules in novae, nova remnants, and observations of novae in other galaxies. The book details knowledge gained from observations across the electromagnetic spectrum, from radio to gamma rays, and discusses some of the most important outstanding problems in classical nova research.

This is the only book devoted solely to the study of classical novae, and as such is an important reference for researchers actively engaged in the subject and graduate students seeking an introduction. The contributors to this book are internationally recognized experts in their field, and present a balanced mix of observation and theory.

MICHAEL BODE is Director of the Astrophysics Research Institute and Professor of Astrophysics at Liverpool John Moores University. His research interests include the study of novae, interstellar and circumstellar dusts, and the use of robotic telescopes to further 'time domain astrophysics' in general.

ANEURIN EVANS is Professor of Astrophysics at Keele University. His research interests include the study of novae, 'born-again' systems like 'Sakurai's Object', the interstellar medium in globular clusters, and mapping the Galactic plane at sub-mm wavelengths.

Cambridge Astrophysics Series

Series editors

Andrew King, Douglas Lin, Stephen Maran, Jim Pringle and Martin Ward

Titles available in this series

10. Quasar Astronomy
by D. W. Weedman

18. Plasma Loops in the Solar Corona
by R. J. Bray, L. E. Cram, C. Durrant and R. E. Loughhead

19. Beams and Jets in Astrophysics
edited by P. A. Hughes

22. Gamma-ray Astronomy, 2nd Edition
by P. V. Ramana Murthy and A. W. Wolfendale

24. Solar and Stellar Activity Cycles
by Peter R. Wilson

25. 3K: The Cosmic Microwave Background Radiation
by R. B. Partridge

26. X-ray Binaries
edited by Walter H. G. Lewin, Jan van Paradijs and Edward P. J. van den Heuvel

27. RR Lyrae Stars
by Horace A. Smith

28. Cataclysmic Variable Stars
by Brian Warner

29. The Magellanic Clouds
by Bengt E. Westerlund

30. Globular Cluster Systems
by Keith M. Ashman and Stephen E. Zepf

32. Accretion Processes in Star Formation
by Lee Hartmann

33. The Origin and Evolution of Planetary Nebulae
by Sun Kwok

34. Solar and Stellar Magnetic Activity
by Carolus J. Schrijver and Cornelius Zwaan

35. The Galaxies of the Local Group
by Stanley van den Bergh

36. Stellar Rotation
by Jean-Louis Tassoul

37. Extreme Ultraviolet Astronomy
by Martin A. Barstow and Jay B. Holberg

38. Pulsar Astronomy, 3rd Edition
by Andrew G. Lyne and Francis Graham-Smith

39. Compact Stellar X-ray Sources
edited by Walter Lewin and Michiel van der Klis

41. The Physics of the Cosmic Microwave Background
by Pavel D. Naselsky, Dmitry I. Novikov and Igor D. Novikov

42. Molecular Collisions in the Interstellar Medium, 2nd Edition
by David Flower

43. Classical Novae, 2nd Edition
edited by Michael Bode and Aneurin Evans

CLASSICAL NOVAE

SECOND EDITION

Edited by

MICHAEL F. BODE
Astrophysics Research Institute, Liverpool John Moores University

ANEURIN EVANS
Astrophysics Group, Keele University

CAMBRIDGE
UNIVERSITY PRESS

CAMBRIDGE UNIVERSITY PRESS
Cambridge, New York, Melbourne, Madrid, Cape Town, Singapore, São Paulo, Delhi

Cambridge University Press
The Edinburgh Building, Cambridge CB2 8RU, UK

Published in the United States of America by Cambridge University Press, New York

www.cambridge.org
Information on this title: www.cambridge.org/9780521843300

First published 1989, Second Edition 2008

Printed in the United Kingdom at the University Press, Cambridge

A catalogue record for this publication is available from the British Library

ISBN 978-0-521-84330-0 hardback

Contents

List of contributors

M. F. Bode	Astrophysics Research Institute, Liverpool John Moores University, Birkenhead, CH41 1LD, UK
Hilmar W. Duerbeck	Vrije Universiteit Brussel, Pleinlaan 2, B-1050 Brussels, Belgium
A. Evans	Astrophysics Group, Keele University, Keele, Staffordshire, ST5 5BG, UK
Masayuki Y. Fujimoto	Department of Physics, Hokkaido University, Sapporo 060-0810, Japan
Robert D. Gehrz	Astronomy Department, University of Minnesota, 116 Church Street S. E., Minneapolis, MN 55455, USA
Peter H. Hauschildt	Hamburger Sternwarte, Gojenbergsweg 112, 21029 Hamburg, Germany
Margarita Hernanz	Institut de Ciencies de l'Espai (ICE/CSIC) and Institut d'Estudis Espacials de Catalunya (IEEC), Campus UAB, Facultat de Ciències, Torre C5 – parell – 2[a] planta, 08193 Bellaterra, Barcelona, Spain
W. Raphael Hix	Physics Division, Oak Ridge National Laboratory, Oak Ridge, TN 37831-6354, USA
Icko Iben Jr.	University of Illinois, Urbana-Champaign, IL 61801, USA
Christian Iliadis	Department of Physics and Astronomy, University of North Carolina, Chapel Hill, NC 27599-3255, USA
Jordi José	Departament de Física i Enginyeria Nuclear, Universitat Politècnica de Catalunya (EUETIB), C. Comte d'Urgell 187, E-08036 Barcelona, Spain and Institut d'Estudis Espacials de Catalunya (IEEC-UPC), Ed. Nexus-201, C. Gran Capitá 2–4, E-08034 Barcelona, Spain
Joachim Krautter	Landessternwarte Königstuhl, D-69117 Heidelberg, Germany
T. J. O'Brien	Jodrell Bank Centre for Astrophysics, Alan Turing Building, University of Manchester, Oxford Road, Manchester, M13 9PL, UK
J. M. C. Rawlings	Department of Physics and Astronomy, University College, Gower Street, London, WC1E 6BT, UK
E. R. Seaquist	Department of Astronomy and Astrophysics, University of Toronto, Toronto, Ontario M5S 1A7, Canada

Allen W. Shafter	Department of Astronomy, San Diego State University, San Diego, CA 92182, USA
Steven N. Shore	Dipartimento di Fisica 'Enrico Fermi', Università di Pisa and INFN, Sezione di Pisa, Largo B. Pontecorvo 3, I-56127 Pisa, Italy
Sumner Starrfield	School of Earth and Space Exploration, Arizona State University, PO Box 871404, Tempe, AZ 85287-1404, USA
Brian Warner	Department of Astronomy, University of Cape Town, Rondebosch 7700, South Africa

Preface to the first edition

Some years ago we blundered, almost by accident, into the field of classical novae. Our prime interest at the time was in their dust formation properties and infrared development; however, it soon became evident that a full understanding of this relatively restricted aspect of the nova outburst could not be achieved without considering *all* aspects of the nova phenomenon. Fortunately, from our point of view, the 1970s was a decade during which several significant advances were made in the understanding of classical novae on both observational and theoretical fronts. Accordingly we were able to take advantage of these advances as they appeared in the research literature. However, with the exception of occasional published conference proceedings, it was apparent that no text existed that covered all aspects – both theoretical and observational – of the classical nova phenomenon.

This book arose out of a casual conversation with Dr Jim Truran during which we bemoaned the fact that there seemed to be no modern equivalent of *the* classic book on the subject, Cecilia Payne-Gaposchkin's *The Galactic Novae*. It seemed to us that such a volume was long overdue. However, it was clear that, with rapid developments in several aspects of the study of novae, no single author could do justice to all the relevant theoretical and multi-wavelength observational material. It was for this reason that we decided to opt for the multi-author approach that the reader will find in this volume. Our initial hope was to produce an up-to-date replacement for *The Galactic Novae*. We now realize, of course, that such an aim was foolhardy, not to say arrogant: Payne-Gaposchkin's book will always remain a classic and we can only hope that present-day and future workers in the field will see fit to use *Classical Novae* to complement *The Galactic Novae*.

Our aim has been to put together a book that presents a balanced mix of observation and theory, without presenting any particular point of view too dogmatically. The book begins with an overview of the general properties of novae, then progresses to discuss the accretion process in nova systems and the physics of the nova outburst. Historically, of course, most of the published observational work on novae has been carried out in the optical, and it is therefore appropriate that a good deal of attention is devoted to a discussion of the available optical data. Also discussed in this context is the way in which optical data relate to major aspects of the nova outburst and to observations at other wavelengths. Subsequent chapters discuss in turn the observational data at radio, infrared, ultraviolet and X-ray wavelengths. The penultimate chapter seeks to place classical novae in the broader family of cataclysmic variables. We hope that the contents will prove useful not only to those familiar with the field of classical novae but also to those in other disciplines who may wish to find a balanced overview of a particular aspect, or who wish to find a particular reference; Chapter 13, on 'Data on Novae', should prove especially valuable in this respect.

This book has had an extremely long gestation period: it has taken far longer from initial conception to delivery than we had originally intended (indeed during this time one of the editors has taken up positions in four different research establishments!). However, even since the first drafts of the various chapters came to hand, there have been a number of significant developments in the field, including the discovery of the remarkable radio remnant associated with the old nova GK Per, the detection of X-ray emission from GQ Mus during its 1983 outburst, the availability of the IRAS satellite infrared data, the realization that the primary in some novae must be O-Ne-Mg white dwarfs and the proposed 'hibernation' of novae between outbursts. Had we kept to our original schedule the book would have become dated rather quickly. However, the time had now come to draw a line and to leave any further developments to future volumes. We would very much like, therefore, to express our heartfelt thanks to the contributing authors, some of whom must have despaired at times of ever seeing their chapters in print. We would also like to record our thanks to various workers and institutions who have granted permission to reproduce material from the published literature.

Finally we must record our gratitude and debt to John Wiley & Sons Ltd and their officers, not only for taking on this venture but also for their ready help and patience during its preparation and production.

MIKE BODE September 1988

NYE EVANS

Preface to the second edition

First conceived around 1981, the first edition of *Classical Novae* was published in 1989, after rather a long gestation period. This was at a time when the International Ultraviolet Explorer observatory was still going strong, the Hubble Space Telescope and the ROSAT X-ray observatory still lay in the future, and observatories that are now delivering data of stunning quality, such as Chandra, XMM and Spitzer, were still on the drawing board. Despite the comment in the preface to the first edition 'had we kept to our original schedule the book would have become dated rather quickly', *Classical Novae* dated *very* quickly, as was inevitable.

We had toyed with the idea of a second edition for some time. It was clear that tinkering at the edges of the first edition would not do: so much had changed since the publication of what we began to refer to as 'CNI'. There were of course the inevitable advances in the quality and nature of the observations' over the entire electromagnetic spectrum, and in our theoretical understanding of the classical nova phenomenon as computing power grew. However, there was also the advent of the NASA Astrophysics Data System (ADS), and the facility to prepare a finding chart at the click of a mouse button (R. A. Downes & M. M. Shara 1989, *PASP* **105**, 127): who could have foreseen this when CNI was being compiled? The latter two rendered the *Data on Novae* chapter of the first edition completely obsolete. There was no alternative but to start what inevitably became known as 'CNII' effectively with a clean sheet.

The catalyst for reinvigorating our enthusiasm for the second edition was the highly successful *Classical Nova Explosions* meeting, held at Sitges, Catalonia, Spain, in May 2002 (M. Hernanz & J. Jose, eds., *American Institute of Physics Conference Proceedings*, Vol. **637**, New York: Melville). This gave us the opportunity to corner several potential authors and invite them to contribute. Even so at least 18 months passed before we were able to negotiate our release from Wiley and to enter discussions with Cambridge University Press.

On the whole the content of CNII differs from that of CNI, and much has changed in this edition (not least the wanderlust of editors). However, several of the authors who contributed to the first edition have also been persuaded to contribute to the second, and the general mix of observation and theory is retained. As far as possible the notation used in the first edition is carried over into the second edition.

We take this opportunity of thanking the authors for their contributions and their willingness to take on board comments and suggestions from the editors, Jacqueline Garget, Vince Higgs, Dawn Preston, and Lindsay Nightingale of Cambridge University Press, and Suresh Kumar and Johnny Sebastian of TEX support, for their advice and support; and the officers of J. Wiley & Sons for releasing us and the contributors to CNI from our obligations with them.

List of symbols

We have attempted to standardize the notation and, as far as possible, the symbols used are the same as those used in the first edition. Included in the right hand column of the listing under 'other symbols' below is the chapter number where each symbol is first defined. Inevitably, some symbols are used for more than one quantity, but where this does occur, the correct meaning should be clear and there should be no ambiguity.

Physical and astronomical constants

c	velocity of light
G	gravitational constant
h	Planck's constant
k	Boltzmann's constant
$L_\odot$	solar luminosity
m_H	mass of hydrogen atom
$M_\odot$	solar mass
N_A	Avogadro's number
$R_\odot$	solar radius
σ	Stefan–Boltzmann constant

Other symbols

Roman characters

a	grain radius	8
a_{grav}	gravitational acceleration	5
a_{max}	maximum grain radius	8
a_n	MMRD constant	2
a_{rad}	radiative acceleration	5
A	amplitude in magnitudes of outburst	2
	atomic mass	11
	constant in the law describing the density profile	7
	semimajor axis of orbit	3
A_V	visual extinction in magnitudes	8
A_0	density parameter at $t = t_0$	7

b_{crit}	impact parameter (of wind accretion)	3
b_n	MMRD constant	2
B	magnetic field	2
	blue magnitude	2
B_λ, B_ν	Planck function in wavelength/frequency units	7, 8
$C(m)$	completeness function	14
D	distance to nova	7
$\mathbf{D}$	gradient of rotational velocity	3
D_{max}	maximum observable distance	3
$E(B-V)$	colour excess due to dust extinction	9
E_{H}	energy released per unit mass by fusion of hydrogen	3
E_{sp}	specific energy	3
f	flux density	7
f_{max}	maximum flux density	7
f_{min}	minimum detectable flux	3
f_{ONe}	fraction of ONe novae	11
f_λ, f_ν	flux in unit wavelength/frequency interval	7, 8
g	acceleration of gravity	5
h_{p}	pressure scale height	6
H_{disk}	scale height of Galactic disk	3
i	inclination to the plane of the sky	12
I	intensity	8
j_{b}	emission measure at time t_{b}	9
J	mean intensity of radiation field	5
J_{orb}	orbital angular momentum	3
J_{s}	spin angular momentum	3
$\dot{J}_{\text{GWR}}$	rate of change of angular momentum due to gravitational radiation	3
$\dot{J}_{\text{MSW}}$	rate of change of angular momentum due to magnetic stellar wind	3
$\dot{J}_{\text{orb}}$	rate of change of orbital angular momentum	3
K	infrared (K) magnitude	14
$K_\rho^{(\text{H})}$, $K_\rho^{(\text{V})}$	material diffusivity in the horizontal and vertical directions	3
ℓ	length scale of emitting region	7
	thickness of ejected shell	8
l_{max}	maximum length scale of eddies	6
	characteristic length scale of hydrodynamic instabilities	3
l_{mix}	mixing length	3
L	luminosity	3
L_{bol}	bolometric luminosity	5

$L_{\rm bp}$	luminosity at blue point of central star of planetary nebula	3
$L_{\rm d}$	luminosity of donor star	3
$L_{\rm Edd}$	Eddington luminosity	8
$L_{\rm H}$	luminosity due to hydrogen burning	3
$L_{\rm He}$	luminosity of helium star	3
$L_{\rm hx}$	hard X-ray luminosity	10
$L_{\rm IR}$	infrared luminosity	8
$L_{\rm max}$	maximum luminosity	9
$L_{\rm RG}$	luminosity of red giant	3
$L_{\rm nova}$	luminosity of nova	3
$L_{\rm peak}$	peak luminosity of TNR	4
$L_{\rm pl}$	plateau luminosity	3
$L_{\rm u}$	Lyman continuum luminosity	7
$L_{\rm WD}$	luminosity of white dwarf	3
L_0	luminosity at outburst	8
m_V	apparent visual magnitude	2
$m_{\rm min,max}$	apparent magnitude at minimum, maximum	14
$m_{\rm pg}$	apparent photographic magnitude	14
$m_{\rm pg}(\rm lim)$	limiting apparent photographic magnitude	14
$\dot{m}_V$	light curve decline rate	14
M	mass	3
M_*	total mass of star	3
$M_{\rm a}$	mass of accretor star	3
$M_{\rm acc}$	mass accreted	3
$M_{\rm bol}$	absolute bolometric magnitude	9
$M_{\rm Ch}$	Chandrasekhar mass	9
$M_{\rm CO}$	mass of electron-degenerate CO or ONe core	3
$M_{\rm crit}$	critical accreted mass for TNR	3
$M_{\rm CS}$	mass of convective shell	3
$M_{\rm d}$	mass of donor star	3
$M_{\rm dust}$	mass of dust shell	8
$M_{\rm dg}$	dredge-up mass	3
$M_{\rm e}$	mass of hydrogen-rich envelope	3
$M_{\rm ej}$	mass of ejecta	7
$M_{\rm gas}$	mass of ejected gas	8
$M_{\rm H}$	mass of hydrogen in shell	10
$M_{\rm He}$	mass of helium core	3
$M_{\rm He}^{\rm init}$	initial mass of helium layer on white dwarf	3
$M_{\rm min,max}$	absolute magnitude at minimum, maximum	2
$M_{\rm pg}$	absolute photographic magnitude	14
$M_{\rm nova}(\rm A)$	Galactic mass isotope A contributed by novae	11
$M_{\rm PN}$	mass of planetary nebula central star	3
$M_{\rm rem}$	net remnant mass	3
$M_{\rm rm}$	critical hydrogen-rich envelope mass	3
$M_{\rm S}$	mass of shocked gas	10

M_t	total binary mass	3
M_V	absolute visual magnitude	2
M_{WD}	mass of white dwarf	3
$\langle M_{15} \rangle$	absolute magnitude 15 days post-maximum	2
M_2	mass of secondary star	2
$\dot{M}$	mass transfer rate in binary	2
	mass loss rate	5
$\dot{M}_{acc}$	mass accretion rate	3
$\dot{M}_{bp}$	mass accretion rate at blue point	3
$\dot{M}_H$	rate of hydrogen burning	3
$\dot{M}_{nova}^{pl}$	rate of burning mass during plateau phase	3
$\dot{M}_{RG}$	mass loss rate from red giant or sub-giant	3
$\dot{M}_{rgw}$	mass loss rate from red giant or AGB star wind	3
$\dot{M}_{MSW}$	mass loss rate due to magnetic stellar wind	3
$\dot{M}_w$	wind mass loss rate from erupting star	3
n	number density	13
n_e	number density of electrons	3
n_H	number density of hydrogen	5
n_i	number density of ions	5
$\mathcal{N}$	Brunt–Väisälä frequency	3
N_X	abundance by number of element X	3
N_H	hydrogen column density	3
N_{GWR}, N_{MSW}	number of interacting binaries in Galaxy below/above period gap	3
N_{gap}	number of CVs in the Galaxy in the period gap	3
N_{nova}	number of novae observed	14
N_{obs}	number of observed systems	3
$\langle N_{fast} \rangle$, $\langle N_{slow} \rangle$	number of fast/slow accretion systems in the Galaxy	3
p	exponent in density law	7
P	pressure	13
P_{crit}	critical pressure for thermonuclear ignition	4
P_{orb}, P_{rot}	orbital/rotation period	2
q	mass ratio M_d/M_a	3
$\langle Q_a \rangle$, $\langle Q_e \rangle$	Planck mean absorptivity/emissivity	8, 13
r	radial coordinate	5
r'	Sloan r' magnitude	14
r_{red}	$= \dot{M}_w/\dot{M}_H$	3
R	apparent R magnitude	9
	radius	3
	spectral resolution	8
$\mathcal{R}$	nova rate	11

R_{bp}	radius at blue point of central star of planetary nebula	3
R_{cond}	dust condensation radius	8
R_d	radius of donor star	3
R_{He}	radius of helium star	3
R_{HeWD}	radius of helium white dwarf	3
R_i	inner radius of shell	7
R_{in}	inner radius of nova atmosphere	5
R_L	Roche lobe radius	3
R_{L1}	radius to L1 point	3
R_{max}	maximum radius	3
R_o	outer radius of shell	7
R_{out}	outer radius of nova atmosphere	5
R_{WD}	radius of white dwarf	3
Rf	flux-Richardson number	3
Ri	Richardson number	3
$R_{\tau_{std}}$	effective radius at τ_{std}	5
S	entropy	3
	superheat of grain	8
t	time	2
$\langle t \rangle$	mean period of nova visibility	14
t_b	time of break in He II emissivity	7
t_{blue}	lifetime of nuclear burning evolution of blue phase	3
t_{cond}	grain condensation time	8
t_d	time of maximum emission from forming dust shell	8
t_H	hydrogen burning time-scale	3
t_i	ionization time-scale of ejected shell	8
t_{max}	time of maximum radio flux	7
t_n	time for light curve to decay n magnitudes from peak	1
t_{PN}	time-scale for hydrogen burning by central stars of planetary nebulae	3
t_r	time between outbursts	2
t_{red}	lifetime of nuclear burning evolution of red phase	3
t_s	time-scale for increase in Strömgren sphere radius	9
t_T	time for spectral energy distribution to depart from black body	8
t_0	turn-off time of photoionizing source	9
	time parameter in wind mass ejection law	7
t_1	time-scale over which mass is ejected	7
T	temperature	4
T_b	temperature at base of envelope	6
	brightness temperature	7
T_{BB}	black body temperature	8
T_{blue}	theoretical prediction of lifetime of blue phase	3
T_{cond}	condensation temperature of grains	8
T_d	dust grain temperature	8
T_e	electron temperature	7
$(T_e)_{bp}$	effective temperature at blue point of central star of planetary nebula	3

T_{eff}	effective temperature	5
T_{gr}	grain colour temperature	8
T_{peak}	peak temperature of TNR	6
T_{rad}	temperature of radiation field	9
T_{red}	theoretical prediction of lifetime of red phase	3
T_{UV}	temperature derived from ultraviolet observations	5
υ_K	luminosity-specific nova rate (at K)	14
V	velocity	3
	visual magnitude	9
V_{conv}	velocity of convection	3
V_d	equatorial velocity of donor star	3
V_{def}	velocity of the deflagration front	6
V_{ej}	velocity of ejecta	2
V_G	volume of the Galaxy	3
V_i, V_o	velocity of inner/outer shell boundary	7
V_{max}	maximum velocity	5
	maximum visual brightness	10
V_{rgw}	velocity of red giant wind	3
V_w	velocity of wind	3
V_∞	terminal velocity of wind	5
	velocity at infinity	3
X	mass fraction of hydrogen	3
X_i	abundance of the ith element	3
Y	mass fraction of helium	3
Z	atomic number	6
	mass fraction of metals	3
Z_{CNO}	mass fraction CNO elements	3

Greek characters

α	recombination coefficient	9
	radio spectral index	7
β	ratio of velocity to velocity of light	5
	spectral index of emissivity of dust	13
γ	electron-to-ion number ratio	7
δ	power law exponent of mass ejection rate	7
ΔM_{He}^{det}	critical mass helium layer for detonation of carbon core of white dwarf	3
Δt_{gap}	time-scale for evolution through period gap	3

Δt_{RG}	time-scale for growth of red giant helium core	3
ΔX_{nuc}	mass fraction of H burnt	3
ΔY_{nuc}	mass fraction of helium produced in convective zone	3
ϵ_{nuc}	nuclear energy generation rate	4
ϵ_{grav}	gravothermal luminosity	3
θ	angular diameter	8
θ_{BB}	black body angular diameter	8
θ_{dust}	angular diameter of dust shell	8
θ_{gas}	angular diameter of ejected gas	8
θ_{max}	angular diameter at maximum radio flux	7
κ	absorption coefficient	7
κ_{ff}	Kramers (free–free) opacity	8
κ_{T}	Thomson scattering opacity	8
λ	wavelength	7
λ_{c}	free–free self-absorption cut-off wavelength	8
μ	mean atomic weight	7
μ_{e}	molecular mass per electron	3
ν	frequency	7
	viscosity	3
ξ	micro-turbulent velocity	5
ρ	mass density	5
ρ_{crit}	critical density for dust condensation	8
ρ_{d}	density of grain material	8
σ_{acc}	wind accretion capture cross-section	3
Σ_{5}	surface brightness at 5 GHz	12
τ	optical depth	5
	e-folding time of radioactive decay	11
τ_{burn}	nuclear burning time-scale	6
τ_{cycle}	duration of outburst cycle	3
τ_{ff}	free–free absorption optical depth	8
τ_{GWR}	time-scale for orbital shrinkage by gravitational wave radiation	3
τ_{H}	hydrogen burning time-scale	3
τ_{hyd}	dynamical time-scale	4
τ_{J}	time-scale for orbital angular momentum loss	3
τ_{lim}	mean nova lifetime	14
τ_{MSW}	lifetime of a long period CV	3
τ_{obs}^{blue}, τ_{obs}^{red}	observed lifetime of blue/red phase	3

τ_{std}	standardized optical depth	5
τ_{th}	gravothermal time-scale	3
τ_{UV}	optical depth in the ultraviolet	9
τ_λ, τ_ν	optical depth at wavelength λ/frequency ν	5, 7
$\tau_{1/2}$	radioactive half-life	4
ϕ	volume filling factor	7
$\omega_{i,max}$	upper bound on growth rates of hydrodynamical instabilities	3
Ω	angular velocity	3
Ω_K	angular velocity of Keplerian rotation	3

1

Novae: an historical perspective

Hilmar W. Duerbeck

1.1 Introduction

Nova, abbreviated from *stella nova*, means *new star* (the plural form is [stellae] novae). Although the Merriam-Webster dictionary indicates its etymological origin to be in New (Renaissance) Latin, the term is in fact found in C. Plinius Secundus, *Naturae Historia*, Book 2, chapter XXIV, written around AD 75 (Pliny, 1855)

> *Idem Hipparchus . . . novam stellam in aevo suo genitam deprehendit; eiusque motu, qua die fulsit, ad dubitationem est adductus, anne hoc saepius fieret moverenturque et eae, quas putamus adfixas*
>
> The same Hipparchus discovered a 'new star' that appeared in his own time and, by observing its motions on the day on which it shone, he was led to doubt whether it does not often happen, that those stars have motion which we suppose to be fixed

although the somewhat obscure text would also permit an identification with a meteor or comet.

Because of the Aristotelian doctrine of the immutability of the *translunar* regions, such an object in the stellar regions would not fit into Aristotle's world view, and other objects now known to be translunar, such as comets, were considered to be atmospheric objects and logically discussed in his book on meteorology (and meteors do indeed belong in that book!). Tycho Brahe was among the first to measure accurately the daily parallaxes of the new star of 1572 and the comet of 1577. His result of immeasurably small parallaxes was a fatal blow to the Aristotelian view of the immutability of the spheres.

1.2 Definition of novae and related stars

The Merriam-Webster dictionary defines a nova as a star that suddenly increases its light output tremendously and then fades away to its former obscurity in a few months or years. This is still modelled after Newton's definition (1726; English translation in Newton, 1729)

> *Hujus generis sunt stellæ fixæ, quæ subito apparent, & sub initio quam maxime splendent, & subinde paulatim evanescunt*
>
> Of this kind are such fixed stars as appear on a sudden, and shine with a wonderful brightness at first, and after vanish by little and little

Classical Novae, 2nd edition, ed. Michael Bode and Aneurin Evans. Published by Cambridge University Press.

(*Principia Mathematica*, 3rd edn, lib. 3, prop. 42, at the end of probl. 22). Hardly any change of definition had occurred at the beginning of the twentieth century (Newcomb 1901, p. 127):

> A distinguishing feature of a star of this class [of new or temporary stars] is that it blazes up, so far as is known, only once in the period of its history, and gradually fades away to its former magnitude, which it commonly retains with ... little or no subsequent variation

And almost the same definition, with first hints on classification, is given by Clerke (1903), p. 275:

> A temporary star is a variable that rises sheer from profound obscurity to a single maximum. The maximum may be prolonged or multiple, but it must be essentially one. The occurrence of a second independent outburst would at once relegate the object to the category of irregular variables. The distinction is perhaps arbitrary, but we can only investigate by dividing.

Until the twentieth century, there was no discrimination between novae and supernovae: because of the poorly known absolute magnitudes and remnants of supernovae, these objects were counted among the novae. From 1917 onwards, serendipitous discoveries of novae in spiral nebulae, mainly by G. W. Ritchey, H. D. Curtis and others, which were soon followed by systematic searches, indicated two distinct groups. An early summary is given by Shapley (1917), who lists nine novae (and two supernovae) in spirals and discusses the implications for the 'island universe' hypothesis, but points out the difficulty of reconciling the findings with 'van Maanen's measures of internal proper motions' of spirals. Such proper motions have long been proven to be illusory. In 1917, however, the cosmic role of the nebulae was unclear, and comparison with Galactic objects was daring. Only believers in the extragalactic nature of nebulae drew the correct conclusions concerning novae and supernovae, e.g. Curtis (1917) in a study which was almost contemporary with that of Shapley.

S Andromedae (the 'nova' of 1885 in M31) and also Z Cen (the 'nova' of 1895 in NGC 5253) were strange outliers, which rivalled their host galaxy in brightness. Lundmark (1920) was the first to speak about *giant novae*, soon followed by Curtis (1921) in what became known as the *Great Debate*: 'It seems certain ... that the dispersion of novae in the spirals, and probably also in our galaxy may reach at least ten absolute magnitudes, as is evidenced by a comparison of S And with the faint novae found recently in this spiral. A division into two magnitude classes is not impossible.' Lundmark (1927) called the supernovae *upper-class Novae*, Baade (1929) *Hauptnovae*, Hubble (1929) *exceptional novae*, Lundmark (1933) *super-Novae*, and again Lundmark (1935) *upper-class Novae* or *super-Novae*. Practically simultaneously, Baade and Zwicky (1934) proposed to designate the luminous stellar explosions *supernovae*.

The other groups of cataclysmic variables, the U Gem and the novalike stars, were usually counted among the common variable stars, as the above remark by Clerke (1903) testifies. *Novae* (to avoid the old-fashioned expression *temporary stars*) were for a long time treated distinctively from the variables, as still exemplified in the review articles in the *Handbuch der Astrophysik* (Ludendorff, 1928; Stratton, 1928). On the other hand, Shapley (1921) already argued that differences between novae and other types of variables were not irreconcilable, and he mentioned the recurrent novae RS Oph, V1017 Sgr and T Pyx, as well as the symbiotic stars RX Pup and Z And which share photometric and/or spectroscopic characteristics with novae.

Lundmark's (1935) study is not only important because it gives an early description of the properties of supernovae. It offers, for the first time, a tripartition of the novae as follows: *upper-class Novae*, or *super-Novae*, with $M_{\rm max}$ at -15, and a frequency of 1 per 50 years in the Milky Way, *middle-class Novae*, or *ordinary Novae*, with $M_{\rm max}$ at -7, and a frequency of 50 per year, and finally the *lower-class Novae*, or *dwarf Novae*, with $M_{\rm max}$ at $+3$ or $+4$.

Lundmark's dwarf novae are not yet completely synonymous with the modern dwarf novae: he had in mind objects like WZ Sge that, at that time, had shown only single outbursts. They were thus clearly separated from SS Cyg and U Gem stars with recurrent outbursts, which were counted among the 'plain' variable stars. Lundmark already speculated that a star has to go several times through the nova stage and that the average interval should be of the order of 400×10^6 years, and also he mentions the repeated outbursts of T Pyx and RS Oph.

According to our present knowledge, Lundmark's *dwarf novae* show the same outburst and structural characteristics as the U Geminorum stars, and thus the terms are nowadays used synonymously. In recent years, Lundmark's group of rarely outbursting large amplitude objects has also been labelled WZ Sge stars (Bailey, 1979) or Tremendous Outburst Amplitude Dwarf novae (TOADs; Howell & Szkody, 1995).

1.3 Theories of novae until the mid twentieth century

The breakthrough in modern nova theory was initiated with Schatzman's (1951) finding that ^{3}He is a trigger of thermonuclear runaways (TNR), and Walker's discovery of the binarity of DQ Her (Walker, 1954). Kraft's series of articles showed that binarity appeared to be a common property of cataclysmic variables in general and novae in particular (Kraft, 1963, 1964), and his idea of explosive hydrogen burning on the surface of the degenerate blue component (Kraft, 1963) was revived by Paczyński (1965). Thus all theoretical research before 1950–1960 should be considered 'historic' or perhaps more properly, 'pre-historic', since basically no references to pre-1950 theoretical articles (unless they also deal with outburst observations) are found in recent papers. Nevertheless, I will give a short overview, which is mainly based on the review chapter 'Old and new attempts to explain novae' in Stein's (1924) book.

Thirteen theories are already quoted in Riccioli's *Almagestum Novum* of 1651, although only a handful of eruptive objects had been found up to that time, or at least studied in some detail (the new stars of 1572 and 1604, P Cyg of 1600, and Mira Ceti, which was also first announced as a new star). The theories appear quite bizarre to modern eyes, and the reader is referred to Riccioli (1651) or Stein (1924) for details.

Later theories with more physical flavour were proposed by Newton (1726) and Maupertuis (1732, 1768). Maupertuis imagined stars highly flattened by rotation, surrounded and 'deranged' by massive planets in highly eccentric orbits of high inclination – an early hint of the existence and properties of extrasolar planets. Newton (1726; English translation 1729) suggested another mechanism which, with hindsight, has an even more pertinent flavour:

> *Sic etiam stellæ fixæ, quae paulatim expirant in lucem & vapores, cometis in ipsas incidentibus refici possunt, & novo alimentae accensæ pro stellis novis haberi*
>
> So fixed stars, that have been gradually wasted by the light and vapours emitted from them for a long time, may be recruited by comets that fall upon them; and from this fresh supply of new fuel those old stars, acquiring new splendor, may pass for new stars

(*Principia Mathematica*, 3rd edn, lib. 3, prop. 42). Old stars accreting fuel (albeit from comets, not from a stellar companion): an idea that would only bear fruit over 200 years later!

A plethora of theories explaining the nova phenomenon appeared at the end of the nineteenth century, especially after the discovery and spectroscopic study of nova T Aur (1892). These were mainly 'collisional' theories, e.g. collisions between two stars, collisions of two meteor streams, or the interaction of a star with an interstellar cloud, where the brightness increase was explained by friction. The latter concept, first put forward by W. H. S. Monck (an

Irish amateur astronomer who also anticipated the Hertzsprung – Russell diagram), was elaborated in a series of papers by Seeliger (1886, 1892) and influenced nova theory for a long time.

Parallel to this, there were attempts to explain the spectroscopic phenomena – fruitless ones explaining line shifts by pressure effects, and in the long run fruitful ones explaining them by the ejection of a shell from the star (Pickering, 1894; Pike, 1929).

While the phenomenological explanation of a nova outburst – the ejection of a gas shell – was securely established by the interpretation of spectra of RR Pic (Hartmann, 1925) at the latest, the cause of the outburst remained fairly obscure, although structural considerations (Biermann, 1939) and nuclear reactions (Schatzman, 1951) were suggested to play a role. Only after the binary nature and the structure of the companions had been clarified could a clearer view of the phenomena be achieved, and hydrodynamic models be developed (e.g. Sparks, 1969).

1.4 Pre-telescopic discoveries, observations and catalogues

Most pre-telescopic discoveries of novae were made in the Far East. From at least as early as 200 BC, officials at the imperial Chinese court maintained a systematic watch on the sky for any unusual celestial events, a habit that spread to Korea around the time of Christ, and in the sixth century AD to Japan. The main motive for celestial observation was astrological.

Historical records from these countries contain references to three main types of 'temporary stars': *xingbo* ('bushy stars'), *huixing* ('broom stars') and *kexing* ('guest stars' or 'visiting stars'); and in addition *liuxing* ('flowing stars'). While the first two almost invariably refer to comets, *liuxing* refers to meteors. In principle, *kexing*, used to describe fixed star-like objects, may be interpreted as encompassing novae and supernovae. However, there are reports about moving *kexing*, and about those that later developed tails; thus caution is warranted when interpreting guest stars as stellar outbursts.

Stephenson and Green (2002) give a detailed overview of Chinese, Korean and Japanese sources, as well as of Far Eastern celestial cartography, including lists of *lunar lodges* (called *mansions* in other publications), and also mention early European and Arab sources.

The earliest compilation of reports of such temporary stars was by the Chinese scholar Ma Dualin in the thirteenth century AD, who listed events from the Han dynasty (from 206 BC onward) to his own time; it has been edited by Biot (1843a,b). Other lists were given by Humboldt (1850) and Zinner (1919). Lundmark (1921) made the first detailed study of the stars in Ma Dualin's list. Important anthologies of Far Eastern reports of comets and guest stars (with English translations of the records) are the following:

(1) Ho Peng Yoke (1962): Chinese, Japanese and Korean sources up to AD 1600;
(2) Ho Peng Yoke and Ang Tian-Se (1970): Chinese sources AD 1368–1911;
(3) Li Qibin (1988): Chinese sources of guest stars only. Based on Zhuang Weifeng's and Wang Lixing's 1987 *Zhongguo Gudai Tianxiang Jilu Zongji* (A unified table of ancient Chinese records of celestial phenomena), he lists 53 'guest stars' between 532 BC and AD 1604, with one additional phenomenon in the fourteenth century BC.

Modern catalogues of Far Eastern sources were made by Hsi Tsê-tsung (1958) and Hsi Tsê-tsung and Po Shu-jen (1966). As for European and Near Eastern sources, the reports are scarce and unreliable; a list of Newton (1972) has been scrutinized by Stephenson (1975), who found that a 'new star', seen in 1245 AD and recorded by the chronicler Albertus of Stade, could be identified with the planet Mars – and, parenthetically, the same identification was already made by Lundmark (1933). This tells a great deal about the quality of at least some of the European chronicles.

Table 1.1. *Catalogue of pre-telescopic Galactic novae and supernovae*

Date	Source	Type	Dur	cl	RA	Decl	CS	Li	Notes
−14th century	C							01	
−531 spring	C	star	–	5	20 50	−10	01	02	
− 203 Aug–Sep	C	po	10 d	5	14 20	+20	02		
− 133 Jun–Jul	C	k'o	–	4	16 00	−25	03	03	HK Sco
− 76 Oct–Nov	C	k'o	–	4	11 10	+75	04	04	
− 75 May–Jun	C	chu	–	5	01 40	+25	05		
− 47 May	C	k'o	–	4	18 40	+25	06		SN??
− 46 Jun–Jul	C	k'o	–	4	04 00	+65	07		
− 4 Mar–Apr	C	hui	>70 d	2	20 20	−15	08	05	
+ 61 Sep 27	C	k'o	70 d	2	14 10	+35	09		
+ 64 May 3	C	k'o	75 d	2	12 20	−05	10	06	SN??
+ 70 Dec–Jan	C	k'o	48 d	1	09 40	+25	11	07	
+ 85 Jun 1	K	k'o	–	5		+65	12		
+ 101 Dec 30	C	k'o	–	4	09 40	+25	13	08	
+ 107 Sep 13	C	k'o	–	4	06 30	+10	14	09	
+ 123									HK Oph
+ 125 Dec–Jan	C	k'o	–	4	17 10	+10	15	10	
+ 126 Mar 23	C	k'o	–	5	12 00	+10	16	11	
+ 126	C							12	Aqr
+ 173									HK Cen
+ 185 Dec 7	C	k'o	600 d	1	14 20	−60	17	13	SN
+ 222 Nov 4	C	k'o	–	4	12 30	0	18	14	
+ 247 Jan 16	C	hui	156 d	2	12 30	−20	19		
+ 290 Apr–May	C	k'o	–	4		+65	20	15	
+ 304 Jun–Jul	C	k'o	–	4	04 20	+15	21	16	
+ 329 Aug–Sep	C	po	23 d	5	12 30	+55	22		
+ 369 Mar–Apr	C	k'o	150 d	1		+65	23	17	HK ? SN??
+ 386 Apr–May	C	k'o	90 d	1	18 30	−25	24	18	HK Sgr SN??
+ 389									HK Aql
+ 393 Feb–Mar	C	k'o	240 d	1	17 10	−40	25	19	HK Sco SN?
+ 396 Jul–Aug	C	star	>50 d	2	04 00	+20	26	20	SN??
+ 402 Nov–Dec	C	k'o	60 d	2	11 10	+10	27		
+ 421 Jan–Feb	C	k'o	–	4	11 30	−15	28	21	
+ 437 Jan 26	C	star	–	5	06 40	+20	29	22	SN??
+ 483 Nov–Dec	C	k'o	–	5	05 30	0	30		SN?
+ 537 Jan–Feb	C	k'o	–	4		+65	31		
+ 541 Feb–Mar	C	k'o	–	4		+65	32	23	
+ 561 Sep 26	C	k'o	–	4	11 30	−15	33	24	
+ 641 Aug 6	C	po	25 d	5	12 20	+20	34		
+ 684 Dec–Jan	J	po	14 d	5	03 40	+25	35		
+ 722 Aug 19	J	k'o	5 d	3	01 00	+60	36		
+ 829 Nov	C	k'o	–	4	07 50	+15	37	25	
+ 837 Apr 29	C	k'o	22 d	3	07 00	+10	38	26	
+ 837 May 3	C	k'o	75 d	1	12 10	+5	39	27	
+ 837 Jun 26	C	k'o	–	5	18 00	−25	40		
+ 877 Feb 11	J	k'o	–	4	23 50	+20	41		
+ 891 May 12	J	k'o	–	4	16 50	−20	42		
+ 900 Feb–Mar	C	k'o	–	5	17 00	+10	43	28	SN??
+ 911 May–Jun	C	k'o	–	4	17 10	+15	44	29	
+ 926 Apr 20	C							30	
+ 945									HK Cep/Cas
+ 1006 Apr 3	ACEJ	k'o	>2 yr	1	15 10	−40	45	31	SN
+ 1011 Feb 8	C	k'o	–	4	19 20	−30	46	32	
+ 1012									HK Ari

Table 1.1. *continued*

Date	Source	Type	Dur	cl	RA	Decl.	CS	Li	Notes
+ 1035 Jan 15	C	star	–	5	01 20	+05	47		
+ 1054 Jul 4	CJ	k'o	660 d	1	05 40	+20	48	33	SN
+ 1065 Sep 11	C	k'o	–	4	09 20	−25	49	34	
+ 1069 Jul 12	C	k'o	11d	3	18 10	−35	50	35	
+ 1070 Dec 25	C	k'o	–	4	02 40	+05	51	36	
+ 1073 Oct 9	K	k'o	–	4	00 10	+05	52		
+ 1074 Aug 19	K	k'o	–	4	00 10	+05	53		
+ 1087 Jul 3	C		–					37	SN?
+ 1138 Jun–Jul	C	k'o	–	4	01 50	+20	54	38	
+ 1139 Mar 23	C	k'o	–	4	14 10	−10	55	39	
+ 1163 Aug 10	K	k'o	–	4	17 30	−20	56		
+ 1166 May 1	C							40	
+ 1175 Aug 10	C	po	5 d	5	15 40	+50	57		
+ 1181 Aug 6	CJ	k'o	185 d	1	01 30	+65	58	41	SN?
+ 1203 Jul 28	C	k'o	9 d	3	17 10	−40	59	42	HK Sco SN??
+ 1224 Jul 11	C	k'o	–	4	17 10	−40	60	43	
+ 1230									HK Oph
+ 1240 Aug 17	C	k'o	–	5	17 10	−40	61	44	
+ 1244 May 14	C							45	SN?
+ 1248	C							46	SN?
+ 1264									HK Cep/Cas
+ 1356 May 3	K	k'o	–	4	05 50	+30	62		
+ 1388 Mar 29	C	star	–	5	00 10	+20	63	47	OE
+ 1399 Jan 5	K	k'o	–	4	18 50	−20	64		
+ 1404 Nov 14	C	star	–	5	19 50	+30	65	48	OE SN?1408
+ 1430 Sep 9	C	star	26 d	4	07 30	+05	66	49	OE SN??
+ 1431 Jan 4	C	star	15 d	5	04 50	−10	67	50	OE SN?
+ 1437 Mar 11	K	k'o	14 d	3	16 50	−40	68		
+ 1460 Feb–Mar	V	star	–	5	11 30	−15	69		
+ 1489 Nov 23	C								OE
+ 1497 Sep 20	C	(star)						51	
+ 1572 Nov 8	CEK	k'o	480 d	1	00 20	+65	70	52	HK Cas,OE SN
+ 1584 Jul 11	C	star	–	5	16 00	−25	71		HK Sco,OE
+ 1592 Nov 28	K	k'o	450 d	1	01 20	−10	72		
+ 1592 Nov 30	K	k'o	120 d	1	00 50	+60	73		
+ 1592 Dec 4	K	k'o	90 d	1	00 00	+60	74		
+ 1604 Oct 8	CEK	k'o	360 d	1	17 30	−20	75	53	HK Oph;OE SN

Date: – refer to years BC, with 0, −1, −2 . . . = 1, 2, 3 . . . BC etc.; + refer to years AD.
Source: A = Arab lands, C = Chinese, E = European, J = Japanese, K = Korean, V = Vietnamese.
Type: k'o = kexing (guest star); po = xingbo (bushy star); hui = huixing (broom star); chu = chuxing (the 'candle star').
Dur = duration: duration of visibility (in days or years).
cl = class: 1 . . . 5 decreasing reliability of being a supernova; Stephenson (1976) suspects that class 3 objects are mainly novae, while in classes 4 and 5, the contamination by comets may be significant.
Coordinates RA, Decl.: Right ascension and declination refer to equinox 1950.0.
CS and *Li*: Number in the catalogues by Clark and Stephenson (1977) and Li Qibin (1988).
Notes: occurrence in the lists of Humboldt (1850), pp. 220 et seq. (= HK), and of Ho Peng Yoke and Ang Tian-Se (1970) (= OE). Humboldt sometimes gives the constellation, which is indicated in three-letter-abbreviation; the classification of an object as a true, likely or possible supernova (SN/SN?/SN??) is based on Li Qibin's (1988) fuzzy classification.

The most convenient catalogue of pre-telescopic novae and supernovae is that given by Stephenson (1976), reprinted in Clark and Stephenson (1977). I have edited it for the purpose of the present volume (Table 1.1) by adding objects from Li Qibin's (1988) more complete survey, which is, however, restricted to Chinese sources. Li Qibin's (1992) maps give a good impression of the 'error circles' of the listed coordinates. It may serve as a first guide to look for interesting objects, but consultation of primary and secondary sources is strongly recommended.

1.5 Modern discoveries, observations and catalogues

The plethora of nova discoveries, starting around 1892 with the discovery of T Aur, was based on the beginning of systematic monitoring of the sky by photographic means. The stations of Harvard College Observatory, both in the northern and southern hemispheres, were later followed by stations in Germany, Russia and elsewhere. After the cessation of systematic sky patrols by the Harvard Observatory in the early 1950s, quite a number of novae were found during objective prism surveys from the 1950s to the 1970s. Since the 1970s, an increasing percentage of novae have been discovered by amateurs. They first visually checked 'stellar patterns' in selected regions by means of binoculars. From about the mid 1970s, photographic and later CCD monitoring became common practice, so that most of the discoveries nowadays are made on films and CCDs. Since the 1980s, discoveries of classical novae have almost totally become a domain of amateur astronomers. A first instructive, although by now somewhat dated, introduction to 'nova hunting' is given by Liller and Mayer (1985).

An overview of early catalogues of novae and variable stars is given by Hagen (1921). Many lists of novae are found in books and articles of novae and variable stars (e.g. in Payne-Gaposchkin, 1957); of course, novae are also included in the General Catalogue of Variable Stars (GCVS; Kholopov, 1985), whose most recent version can be found on the web:

http://www.sai.msu.su/groups/cluster/gcvs/gcvs/

This includes a query form which allows searches for objects of a specified type of variability.

Among those nova catalogues that include finding charts, we should mention Duerbeck (1987), Bode and Evans (1989), Downes and Shara (1993) and Downes *et al.* (1997). For the most up-to-date information, the reader is referred to the electronic data base

http://icarus.stsci.edu/~downes/cvcat/

(Downes *et al.*, 2001), which gives the status as of early 2006, and is also available on CD (Downes *et al.*, 2005).

1.6 Photometric and spectroscopic properties of novae

1.6.1 Classification of nova light curves

It goes without saying that novae, from the beginning, were accepted as unique objects. Contrary to most other variable stars, whose light variations often showed similarities that permitted them to be sorted into various classes (Mira stars, cepheids, eclipsing binaries ...), each carefully studied nova exhibited unique characteristics, fluctuations, pulsations, deep brightness drops etc. After sufficient material was collected, in the 1930s, suggestions of light-curve classes were proposed. Almost simultaneously, and probably independently, Lundmark (1935) and Gerasimovic (1936) proposed very similar schemes. Gerasimovic introduced two principal groups, (I) slow novae and (II) flashing novae. The first group was subdivided into four subgroups according to the degree of slowness, from permanent novae (P Cyg) through very slow novae (η Car) to slow ones (RR Pic, DQ Her), while the second group was subdivided into fast novae without and with a brightness recovery (V 603 Aql and T CrB, respectively). Lundmark formed five groups arranged according to speed and additional characteristics: flash novae proper (T CrB), flash and oscillation novae (V603 Aql), wave novae (T Aur), wave-oscillation novae (P Cyg) and 'jump novae' whose light curve resembled 'the temperature curve of a malaria patient', for which he could only cite examples in M31 (Novae 26, 38, 40, 53).

After a number of years, another scheme was suggested by Woronzow-Weljaminow (1953), building upon the previous ones. Calling the fast 'flash' novae R (for rapid), and the slow novae S, he added subclasses such as o (for oscillation), d (for depression), and s (for enhanced slowness). Thus his types were Rs (CP Pup), Ro (GK Per), Rd (T CrB), Ss (V841 Oph), So (RR Pic), Sd (DQ Her) and Sss (RT Ser).

Finally, Duerbeck (1981), having at first overlooked previous attempts in the field, introduced classes A/Ao (rapid novae with smooth light curves, with/without oscillations), Ba/Bb (fairly rapid light curve with some weak/expressed fine structure), Ca/Cb (flat/declining maximum, followed by a marked drop due to dust formation), and D (slow novae with structured light curves). At the same time, he drew attention to the fact that only the novae with A-type light curves have super-Eddington luminosities, while novae of light-curve types B,C and D radiate for an extended time near the Eddington limit; their light fluctuations can be explained by variations in the photospheric radius, i.e. irregularities in mass loss.

1.6.2 Dividing nova light curves: the speed classes

As regards the speed classes, once again it was Gerasimovic (1936) who started to divide novae into several groups. This first attempt does not appear to be well designed; it nevertheless comprises the two groups *slow novae* and *flashing novae*, the latter being more or less synonymous with fast novae. A refinement was made by McLaughlin (1939), who introduced the speed classes: fast, average, slow, RT Ser, as derived from their light-curve decay times t_2 and t_3, the times to decay by 2 and 3 magnitudes respectively. The next step by McLaughlin (1945) was the establishment of a light-curve–luminosity relation, where the group of very fast novae was introduced, and also the question of the existence of subluminous novae (Lundmark's (1927, 1935) *dwarf novae*) was discussed. Finally, the speed classes were slightly redefined on the basis of their t_2-times by Payne-Gaposchkin (1957), and this definition was repeated and thus 'canonized' by Warner (1995, p. 263; see also Chapter 2).

Using a set of light curves from the compilation of Downes, Duerbeck and Delahodde (2001), we derived simple relations between t_2 and t_3, disregarding whether the light curve was based on a visual or blue/photographic passband. For all novae, the relation $t_3 = 1.75t_2$ holds; the slope is, however, dominated by the slow objects. Splitting the sample in two groups, (a) very fast and fast, and (b) moderately fast, slow and very slow objects, according to Payne-Gaposchkin's (1957) classification, yields the conversion rules-of-thumb for (a) $t_3 = 2.10t_2$ and for (b) $t_3 = 1.75t_2$.

In Table 1.2, the boldface numbers are those given by the respective authors, or estimated from their tables and diagrams. Values of t_2 and t_3 missing from their papers were calculated according to the rules-of-thumb given in the previous paragraph, with some generous rounding to define suitable intervals.

It is clearly seen that McLaughlin's (1939) 'fast' is equivalent to Bertaud's (1951) 'rapide', and McLaughlin's (1945) 'very fast' and 'fast' correspond to those of Payne-Gaposchkin

Table 1.2. *Definitions of speed classes of classical novae*

Author	Class	t_2	t_3
Gerasimovic (1936)	flashing		
	slow		
McLaughlin (1939)	fast	**<29**	**<49**
	average	**30–49**	**50–84**
	slow	**50–299**	**85–499**
	RT Ser	**>300**	**>500**
McLaughlin (1945)	very fast	<7	**<15** (10)
	fast	8–24	**15–45** (30)
	average	25–49	**50–84** (60)
	slow	50–250	**85–449** (200)
	RT Ser	>570	**>1000**
Bertaud (1951)	rapide		**<50**
	lente		**>50**
Payne-Gaposchkin (1957)	very fast	**<10**	<20
	fast	**11–25**	21–49
	moderately fast	**26–80**	50–140
	slow	**81–150**	141–264
	very slow	**151–250**	265–440

(1957). McLaughlin's 'average', 'slow' and RT Ser undergo no change between 1939 and 1945, and are summarized under Bertaud's 'nova lente' group. As can be seen by comparing McLaughlin (1945) and Payne-Gaposchkin (1957), there are differences – the latter's 'moderately fast' group has a larger extent than McLaughlin's 'average' group, while Payne-Gaposchkin's 'slow' group has narrower limits than McLaughlin's. Since very slow or RT Ser novae are extremely rare, a suitable classification of such objects is problematic.

Finally, it should be mentioned that the *General Catalogue of Variable Stars* (Kholopov, 1985) divides novae into 'fast' (NA, $t_3 < 100$ days), 'slow' (NB, $t_3 > 150$ days) and 'very slow' (NC, more than a decade at maximum light). This somewhat inconsistent classification has never enjoyed extensive use outside the GCVS.

1.6.3 Spectroscopic classification of novae

Two novae which were very thoroughly observed spectroscopically, DN Gem (1912) and DQ Her (1934), were instrumental in the development of spectral classification of novae. The monograph by Stratton (1920) led to the isolation of seven spectral stages of novae, and soon afterwards the International Astronomical Union proposed to expand a preliminary Harvard designation – Q – into a sequence Qa...Qc, Qu...Qz (Adams, 1923). Another well-observed nova, DQ Her, for which a detailed atlas was prepared (Stratton & Manning, 1939), led another ardent student of novae, D. B. McLaughlin, to divide the evolution of nova spectra into various stages. He divided Harvard class Q into 10 or 11 subclasses (Q0...Q9.5), traversed by a nova in the course of its outburst (McLaughlin, 1938, 1946). In addition, he defined the stages (McLaughlin, 1942b), and presented a sequence of 'synthetic' spectra of a typical nova outburst (McLaughlin, 1944a), showing pre-maximum, maximum and post-maximum spectra at magnitude intervals of about 1, and spectra in the nebular and post-nova stage.

McLaughlin (1944b) was also the first to draw attention to a remarkable difference in the strength of the [Ne III] 3869, 3968 Å lines in the spectra of novae in the nebular stage. Only in the 1980s did this finding lead to the concept of neon novae, i.e. outbursts occurring on massive white dwarfs that show ejecta enriched in O, Mg and Ne.

The refined Q classification of nova spectra never enjoyed widespread use. A more physical classification of nova spectra, which is well suited for modern observations with linear receivers, covering a wide wavelength range, was proposed by Williams *et al.* (1991).

1.6.4 Novae as distance indicators

The occurrence of novae in spiral nebulae spurred interest in determining their absolute magnitude, in order to derive the distances of nebulae. Knut Lundmark was undoubtedly the pioneer of such studies, and he published half a dozen calibrations between 1919 and 1939. His first attempt (Lundmark, 1920) led to a mean absolute magnitude at maximum, $\langle M \rangle = -4.0$. In subsequent papers (e.g. Lundmark, 1923), he gave an absolute magnitude averaged from the results of various methods (trigonometric and secular parallaxes, magnitude of novae near the Galactic centre, and Galactic distribution, compared with B stars). Over the years, his absolute magnitude calibration converged to $\langle M \rangle = -7.0$ (Lundmark, 1941).

When Lundmark terminated his studies, McLaughlin took over. After having investigated light-curve forms and connections between rate of decline and ejection velocities, he started to investigate the luminosities. In addition he started to derive average luminosities of novae, using parallaxes, proper motions, nebular expansion parallaxes, interstellar line strengths and other distance determinations (McLaughlin, 1942a). In his 1942 study, he also obtained an average value of $\langle M \rangle = -7.0 \pm 0.2$, but his most important result was that slow novae are, on the average, 1.8 magnitudes fainter than fast ones. His suggestion that there exists a connection between luminosity and rate of decline, or a 'life–luminosity relation', was elaborated three years later, when McLaughlin (1945) published his diagram where he plotted M versus t_3, and also drew attention to a group of dwarf novae – the prototype being WZ Sge – with noticeably fainter absolute magnitudes. This 'maximum magnitude–rate of decline' relation has been recalibrated several times in recent decades (see Chapters 2 and 14).

1.7 Novae and related stars

In the *General Catalogue of Variable Stars* (Kholopov, 1985), the novae (split into three hardly used groups according to speed class) are counted among the cataclysmic variables. Other types in this category are the novalike stars, the recurrent novae, the supernovae (Types I and II), the U Gem stars (subdivided into SS Cyg, SU UMa and Z Cam stars), and finally the Z And or symbiotic stars. It was already mentioned that in the beginning of variable star research, novae formed a class of their own, aside from (and even before!) the group of variable stars. Early in the twentieth century the group of novae split into supernovae, novae and dwarf novae, although in the beginning, dwarf novae were not synonymous with U Gem stars in general, but included objects that are now identified as WZ-Sge-type systems with rare outbursts.

The term *classical nova* was possibly first used by Gerasimovic (1934), when he discussed cycle lengths of SS Cyg variables and linked them with T Pyx and classical novae whose outburst cycles would be longer than 300 years; it was also used by Payne-Gaposchkin (1954), but only after a paper by Warner (1972) did the term come into general use.

1.7.1 Novae and other stellar eruptions

Some novae do not qualify as *classical* novae. One is the shell star P Cyg, called a 'permanent nova' by Gerasimovic (1936), another is the luminous variable η Car, which was listed by him as a *semi-permanent nova*; both objects were counted among the slow novae by Gerasimovic.

Other unusual objects can be confused with novae, especially if no spectroscopic observations or distance estimates are available. Distant relatives of classical novae are those symbiotic stars showing phases of brightening, such as RR Tel, PU Vul or RT Ser, which are also called *symbiotic novae*. A list of symbiotic novae is given by Warner (1995, p. 304). Other objects belonging to late stellar evolution are V838 Mon and possibly related objects like V4332 Sgr.

The peculiar Nova Aql 1919 (V605 Aql) has been linked to the R CrB stars, and FG Sge was considered as a 'unique variable'. The discovery (in 1996) and intense observation of Sakurai's object (V4334 Sgr) has greatly clarified the behaviour of these final helium flash objects, which are highly evolved single stars of intermediate mass on their way to the white dwarf region, which experience their last thermal pulse very late on their evolutionary track to the white dwarf region. The proceedings of the workshop on Sakurai's Object (Evans & Smalley, 2002) give a thorough discussion of properties and interpretations of such objects. There are good reasons for supposing that one of the earliest *new stars* discovered in the west, CK Vul (nova Vul 1670) might also have been a final helium flash (Duerbeck, 1990; Evans *et al.*, 2002).

Another 'nova' which has perplexed astronomers for quite some time is FU Ori, the prototype of the FU Ori stars or FUORS. These, like V1073 Cyg, V1515 Cyg and other objects, are pre-main-sequence objects which are found in star-forming regions.

1.7.2 Novae and other cataclysmic stars: evolutionary connections

Early in the twentieth century, astronomers had already tried to establish evolutionary connections between types of variables. A good modern example is the possible connection between Type I supernovae and recurrent novae – if recurrent nova explosions occur on massive white dwarfs, and if these show a long-term gain in mass, they might undergo a carbon deflagration and become a Type I supernova. Another scenario connects classical novae with novalikes, dwarf novae (or even detached systems; see also Chapter 3). This idea, popularized by Shara *et al.* (1986) as the *hibernation scenario*, had, however, already been considered in papers by Vogt (1981, 1982) and Duerbeck (1984), although *deep hibernation* was not considered by these authors.

Crucial in this respect is the question of the recurrence times of classical novae. Are there millennia between the outbursts (Case 1), thus giving us no chance to see 'in real time' the photometric and spectroscopic behaviour between outbursts? Then identifications of the remnants of very old novae, e.g. from the Far Eastern lists, would at least permit us a glimpse of the state of a nova between outbursts. Or are there only a few hundred or thousand years between outbursts (Case 2), and do we have a chance to identify a present-day nova outburst with another one on the Far Eastern lists? Or are novae recurrent on even shorter time-scales (Case 3)?

With regard to nova Sco 1437, Shara *et al.* (1990, 1993) have argued for Case 1, claiming that a well-observed bright nova had no post-nova in the appropriate magnitude range, and thus must be in deep hibernation (also meaning low mass transfer rates and long outburst intervals; see Chapter 2 for further discussion of the hibernation hypothesis).

Recently Shara *et al.* (2007) have reported the recovery of a shell around a dwarf nova, suggesting that this object underwent a classical nova outburst ~ 1000 years ago – an example for a Case 1 or 2 scenario.

An example for Case 2 may be HT Cas, which is a SU UMa type dwarf nova, but with physical characteristics which are quite similar to those of CP Pup (= Nova Pup 1942), except that the accretion rate is about a factor of 5 less (Duerbeck, 1993). In this region, a guest star was observed on AD 722 August 19 which disappeared after five days (Stephenson, 1976, object 36).

An example for Case 3 is IM Nor, which erupted in 1920 and 2002. Although it now qualifies as a 'recurrent nova', its spectral evolution in 2002 resembled that of a classical nova. Clearly there are some unusual novae that recur on short time-scales – the *recurrent novae*. But it is obvious that classical novae have a wide range of outburst intervals, and perhaps, with IM Nor, we are just beginning to see the 'short-period tip-of-the-iceberg' (Duerbeck, 2002).

1.8 Concluding remarks

We have attempted to outline the evolution of the concept of 'new stars' or *novae* over the ages. Only in the twentieth century have their physical properties been established to such a degree that they could be clearly separated from the apparently similar phenomena of supernovae and dwarf novae. The discovery in the 1950s and 1960s that some novae are interacting binary stars with a white dwarf component, and the understanding of the process of explosive hydrogen burning led to a blossoming of this field of research, which is manifested in the following pages.

Acknowledgements

I thank O. Osterbrock (Lick Observatory) and B. Warner (Cape Town) for advice, and B. Corbin and G. Shelton (USNO) for bibliographical help. This research made use of NASA's Astrophysics Data System Bibliographic Services.

References

Adams, W. S., 1923, *Ap. J.*, **57**, 65.

Baade, W., 1929, Inaugural lecture, Hamburg, Jan. 30, 1929, quoted by D. E. Osterbrock (2001), *Walter Baade*. Princeton: Princeton University Press, p. 57.

Baade, W., & Zwicky, F., 1934, *PNAS*, **20**, 254.

Bailey, J., 1979, *MNRAS*, **189**, 41.

Bertaud, C., 1951, *Ann. Astrophys.*, **14**, 199.

Biermann, L., 1939, *Z. Astrophys.*, **18**, 344.

Biot, E., 1843a, *Catalogue des cometes observées en Chine depuis l'an 1230 jusqu'à l'an 1640 de notre ère (extrait du supplement du Wen-hian-thoung-khao, et de la grande Collection des vingt-cinq historiens de la Chine). Connaissance des temps...pour l'an 1846*. Paris: Bureau des Longitudes, Bachelier Imprimeur-Libraire, Add. p. 44.

Biot, E., 1843b, *Catalogue des étoiles extraordinaires observées en Chine depuis les temps anciens jusqu'à l'an 1203 de notre ère (extrait du livre 294 de la grande collection de Ma-touan-lin): Connaissance des temps...pour l'an 1846*. Paris: Bureau des Longitudes, Bachelier Imprimeur-Libraire, Add. p. 60.

Bode, M., Duerbeck, H. W., & Evans, A., 1989, in *Classical Novae*, 1st edn, ed. M. F. Bode, & A. Evans. New York and Chichester: Wiley, p. 249.

Clark, D. H., & Stephenson, F. R., 1977, in *The Historical Supernovae*. Oxford: Pergamon Press, p. 46.

Clerke, A. M., 1903, *Problems in Astrophysics*. London: A. & C. Black.

Curtis, H. D., 1917, *PASP*, **29**, 206.

Curtis, H. D., 1921, *Bull. Nat. Res. Council*, **2**, pt. 3, 171.

Downes, R. A., & Shara, M. M., 1993, *PASP*, **105**, 127.

Downes, R. A., Webbink, R. F., & Shara, M. M. 1997, *PASP*, **109**, 345.

Downes, R. A., Duerbeck, H. W., & Delahodde, C. E., 2001, *J. Astron. Data*, **7**, No. 6.

Downes, R. A., Webbink, R. F., Shara, M. M. *et al.*, 2001, *PASP*, **113**, 764.

Downes, R. A., Webbink, R. F., Shara, M. M. *et al.*, 2005, *J. Astron. Data*, **11**, 2.

Duerbeck, H. W. 1981, *PASP*, **93**, 165.

Duerbeck, H. W., 1984, *Ap. &SS*, **99**, 363.

Duerbeck, H. W., 1987, *Space Sci. Rev.*, **45**, 1.

Duerbeck, H. W., 1990, in *Physics of Classical Novae*, IAU Colloquium 122, ed. A. Cassatella, & R. Viotti. Berlin: Springer, p. 34.

Duerbeck, H. W., 1993, in *Cataclysmic Variables and Related Physics*, 2nd Technion Haifa Conference, ed. O. Regev, & G. Shaviv. *Annals Israel Physical Society*, **10**. Bristol: Institute of Physics Publishing, p. 77.

Duerbeck, H. W., 2002, in *Classical Nova Explosions*, ed. M. Hernanz, & J. José. New York: American Institute of Physics, p. 514.

Evans, A., & Smalley, B., 2002, *Ap. &SS*, **279**, 1.

Evans, A., van Loon, J. Th., Zijlstra, A. A. *et al.*, 2002, *MNRAS*, **332**, L35.

Gerasimovic, B. P., 1934, *AN*, **251**, 255.

Gerasimovic, B. P., 1936, *Poulkovo Observatory Circular* No. 16, pp. 5–10 = *Pop. Astr.*, **44**, 78.

Hagen S. J., J. G., 1921, *Die Veränderlichen Sterne*, Erster Band: Geschichtlich-technischer Teil. Freiburg im Breisgau: Herder Verlagsbuchhandlung.

Hartmann, J., 1925, *AN*, **226**, 63.

Ho Peng Yoke, 1962, *Vistas Astron.*, **5**, 127.

Ho Peng Yoke, & Ang Tian-Se, 1970, *Oriens Extremus, Zeitschrift für Sprache, Kunst und Kultur der Länder des Fernen Ostens*, **17**, 63.
Howell, S. B., & Szkody, P., 1995, in *Cataclysmic Variables*, ed. A. Bianchini, M. della Valle, & M. Orio. Dordrecht: Kluwer Academic Publishers, p. 335.
Hsi Tsê-tsung, 1958, *Smithsonian Contributions to Astrophysics*, **2**, 109 (also published in *Astr. Zhurnal*, **34**, 159 = *Sov. Astr. – AJ*, **1**, 161).
Hsi Tsê-tsung [Xi Ze-zong], & Po Shu-jen, 1966, *Science*, **154**, 597.
Hubble, E., 1929, *Ap. J.*, **69**, 103.
Humboldt, A. von, 1850, *Kosmos. Entwurf einer physischen Weltbeschreibung*, Dritter Band. Stuttgart and Augsburg: J. G. Cotta, p. 220. English edn *Cosmos*, Vol. 3, translated by E. C. Otte. London: Bohn, p. 204.
Kholopov, P. N., 1985, *General Catalogue of Variable Stars*, 4th edn, Vol. 1. Moscow: Nauka.
Kraft, R. P., 1963, in *Advances in Astronomy and Astrophysics*, Vol. 2, ed. Z. Kopal. New York and London: Academic Press, p. 43.
Kraft, R. P., 1964, *Ap. J.*, **139**, 457.
Li Qibin, 1988, in *High Energy Astrophysics*, ed. G. Börner. Berlin and New York: Springer, p. 2.
Li Qibin, 1992, in *Supernovae and their Remnants*, ed. Li Qibin, Ma Er, & Li Zongwei. Beijing: International Academic Publishers, p. 57.
Liller, B., & Mayer, B., 1985, *The Cambridge Astronomy Guide: A Practical Introduction to Astronomy*. Cambridge: Cambridge University Press.
Ludendorff, H., 1928, in *Handbuch der Astrophysik*, Vol. 6, ed. G. Eberhard, A. Kohlschütter, & H. Ludendorff. Berlin: Springer, p. 49.
Lundmark, K., 1920, *Kungl. Svensk Vetensk. Hand.*, **60**, No. 8.
Lundmark, K., 1921, *PASP*, **33**, 225.
Lundmark, K., 1923, *PASP*, **35**, 95.
Lundmark, K., 1927, *Nova Acta Reg. Soc. Sci. Upsal.*, Vol. extraordinem.
Lundmark, K., 1933, *Lund Obs. Circ.*, **8**, 216.
Lundmark, K., 1935, *Medd. Lunds Astr. Obs.*, Ser. II, Band VIII, No. 74.
Lundmark, K., 1941, in *Colloque International d'Astrophysique*, 1939, Tome I: *Observation des novae*, ed. A. J. Shaler. Paris: Hermann & Cie, p. 11.
McLaughlin, D. B., 1938, *Pop. Astr.*, **46**, 373.
McLaughlin, D. B., 1939, *Pop. Astr.*, **47**, 410, 481, 538.
McLaughlin, D. B., 1942a, *Pop. Astr.*, **50**, 233.
McLaughlin, D. B., 1942b, *Ap. J.*, **95**, 428.
McLaughlin, D. B., 1944a. *Pop. Astr.*, **52**, 109.
McLaughlin, D. B., 1944b, *AJ*, **51**, 20.
McLaughlin, D. B., 1945, *PASP*, **57**, 69.
McLaughlin, D. B., 1946, *AJ*, **52**, 46.
Maupertuis, P. L. M de, 1732, *Discours sur les différentes figures des astres, où l'on tire des conjectures sur les étoiles qui paroissent changer de grandeur*... Paris: Imprimerie Royale.
Maupertuis, P. L. M de, 1768, *Discours sur les différentes figures des astres, où l'on essaie d'expliquer les principaux phénomenes du ciel, Oeuvres*, Tome I. Lyon: Bruyset.
Newcomb, S., 1901, *The Stars: A Study of the Universe*. New York: Putnam.
Newton, I., 1726, *Philosophiae Naturalis Principia Mathematica*, 3rd edn (1726), ed. A. Koyré, & J. B. Cohen, Cambridge: Harvard University Press.
Newton, I., 1729, *The Mathematical Principles of Natural Philosophy* (translated by A. Motte). London: B. Motte, 1729. First American edn: New York: D. Adee, 1848.
Newton, R. R., 1972, *Medieval Chronicles and the Rotation of the Earth*. Baltimore and London: Johns Hopkins University Press.
Paczyński, B., 1965, *Acta Astron.*, **15**, 197.
Payne-Gaposchkin, C., 1954, *Variable Stars and Galactic Structure*. London: The Athlone Press.
Payne-Gaposchkin, C., 1957, *The Galactic Novae*. Amsterdam: North-Holland.
Pickering, W. H., 1894, *A&A*, **13**, 201.
Pike, S. R., 1929, *MNRAS*, **84**, 538.
Plinius, C. (Pliny the Elder), 1855, *The Natural History of Pliny*, transl. and ed. J. Bostock, & H. T. Riley. London: H. G. Bohn.

Riccioli, G. B., 1651, *Almagestum novum: astronomiam veterem novamque complectens observationibus aliorum et propriis nouisque theorematibus, problematibus, ac tabulis promotam*. Ex typographia haeredis Victorij Benatij, Bononiae [digital copy available from: www.gallica.fr].
Schatzman, E., 1951, *Ann. Astrophys.*, **14**, 294.
Seeliger, H., 1886, *AN*, **113**, 353.
Seeliger, H., 1892, *AN*, **130**, 393.
Shapley, H., 1917, *PASP*, **29**, 213.
Shapley, H., 1921, *PASP*, **33**, 185.
Shara, M. M., Livio, M., Moffat, A. F. J., & Orio, M., 1986, *Ap. J.*, **311**, 163.
Shara, M. M., Potter, M., Moffat, A. F. J., Bode, M., & Stephenson, F. R., 1990, in *Physics of Classical Novae*, ed. A. Cassatella, & R. Viotti. Berlin: Springer, p. 57.
Shara, M. M., Moffat, A. F. J., Potter, M., Bode, M., & Stephenson, F. R., 1993, in *Cataclysmic Variables and Related Physics. Annals Israel Physical Society*, **10**. Bristol: Institute of Physics Publishing, p. 84.
Shara, M. M., Martin, C. D., Seibert, M. *et al.*, 2007, *Nature*, **446**, 159.
Sparks, W. M., 1969, *Ap. J.*, **156**, 569.
Stein S.J., J., 1924, *Die veränderlichen Sterne*. Zweiter Band: Mathematisch-physikalischer Teil. Freiburg im Breisgau: Herder Verlagsbuchhandlung.
Stephenson, F. R., 1975, in *Origin of Cosmic Rays*, ed. J. L. Osborne, & A. W. Wolfendale. Dordrecht: Reidel, p. 339.
Stephenson, F. R., 1976, *QJRAS*, **17**, 121. Also included in Clark & Stephenson (1977).
Stephenson, F. R., & Green, D. A., 2002, *Historical Supernovae and their Remnants*. Oxford: Clarendon Press.
Stratton, F. J. M., 1920, *Ann. Solar Phys. Obs., Cambridge*, **4**, Part I.
Stratton, F. J. M., 1928, in *Handbuch der Astrophysik*, Vol. 6, ed. G. Eberhard, A. Kohlschütter, & H. Ludendorff. Berlin: Springer, p. 251.
Stratton, F. J. M., & Manning, W. H., 1939, *Atlas of Spectra of Nova Herculis 1934*. Cambridge: Solar Phys. Obs.
Vogt, N., 1981, *SU-UMa-Sterne und andere Zwergnovae. Eine Untersuchung ihrer Eruptionsmechanismen, ihrer Struktur und entwicklungsgeschichtlichen Stellung unter den kataklysmischen Doppelsternen*. Habilitation thesis, Ruhr-Universität, Bochum.
Vogt, N., 1982, *Mitt. Astr. Ges.*, **57**, 79.
Walker, M. F., 1954, *PASP*, **66**, 230.
Warner, B. 1972, *MNRAS*, **160**, 35P.
Warner, B., 1995, *Cataclysmic Variable Stars*. Cambridge: Cambridge University Press.
Williams, R. E., Hamuy, M., Phillips, M. M. *et al.* 1991, *Ap. J.*, **376**, 721.
Woronzow-Weljaminow, B. A., 1953, *Gasnebel und Neue Sterne*. Berlin: Kultur und Fortschritt (Russian original published in 1948).
Zinner, E., 1919, *Sirius*, **52**, 25, 127, 152.

2

Properties of novae: an overview

Brian Warner

2.1 Introduction

Although no two novae show exactly the same properties or development during eruption, there are systematics in the overall population of novae that assist in the interpretation of the physics underlying their behaviour and their relationship to other systems of stars. Furthermore, some of these properties are of value in other areas, such as distance indicators for extragalactic research. Here we give a general introductory overview of the general properties of novae, derived largely from ground-based observations; later chapters add more detail and extend the wavelength range.

2.2 Frequency and Galactic distribution of novae

The frequency of classical nova discoveries over the past century, corrected for non-uniformity of coverage in time, is shown in Table 2.1 (Duerbeck, 1990). In the range $4 \leq m_V \leq 6$ the values do not increase as fast as expected, showing that many bright novae go undetected. At fainter magnitudes the increase is due to the contribution of novae in the Galactic bulge. The total mean detected nova rate is $\sim 3\,\mathrm{yr}^{-1}$. There are so many factors that lead to incompleteness in nova searches (seasonal, weather, sky coverage biased towards the Galactic bulge, missed fast novae) that it is possible to conclude that Table 2.1 shows the few novae that were actually detected from an observable number of $\sim 12\,\mathrm{yr}^{-1}$ (Liller & Mayer, 1987). Only when new all-sky monitors such as the Large Synoptic Survey Telescope are in operation will more complete coverage become possible.

Further allowance for incompleteness, selection effects and interstellar extinction produces estimates for a total Galactic nova rate of $73 \pm 24\,\mathrm{yr}^{-1}$ (Liller & Mayer, 1987), $29 \pm 17\,\mathrm{yr}^{-1}$ (Ciardullo *et al.*, 1990), $30 \pm 10\,\mathrm{yr}^{-1}$ (Shafter, 2002) and $34^{+15}_{-12}\,\mathrm{yr}^{-1}$ (Darnley *et al.*, 2006).

The apparent Galactic distribution of novae shows a strong concentration towards the plane and bulge (Figure 2.1). At higher spatial resolution the novae are seen preferentially in areas of low interstellar extinction (Plaut, 1965). The Galactic distribution is clearly predominantly that of disc and bulge populations, but the existence of at least one nova in a globular cluster (Shara & Drissen, 1995) shows that there can be an extreme Population II component. Looking in more detail, Della Valle *et al.* (1992) find evidence that fast novae (defined below) are far

Classical Novae, 2nd edition, ed. Michael Bode and Aneurin Evans. Published by Cambridge University Press.

Table 2.1. *Average rates of nova discovery*

m_V (max) (mag)	Discovery rate (yr^{-1})	m_V (max) (mag)	Discovery rate (yr^{-1})
< 1	0.04	4–5	0.05
1–2	0.02	5–6	0.14
2–3	0.04	6–7	0.47
3–4	0.03	7–8	0.58

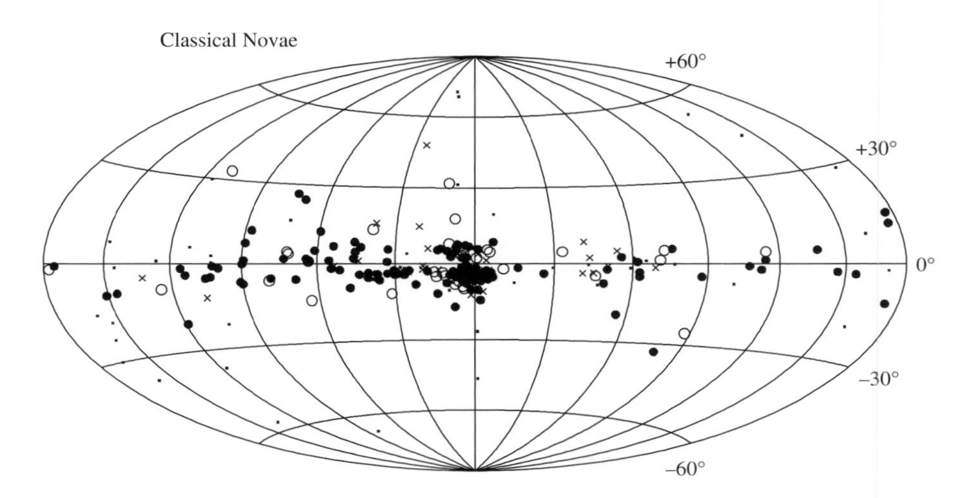

Fig. 2.1. The distribution in Galactic coordinates of classical novae (an Aitoff projection). The Galactic Centre is located at the centre of the diagram, with increasing Galactic longitude towards the left. Filled circles mark 132 fast novae (NA/NA:), 40 slow novae (NB/NB:) are plotted as open circles, the crosses mark 58 uncategorized novae (N) and the small dots correspond to 58 uncertain novae (N:/N::). All data are from the Downes *et al.* CV catalogue (`http://icarus.stsci.edu/~downes/cvcat/`); see also Downes *et al.* (2005).

more concentrated (vertical height $z < 100$ pc) towards the Galactic plane than slow novae (which extend up to 1000 pc and are associated with the Galactic bulge), and this shows in the predominance of the former towards the Galactic anti-centre and of the latter towards the bulge. More details are given in Chapter 14.

Any classical nova that shows more than one eruption becomes, by definition, a recurrent nova. No doubt all novae are recurrent, but their intervals, t_r, between eruptions are in general greater than our observational baseline; none of the novae found in ancient records (mostly oriental, from one to two thousand years ago; see Chapter 1) have been seen to recur, so for most novae $t_r \gtrsim 10^3$ yr, but there is the appearance of a tail of the distribution of nova recurrence times that we are able to observe where $t_r \sim 100$ yr. However, the observed recurrent novae do not resemble the normal classical novae (see Section 2.7 below).

Table 2.2. *Classification of nova light curves*

Speed class	t_2 (days)	$\dot{m}_V$ (mag d^{-1})
Very fast	<10	>0.20
Fast	11–25	0.18–0.08
Moderately fast	26–80	0.07–0.025
Slow	81–150	0.024–0.013
Very slow	151–250	0.013–0.008

2.3 Light curves during eruption

Modern studies of nova light curves still depend largely on the descriptions introduced 50 or more years ago. Almost all novae rise rapidly (1–3 d) and are not then sufficiently well observed to warrant division into types. To quantify the very different rates of decline from maximum light, Payne-Gaposchkin (1957) introduced the 'speed classes' listed in Table 2.2 (where it will be noted that there are no 'normal' novae, they are either fast or slow; cf. dwarf and giant stars). The notation t_n is used to designate the number of days, t, that the nova took to fall n magnitudes from maximum. The implied rate of decline, dV/dt, is also listed. From a correlation of measured values, Warner (1995) finds that $t_3 \approx 2.75\, t_2^{0.88}$ (see also Chapters 1 and 14).

In many compilations, including the Downes and Shara (1993) catalogue and its later versions, the speed classes are contracted into the notation NA, NB and NC, meaning fast, slow and very slow. Recurrent novae are designated NRA if they have giant secondaries, and NRB if they have non-giant secondaries.

By compression of their time-scales McLaughlin (1939, 1960) found that all nova optical light curves can be made to resemble each other, and that the spectral evolutions are at the same time brought into better accord. The idealized light curve is shown in Figure 2.2.

2.3.1 Initial rise, pre-maximum halt and final rise

Few novae have been caught on the initial rise – though it is sometimes possible to construct at least part of the early rise from patrol images of the sky taken by amateurs (V1500 Cyg 1975 is a good example: Liller *et al.*, 1975). The brightening to within about 2 magnitudes from maximum takes at most 3 days, even for the slowest novae. In many novae there is a pause, ranging from a few hours in fast novae to a few days in slow novae, about 2 magnitudes below maximum. The nova then brightens to maximum, taking 1–2 days for fast novae and several weeks for the slowest novae. The maximum phase itself is relatively short-lived, being only hours for very fast novae and only a few days for slow novae.

2.3.2 Early decline and transition

The initial fall from maximum is usually smooth, except for slow novae, which have brightness variations on time-scales of 1–20 days with amplitudes up to 2 mag. At 3 to 4 magnitudes below maximum three distinct behaviours are possible: (i) some novae fall into a minimum 7 to 10 mag deep and lasting for months or even years, after which the star recovers and follows an extrapolated decline; (ii) other novae start large amplitude oscillations with quasi-periods of ~5–15 days and amplitudes of up to 1.5 mag; (iii) a few novae, comprising

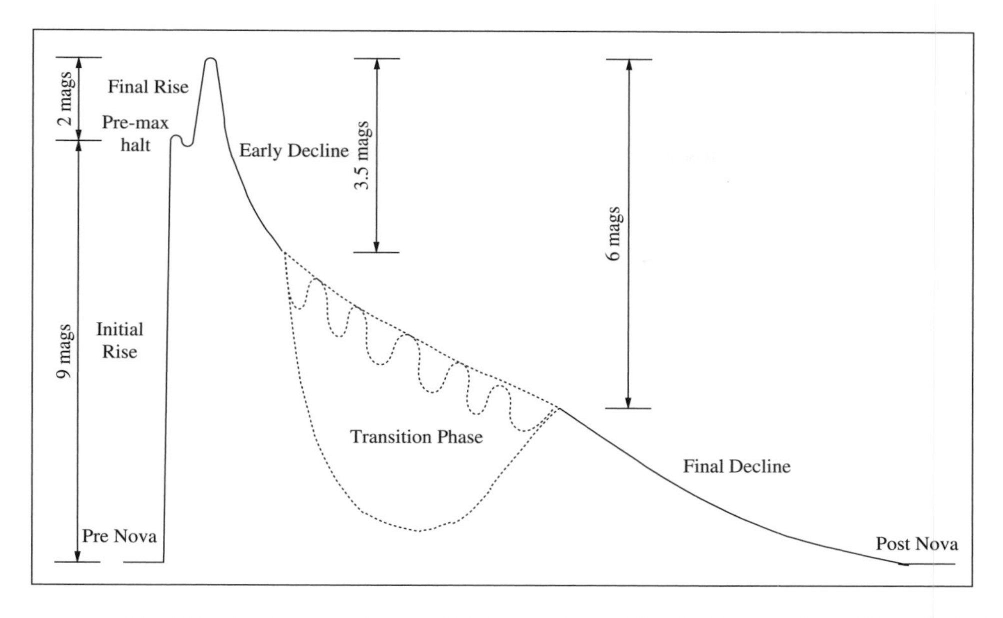

Fig. 2.2. Morphology of a nova light curve. From *Classical Novae*, first edition (Bode & Evans, 1989).

about one-third of the fast and very fast systems plus a few of the slower ones, pass through this transition region without any noticeable peculiarities.

The deep minimum that occurs on the decline of some novae is due to the formation of dust in the gas ejected by the eruption. This was first made clear by infrared photometry of FH Ser, which developed an infrared excess when the optical light curve began to fall into its 5 mag minimum (Hyland & Neugebauer, 1970; Geisel, Kleinmann & Low, 1970; see also Chapters 1 and 13). The sum of the optical and infrared energies remained approximately constant at $\sim 1.5 \times 10^4\,L_\odot$. A second indication that a nova eruption is roughly a constant bolometric luminosity event came soon after – the realization that in the early stages of decline the fall in optical flux is counterbalanced by an increase in ultraviolet flux was a result of OAO-2 satellite measurements (Gallagher & Code, 1974). Here the reason is that as the ejected shell becomes optically thinner we see radiation coming from deeper in towards the central binary with its heated primary (see Chapters 4 and 5).

The modern understanding of nova eruptions is that they have bolometric luminosities that are all at or somewhat above the Eddington luminosity, and the total radiant output lies in the range 10^{45}–10^{46} erg. The parameter determining the speed class is the initial rate of generation of energy, which itself is largely a function of the mass of the primary.

2.4 Relationships among amplitude, rate of decline, absolute magnitude and expansion velocity

The most direct correlation of nova properties, independent of distance and reddening, is that of amplitude of eruption versus t_n. For both parameters there can be problems of measurement, especially for the fainter novae: t_n requires good photometric coverage of the maximum stages (and the maximum itself may often be missed, even for the brightest novae), and the amplitude (or range) of eruption, $A = m_{\rm min} - m_{\rm max}$, can only be measured if and when the nova has settled down as an identifiable remnant. For older novae, with poorly determined positions, identification of the remnant has sometimes been with an unrelated field star that is much brighter than the real minimum state (e.g. Warner (1986), where all of the suggested misidentifications have later been verified), which reduces the apparent amplitude.

The A/t_2 correlation is shown in Figure 2.3, and extends from $A \sim 15$ mag for the fastest novae to $A \sim 7$ mag for the slowest. The added lines explain part of the vertical scatter, which is due to a range of $M_{\rm min}$ (see below), but if novae at minimum are at all similar to each other it is clear that there must be a strong correlation between $M_{\rm max}$ and t_n.

Edwin Hubble (1929), from studies of novae in M31, noticed that the brighter novae fade faster, and McLaughlin (1945), using absolute magnitudes from expansion parallaxes, interstellar line strengths and Galactic rotation, found the same correlation between maximum luminosity and rate of decline for Galactic novae. Many other studies have been made of novae in M31, the Magellanic Clouds and the Galaxy; an overview of these early studies is given in Chapter 14.

The importance of the maximum magnitude/rate of decline (MMRD) relationship as an extragalactic distance indicator is obvious and has resulted in continued discussion and updating. Probably the most influential early determinations were those of van den Bergh (1975) and de Vaucouleurs (1978), where the former adopted a non-linear relationship and the latter a linear one. The most recent work maintains this division of opinion, but as it affects only the fainter (slower) novae it has not been of consequence for extragalactic distance determinations.

Capaccioli *et al.* (1989) concluded that the MMRD relationship is the same for galaxies of all Hubble types, which means that all relevant observations can be pooled for analysis. This has been done by Della Valle and Livio (1995), whose observational MMRD results are the most comprehensive to date. However, they choose to fit a non-linear relationship (an S-curve) in the same manner as van den Bergh. This choice is driven by a theoretical MMRD relationship derived by Livio (1992) for nova eruptions on white dwarfs with $M = 0.6\,{\rm M}_\odot$, which therefore omits the slowest novae, and is also affected by the paucity of observations of extragalactic slow novae ($t_2 > 50$ d) which are intrinsically much fainter. This can be further seen (Warner, 1995) by the incompatibility of the S-curve with the A/t_2 relationship (Figure 2.3), which is derived for Galactic novae and therefore includes slow novae. We conclude that the inclusion of slow novae points to a more linear MMRD correlation, and this allows the use of simple fitting coefficients, as discussed below.

The discussion by Downes and Duerbeck (2000) of 28 Galactic novae for which expansion parallaxes are available finds a linear MMRD relationship to be adequate, with no significant differences between novae of different composition types. They conclude that a typical $M_{\rm V}$ uncertainty when using this method is ~ 0.5 mag. Part of this scatter may be due to the effect of assuming that nova shells expand with spherical symmetry, whereas they are usually prolate (Gill & O'Brien, 2000; Wade, Harlow and Ciardullo, 2000; see also Chapter 12), and part is due to a spread in nova properties (see below). The axial ratio of a nova shell is correlated with speed class, such that slower novae are more elongated (Slavin, O'Brien & Dunlop, 1995) – an extreme example is HR Del, which has the most bipolar shape and is one of the slowest novae (Harman & O'Brien, 2003).

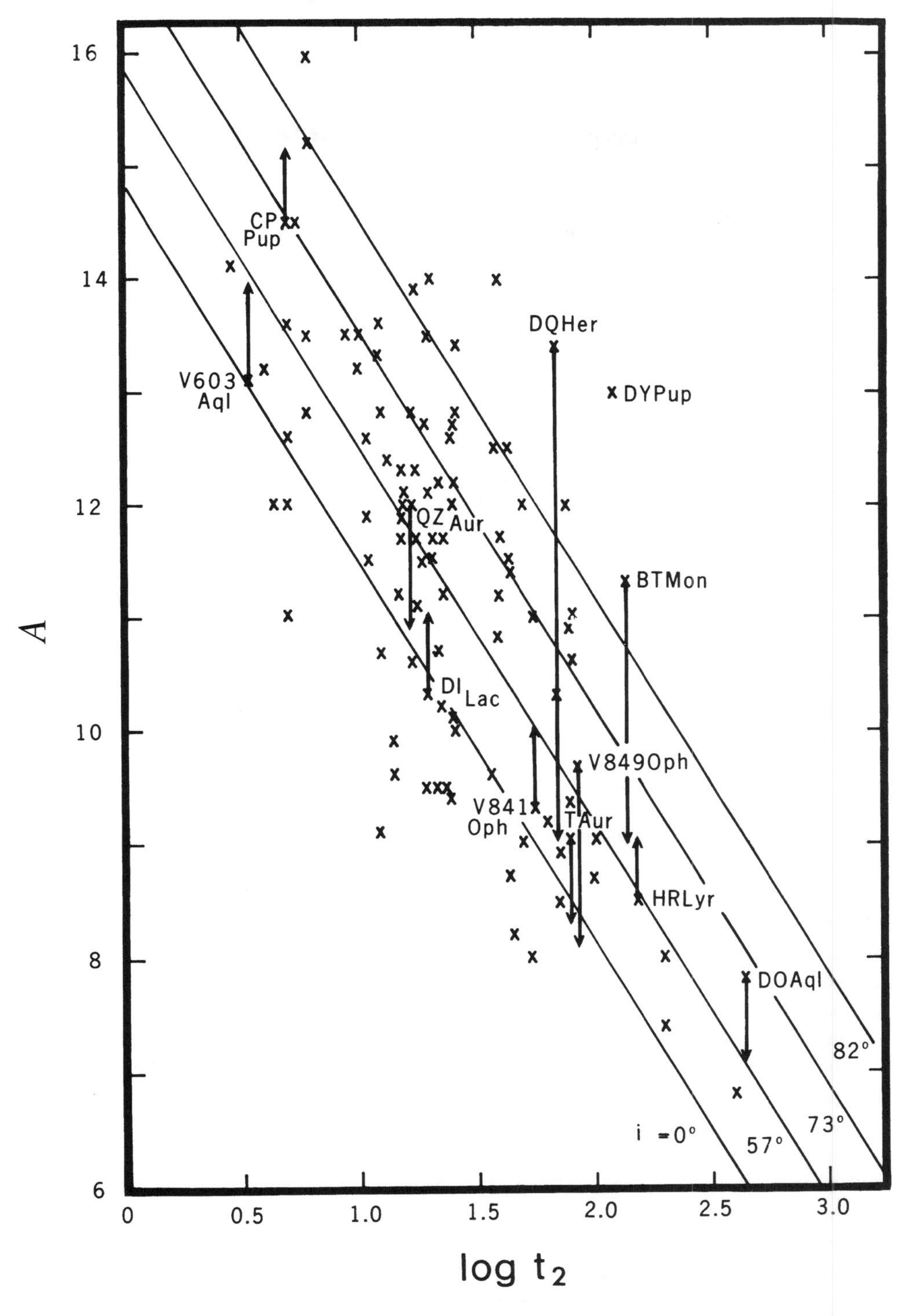

Fig. 2.3. Amplitude (A) versus rate of decline for classical novae. From Warner (1995). The sloping lines show the effects of different inclinations. The vertical lines show the effect on amplitude of correcting for this effect on novae of known inclination.

Table 2.3. *Typical values of MMRD constants*

M	n	a_n	b_n	Reference
pg	3	−11.3	2.4	1
B	3	−10.67 (± 0.30)	1.80 (± 0.20)	2
V	2	−10.70 (± 0.30)	2.41 (± 0.23)	3
V	2	−11.32 (± 0.44)	2.55 (± 0.32)	4
V	3	−11.99 (± 0.56)	2.54 (± 0.35)	4

References: 1. de Vaucouleurs (1978); 2. Pfau (1976); 3. Cohen (1985); 4. Downes & Duerbeck (2000).

Table 2.4. *Determinations of* M_{15}

Colour	M_{15}	Reference
pg	−5.86	Schmidt-Kaler (1957)
pg	−5.5 (± 0.18)	de Vaucouleurs (1978)
V	−5.2 (± 0.1)	Buscombe & de Vaucouleurs (1955)
V	−5.60 (± 0.45)	Cohen (1985)
V	−6.05 (± 0.44)	Downes & Duerbeck (2000)

Della Valle (1991), however, finds that outside of the 0.5 spread in magnitude there is evidence for a few ($< 10\%$) super-bright novae, standing a magnitude or more above the usual relationship.

The linear MMRD relationship is usually expressed in the form

$$M = b_n \log t_n(d) + a_n \tag{2.1}$$

where M is the absolute magnitude (variously V, B or pg) and n = 2 or 3. A selection of typical values of a_n and b_n, including some earlier to compare with the most recent, is given in Table 2.3; see also Equation (14.6). A more complex version is given in Chapter 14.

Another empirical relationship that is useful as an extragalactic distance indicator is the fact that M_{15}, the absolute magnitude 15 days after maximum, is the same for all speed classes (Buscombe & de Vaucouleurs, 1955). A selection of values is given in Table 2.4. The use of classical novae as extragalactic distance indicators, via these relationships, is discussed in Chapter 14, particularly Section 14.4.1.

A further empirical relationship is that relating t_n to the mean expansion velocity $V_{\rm ej}$ of the ejecta. The two sets of absorption lines typically seen in the early stages of the nova eruption have different velocities, and both increase with time. These relationships are listed in Sections 2.5.2 and 2.5.3.

The use of classical novae, via these relationships, as extragalactic distance indicators is discussed in Chapter 14.

The above general empirical relationships can be explained theoretically through the continuous ejection model of nova explosions. More details are given in Chapters 5 and 4, but the underlying physics was provided by Shara (1981a,b) who showed that deviations of up to 0.5 mag in the MMRD relationship arise from the spread of white dwarf luminosities before eruption, that the same physics results in a convergence of M_V about 15 days after maximum, and that the V_{ej}/t_n relationships arise through approximate equipartition of energy between the radiation field and the ejected matter.

2.5 Spectral evolution during eruption

There is a wealth of literature on the dramatic changes that take place in the spectra of novae from maximum down to eventual quiescence. Modern analyses continue to refer to the various taxonomic classes that were first introduced by McLaughlin (1942, 1944) and emphasized by Payne-Gaposchkin (1957), but there is now also a modern classification scheme for the emission line evolution which is more physically based (Williams, 1990, 1992). It remains an extraordinary fact that despite the complexities in evolution of their spectra during eruption, almost all novae follow the sequence of four successive systems of absorption lines and the five overlapping systems of emission lines. Here we give only a brief outline (see also Chapters 8 and 9).

2.5.1 Pre-maximum spectrum

Few spectra have been obtained on the fast rise to maximum, but the pre-maximum halt and final rise (Figure 2.2) provide better opportunities, especially for the slower novae. The spectra are dominated by broad, blue-shifted absorption lines, resembling those of early type stars, often with the addition of P Cygni profiles. Spectral types at maximum are usually in the range B5 to F5, with later types and smaller blue shifts associated with slower novae. The widths of the lines are approximately equal to the expansion velocities, which range from -1300 km s^{-1} for very fast to ~ -100 km s^{-1} for the slowest novae.

These pre-maximum spectra are characteristic of a uniformly expanding optically thick envelope, cooling as it expands.

2.5.2 Principal spectrum

The principal spectrum appears at visual maximum, has stronger and more blue-shifted absorption lines than those at pre-maximum and resembles that of an A or F supergiant with enhancements of the lines of C, N or O. The absorption lines often show multiple substructure, the velocity evolutions of which vary greatly from component to component and from nova to nova; these components evolve into the emission lines of the nebular shell (see below). The average velocities of the absorption lines correlate with speed class, being up to 1000 km s^{-1} for the fastest novae down to 150 km s^{-1} for the slowest (see Tables 10.2 and 10.3 of (Payne-Gaposchkin, 1957). McLaughlin (1960) found that the velocities can be represented by

$$\log V_{ej}\,(\mathrm{km\,s^{-1}}) = 3.70 - 0.5 \log t_3\,(\mathrm{d}) = 3.57 - 0.5 \log t_2\,(\mathrm{d}). \qquad (2.2)$$

At, or immediately after, maximum the multiple absorption lines develop P Cyg profiles, the strongest lines are of H I, Ca II, Na I and Fe II. A few days after maximum emission lines of [O I] and [N II] appear, followed by [O III]. In the ultraviolet, the spectrum contains strong emission lines, primarily resonance and low-lying intercombination lines of abundant ions of He, C, N, O, Mg, Al and Si.

2.5.3 Diffuse enhanced spectrum

The third distinctive suite of absorption lines, similar to that of the principal spectrum but about twice as broad and blue-shifted (again correlated with speed class), appears shortly after maximum light (typically from 1 to 20 days, according to speed class). This is the diffuse enhanced system, which also has P Cyg profiles (the broad emission components underlying those of the principal spectrum), and lasts for about two weeks in very fast novae to three or four months in typical slow novae. In the later stages the broad absorptions often show multiple narrow structures. For the average velocities McLaughlin (1960) found

$$\log V_{ej}\,(\mathrm{km\,s^{-1}}) = 3.81 - 0.41 \log t_3\,(\mathrm{d}) = 3.71 - 0.4 \log t_2\,(\mathrm{d}). \tag{2.3}$$

2.5.4 *Orion spectrum*

As if not already complicated enough, after novae have declined by one to two magnitudes from maximum a further system appears – the Orion spectrum, named for its similarity to the stellar wind lines in luminous OB stars. Its absorption lines are usually single, diffuse and blue-shifted by at least as much as the diffuse enhanced system, from –2700 to –1000 km s^{-1} according to speed class. These velocities become steadily larger until the Orion spectrum disappears (about 2 mag down from maximum for slow novae and 4 mag down for fast novae). The lines in the optical are at first predominantly from He I, C II, N II and O II (not always with accompanying H I), and later emission lines of N III and N V appear. There are occasional dramatic changes in line strength, both in absorption and emission components. Particular species can become very strong; 'nitrogen flaring' is an example, with exceptional enhancement of the N III multiplet near 4640 Å and other N II and N III lines. Even before the Orion spectral phase an O I 'flash' can occur, where the O I 7772, 8448 Å lines rival Hα in strength.

In the ultraviolet, the emission lines do not change greatly during the first three of these spectral stages, other than by increasing in equivalent width as the continuum decays. The sequence of events is a result of reduction in opacity of the outer expanding layers, resulting in the photosphere moving inwards. In the initial optically thick phase the envelope is ionization bounded and the outermost gas can produce neutral and low ionization lines. Later, ionizing radiation from the hot interior escapes and, with the additional effect of decreasing densities, steadily increases the ionization state of the outer layers. The multiple line components arise from the variety of blobs of gas being ejected at different speeds in different directions (often with systematically different speeds in equatorial and polar directions).

2.5.5 *The nebular spectrum*

The final recognized distinctive stage is the development of an entirely emission spectrum, at first retaining the [O I] and [N II] components of the principal spectrum, and then producing [O III] and [Ne III], steadily evolving towards the appearance of the spectrum of a planetary nebula. Eventually 'coronal lines' develop if the ionizing radiation reaches temperatures over 10^6 K (see Chapter 8), with lines from successive ionization states up to [Fe XIV] and other high ions from abundant elements. Such lines are also seen in the ultraviolet.

Some novae show extremely strong lines of neon, especially [Ne III] and [Ne V], particularly prominently in the nebular phase, which has led to a class known as 'neon novae'. These come predominantly from the fast and very fast categories.

2.5.6 *The post-nova spectrum*

After some years to decades, the ejected shell may become spatially resolvable, so spectra of individual components of the ejecta, and images of the expanding shell in the light of individual spectral lines, can be obtained (see Chapter 12). The emission spectra of these shells are not like those of other gaseous nebulae – the lines are principally of permitted recombination lines of H, He, C, N and O, with the addition of [N III] and strong Balmer continuum emission.

2.5.7 *Modern spectral classification*

Williams (1990, 1992) has proposed a classification scheme for the emission line spectra of novae that takes into account modern understanding of the changing photo-ionization caused by expansion of the shell and hardening of the central radiation source. The high densities in the early stages restrict the lines to permitted transitions, of two classes – 'Fe II' and 'He/N' according to which species dominate. The Fe II novae, constituting about 60% of the total, have $1000 < V_{ej}\ (\mathrm{km\,s^{-1}}) < 3500$ and, in general, evolve into either 'standard' novae ($\sim$45% of all novae) or neon novae. The He/N novae have $V_{ej} > 2500\ \mathrm{km\,s^{-1}}$. All novae with $V_{ej} > 2500\ \mathrm{km\,s^{-1}}$ tend to evolve into either neon novae (30% of all novae), or into the 'coronal' class (15% of the total). The remaining 10% fall into a class having no forbidden lines.

2.6 Novae as close binaries

Most, but not all, of the observed aspects of erupting novae carry no imprint of their underlying binarity. It was Walker's discovery (Walker, 1954) of eclipses in DQ Her (Nova Her 1934) and Kraft's classic paper 10 years later (Kraft, 1964) that showed that it is virtually certain that all novae (and, more generally, all cataclysmic variables – CVs) are close binaries, in general with short orbital periods (typically hours). Here we will list some of the basic data for these systems and point out features of relevance to eruptions and some of the interconnections between CVs of different types (see also Chapter 3).

The dimensions of a binary system are set by its orbital period P_{orb} and by the masses of its components. For Roche-lobe-filling main sequence secondaries the mass is uniquely determined by P_{orb} and is approximately

$$M_2/\mathrm{M}_\odot = 0.065 P_{orb}^{5/4}\ (\mathrm{h})$$

(Warner, 1995). The masses of the white dwarf primaries can have a wide range up to the Chandrasekhar Limit, but are usually found to be well above $\sim 0.5\,\mathrm{M}_\odot$ in novae, which is an observational selection effect resulting from the higher frequency of eruptions at higher mass. The frequency distribution of P_{orb} is a useful clue to the orbital evolution (Nelson, MacCannell & Dubeau, 2004), so we give here (Table 2.5) a list of known orbital periods (Warner, 2002), supplemented with later published values and work by Woudt and Warner. Any system with $P_{orb} \gtrsim 8\,\mathrm{h}$ will have a secondary that is above the main sequence; in the two longest period systems the secondary is a red giant. An attempt to correct the observed distribution for various selection effects has been made by Diaz and Bruch (1997).

For the deeply eclipsing BT Mon, P_{orb} is found from archival photographic plates to have been smaller by 4 parts in 10^5 before eruption, which gives a dynamical estimate of $3 \times 10^{-5}\,\mathrm{M}_\odot$ for the mass of the ejecta (Schaefer & Patterson, 1983) (they discount the only other claimed P_{orb} change, in DQ Her by Ahnert (1960)). The recurrent nova CI Aql was observed extensively before and after its 2000 outburst and showed a P_{orb} increase of 4 parts in 10^6 (Lederle & Kimeswenger, 2003).

Table 2.5. *Orbital periods of novae*

Nova	Year[a]	Magnitude (mag)	P_{orb} (h)	Nova	Year	Magnitude (mag)	P_{orb} (h)
RW UMi	1956	6–18.5	1.418	DO Aql	1925	8.7–16.5	4.026
GQ Mus	1983	7.2–18.3	1.425	V849 Oph	1919	7.3–17	4.146
CP Pup	1942	0.5–15.2	1.474	GI Mon	1918	5.6–16.6	4.32
T Pyx	1902R	6.5–15.3	1.829	V697 Sco	1941	10.2–19.7	4.49
V1974 Cyg	1992	4.2–15.2	1.950	DQ Her	1934	1.3–14.6	4.647
RS Car	1895	7.0–18.5	1.980	CT Ser	1948	7.9–16.6	4.68
DD Cir	1999	7.7–20.2	2.340	T Aur	1891	4.2–15.2	4.906
IM Nor	1920R	7.8–22.0	2.46	V446 Her	1960	3.0–17.8	4.97
V Per	1887	9.2–18.5	2.571	V533 Her	1963	3.0–15.0	5.04
QU Vul	1984	5.6–17.5	2.682	HZ Pup	1963	7.7–17.0	5.11
V2214 Oph	1988	8.5–20.5	2.804	AP Cru	1936	10.7–18.0	5.12
V630 Sgr	1936	1.6–17.6	2.831	HR Del	1967	3.5–12.3	5.140
V351 Pup	1991	6.4–19.0	2.837	V1425 Aql	1995	7.5–>19	6.14
V4633 Sgr	1998	7.4–>20	3.014	V4743 Sgr	2002	5.0–16.8	6.74
DN Gem	1912	3.5–16.0	3.068	BY Cir	1995	7.2–17.9	6.76
V1494 Aql	1999	4.0–>16	3.232	V838 Her	1991	5.4–15.4	7.143
V1668 Cyg	1978	6.7–19.8	3.322	BT Mon	1939	8.5–16.1	8.012
V603 Aql	1918	−1.1–11.8	3.324	V368 Aql	1936	5.0–15.4	8.285
DY Pup	1902	7.0–19.6	3.336	QZ Aur	1964	6.0–17.5	8.580
V1500 Cyg	1975	2.2–18.0	3.351	Q Cyg	1876	3.0–15.6	10.08
V909 Sgr	1941	6.8–20	3.36	DI Lac	1910	4.6–15.0	13.050
RR Cha	1953	7.1–18.4	3.370	V841 Oph	1848	4.2–13.5	14.50
RR Pic	1925	1.0–12.1	3.481	CI Aql	1917R	8.8–15.6	14.84
V500 Aql	1943	6.6–17.8	3.485	V723 Cas	1995	4.2–13.5	16.638
V533 Her	1963	3.0–14.8	3.53	CP Cru	1996	9.2–19.6	22.7
V992 Sco	1992	8.3–17.2	3.686	U Sco	1863R	8.8–19.5	29.53
WY Sge	1783	5.4–20.7	3.687	X Ser	1903	8.9–18.3	35.5
OY Ara	1910	6.0–17.5	3.731	GK Per	1901	0.2–13.9	47.923
V1493 Aql	1999	10.4–>21	3.74	V1017 Sgr	1919	7.2–13.7	137.1
V382 Vel	1999	2.7–16.6	3.795	T CrB	1866R	2.0–11.3	227.57 d
V4077 Sgr	1982	8.0–22	3.84	RS Oph	1898R	4.3–12.5	455.72 d

[a] R = recurrent.

The binary nature of novae is evinced during and after eruption through its imprint on the ejecta, arising from a bipolar wind and/or anisotropic irradiation of the shell. In addition, the transfer of energy and angular momentum between the expanding shell and the secondary is able to account for the observed axial symmetry and clumpiness of the ejecta (Lloyd, O'Brien & Bode, 1997) – see Chapter 12.

2.7 Recurrent novae

With recognition of the binary nature of novae we can now look at the nature of recurrent novae. Although there are only nine definite (and a few possible) recurrent novae, they group themselves into three very distinct subtypes:

(1) The T CrB type have giant secondaries; their high frequency of eruptions is attributable to the high rate of mass transfer $\dot{M}$ generated by giants. The only definite members of the class are T CrB, V745 Sco, V3890 Sgr and RS Oph. There is some possible overlap or confusion between this class and the symbiotic novae, which have giant secondaries underfilling their Roche lobes.

(2) The U Sco type, comprising U Sco, V394 CrA, CI Aql and LMC-RN, also have evolved secondaries, if U Sco and CI Aql with their periods ~ 1 d are typical of the class. The visibility of the secondary's spectrum in each supports this. They are among the fastest novae observed, making it probable that some of their eruptions have been overlooked. In quiescence their spectra, unlike normal novae, are dominated by He lines.

(3) The T Pyx type has been a class with only one member, T Pyx itself, which shows a relatively slow decay ($t_3 = 88$ d) with oscillations in the transition region. Its short $P_{\rm orb}$ (1.83 h) also put it in a class by itself, but the recently recognized recurrent nova IM Nor, with $P_{\rm orb} = 2.46$ h and $t_3 = 50$ d allies it to T Pyx. T Pyx itself, has a quiescent bolometric luminosity an order of magnitude greater than expected, and is suspected of being in a rapid and possibly terminal phase of evolution (Knigge, King & Patterson, 2000).

The three recurrent nova subtypes are therefore variously different from normal classical novae, which suggests that their short intervals between eruptions are not simply the observable end of a wide distribution of $t_{\rm r}$.

2.8 Novae in quiescence and before eruption

In most cases the immediate pre-eruption and eventual post-eruption magnitudes of novae are found to be very similar (Robinson, 1975; but see below): a renegade, BT Mon, was later found to conform (Schaefer & Patterson, 1983). However, notable exceptions are GQ Mus (1983), CP Pup (1942) and V1500 Cyg (1975), and to a lesser extent V2214 Oph, V1974 Cyg and RW UMi, all of which rose from exceptionally faint magnitudes (but they have returned to more normal post-eruption magnitudes). These stars were evidently in very low states of $\dot{M}$ prior to eruption, though V1500 Cyg rose to a higher state about a week before eruption.

Anticipation of an eruption is not uncommon (Robinson, 1975): pre-eruption light curves of nearly half of the novae with good coverage showed slow rises of 0.25–1.5 mag in the 1 to 15 years prior to eruption, the most dramatic being V533 Her (Figure 2.4).

At maximum brightness a nova has roughly spherical symmetry, but in post-eruption quiescence it is powered by accretion luminosity, arising largely from the almost two-dimensional accretion disc. There is therefore a strong dependence of quiescent magnitude on orbital inclination, which transfers to the amplitude A (Warner, 1986). In Figure 2.3 this is shown by

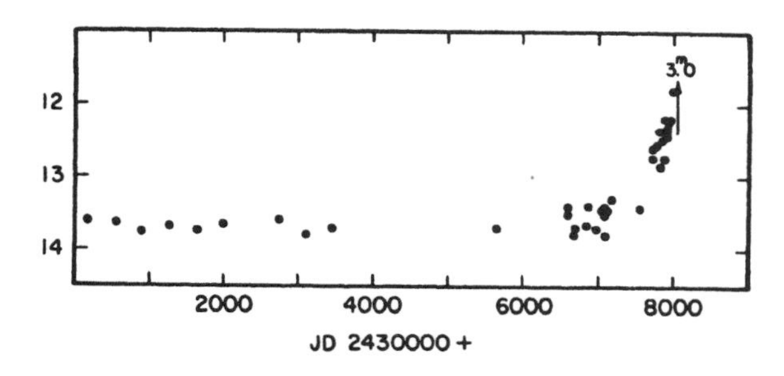

Fig. 2.4. Pre-eruption light curve of V533 Her. From Robinson (1975).

the lines drawn for a range of inclinations. In particular, this fully accounts for the very large amplitudes of outburst of such high inclination novae as DQ Her and T Aur, which caused Payne-Gaposchkin (1957) to wonder whether they belonged to a different class. In fact, the inclination effect on the M_V of novae is one of the most direct ways of demonstrating the existence of thin discs (Figure 2 of Warner, 1987) and has been seen also in IUE observations of the ultraviolet luminosities of quiescent novae (Selvelli, 2004).

A selection of light curves of novae at or approaching quiescence is shown in Figure 2.5. These cover a limited range of relatively high inclinations because they are chosen to show orbital modulations. The light curves of quiescent novae typically show a great deal of activity on short time-scales (seconds to minutes).

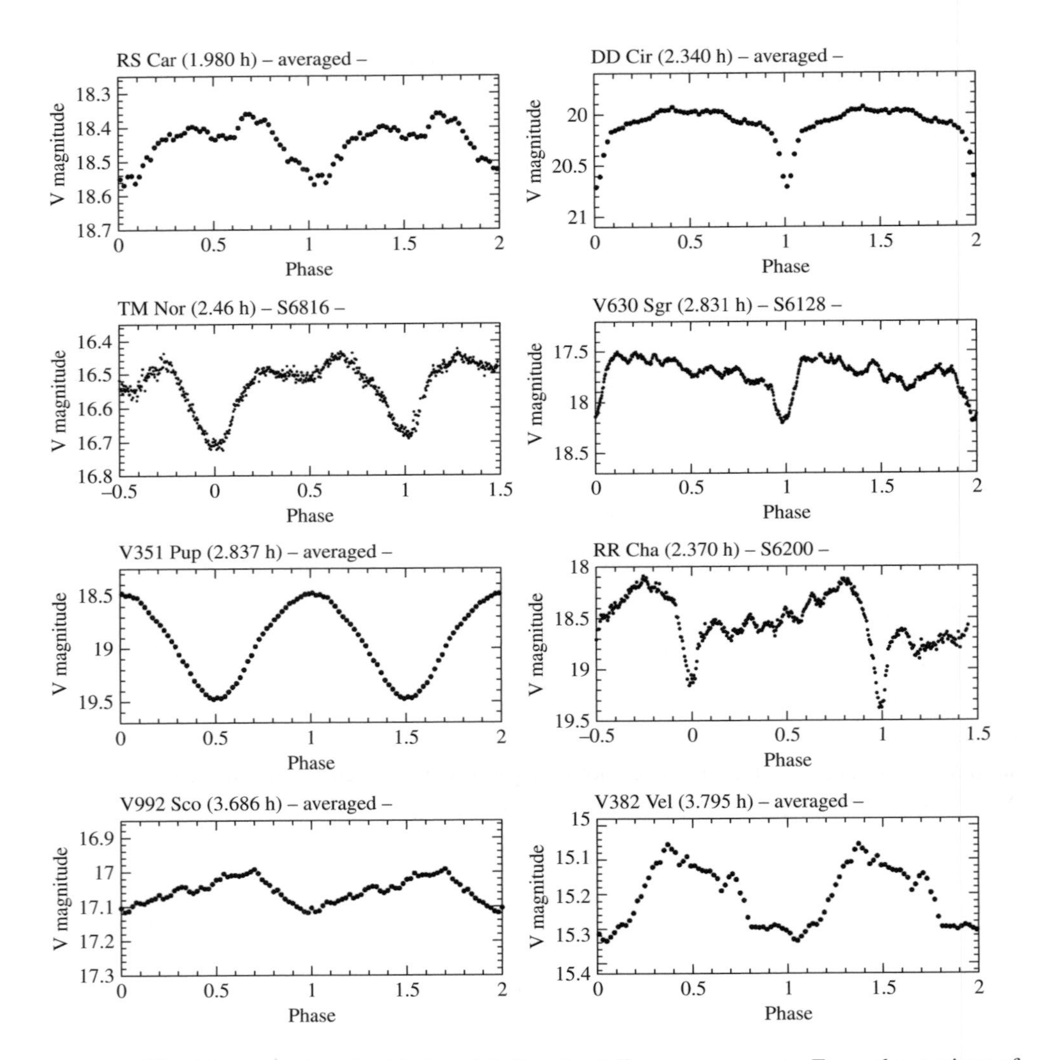

Fig. 2.5. Examples of orbital modulations in stellar nova remnants. From observations of P. A. Woudt and B. Warner.

Spectra of quiescent novae are typical of high $\dot{M}$ discs: blue continua arising from the optically thick disc and either or both of broad Balmer absorption or emission lines. In a few systems the lines are very weak: there is a strong correlation between equivalent width and inclination, with the face-on discs having very weak lines because the continuum is then at its strongest (Warner, 1986, 1987). Generally He II 4686 Å and C III/N III 4650 Å are in emission, which distinguishes these from dwarf novae and is probably the result of ionization by the strong ultraviolet radiation from the primary, which was heated during the nova eruption. There are many CVs classified as 'nova-like' variables because of their similarity in binarity and spectra to quiescent novae; these are further examples of high $\dot{M}$ systems. Among them will be post-novae whose eruptions went unrecorded, and novae waiting to happen. The slow rise in brightness before eruption, mentioned above, is a signature to watch for among the nova-likes.

The theory of low $\dot{M}$ discs postulates an instability that results in storage of mass in the disc until a thermal instability sets in and changes $\dot{M}$ through the disc to a higher value, dumping the accumulated mass onto the primary. The result is recurrent outbursts that produce the dwarf nova class of CVs. The high $\dot{M}$ state of post-nova discs prohibits this instability, but its eventual decrease allows a transformation from stable disc to full amplitude dwarf nova outbursts. This has been seen in only a few novae – V446 Her, GK Per and V1017 Sgr, all of which have quite long orbital periods and hence large separations. This is relevant because irradiation from the hot primary should keep smaller discs entirely above the critical temperature (and hence $\dot{M}$) at which instability sets in. If only the inner regions of the disc are maintained above the critical point then 'stunted' dwarf nova outbursts (involving only the outer annuli) could be possible – and are indeed seen in some old novae (Honeycutt, Robertson & Turner, 1998). Claims of possible outbursts seen in pre-novae V446 Her (Robinson, 1975) and V3890 Sgr (Dinerstein, 1973) have been rejected as being unlike true dwarf novae (Robinson, 1975; Warner, 2002). But in fact, most high $\dot{M}$ discs vary in brightness by a magnitude or more, sometimes cyclically with time-scales typical of dwarf nova outbursts (Honeycutt, 2001), and are probably 'stunted' outbursts.

The nature of novae centuries after eruption is an area of some dispute (see also Chapter 1). The effect on the secondary of irradiation during eruption leads to enhanced $\dot{M}$, which should be maintained for a century or so until the primary has cooled down (Prialnik, 1986). To maintain the long-term average $\dot{M}$ (which is determined by mechanisms of angular momentum loss from the orbit) the excesses during eruption should be followed by a long phase of low $\dot{M}$ – perhaps with $\dot{M}$ effectively becoming zero, known as 'hibernation' (Kovetz, Prialnik & Shara, 1988). The question has been whether hibernation has any observational support. There is a case that this does not take place within ~ 200 yr (Somers, Ringwald & Naylor, 1997). But high $\dot{M}$ may be sustained for centuries by the positive feedback effect of irradiation (Warner, 2002), which will be stronger for smaller separations and thus may account for the recent finding that novae with $P_{\mathrm{orb}} \lesssim 5$ h are much brighter after than before eruption (Duerbeck, 2003).

The best observational evidence for the disappearance of old novae is that none of the remnants of very bright novae noted in the Oriental records of one or two millennia ago, many of which would have become nova-likes at $m_V \sim 14{-}16$, are observable today (Shara, 1989); see Chapter 1 for further details. Furthermore, two binaries in which the secondaries just fill their Roche lobes and have all the appearance of hibernating novae have recently been found: BPM 71214 at $m_V = 13.6$ and EC 13471-1268 at $m_V = 14.8$, with orbital periods of

Table 2.6. *Novae that are intermediate polars*

Nova	Eruption	P_{orb} (h)	P_{rot} (s)	Reference
V533 Her	1963	5.04	64	1
DQ Her	1934	4.65	71	2
V373 Sct	1975	–	258	3
GK Per	1901	47.92	351	4
DD Cir	1999	2.34	670	3
HZ Pup	1963	5.11	1212	5
AP Cru	1936	5.12	1837	6
RR Cha	1953	3.37	1950	6
GI Mon	1918	4.32	2916	7
V1425 Aql	1995	5.42	5188	8
V697 Sco	1941	4.49	11916	9

References: 1. Patterson (1979); 2. Walker (1957); 3. Woudt & Warner (2003); 4. Watson, King & Osborne (1985); 5. Abbott & Shafter (1997); 6. Woudt & Warner (2002); 7. Woudt, Warner & Pretorius (2004); 8. Retter, Leibowitz & Kovo-Kariti (1998); 9. Warner & Woudt (2002a).

4.84 h and 3.62 h respectively (Kawaka & Vennes, 2003; O'Donoghue *et al.*, 2003). Another star, LTT 329, has also long been thought to be a candidate (Bragaglia *et al.*, 1990).

There is therefore an increasing case to be made for the idea of 'cyclical evolution' (Shara *et al.*, 1986; Warner, 1987, 1995), where all subtypes of CVs are simply different states of the same basic objects, where $\dot{M}$ is the dominant determinant of behaviour and hence of CV type, and where transitions between them are driven by nova eruptions on time-scales of $10^3 - 10^5$ yr (see Chapter 3).

2.9 Magnetic novae

To the above picture must be added novae that occur in CVs where the primary has a strong magnetic field. At least one-quarter of CVs have magnetic fields strong enough to determine the path of mass transfer – for the strongest ($B \gtrsim 10^7$ G), known as polars, the field lines connect to the inter-star stream and prevent any accretion disc from forming. In the intermediate polars ($10^6 \lesssim B \lesssim 10^7$ G) a disc forms but the inner parts are truncated where the primary's magnetosphere controls the flow (see Warner, 1995, for a review of these structures). Observed nova eruptions have occurred in the polar V1500 Cyg and the intermediate polars listed in Table 2.6. In all cases the magnetic nature of the primary was only discovered from observations made after the eruption. In addition, analysis of the eclipses in V Per (nova in 1887) indicates that the central parts of its disc are missing (Shafter & Abbott, 1989), which is a signature of an intermediate polar.

In polars the rotation of the primary is normally synchronized (through magnetic linkage to the secondary) with orbital revolution, but V1500 Cyg was found to have $P_{orb} = 201.04$ min and $P_{rot} = 197.55$ min, with the latter increasing at a rate such that synchronism will occur in ~ 185 yr (Schmidt *et al.*, 1995). V1500 Cyg is therefore regarded as a polar temporarily de-synchronized by angular momentum transfer during the 1975 eruption. The existence of at least three other polars (V1432 Aql, BY Cam and CD Ind) with P_{orb} and P_{rot} differing by $\sim 1\%$ argues for their also having been novae in the relatively recent past.

See Warner (2002) for a review of magnetism in novae.

2.10 Rapid photometric oscillations in novae

Rapid non-periodic flickering on time-scales of minutes or less was found by Walker (1957) to be a common property of novae in quiescence. Flickering is now known to be ubiquitous in CVs and is associated with the mass-transfer process, originating in the impact region of the stream on the disc, or in probably flare-like magnetic processes in the disc itself.

During his survey of old novae Walker (1956) found the earliest, but at the time unrecognized, evidence for a magnetic primary in a nova by his discovery of a highly stable photometric modulation in DQ Her with a period of 71.1 s. This is now realized to be a rapidly rotating intermediate polar; V533 Her also for a few years showed a 63.6 s modulation (Patterson, 1979).

Another common feature of short time-scale photometry of CVs, especially among those with high $\dot{M}$ discs, is the occasional presence of quasi-periodic brightness variations. There are two principal types – those known as dwarf nova oscillations (DNOs) with periods in the range 3–100 s and moderate stability, and quasi-periodic oscillations (QPOs) of low stability and periods typically of 80–2000 s (reviewed in Warner, 2004). Although commonly seen in dwarf novae in outburst and in nova-likes, for an unknown reason they are rare among novae, especially DNOs, where the only known are in RR Pic (20–40 s). As the interpretation of DNOs involves a model similar to that of intermediate polars (Warner & Woudt, 2002b) it is possible that the very high post-eruptive $\dot{M}$ in a nova in general crushes the primary's magnetic field to the surface, preventing magnetically controlled accretion. QPOs have been seen at 1400 s in V533 Her, at 1800 s in BT Mon, at 5000 s in GK Per and at 750–1300 s in V842 Cen.

2.11 Conclusions

This chapter gives only an initial taste of the intricacies and value of nova research. Novae are observable over a very wide range of wavelengths, from gamma rays to radio waves, and discussion of these constitutes the majority of the later chapters in this book. Furthermore, novae are important events in the lives of CVs, which themselves are optically the most easily observable of the general class of interacting binaries. The role that nova explosions play in the evolution of CVs is still being worked out. Models of the explosions themselves involve physics at the forefront of knowledge (see Chapter 6), and therefore novae can act as test beds for such branches of physics as nuclear reactions, propagation of shock waves, radiative transfer in moving media, hydrodynamics and the structure of white dwarfs.

Acknowledgements

The author's research is supported financially by the University of Cape Town. Comments by Hilmar Duerbeck and the assistance of Patrick Woudt in preparation of the final manuscript are gratefully acknowledged. This chapter was prepared while the author was a visitor at the Nicolaus Copernicus Astronomical Center in Warsaw.

References

Abbott, T. M. C., & Shafter, A. W., 1997, in *Accretion Phenomena and Related Outflows*, ed. D. T. Wickramasinghe, G. V. Bicknell, & L. Ferrario. ASP Conf. Ser. **121**, p. 679.
Ahnert, P., 1960, *AN*, **285**, 191.
Bode, M. F., & Evans, A. (eds.), *Classical Novae*, 1st edn. New York and Chichester: Wiley.
Bragaglia, A., Greggio, L., Renzini, A., & D'Odorico, S., 1990, *Ap. J.*, **365**, L13.
Buscombe, W., & de Vaucouleurs, G., 1955, *Observatory*, **75**, 170.
Capaccioli, M., Della Valle, M., Rosino, L., & D'Onofrio, M., 1989, *AJ*, **97**, 1622.
Ciardullo, R., Tamblyn, P., Jacoby, G. H., Ford, H. C., & Williams, R. E., 1990, *AJ*, **99**, 1079.
Cohen, J. G., 1985, *Ap. J.*, **292**, 90.
Darnley, M. J., Bode, M. F., Kerins, E. J. *et al.*, 2006, *MNRAS*, **369**, 257.
de Vaucouleurs, G., 1978, *Ap. J.*, **223**, 351.
Della Valle, M., 1991, *A&A*, **252**, L9.
Della Valle, M., & Livio, M., 1995, *Ap. J.*, **452**, 704.
Della Valle, M., Bianchini, A., Livio, M., & Orio, M., 1992, *A&A*, **266**, 232.
Diaz, M. P., & Bruch, A., 1997, *A&A*, **332**, 807.
Dinerstein, H. L., 1973, *J. Am. Assoc. Variable Star Observers*, **2**, 71.
Downes, R. A., & Duerbeck, H. W., 2000, *AJ*, **120**, 2007.
Downes, R. A., & Shara, M. M., 1993, *PASP*, **105**, 127.
Downes, R. A., Webbink, R. F., Shara, M. M. *et al.*, 2005, *J. Astron. Data*, **11**, 2.
Duerbeck, H. W., 2003, in *Interplay of Periodic, Cyclic and Stochastic Variability in Selected Areas of the H–R Diagram*, ed. C. Sterken. Astronomical Society of the Pacific, ASP Conf. Ser. **292**, p. 323.
Gallagher, J. S., & Code, A. D., 1974, *Ap. J.*, **189**, 303.
Geisel, S. L., Kleinmann, D. E., & Low, F. J., 1970, *Ap. J.*, **161**, L101.
Gill, C. D., & O'Brien, T. J., 2000, *MNRAS*, **314**, 175.
Harman, D. J., & O'Brien, T. J., 2003, *MNRAS*, **344**, 1219.
Honeycutt, R. K., 2001, *PASP*, **113**, 473.
Honeycutt, R. K., Robertson, J. W., & Turner, G. W., 1998, *AJ*, **115**, 2427.
Hubble, E. P., 1929, *Ap. J.*, **69**, 103.
Hyland, A. R., & Neugebauer, G., 1970, *Ap. J.*, **160**, L177.
Kawaka, A., & Vennes, S., 2003, *Ap. &SS*, **296**, 481.
Knigge, C., King, A. R., & Patterson, J., 2000, *A&A*, **364**, L75.
Kovetz, A., Prialnik, D., & Shara, M. M., 1988, *Ap. J.*, **325**, 828.
Kraft, R. P., 1964, *Ap. J.*, **139**, 457.
Lederle, C., & Kimeswenger, S., 2003, *A&A*, **397**, 951.
Liller, W., & Mayer, B., 1987, *PASP*, **99**, 606.
Liller, W., Shao, C. Y., Mayer, B. *et al.*, 1975, IAUC 2848.
Livio, M., 1992, *Ap. J.*, **393**, 516.
Lloyd, H. M., O'Brien, T. J., & Bode, M. F., 1997, *MNRAS*, **284**, 137.
McLaughlin, D. B., 1939, *Pop. Astr.*, **47**, 410, 481, 538.
McLaughlin, D. B., 1942, *Pop. Astr.*, **52**, 109.
McLaughlin, D. B., 1944, *Publ. Obs. Univ. Michigan*, **8**, 149.
McLaughlin, D. B., 1945, *PASP*, **57**, 69.

McLaughlin, D. B., 1960, in *Stellar Atmospheres*, ed. J. L. Greenstein. Chicago: University of Chicago Press.
Nelson, L. A., MacCannell, K. A., & Dubeau, E., 2004, *Ap. J.*, **602**, 938.
O'Donoghue, D., Koen, C., Kilkenny, D. *et al.*, 2003, *MNRAS*, **345**, 506.
Patterson, J., 1979, *Ap. J.*, **233**, L13.
Payne-Gaposchkin, C., 1957, *The Galactic Novae*. Amsterdam: North Holland Publishing Co.
Pfau, W., 1976, *A&A*, **50**, 113.
Plaut, L., 1965, in *Galactic Structure*, ed. A. Blaauw, & M. Schmidt. Chicago: University of Chicago Press, p. 311.
Prialnik, D., 1986, *Ap. J.*, **310**, 222.
Retter, A., Leibowitz, E. M., & Kovo-Kariti, O., 1998, *MNRAS*, **293**, 145.
Robinson, E. L., 1975, *AJ*, **80**, 515.
Schaefer, B. E., & Patterson, J., 1983, *Ap. J.*, **268**, 710.
Schmidt, G. D., Liebert, J., & Stockman, H. S., 1995, *Ap. J.*, **441**, 414.
Schmidt-Kaler, T., 1957, *Z. Astrophys.*, **41**, 182.
Selvelli, P., 2004, *Baltic Astron.* **13**. 90.
Shafter, A. W., 2002, in *Classical Nova Explosions*, ed. M. Hernanz, & J. José. New York: American Institute of Physics, p. 462.
Shafter, A. W., & Abbott, T. M. C., 1989, *Ap. J.*, **339**, L75.
Shara, M. M., 1981a, *Ap. J.*, **243**, 926.
Shara, M. M., 1981b, *Ap. J.*, **243**, 268.
Shara, M. M., 1989, *PASP*, **101**, 5.
Shara, M. M., & Drissen, L., 1995, *Ap. J.*, **448**, 203.
Shara, M. M., Livio, M., Moffat, A. F. J., & Orio, M., 1986, *Ap. J.*, **311**, 163.
Slavin, A. J., O'Brien, T. J., & Dunlop, J. S., 1995, *MNRAS*, **276**, 353.
Somers, M. W., Ringwald, F. A., & Naylor, T., 1997, *MNRAS*, **284**, 359.
van den Bergh, S., 1975, in *Galaxies and the Universe*, ed. A. Sandage, M. Sandage, & J. Kristian. Chicago: University of Chicago Press, p. 509.
Wade, R. A., Harlow, J. J. B., & Ciardullo, R., 2000, *PASP*, **112**, 614.
Walker, M. F., 1954, *PASP*, **66**, 230.
Walker, M. F., 1956, *Ap. J.*, **123**, 68.
Walker, M. F., 1957, in *Non-stable Stars*. ed. G. H. Herbig. Cambridge: Cambridge University Press, p. 46
Warner, B., 1986, *MNRAS*, **222**, 11.
Warner, B., 1987, *MNRAS*, **227**, 23.
Warner, B., 1995, *Cataclysmic Variable Stars*. Cambridge: Cambridge University Press.
Warner, B., 2002, in *Classical Nova Explosions*, ed. M. Hernanz, & J. José. New York: American Institute of Physics, p. 3.
Warner, B., 2004, *PASP*, **116**, 115.
Warner, B., & Woudt, P. A., 2002a, *PASP*, **114**, 1222.
Warner, B., & Woudt, P. A., 2002b, *MNRAS*, **335**, 84.
Watson, M. G., King, A. R., & Osborne, J., 1985, *MNRAS*, **212**, 917.
Williams, R. E., 1990, in *Physics of Classical Novae*, ed. A. Cassatella, & R. Viotti. New York and Berlin: Springer, p. 215.
Williams, R. E., 1992, *AJ*, **104**, 725.
Woudt, P. A., & Warner, B., 2002, *MNRAS*, **335**, 44.
Woudt, P. A., & Warner, B., 2003, *MNRAS*, **340**, 1011.
Woudt, P. A., Warner, B., & Pretorius, M. L. (2004). *MNRAS*, **351**, 1015.

3

The evolution of nova-producing binary stars

Icko Iben Jr. and Masayuki Y. Fujimoto

3.1 Introduction

We discuss the evolution of both stellar components of cataclysmic variables (CVs) and symbiotic stars from formation to termination, identifying the modes and estimating the rates of mass tranfer as functions of the period of the system, and suggesting how the composition of the nova ejecta depends on the rate of mass transfer, the processes of mixing between accreted material and material in the underlying white dwarf, and the mass-transfer, mixing, and prior outburst history of the system.

Left completely out of the discussion is the disk component which, in CVs and perhaps in some symbiotic stars, mediates mass transfer between the mass donor and the white dwarf accretor. We begin with an outline of the topics to be discussed.

(1) Definitions: CVs are here defined as close binary systems in which one component is a CO or ONe white dwarf and the other is a Roche-lobe-filling (or nearly Roche-lobe-filling) star which can be a main-sequence star, a small hydrogen-rich white dwarf, a red giant (Algol-like CV) or a helium white dwarf or helium main sequence star (helium CV). In symbiotic stars, the companion of the white dwarf is a red giant or Asymptotic Giant Branch (AGB) star which, in general, does not fill its Roche lobe.

(2) Precursors of CVs are binary star systems with the following properties: (a) the initial primary is massive enough to develop a hydrogen- and/or helium-exhausted core in less than a Hubble time and orbital separation is initially large enough that the primary develops this core before it fills its Roche lobe; (b) the initial orbital separation is small enough and the initial mass ratio is large enough that, when the primary fills its Roche lobe, a common envelope is formed, leading to orbital shrinkage sufficiently extensive that mass transfer from the initial secondary to the white dwarf remnant of the initial primary can occur, leading to nova outbursts in less than a Hubble time. Precursors of symbiotic stars differ from precursors of CVs by being much wider systems in which either (a) the primary does not fill its Roche lobe before it becomes a white dwarf; (b) the primary becomes a thermally pulsing AGB star with a dusty wind and common-envelope

Classical Novae, 2nd edition, ed. Michael Bode and Aneurin Evans. Published by Cambridge University Press.

formation is avoided; or (c) the initial components are of comparable mass, so that Roche-lobe filling does not lead to extensive orbital shrinkage.

(3) The condition that nova outbursts occur in less than a Hubble time requires that mechanisms which can drive mass transfer at a rate in excess of $\sim 10^{-11}\,M_\odot\,yr^{-1}$ exist and are activated. In CVs, if the donor is a main-sequence star of low mass, a magnetic stellar wind (MSW) and/or gravitational wave radiation (GWR) are thought to operate. If the donor is a Roche-lobe-filling red giant or sub-giant with a large electron-degenerate helium core, the mass-driving mechanism is envelope expansion due to the increase with time in the luminosity of the hydrogen-burning shell at the edge of the core. In symbiotic stars, accretion from the wind of the red giant or AGB component is the mechanism of mass transfer.

(4) A zeroth-order model can be constructed to describe the rate of evolution of a typical CV first through (a) a long-period orbit-shrinking phase ($3\,h \lesssim P_{orb} \lesssim 10$ h) during which a MSW forces mass transfer at rates of $\dot{M}_{acc} \sim 10^{-8} - 10^{-9}\,M_\odot\,yr^{-1}$, then through (b) a 'period gap' phase ($2\,h \lesssim P_{orb} \lesssim 3\,h$) when no mass transfer takes place and angular momentum loss due to GWR causes orbital shrinkage, next through (c) a short-period orbit-shrinking phase ($80\,min \lesssim P_{orb} \lesssim 2\,h$) during which GWR forces mass transfer at rates of $\dot{M}_{acc} \leq 2 \times 10^{-10}\,M_\odot\,yr^{-1}$, and finally through (d) a period-increasing phase which begins when the hydrogen-rich mass donor which has ceased to burn nuclear fuel becomes a very low mass white dwarf whose radius increases as it loses mass. The model predicts that novae occur $\sim$30–70 times more frequently in long-period than in short-period systems, despite the fact that there are perhaps 50–100 times as many short-period systems as long-period systems in the Galaxy.

(5) In the immediate precursor binary system, a helium buffer layer exists at the surface of the white dwarf and, after transfer of hydrogen-rich matter to the white dwarf begins, this layer may, in some systems, inhibit or even prevent mixing between heavy-element-rich white dwarf matter and accreted matter. The buffer layer may grow in mass as accreted hydrogen-rich matter burns to helium during ordinary nova oubursts; if it becomes massive enough for the ignition of helium, a helium-burning nova outburst is triggered. If the outburst is strong enough and the buffer layer attempts to expand to giant dimensions, most of the buffer layer is removed by a wind.

Prior to the removal of the buffer layer, mixing between buffer-layer helium and accreted matter can lead to large enhancements in the abundance of helium in the nova ejecta. If the mass of the buffer layer becomes small enough, mixing through the buffer layer between hydrogen-rich accreted matter and the heavy-element-rich matter in the interior of the white dwarf can lead to large overabundances of heavy elements in the nova ejecta.

Particle diffusion and thermal convection are two straightforward mixing processes whose consequences have been extensively explored in nova-modelling experiments which demonstrate that, with the right choice of parameters, the large overabundances in heavy elements seen in many observed nova spectra can be accounted for.

(6) A simple model can be constructed to follow the compositional history of the surface layers of the accretor, after many outbursts have led to the establishment of a cyclical 'steady state'. The model allows for an arbitrary degree of mixing and may be used to deduce the degree of mixing required to explain abundances observed in real novae. The model suggests that, in short-period CVs, particle diffusion and thermal convection lead to sufficiently extensive mixing between accreted material and the heavy-element-enriched matter of the white dwarf that heavy elements are enhanced during a nova outburst by

an order of magnitude (relative to solar) for all accretion rates of relevance. However, in long-period CVs in which there is much less time between outbursts for mixing to operate, mixing by particle diffusion and thermal convection do not appear to be capable of accounting for observed overabundances of heavy elements.

(7) The missing mechanism may be due to instabilities associated with differential rotation, which is expected to be present as the white dwarf is spun up by accretion of angular momentum; a baroclinic instability has been found to be effective in causing mixing on a dynamical time-scale. If the white dwarf is corotating with the orbital angular velocity, other forms of mixing such as Eddington–Sweet circulation may play a role.

(8) From the standpoint of stellar evolution, there are two major differences between CVs and symbiotic stars. The first difference is that, in symbiotics, neither the donor nor the accretor, even at maximum radius after a nova outburst, fills its Roche lobe. In CVs, the donor always fills its Roche lobe and, shortly after the start of an eruption, matter above the burning shell in the accretor expands far beyond the orbit defined by the donor and the core of the accretor, and this matter is lost from the system. Thus, all other things being equal, a symbiotic system can retain more accreted material than can a CV, adding helium to the helium buffer layer. The second major difference is that, in symbiotic stars, the mass-transfer mechanism is accretion from the wind of the red giant or AGB component. In general, since accretion from a wind can be inefficient, the total duration of the symbiotic system phase is expected in many cases to be too short for the buffer layer to achieve the critical mass for a helium flash to occur, so, although a large helium overabundance can appear in the nova ejecta, as observed, no large overabundances of heavy elements are expected, in agreement with extant observations.

(9) Neither CVs nor symbiotic stars are major precursors of Type Ia supernovae (SNe Ia): (a) the formation frequency of CVs is of the order of 10 times smaller than the frequency of SNe Ia in the Galaxy; (b) in CVs which produce expanding shells exhibiting large overabundances of heavy elements, the white dwarf loses more mass during outburst than it gains through accretion between outbursts, preventing the growth of a white dwarf massive enough to experience an explosion of supernova proportions; (c) in symbiotic stars, most of the matter in the wind flowing from the donor escapes from the binary system and, even though the white dwarf component may temporarily grow in mass by adding helium to the initial helium buffer layer, there is insufficient time and/or not enough accreted mass in most cases for the buffer layer to reach the critical mass needed for triggering a helium flash. If, in some systems, the critical helium layer mass is reached, a helium flash driven outburst forces the layer to expand to giant dimensions and wind-driven mass-loss removes most of the layer, again preventing the growth of a CO white dwarf core massive enough to experience a type Ia supernova explosion.

3.2 Origin and evolution: overview

Precursors of CVs are thought to be binaries consisting of a low mass secondary and a primary which fills its Roche lobe after developing an electron-degenerate core composed of carbon and oxygen or of oxygen, neon and a modest abundance of heavier elements. After a common envelope event which reduces the orbital separation until the secondary main-sequence component nearly fills its Roche lobe, the primary is converted into a white dwarf. In consequence of a MSW, the orbital separation continues to decrease until the secondary begins to transfer hydrogen-rich matter to the white dwarf, which experiences a hydrogen-burning thermonuclear runaway whenever the accreted mass exceeds a critical value of the order of $10^{-5} - 10^{-4}\,M_\odot$. Once the mass of the donor decreases below $\sim 0.22 - 0.3\,M_\odot$, GWR replaces the MSW as the main driver of mass transfer.

The primary in the precursor main-sequence system is of mass in the range 2–10 $M_\odot$ and the orbital separation is large enough that the primary does not fill its Roche lobe until it has exhausted either hydrogen or helium at its center and developed a hot white-dwarf core of mass in the range $\sim 0.5 - 1.3\,M_\odot$. When the primary overflows its Roche lobe, a common envelope is formed and, if the mass of the secondary is much smaller than the initial mass of the primary, most of the mass ejected by the primary into the common envelope leaves the system. The large difference in mass between the initial primary and its remnant means that considerable orbital energy is used up in expelling the common envelope, and this translates into a large decrease in orbital separation. After expulsion of the common envelope and some short-lived residual burning of hydrogen or helium, or both, the remnant becomes a cooling white dwarf.

In many instances, the initial secondary does not fill its Roche lobe immediately after the common envelope phase. However, if it is of small enough mass ($\lesssim 1.5\,M_\odot$) and is rotating rapidly enough, it will exhibit magnetic activity and support a stellar wind which carries off spin angular momentum in consequence of torques between the magnetic field, which is anchored in the secondary, and ionized particles in the wind. If the orbital shrinkage during the preceding common envelope phase has been sufficient, tidal torques exerted by the white dwarf component will keep the secondary spinning at a frequency nearly equal to the orbital frequency. The spin angular momentum of the secondary is maintained at the expense of orbital angular momentum and the orbital separation therefore decreases.

Ultimately, the secondary fills its Roche lobe, and the MSW drives mass transfer. The orbital period decreases from an initial value $P_{\rm orb} \sim 6 - 10$ h to $P_{\rm orb} \sim 3$ h. In the number–period distribution defined by known CVs, the density (number per period interval) of systems in the period interval 2–3 h is significantly smaller than the density on either side of the period 'gap'. This has been interpreted as being due to a sharp decline in the efficacy of the MSW at $P_{\rm orb} \sim 3$ h. A popular explanation is that the secondary becomes completely convective, and this leads to a re-arrangement in the geometry of the magnetic field which considerably reduces the torquing action between anchored magnetic field and ionized wind particles. In any case, since mass transfer effectively ceases, it must be that, during the preceding mass-transfer phase, the radius of the secondary is larger than that of an isolated star of the same mass. We infer that a low mass star is bloated in response to mass-loss and/or because magnetic activity creates a chromosphere which extends beyond the Roche lobe, or that some combination of these possibilities is operating. Experiments with stellar models show that mass-loss does indeed induce bloating, and the degree of bloating increases with increasing mass-loss rate.

By the time the period is as small as 3 h, GWR has become an effective way of decreasing the orbital angular momentum. After $\sim 5–8 \times 10^8$ yr of GWR, the orbital period has decreased to ~ 2 h; the secondary once again comes into contact with its Roche lobe and mass transfer resumes, this time driven by the loss of angular momentum by GWR.

Nova outbursts occur whenever the mass of hydrogen-rich material above the hydrogen-deficient interior of the white dwarf component exceeds a critical value $M_{\rm crit}$, which depends on the mass-accretion rate and on the mass and interior temperature of the white dwarf. During the early evolution of a CV, it also depends on the mass of a helium surface layer which the white dwarf posesses before mass transfer begins. In general, as a given CV evolves toward shorter periods, the mean mass-transfer rate decreases, the amount of mass accreted and the time between outbursts increase. Further, the interior temperature of the white dwarf decreases and the character of mixing of elements between the accreted layer and the interior changes, leading to a variation with time in the abundance patterns that appear in the nova ejecta. A change with time in the average mass of the white dwarf also contributes to the variation in abundance patterns.

3.3 Mean mass transfer rates in CVs and symbiotic stars

The first order of business is an estimate of long-term, time-averaged, mass-transfer rates. We bypass the complications associated with the formation of an accretion disk and of the dwarf-nova discharges of this disk, focusing exclusively on the accretion rate averaged over many classical nova-producing cycles.

3.3.1 Mass transfer driven by a magnetic stellar wind

The development here follows that of Tutukov (1983; see also Iben & Tutukov, 1984b; Iben, Fujimoto & MacDonald, 1992a). The Skumanich (1972) 'law' for magnetically active low-mass single stars is

$$V \sim 2\ \mathrm{km\,s^{-1}} \left(\frac{5\times10^{9}\ \mathrm{yr}}{t}\right)^{1/2},$$

where V is the equatorial velocity, and t is the age of the star. The spin angular momentum J_s of a single main-sequence star rotating as a solid body is

$$J_\mathrm{s} = 0.1\alpha MRV,$$

where M is the mass, R is the radius, and α is a parameter of order unity. It follows from these two equations that the rate of spin angular momentum loss is

$$\left(\frac{1}{J_s}\frac{\mathrm{d}J_\mathrm{s}}{\mathrm{d}t}\right)_{\text{single star}} = -10^{-10}\,\mathrm{yr}^{-1}\left(\frac{V}{2\ \mathrm{km\,s^{-1}}}\right)^{2}. \tag{3.1}$$

The orbital angular momentum of a binary system is given by

$$J_\mathrm{orb} = M_\mathrm{d}M_\mathrm{a}\left(\frac{GA}{M_\mathrm{t}}\right)^{1/2}, \tag{3.2}$$

where G is Newton's gravitational constant, A is the semimajor axis, M_d is the mass of the donor, M_a is the mass of the accretor, and $M_\mathrm{t} = M_\mathrm{a} + M_\mathrm{d}$ is the total mass of the system. The relationship between the semimajor axis A and the effective radius R_L of the Roche lobe enclosing the donor may be approximated by (Iben & Tutukov, 1984a,b)

$$R_\mathrm{L} = 0.52\left(\frac{M_\mathrm{d}}{M_\mathrm{t}}\right)^{0.44} A, \tag{3.3}$$

and the relationship between the radius and mass of the donor may be written as

$$\frac{R_\mathrm{d}}{\mathrm{R}_\odot} = \rho\left(\frac{M_\mathrm{d}}{\mathrm{M}_\odot}\right)^{1-\epsilon}, \tag{3.4}$$

where ρ and ϵ are parameters obtained by fitting to the observations. Setting $R_{\rm L} = R_{\rm d}$, and assuming that all of the mass lost by the donor is accreted by the white dwarf, Equations (3.2)–(3.4) give

$$\frac{1}{J_{\rm orb}}\frac{{\rm d}J_{\rm orb}}{{\rm d}t} = \frac{1}{M_{\rm d}}\frac{{\rm d}M_{\rm d}}{{\rm d}t}\left(1.28 - \frac{\epsilon}{2} - \frac{M_{\rm d}}{M_{\rm a}}\right). \tag{3.5}$$

If tidal forces maintain the spin period of the donor equal to the orbital period, the equatorial velocity $V_{\rm d}$ of the donor satisfies

$$V_{\rm d} = \left(\frac{R_{\rm d}}{A}\right)\left(\frac{GM_{\rm t}}{A}\right)^{1/2}. \tag{3.6}$$

Replacing V in Equation (3.1) by $V_{\rm d}$ from Equation (3.6), setting $\dot{J}_{\rm orb}/J_{\rm orb}$ given by Equation (3.5) equal to $\dot{J}_{\rm s}/J_{\rm s}$ in Equation (3.1), and using Equation (3.4), we obtain the mass-transfer rate

$$\dot{M}_{\rm MSW} = -\dot{M}_{\rm d} = \frac{\alpha}{\rho}\,\frac{M_{\rm d}^{\epsilon}\, q^{2.2} M_{\rm a}\,(M_{\rm a}/M_{\rm t})^{0.2}}{(1.28 - \epsilon/2 - q)}\; 1.8 \times 10^{-8}\,{\rm M}_\odot\,{\rm yr}^{-1},$$

where $\dot{M}_{\rm d} = {\rm d}M_{\rm d}/{\rm d}t, q = M_{\rm d}/M_{\rm a}$, and $M_{\rm a}$ and $M_{\rm d}$ are in solar units. For long-period systems ($P_{\rm orb}$ = 3–10 h), the choices $\rho \sim 0.86$ and $\epsilon \sim 1/6$ are consistent with models by Grossman, Hays and Graboske (1974) and by Iben and Tutukov (1984b) and, for $(M_{\rm d}, M_{\rm a}) = (0.3, 1)$, give $P_{\rm orb} = 3$ h and

$$\dot{M}_{\rm MSW} \sim \alpha\left(10^{-7.28}\, M_{\rm d}^{3.15}\right){\rm M}_\odot\,{\rm yr}^{-1}. \tag{3.7}$$

So, the time-scale for mass-loss by the MSW is

$$\tau_{\rm MSW} = \frac{M_{\rm d}}{\dot{M}_{\rm MSW}} = \frac{1}{\alpha}\,\frac{10^{7.28}}{M_{\rm d}^{2.15}}\ {\rm yr}.$$

With $\alpha = 1$ and $M_{\rm d}$ = 0.22–0.3 $\rm M_\odot$,

$$(\tau_{\rm MSW})_{P_{\rm orb}=3\,{\rm h}} \sim (2.5\text{–}4.9) \times 10^8\ {\rm yr},$$

with the longer time-scale being associated with a smaller donor mass.

For CVs with periods near $P_{\rm orb} = 3$ h, the mean of observationally estimated mass-transfer rates (Patterson, 1984, Figure 7) is about an order of magnitude smaller than suggested by Equation (3.7). However, the scatter in the estimated rates is two orders of magnitude, so a comparison is not very illuminating. Further, there is no assurance that mass-transfer rates shortly before or shortly after a nova outburst reflect the secular mass-transfer rate averaged over many outburst cycles. We could guess that there is an observational selection effect that favors finding CVs that have recently experienced an outburst and that mass-loss during outburst leads to a small adjustment in orbital characteristics and a temporary alteration in the mass-transfer rate.

3.3.2 *Inflation of low mass stellar models in response to mass-loss and the origin of the period gap*

A better understanding of the factors involved in producing the period gap requires, at the very least, an understanding of the interior properties of and the mass-radius relationship for low mass stars that are losing mass and have masses in the range $0.25 \pm 0.1\,M_\odot$. After Kumar (1963) first constructed models of very low mass stars, almost three decades passed before such stars attracted widespread theoretical attention, with observational searches for 'brown dwarfs' (stellar objects not massive enough for nuclear fuel to be their major source of luminosity) being one of the factors responsible for the revival of interest (Burrows & Liebert, 1994). With one notable exception (D'Antona & Mazzitelli (1994), who went to higher masses) much of the theoretical work in recent years has centered on the brown dwarf question, with attention being concentrated on stars less massive than $0.2M_\odot$ (e.g. Laughlin & Bodenheimer, 1993; Burrows *et al.*, 1993; Laughlin, Bodenheimer & Adams, 1997) and with minimal attention given to interior properties and no attention given to the effect of mass-loss on the mass-radius relationship.

Model-based calculations of CV orbital evolution take into account the response of the donor radius to mass-loss, but typically this response is not made explicit (e.g. Iben & Tutukov, 1984b). We have therefore conducted experiments to examine the response of a low mass star to mass-loss at different rates. We find that, when convection first extends all the way from the surface to the center, the larger the mass-loss rate, the smaller is the model mass, being 0.325, 0.28, and $0.18\,M_\odot$, respectively, for the mass-loss rates 10^{-10}, 10^{-9}, and $10^{-8}\,M_\odot\,\mathrm{yr}^{-1}$. When a model becomes fully convective, its radius is larger than that of a constant-mass model by 1%, 15%, and 97%, respectively, for the same three mass-loss rates. For a given mass-loss rate, the ratio of $R(M, \dot{M}_{\rm acc} < 0)$ to $R(M, \dot{M}_{\rm acc} = 0)$ increases with decreasing mass. For example, for $\dot{M}_{\rm acc} = 10^{-9}\,M_\odot\,\mathrm{yr}^{-1}$, $R(M, \dot{M}_{\rm acc} < 0)/R(M, \dot{M}_{\rm acc} = 0)$ is 1.1, 1.15, and 1.3 for $M = 0.3$, 0.25 and $0.2\,M_\odot$, respectively. Bloating occurs because, at any point in the model, the pressure becomes smaller as the weight of matter beyond this point becomes smaller.

Thus, a low mass star is indeed bloated when it is forced to transfer mass. However, many low mass stars have very active chromospheres, and it is quite possible that, in CVs, considerable matter is thrown out beyond the donor's Roche lobe by flares. It may be that, when magnetic activity dies down in CVs at the upper end of the period gap, both a decrease in donor radius and a diminution in flare activity contribute to the decrease in the mass-transfer rate. In any case, the reduction in radius when mass transfer ceases is supported by theoretical models. In order for the donor not to fill its Roche lobe again and resume mass transfer until the orbital period reaches $P_{\rm orb} \sim 2$ h, the donor radius must become ~31% smaller than the radius of the Roche lobe when $P_{\rm orb} \sim 3$ h. If our models could be trusted, i.e., were constructed with completely correct input physics, and if we could assert that flaring activity had nothing to do with mass transfer (other than being a part of the MSW activity), we could estimate the mass-transfer rate from the requirement that, when it first becomes fully convective, the donor radius is larger by ~31% than that of a model of the same mass which is not losing mass. For example, using this criterion and accepting our model results at face value, we would guess that, at the end of the long-period phase, $\dot{M}_{\rm acc} \sim 2 \times 10^{-9}\,M_\odot\,\mathrm{yr}^{-1}$ and donor mass $M_{\rm d} \sim 0.22\,M_\odot$. On the other hand, if we suppose that flaring activity accounts for half of the effective bloating, we could argue that convection extends to the stellar center when donor mass is $M_{\rm d} \sim 0.28\,M_\odot$ and that, for the last $\sim 6 \times 10^7$ yr of the long-period phase, the donor loses mass at the rate $\dot{M}_{\rm acc} \sim 10^{-9}\,M_\odot\,\mathrm{yr}^{-1}$.

As an aside, we note that the interior behavior of mass-losing models is quite different from that of constant-mass models. For example, for a mass-loss rate above a critical value (which is smaller than $10^{-10}\,M_\odot\,\mathrm{yr}^{-1}$ in our experiments), the entropy S of a model decreases with time, but the magnitude of the decrease is smaller, the larger the mass-loss rate. For any given model mass, the larger the mass-loss rate, the larger is the entropy; in fact, for the highest mass-loss rate examined, the model entropy remains close to an adiabat. And yet, at any given mass, the higher the mass-loss rate, the smaller are both the central temperature and the central density. Thus, as they expand, the mass-losing models also cool, and gravothermal energy makes a significant contribution to the model luminosity. For example, when the model which loses mass at the rate $10^{-9}\,\mathrm{M}_\odot\,\mathrm{yr}^{-1}$ becomes fully convective, its luminosity is $\sim 0.016\,\mathrm{L}_\odot$, but its nuclear burning luminosity is only $\sim 0.006\,\mathrm{L}_\odot$; over 60% of its surface luminosity is due to the release of internal thermal and gravitational potential energy. Paradoxically, even though its entropy is larger than that of a model of the same mass losing mass more slowly, the gravothermal luminosity of the model losing mass more rapidly is greater than that of the model losing mass less rapidly. The resolution of the apparent paradox lies with the denominator Δt in $\epsilon_{\mathrm{grav}} = -T\Delta S/\Delta t$.

3.3.3 Angular momentum loss due to gravitational wave radiation

For a binary system in a circular orbit, the rate of orbital angular momentum loss due to GWR is given by (e.g. Landau & Lifshitz, 1951)

$$-\frac{\dot{J}_{\mathrm{GWR}}}{J_{\mathrm{GWR}}} = \frac{1}{\tau_J} = \frac{32}{5}\,\frac{M_{\mathrm{a}}\,M_{\mathrm{d}}\,M_{\mathrm{t}}}{A^4}\left[\left(\frac{G\,\mathrm{M}_\odot}{\mathrm{R}_\odot\,c^2}\right)^3\frac{c}{\mathrm{R}_\odot}\right]$$
$$= \frac{1}{1.2\times 10^9\,\mathrm{yr}}\,\frac{M_{\mathrm{a}}\,M_{\mathrm{d}}\,M_{\mathrm{t}}}{A^4}, \tag{3.8}$$

where masses and orbital separation A in the last equality are in solar units. Angular momentum loss by GWR presumably carries the progenitors of short-period CVs through the period gap. The paucity of known variables in the period gap suggests that, on reaching the long-period side of the gap, the main-sequence donor shrinks within its Roche lobe on a time-scale shorter than the time-scale for orbital shrinkage due to GWR. This latter time-scale is found by first noting, from Equations (3.2) and (3.8), that

$$\frac{1}{8}\frac{\mathrm{d}A^4}{\mathrm{d}t} = -\frac{1}{\tau_{\mathrm{J}}(A)}, \tag{3.9}$$

where, from Equation (3.8), $\tau_J(A)$ is

$$\tau_J(A) = 1.2\times 10^9\,\mathrm{yr}\,\frac{A^4}{M_{\mathrm{d}}\,M_{\mathrm{a}}\,M_{\mathrm{t}}}.$$

Integrating Equation (3.9), we have that the time for the orbital semimajor axis to shrink from A to A_{f} is

$$\Delta t = \tau_{\mathrm{GWR}}(A)\left[1-\left(\frac{A_{\mathrm{f}}}{A}\right)^4\right], \tag{3.10}$$

where

$$\tau_{\mathrm{GWR}}(A) = \frac{\tau_{\mathrm{J}}(A)}{8} = 1.5\times 10^8\,\mathrm{yr}\,\frac{A^4}{M_{\mathrm{a}}\,M_{\mathrm{d}}\,M_{\mathrm{t}}}$$

can be defined as the GWR time-scale. Adopting $M_{\mathrm{d}} = 0.3\,\mathrm{M}_\odot$ and $M_{\mathrm{a}} = 1.0\,\mathrm{M}_\odot$ for CVs at the upper end of the period gap, we have $A \sim 1.15$ and $\tau_{\mathrm{GWR}} \sim 6.8\times 10^8$ yr. As the system

Table 3.1. $\tau_{\rm GWR}$ *for various values of* $M_{\rm d}$, $M_{\rm a}$ *and* A

$M_{\rm d}$ ($\rm M_\odot$)	$M_{\rm a}$ ($\rm M_\odot$)	A ($\rm R_\odot$)	$\tau_{\rm GWR}$ (10^8 yr)
0.22	0.6	0.99	13.3
0.3	1.0	1.15	6.8
0.3	0.6	1.02	10.0
0.22	1.0	1.13	9.1

evolves across the period gap, its orbital separation decreases by a factor $A_{\rm f}/A = (2/3)^{2/3}$ in a time given by Equation (3.10) as $\Delta t = 0.661\ \tau_{\rm GWR}(A) \sim 4.5 \times 10^8$ yr. Repeating this exercise with various values of $M_{\rm d}$ and $M_{\rm a}$ leads to the A and $\tau_{\rm GWR}$ values in Table 3.1.

In summary, the time for a typical system to evolve through the period gap is

$$\Delta t_{\rm gap} = 0.661(\tau_{\rm GWR})_{P_{\rm orb}=3\,{\rm h}} \sim (5.5\text{–}7.5) \times 10^8\ {\rm yr},$$

with the longer transit time being associated with a smaller donor mass.

The gravothermal time-scale of a constant mass, homogeneous, main-sequence star can be defined by

$$\tau_{\rm th} \sim 3 \times 10^7\ {\rm yr}\ \frac{M_{\rm d}^2}{L_{\rm d} R_{\rm d}}.$$

The slope of the mean mass-luminosity relationship for low mass stars in wide binaries is $d\log L/d\log M \sim 2.2$ (where again both L and M are in solar units). Adopting $L_{\rm d} \sim M_{\rm d}^2$ and $R_{\rm d} \sim 0.86 M_{\rm d}^{0.833}$, we have

$$(\tau_{\rm th})_{P_{\rm orb}=3\,{\rm h}} \sim 10^8\ {\rm yr}$$

for mass donors in CVs near the upper edge of the period gap. Thus, in agreement with expectations, $\tau_{\rm th}$ is smaller than $\tau_{\rm GWR}$.

3.3.4 *Mass transfer driven by gravitational wave radiation*

Assuming that, when the CV reaches the short-period end of the period gap, the donor star fills its Roche lobe, equating the angular momentum loss rates given by Equations (3.5) and (3.8) produces

$$\dot{M}_{\rm d} \sim -\left[1.28 - 0.5\epsilon - \frac{M_{\rm d}}{M_{\rm a}}\right]^{-1} \frac{M_{\rm a}^2\, M_{\rm d}\, M_{\rm t}}{A^4}\ \frac{\rm M_\odot}{1.2 \times 10^9\ {\rm yr}}.$$

Eliminating A from this equation with the help of Equations (3.3) and (3.4) ($R_{\rm L} = R_{\rm d}$), we have further that

$$\dot{M}_{\rm GWR} = -\dot{M}_{\rm d} = \frac{1}{\rho^4}\ \frac{M_{\rm d}^{-0.24+4\epsilon}\ M_{\rm a}}{M_{\rm t}^{0.76}}\ \frac{10^{-10.21}}{1.28 - \epsilon/2 - M_{\rm d}/M_{\rm a}}\ {\rm M_\odot\ yr^{-1}}.$$

For short-period systems ($P_{\rm orb} = 1.33$–2 h), the choices $\rho \sim 0.46$ and $\epsilon \sim 0.45$ produce a mass versus mean density curve that passes through the points derived by Webbink (1990) from the observations and give $P_{\rm orb} = 2\,{\rm h}$ when $(M_{\rm d}, M_{\rm a}) = (0.3, 1)$ and $P_{\rm orb} = 1.33\,{\rm h}$ when $(M_{\rm d}, M_{\rm a}) = (0.03, 1)$. So, for $M_{\rm a} = 1.0$,

$$\dot{M}_{\rm GWR} \sim 10^{-8.80} M_{\rm d}^{1.63}\,{\rm M_{\odot}\,yr^{-1}}\,, \tag{3.11}$$

corresponding to an accretion rate of $\dot{M}_{\rm GWR} \sim 10^{-9.65}\,{\rm M_{\odot}\,yr^{-1}}$ at $P_{\rm orb} = 2\,{\rm h}$, and of $\dot{M}_{\rm GWR} \sim 10^{-11.28}\,{\rm M_{\odot}\,yr^{-1}}$ at $P_{\rm orb} = 80\,{\rm min}$. Observational estimates (Patterson, 1984, Figure 7) suggest a variation from a mean of $\langle \dot{M}_{\rm obs} \rangle \sim 10^{-10}\,{\rm M_{\odot}\,yr^{-1}}$ at $P_{\rm orb} = 2\,{\rm h}$ to $\langle \dot{M}_{\rm obs} \rangle \sim 10^{-11}\,{\rm M_{\odot}\,yr^{-1}}$ at $P_{\rm orb} = 80\,{\rm min}$. But again, the scatter in observational estimates is large, being of the order of a factor of 10 at a given period.

The choices of $\rho \sim 0.46$ and $\epsilon \sim 0.45$ as parameters in the mass–radius relationship might seem rather odd, but it must be remembered that the equation of state in the interiors of very low mass stars is complicated; in regions which are only partially ionized, free electrons are nevertheless partially degenerate. Furthermore, as its mass decreases, the donor makes a transition from a main-sequence like object to a white dwarf like object. For these reasons, it makes sense to rely on an observationally grounded mass–radius relationship rather than on one given by theoretical stellar models.

The mass of $\sim 0.03\,{\rm M_{\odot}}$ that our approximation gives for the donor at $P_{\rm orb} = 80\,{\rm min}$ suggests that the mass–radius relationship we have adopted is inappropriate at this period; for such a small mass, the donor must be a white dwarf. In fact, it is the transformation of the donor into a white dwarf which can account for the termination of the CV distribution at 80 min (Paczyński & Sienkievicz, 1983). Since the radius of a white dwarf increases with decreasing mass, it is clear that, as a white dwarf component which fills its Roche lobe loses mass, the orbital period must increase. Paczyński and Sienkievicz (1983) construct models to demonstrate in detail how this occurs.

To a first approximation, the radius (in solar radii) of a low mass white dwarf is related to stellar mass (in ${\rm M_{\odot}}$) by

$$R_{\rm WD} \sim 0.01 \left(\frac{2}{\mu_{\rm e}}\right)^{5/3} \frac{1}{M_{\rm WD}^{1/3}}, \tag{3.12}$$

where $\mu_{\rm e}$ is the molecular mass per electron. Adopting

$$\mu_e = \frac{2}{1+X} = \frac{2}{1+0.71} = 1.15$$

and using Equations (3.3) and (3.12) in Kepler's law,

$$P_{\rm orb}({\rm h}) = 2.76\,\frac{A^{3/2}}{M_t^{1/2}},$$

we find $M_{\rm WD} = 0.035\,{\rm M_{\odot}}$ at $P_{\rm orb} = 80\,{\rm min}$, not unlike the mass obtained with the power-law approximation to the observationally based mass–radius relationship. For the adopted composition (mass fraction of hydrogen $X = 0.71$), the mass-transfer rate due to GWR in a system with a low mass hydrogen-rich white dwarf donor is

$$\dot{M}_{\rm d} \sim -\frac{M_{\rm WD}^{5.09}\,M_{\rm a}}{M_{\rm t}^{0.76}}\,\frac{1.84\times 10^{-4}}{0.61 - M_{\rm WD}/M_{\rm a}}\,{\rm M_{\odot}\,yr^{-1}},$$

suggesting that $\dot{M}_{\rm d} \sim 7.1\times 10^{-12}\,{\rm M_{\odot}\,yr^{-1}}$ for the shortest-period CVs with hydrogen-rich white dwarf donors.

3.3.5 *Mass transfer due to nuclear evolution in Algol-like CVs*

For CVs with periods larger than about 10 hours, the model of a relatively homogeneous main-sequence mass donor breaks down. As cases in point, consider the systems BV Cen ($P_{\rm orb} = 14.6$ hours), GK Per ($P_{\rm orb} = 47.9$ hours) and V1017 Sgr ($P_{\rm orb} = 137$ hours). From Kepler's law, the semimajor axis for these systems is $A = 3.03 M_{\rm t}^{1/3}$, $6.70 M_{\rm t}^{1/3}$, and $13.5 M_{\rm t}^{1/3}$, respectively. Equation (3.3) tells us that the radius of the Roche lobe about the mass donor satisfies $R_{\rm Ld} M_{\rm d}^{-0.44} M_{\rm t}^{0.11} = 1.57, 3.48$, and 7.02, respectively. If the donor were a homogeneous main-sequence star, its mass (in $M_\odot$) would be about $M_{\rm d} = 2$, 8, and 33, respectively. For at least two reasons, these masses are unacceptably large. First, they are too large to permit stable mass transfer, even for a white dwarf component of mass near the Chandrasekhar mass. Second, main-sequence stars of such large masses do not support a MSW.

During the main-sequence phase, single stars approximately double their radius. This means that some of the longer-period CVs contain mass donors which are evolved main-sequence stars and modeling such systems properly requires detailed evolutionary calculations such as those performed by Iben and Tutukov (1984b).

For the longest-period systems, one must turn to models in which the donor has evolved beyond the main sequence into a subgiant or giant with an electron-degenerate helium core. When the mass of the hydrogen-rich envelope above the electron-degenerate helium core is more massive than several hundredths of a solar mass, the radius of the donor is a function primarily of the mass $M_{\rm He}$ of the helium core. For a Population I composition (Iben & Tutukov, 1984a,b),

$$R_{\rm d} \sim 10^{3.38} M_{\rm He}^4 . \tag{3.13}$$

Differentiating Equations (3.13), (3.2), and (3.4) and setting $\dot{R}_{\rm L}/R_{\rm Ld} = \dot{R}_{\rm d}/R_{\rm d}$ gives

$$\frac{\dot{M}_{\rm d}}{M_{\rm d}} \sim -\frac{1}{0.78 - M_{\rm d}/M_{\rm a}} \left[\frac{\dot{J}_{\rm orb}}{J_{\rm orb}} - 2\,\frac{\dot{M}_{\rm He}}{M_{\rm He}} \right].$$

Assuming that the hydrogen abundance by mass in accreted material is $X = 0.7$, one has that

$$\dot{M}_{\rm He} = 1.33 \times 10^{-11}\, L_{\rm RG}\, M_\odot\, {\rm yr}^{-1},$$

where $L_{\rm RG}$ is in units of $L_\odot$. The luminosity of the giant or sub-giant is related to its core mass by (Iben & Tutukov, 1984a)

$$L_{\rm RG} \sim 10^{5.6} M_{\rm He}^{6.5}, \tag{3.14}$$

where $M_{\rm He}$ is in units of $M_\odot$ and $L_{\rm RG}$ is in units of $L_\odot$. Thus,

$$\dot{M}_{\rm He} \sim 10^{-5.276} M_{\rm He}^{6.5}\ M_\odot\, {\rm yr}^{-1}, \tag{3.15}$$

where $M_{\rm He}$ is in units of $M_\odot$.

There is no reason to suppose that the MSW ceases to operate when the star is a red giant and, assuming Roche-lobe filling and setting $\dot{J}_{\rm orb} = \dot{J}_{\rm GWR} + \dot{J}_{\rm MSW}$, we have

$$\dot{M}_{\rm RG} = \dot{M}_{\rm d} = -\dot{M}_{\rm evo}(1 + \gamma), \tag{3.16}$$

Table 3.2. *Estimates of mass-transfer rates for long-period CVs*

System	P_{orb} (h)	M_{He} ($M_\odot$)	$\dot{M}_{evo}$ ($M_\odot$ yr^{-1})	$\dot{M}_{MSW}/\dot{M}_{evo}$	$-\dot{M}_{RG}$ ($M_\odot$ yr^{-1})
BV Cen	14.6	0.16	0.79×10^{-10}	9.5	8.3×10^{-9}
GK Per	47.9	0.20	2.7×10^{-9}	1.27	6.1×10^{-9}
V1017 Sgr	137	0.23	5.8×10^{-9}	0.36	7.9×10^{-9}
T CrB	5460	0.39	2.2×10^{-7}	0.0036	2.2×10^{-7}

where

$$\dot{M}_{evo} \sim 5.18 \times 10^{-9} \frac{M_d}{0.78 - M_d/M_a} \left(\frac{M_{He}}{0.25}\right)^{5.5} M_\odot \text{ yr}^{-1} \tag{3.17}$$

and

$$\gamma = 0.37 \frac{M_d}{M_t^{0.2} M_a} \left(\frac{0.25}{M_{He}}\right)^9 . \tag{3.18}$$

The quantity γ gives the contribution of the MSW to the mass-loss rate relative to the contribution of the evolution of the donor.

The orbital period of the system is related to the component masses by

$$P_{orb}(\text{hours}) \sim 0.865 \frac{M_t^{0.16}}{M_d^{0.66}} (10 M_{He})^6 . \tag{3.19}$$

Solving this equation for the mass of the core of the donor, we have

$$M_{He} \sim 0.1 \frac{M_d^{0.11}}{M_t^{0.0267}} \left(\frac{P_{orb}}{0.865 \text{ hours}}\right)^{1/6} . \tag{3.20}$$

When applied to the three longest-period CVs (BV Cen, GK Per, and V1017 Sgr), Equation (3.20) gives the core masses shown in the third column of Table 3.2. Assuming $M_a \sim 1$ and $M_d/M_a \sim 0.5$, the quantity $\dot{M}_{evo}$ defined by Equations (3.17) for these systems is shown in the fourth column, the ratio γ given by Equation (3.18) is shown in the fifth column, and the total estimated mass-loss rate defined by Equation (3.16) is given in the last column.

It is interesting to estimate the amount of mass which can be transferred while the donor evolves as a red giant. Integration of Equation (3.15) gives

$$\Delta t_{RG} \sim 1.65 \times 10^7 \text{ yr} \left(\frac{0.25}{M_{He}}\right)^{5.5} \left[1 - \left(\frac{M_{He}}{M_{He,f}}\right)^{5.5}\right]$$

as the time for the mass of the helium core to grow from an initial value of M_{He} to a final value of $M_{He,f}$. For example, the time for core mass to grow from $0.16 M_\odot$ to $0.23 M_\odot$ (BV Cen to V1017 Sgr) is $\sim 1.66 \times 10^8$ yr. Approximating the mass-loss rate from the last column of Table 3.2 by $\sim 7 \times 10^{-9} M_\odot$ yr^{-1}, we guess that a total mass of $\sim 0.12 M_\odot$ is lost by the donor as its core grows from $0.16 M_\odot$ to $0.23 M_\odot$.

Another system which may fall into the category of CVs with Roche-lobe-filling giants is the recurrent nova T CrB. The orbital period of this system is $P_{orb} = 227.6$ days and the masses of the white dwarf and red giant components are estimated to be $M_a \sim 1.2$ and $M_d \sim 0.7$, respectively (Mikołajewska, 2003), giving a semimajor axis $A \sim 195\,R_\odot$ and a Roche-lobe radius about the red giant component of $\sim 65\,R_\odot$. Equation (3.13) suggests $M_{He} \sim 0.41\,M_\odot$. The luminosity of the red giant component is estimated to be $L_d \sim 650\,L_\odot$ and Equation (3.14) gives the estimate of $M_{He} \sim 0.37\,M_\odot$. With these estimates of core mass, the MSW can be neglected, and the mass-transfer rate given by Equation (3.17) is

$$\dot{M}_d \sim -(1.6 - 2.8) \times 10^{-7}\,M_\odot\,yr^{-1}.$$

These results are summarized in the last row of Table 3.2.

3.3.6 *Mass transfer in symbiotic stars due to accretion from a wind*

In the Hoyle–Lyttleton (1939) treatment of accretion from a plane parallel wind, there is a critical radius or 'impact parameter' b_{crit} such that all wind matter flowing (at large distances from the accretor) in a cylinder of this radius will be focused by gravity on to the accretor. They approximate the critical radius by

$$b_{crit} = \frac{2GM}{v_\infty^2},$$

where v_∞ is the speed of the wind relative to the speed of the accretor at large distances from the accretor. Numerically,

$$b_{crit} = 611\,\frac{M_{acc}}{v_{25}^2}\,R_\odot,$$

where the mass M_{acc} of the accretor is in solar units and v_{25} is v_∞ in units of 25 km s^{-1}. The cross-section for capture by the accretor is

$$\sigma_{acc} = \pi b_{crit}^2.$$

In the context of symbiotic stars, a zeroth order approximation to the rate at which mass is accreted from the wind of the giant companion is (Tutukov & Yungelson, 1976)

$$\dot{M}_{acc} = \frac{\dot{M}_{rgw}}{4\pi A^2}\,\sigma_{acc} = \frac{1}{4}\left(\frac{b_{crit}}{A}\right)^2 \dot{M}_{rgw}, \tag{3.21}$$

where $\dot{M}_{rgw}$ is the red giant (or AGB star) wind mass-loss rate.

Adopting as typical parameters $M_a \sim 0.6\,M_\odot$, $M_t \sim 3\,M_\odot$, and $v_{rgw} \sim 25$ km s^{-1}, Iben & Tutukov (1996) approximate this equation by

$$\dot{M}_{acc} \sim \dot{M}_{rgw}\left(\frac{141\text{ days}}{P_{orb}}\right)^{4/3}. \tag{3.22}$$

This approximation is obviously a considerable overestimate when P_{orb} is close to 141 days and it could also be an overestimate for periods substantially larger than this. On the other hand, the Roche-lobe geometry and the tidal distortions of the giant component may lead to an accretion stream which is concentrated in the orbital plane, and, at large P_{orb} ($\gg$141 days), to an effective accretion rate larger than that given by Equation (3.22).

3.4 A zeroth-order model for CV evolution

Assuming that the mass of a typical white dwarf component is *and remains* constant at $1\,M_\odot$ and that the birthrate of CVs is independent of the initial mass of the donor for initial donor mass $M_d < 1.54\,M_\odot$, a simple but illustrative model of CV evolution can be constructed (Iben, Fujimoto & MacDonald, 1992a). This model makes use of results described in Section 3.6.

3.4.1 *CV number density versus mean accretion rate*

At time t, let $n(M_d, t)\,dM_d$ be the number in the entire Galaxy of CVs for which donor mass lies between M_d and $M_d + dM_d$, let $(dn/dt)_b\,dM_d$ be the rate at which CVs are born in the interval dM_d, and let $(\partial n/\partial t)\,dM_d$ be the rate at which the number of CVs in the interval dM_d actually changes with time.

Number conservation requires that

$$\left(\frac{dn}{dt}\right)_b = \frac{\partial n}{\partial t} + \frac{\partial}{\partial M_d}\left(n\frac{dM_d}{dt}\right). \tag{3.23}$$

If a steady state exists, $\partial n/\partial t = 0$ and $n = n(M_d)$. As a birthrate, choose

$$\left(\frac{dn}{dt}\right)_b = 0.001\beta\ \mathrm{yr}^{-1}\,M_\odot{}^{-1} \tag{3.24}$$

for $M_d = 0.03 \rightarrow 1.5$, and

$$\left(\frac{dn}{dt}\right)_b = 0 \tag{3.25}$$

for $M_d > 1.5\,M_\odot$.

Since mass transfer can be stable only if the mass ratio $M_d/M_a = q < 0.8$, a common envelope is expected to be formed in all systems for which $q > 0.8$ when contact is first achieved, and the mass that is transferred by the donor into the common envelope is also lost rapidly from the system until $q \leq 0.8$. Since it has been assumed that all accretors are of mass $\sim 1\,M_\odot$, all systems for which initially $0.8 < M_d/M_\odot < 1.5$ evolve rapidly (after Roche-lobe contact has been established) into systems with $M_d \leq 0.8\,M_\odot$. Hence, the flux of new systems into the pool at $M_d < 0.8\,M_\odot$ is

$$n(M_d)\,\dot{M}_d = -0.0007\beta\ \mathrm{yr}^{-1}.$$

Setting $\partial n/\partial t = 0$, and using Equations (3.23)–(3.25), it follows that

$$n(M_d)\,\dot{M}_d = -0.001\beta\,(1.5 - M_d). \tag{3.26}$$

For $P_{orb} > 3\,h$, Equations (3.7) and (3.26) give

$$n(M_d) = 10^{4.28}\,\frac{\beta}{\alpha}\,(1.5 - M_d)M_d^{-3.15}. \tag{3.27}$$

Assuming that the mass of a donor at the long-period end of the period gap is $0.3\,M_\odot$, integration from $M_d = 0.3$ to 0.8 gives, as the total number of CVs in the Galaxy with $P_{orb} > 3\,h$, $N_{MSW} = 1.11 \times 10^5\,\beta/\alpha$. From this estimate and Equation (3.24) it follows that the lifetime of a typical long-period CV is $\tau_{MSW} \sim 0.9 \times 10^8\,\alpha^{-1}$ yr. If, instead, we assume $M_d = 0.22$ at the long-period end of the period gap, $N_{MSW} = 2.5 \times 10^5 \beta/\alpha$ and $\tau_{MSW} \sim 2 \times 10^8 \alpha^{-1}$ yr. Altogether, then, we estimate

$$N_{MSW} = (1.1\text{–}2.5) \times 10^5 \frac{\beta}{\alpha}.$$

For $P_{orb} < 2$ h, Equations (3.11) and (3.26) give

$$n(M_d) = 10^{5.80} \beta (1.5 - M_d) M_d^{-1.63}.$$

Integration from $M_d = 0.03\,M_\odot$ to $0.3\,M_\odot$ gives, as the total number of CVs in the Galaxy with $P_{orb} < 2$ h, $N_{GWR} = 1.32 \times 10^7 \beta$. From this result and Equation (3.25), it follows that the lifetime of a typical short-period CV is $\tau_{GWR} \sim 0.67 \times 10^{10}$ yr. Assuming, instead, that $M_d = 0.22\,M_\odot$ at the long-period end of the period gap $N_{GWR} = 0.93 \times 10^7 \beta$ and $\tau_{GWR} \sim 0.47 \times 10^{10}$ yr. Thus, we estimate

$$N_{GWR} \sim (0.9 - 1.3) \times 10^7 \beta.$$

Using the estimates made in Section 3.3.3, the number of systems in the period gap is predicted to be

$$\begin{aligned} N_{gap} &\sim 0.001\,\beta \times (1.5\text{–}0.03) \times (5.5\text{–}7.5) \times 10^8 \\ &\sim (0.8\text{–}1.1) \times 10^6 \beta. \end{aligned}$$

In summary, the simple model suggests that there are approximately 50–100 times as many short-period CVs as there are long-period CVs, and corresponding lifetimes are in the inverse ratio. There are predicted to be about 4–7 times as many systems in the period gap as there are long-period CVs. However, the lack of a disk and hot spot presents a huge observational selection obstacle for these stars, so a prediction is not very useful.

3.4.2 Nova rate versus mean accretion rate

The differential rate at which novae occur per unit time is

$$\frac{d\mathcal{R}}{dM_d} = -\frac{n(M_d)}{(M_{acc}/\dot{M}_d)},$$

where $d\mathcal{R}$ is the number of novae occurring per year in CVs with donor masses in the range $M_d \to M_d + dM_d$ and M_{acc} is the mass accreted between outbursts for such CVs accreting at the rate $\dot{M}_d$.

Fitting to values of M_{acc} appropriate for a $1\,M_\odot$ accretor with a steady state thermal distribution achieved after many pulses (Section 3.6) gives (Iben, Fujimoto & MacDonald, 1992a, b)

$$M_{acc} \sim 10^{-7.94} \dot{M}_{MSW}^{-0.34}, \tag{3.28}$$

and using Equations (3.7) and (3.27),

$$\left(\frac{d\mathcal{R}}{dM_d}\right)_{MSW} = 10^{2.46} \beta \alpha^{0.34} (1.5 - M_d) M_d^{1.07}$$

for $P_{orb} > 3$ h. Integration from $M_d = 0.3\,M_\odot$ to $0.8\,M_\odot$ gives $\mathcal{R}_{MSW} = 70 \beta \alpha^{0.34}$ yr^{-1}. Doing the integration from $M_d = 0.22\,M_\odot$ to $0.8\,M_\odot$ increases this estimate by only 10%. Altogether, we estimate

$$\mathcal{R}_{MSW} \sim (70\text{–}77) \beta \alpha^{0.34} \text{ yr}^{-1}$$

as the frequency with which novae are produced by CVs with $P_{\rm orb} > 3\,{\rm h}$.

Similarly, Equations (3.12) and (3.28), along with Iben, Fujimoto and MacDonald (1992a,b):

$$M_{\rm acc} \sim 10^{-14.24}\ \dot{M}_{\rm GWR}^{-1.04}$$

give

$$\left(\frac{{\rm d}\mathcal{R}}{{\rm d}M_{\rm d}}\right)_{\rm GWR} = 10^{2.09}\ \beta\ (1.5 - M_{\rm d})\ M_{\rm d}^{1.70},$$

and integration over the interval $M_{\rm d} = 0.03\,{\rm M}_\odot$ to $0.3\,{\rm M}_\odot$ gives $\mathcal{R}_{\rm GWR} = 2.26\ \beta\ {\rm yr}^{-1}$. Doing the integration from $M_{\rm d} = 0.03\,{\rm M}_\odot$ to $0.22\,{\rm M}_\odot$ decreases this estimate by a factor of 2.2. Thus, we estimate

$$\mathcal{R}_{\rm GWR} = (1 - 2.3)\beta\ {\rm yr}^{-1}$$

as the frequency with which novae are produced by CVs with $P_{\rm orb} < 2\,{\rm h}$.

In summary, we estimate the frequency of novae produced by long-period CVs to be between $\sim$30 and $\sim$70 times the frequency of novae produced by short-period CVs and guess the total frequency of novae to be $\sim 70\beta\ {\rm yr}^{-1}$ (setting $\alpha \sim 1$). The observed classical nova frequency in M31 ($\sim 65\ {\rm yr}^{-1}$, cf. Chapter 14), and a frequency in the Milky Way of around 30 ${\rm yr}^{-1}$ (cf. Chapters 2 and 14) may be taken as a modest justification for the choice $\beta \sim 0.5$–1.

3.4.3 *Predicted observational CV distribution functions*

If $f_{\rm min}$ is the smallest flux which can be detected in a given band, a system of intrinsic luminosity L in that band can be seen out to a distance $D_{\rm max}(L) = (L/4\pi f_{\rm min})^{1/2}$. If $D_{\rm max}$ for the system of smallest intrinsic luminosity is large compared with the scale height $H_{\rm disk}$ of the Galactic disk (the disk case), the number of observed systems ${\rm d}N_{\rm obs}(L)$ with luminosities in the range L to $L + {\rm d}L$ is related to the total number of systems ${\rm d}N(L)$ in this luminosity interval by

$${\rm d}N_{\rm obs}(L) = \left(\frac{{\rm d}N(l)}{V_{\rm G}}\right)\ 2\pi\ H_{\rm disk}\,D_{\rm max}^2(L) \propto {\rm d}N(L)L,$$

where $V_{\rm G}$ is the volume of the Galaxy. If the reverse is true (the spherical case), then

$${\rm d}N_{\rm obs}(L) = \left(\frac{{\rm d}N(L)}{VG}\right)\left(\frac{4\pi}{3}\right) D_{\rm max}^3(L) \propto {\rm d}N(L)L^{3/2}.$$

To obtain normalized 'observational' distribution functions $\bar{n}_{\rm obs}(M) = \dfrac{{\rm d}\bar{N}_{\rm obs}}{{\rm d}\log \dot{M}_{\rm acc}}$, assume that $L \propto \left[\dfrac{\dot{M}_{\rm acc}(M)}{\dot{M}_{\rm acc}(0.8)}\right]^{1-\delta}$, set ${\rm d}\bar{N}(L) = \left[\dfrac{n(M)}{n(0.8)}\right]{\rm d}M$, and use ${\rm d}M = 2.3026\,M\ \dfrac{{\rm d}\log \dot{M}_{\rm acc}}{{\rm d}\ln \dot{M}_{\rm acc}/{\rm d}\ln M}$ to find

$$\bar{n}_{\rm obs}(M) = 2.3026\,M\left(\frac{{\rm d}\ln \dot{M}_{\rm acc}}{{\rm d}\ln M}\right)^{-1}\left[\frac{n(M)}{n(0.8)}\right]\left[\frac{\dot{M}_{\rm acc}(M)}{\dot{M}_{\rm acc}(0.8)}\right]^{m(1-\delta)}, \qquad (3.29)$$

where $m = 1$ in the disk case and $m = 3/2$ in the spherical case. The subscript d has been dropped from $M_{\rm d}$. Finally,

$$\bar{n}_{\rm obs}(M) = 2.3026M \left(\frac{\mathrm{d}\ln \dot{M}_{\rm acc}}{\mathrm{d}\ln M}\right)^{-1} \left(\frac{1.5 - M}{0.7}\right) \left[\frac{\dot{M}_{\rm acc}(M)}{\dot{M}_{\rm acc}(0.8)}\right]^{(m-1-m\delta)}. \quad (3.30)$$

We continue the discussion on the assumption that the typical mass of the donor at the long-period end of the period gap is $0.3\,\mathrm{M}_\odot$. In the disk case ($m = 1$), setting $\delta = 0$ gives

$$\bar{n}_{\rm obs} = 1.04\,(1.5 - M)M,$$

when $M > 0.3\,\mathrm{M}_\odot$, and

$$\bar{n}_{\rm obs} = 2.02\,(1.5 - M)M,$$

when $M < 0.3\,\mathrm{M}_\odot$. The ratio of slow accretion systems to fast accretion systems is

$$\left(\frac{\bar{N}_{\rm slow}}{\bar{N}_{\rm fast}}\right)_{m=1,\delta=0} \sim 0.8.$$

The observational distribution functions for the spherical case ($m = 3/2$) when $\delta = 0$ are

$$\bar{n}_{\rm obs} = 1.48\,(1.5 - M)M^{2.58},$$

when $M > 0.3$, and

$$\bar{n}_{\rm obs} = \frac{0.50}{\alpha^{1/2}}\,(1.5 - M)M^{1.81},$$

when $M < 0.3$. The ratio of slow accretion systems to fast accretion systems is

$$\left(\frac{\bar{N}_{\rm slow}}{\bar{N}_{\rm fast}}\right)_{m=3/2,\delta=0} \sim \frac{0.08}{\alpha^{1/2}}.$$

Observed numbers of short-period and long-period systems are comparable, and one might be tempted to infer that the spherical case is inappropriate for the observed sample. However, if the sample is dominated by systems studied primarily in optical bands, one must take into account the fact that the mean temperature of an accretion disk increases with increasing accretion rate, with the result that the luminosity at optical wavelengths decreases with increasing accretion rate relative to the total luminosity. This effect can be modeled by setting $\delta > 0$ in Equations (3.29) and (3.30).

Estimates of δ can be derived from theoretical models of accretion disks. The results are $\delta \sim 0.33$ for long-period systems and $\delta \sim 0.37$ for short-period systems (Iben, Fujimoto & MacDonald, 1992a).

The spherical case with $\delta = 1/3$ gives precisely the same distribution as does the disk case with $\delta = 0$ and thus predicts a ratio of slow to fast accretors not unlike that which is observed. The disk case with $\delta = 1/3$ gives a ratio of slow to fast accretion systems which is an order of magnitude too large. Thus, the spherical case (with $\delta \sim 1/3$) is the appropriate one.

In summary, the model predicts that (1) the frequency with which classical nova outbursts occur varies with the time-averaged mass-transfer rate in CV nova precursors in such a way that the total frequency for long-period, high mass-transfer rate CVs is ~30–70 times larger than it is for short-period, small mass-transfer rate CVs, even though short-period systems outnumber long-period systems by a factor of ~50–100 and that (2) if the probability for detection increases at least linearly with the mass-transfer rate, the number of observed short-period, small mass-transfer rate CVs in a magnitude limited sample is comparable with the number of observed long-period high mass-transfer rate CVs in this sample. This appears to agree with the observations.

3.4.4 Helium white dwarf CVs: AM CVn stars

In AM CVn stars, of which there are at least seven established examples, the donor is probably a helium white dwarf, so that

$$R_{\rm HeWD} \sim \frac{0.01}{M_{\rm HeWD}^{1/3}}.$$

Setting $R_{\rm HeWD} = R_{\rm L1}$ in Equation (3.3) and using Kepler's law gives

$$P_{\rm orb}(\text{minutes}) \sim 0.44\,\frac{M_{\rm t}^{0.16}}{M_{\rm HeWD}^{1.16}}. \tag{3.31}$$

Solving Equation (3.31) for $M_{\rm HeWD}$ gives donor masses of $\sim$0.018 $\rm M_{\odot}$for GP Com ($P_{\rm orb} =$ 46.5 min), $\sim$0.041 $\rm M_{\odot}$ for AM CVn ($P_{\rm orb} = 17.5$ min) and $\sim$0.12 $\rm M_{\odot}$ for RXJ0806.3+1527 ($P_{\rm orb} \sim 5.35$ min).

If angular momentum loss is by GWR,

$$\dot{M}_{\rm d} = -\dot{M}_{\rm GWR} \sim -\frac{M_{\rm d}^{5.09}\,M_{\rm a}}{M_{\rm t}^{0.76}}\,\frac{6.09 \times 10^{-3}}{0.61 - M_{\rm d}/M_{\rm a}}\,\rm M_{\odot}\,yr^{-1},$$

suggesting that $|\dot{M}_{\rm d}| = 2.5\times10^{-7}\,\rm M_{\odot}\,yr^{-1}$, $0.93\times10^{-9}\,\rm M_{\odot}\,yr^{-1}$, and $1.4\times10^{-11}\,\rm M_{\odot}\,yr^{-1}$, for RXJ0806.3+1527, AM CVn, and GP Com, respectively.

3.4.5 Helium star CVs

For completeness, we comment on some properties of systems in which the donor is a helium star. Such systems are, of course, the consequence of close binary star evolution during which common envelope formation occurs and leads to sufficient orbital shrinkage that the helium-star component can eventually make Roche-lobe contact. Iben *et al.* (1987) find that, typically, mass-transfer rates of the order of $\dot{M}_{\rm acc} \sim 3 \times 10^{-8}\,\rm M_{\odot}\,yr^{-1}$ are to be expected. For accretion at this rate, the helium-burning flash which occurs when a helium layer of critical mass has been built up is expected to be violent enough that, when an outburst occurs, the helium layer will evolve to red giant proportions (see Section 3.5.1) and most of the matter accreted between outbursts will be lost in a wind (e.g. Kato & Hachisu, 2004).

Savonije, de Kool and van den Heuvel (1986) calculate the evolution of an X-ray binary in which the donor is assumed to be a helium star. As its mass decreases from 0.6 $\rm M_{\odot}$ to 0.23 $\rm M_{\odot}$, the radius and mass of the donor are related approximately by

$$R_{\rm He} \sim 0.256 M_{\rm He}^{1.15}\,\rm R_{\odot}. \tag{3.32}$$

Setting $R_{\rm He}$ in Equation (3.32) to $R_{\rm Ld}$ in Equation (3.3) and using Kepler's law gives

$$P_{\rm orb}(\text{minutes}) \sim 479 M_{\rm He}^{1.065} M_{\rm t}^{0.16}.$$

For an unevolved helium star, mass and luminosity are related by (Iben & Tutukov, 1991)

$$L_{\rm He} \sim 234 M_{\rm d}^{3.98}\,\rm L_{\odot},$$

whereas, for an evolved core helium-burning star (Iben, 1990),

$$L_{\mathrm{He}} \sim 655 M_{\mathrm{d}}^{4.81}\,\mathrm{L}_{\odot}.$$

Iben and Tutukov (1991) examine situations in which the donor has a CO core and is burning helium in a shell, the initial mass of which is given by Equation (3.35). The results can be applied to understanding systems such as U Sco which experience recurrent nova outbursts in which the ejecta appear to be primarily helium. Because the initial helium layer is small, helium burning is soon over and it is not likely that such systems evolve into supernovae.

3.5 Hydrogen-burning outbursts and critical helium-layer masses

In early studies (e.g. Giannone & Weigert, 1967; Starrfield, Sparks & Truran, 1974a,b), just one cycle was followed prior to a hydrogen-burning thermonuclear outburst. Paczyński and Żytkow (1978) followed the development of several successive flashes for a range of mass-accretion rates, showing that outburst strength increases with each cycle. They also identified mass-accretion rates that lead to steady state hydrogen burning at high surface temperatures. In their work, and in similar work by others (e.g. Sion, Acierno & Tomczyk, 1979; Iben, 1982; Sion & Starrfield, 1986; and Livio, Prialnik & Regev, 1989), no particular attention was paid to the consequences of building up a helium layer, although Iben (1982) carried the evolution of one model to the ignition of a helium shell flash. Fujimoto (1982a,b) identified three regions in the $\dot{M}_{\mathrm{acc}}$–M_{WD} plane where one expects (1) steady state burning, (2) weak hydrogen-burning flashes and (3) strong hydrogen-burning flashes. The first systematic study of the long term evolution of hydrogen-accreting white dwarfs as a function of accretion rate is that of José, Hernanz and Isern (1993), who used a semi-analytical computational technique in plane-parallel geometry and followed evolution through a large number of hydrogen shell flashes and also some consequent helium shell flashes. Cassisi, Iben and Tornambé (1998) and Piersanti *et al.* (1999, 2000) conducted similar work, but in the spherically symmetric quasistatic approximation.

The dependence on white dwarf mass and on accretion rate of the characteristics of hydrogen-burning flashes studied by the latter authors (accretion onto a CO white dwarf of initial masses $M_{\mathrm{WD}} = 0.516$ and $0.8\,\mathrm{M}_{\odot}$) and by Livio, Prialnik and Regev (1989) (accretion onto a white dwarf of initial mass $M_{\mathrm{WD}} = 1\ \mathrm{M}_{\odot}$) delineate the three regions identified by Fujimoto (1982a,b). Accretion of hydrogen-rich matter at rates slightly larger than given by

$$\log \dot{M}_{\mathrm{acc}}\left(\mathrm{M}_{\odot}\,\mathrm{yr}^{-1}\right) = -6.6 + 1.10\ (M_{\mathrm{WD}} - 0.8)(M_{\mathrm{WD}} - 0.8) \qquad (3.33)$$

produces red giants or AGB stars. Mild hydrogen shell flashes occur for accretion rates slightly smaller than given by

$$\log \dot{M}_{\mathrm{acc}}\left(\mathrm{M}_{\odot}\,\mathrm{yr}^{-1}\right) = -6.9 + 2.09\ (M_{\mathrm{WD}} - 0.8). \qquad (3.34)$$

For accretion rates in the region with borders defined by Equations (3.33) and (3.34), steady state hydrogen burning at high surface temperature takes place. Finally, strong hydrogen shell flashes occur for accretion rates below those given by

$$\log \dot{M}_{\mathrm{acc}}\left(\mathrm{M}_{\odot}\,\mathrm{yr}^{-1}\right) = -8.2 + 1.48\ (M_{\mathrm{WD}} - 0.8).$$

Cassisi, Iben and Tornambé (1998) and Piersanti *et al.* (1999, 2000) demonstrate that, when flashes are relatively mild (defined as being amenable to quasistatic calculations), outburst strength approaches asymptotically a locally unique value and reaches this value after typically 10–20 cycles. This means that, after a relatively small number of nova outbursts, the steady state thermal distribution as defined by Fujimoto and Sugimoto (1979) has been established for some distance below the accreted layer.

3.5.1 Critical helium-layer masses without mass-loss and diffusion

After a common envelope event (Paczyński, 1976), if the initial mass of the primary in a CV precursor is less than $\sim$8 $M_\odot$, the remnant of the initial primary that emerges from the common envelope is a CO white dwarf (Iben & Tutukov, 1985, 1993). If the initial mass of the primary is between $\sim$8 $M_\odot$ and $\sim$10.5 $M_\odot$, it is an ONe white dwarf (García-Berro & Iben, 1994; Ritossa, García-Berro & Iben, 1996; García-Berro, Ritossa & Iben, 1997; Iben, Ritossa & García-Berro, 1997; Ritossa, García-Berro & Iben, 1999; Gil-Pons & García-Berro, 2001, 2002). Above the CO or ONe matter there exists a layer of nearly pure helium which has a mass $M_{\rm He}^{\rm init}$ which is essentially the same as the mass in the helium shell of a single AGB star at the termination of helium burning in a thermal pulse cycle (Iben, 1977). This latter is given approximately by (Iben & Tutukov, 1996)

$$\log \mathrm{M}_{\rm He}^{\rm init} = -1.835 - 2.67\, M_{\rm WD}\, (M_{\rm WD} - 0.648), \tag{3.35}$$

where $M_{\rm WD}$, the mass of the white dwarf, and $M_{\rm He}^{\rm init}$ are in solar units.

An important question is whether or not, in the process of accreting hydrogen-rich matter, the white dwarf in a CV or symbiotic star system can add to this 'buffer' layer. In those systems which experience nova outbursts with spectra showing large enhancements of heavy elements, it is clear that mixing between white dwarf material and accreted matter has taken place and that, currently at least, the white dwarf loses more matter in an outburst than it gains between outbursts. Whether the mass-loss is due to common envelope interaction between the core of the bursting star and its companion, to dynamical acceleration powered by the nuclear energy injected during outburst, to wind mass-loss from the expanding envelope of the bursting star, or to some combination of these processes is irrelevant to the question. Nature has determined that the white dwarf which experiences these outbursts loses mass on a secular time-scale encompassing many nova outburst cycles.

Over the years, a number of experiments have been performed by accreting pure helium, thereby circumventing the complication of following hydrogen shell flashes. The results are, of course, not the same as when hydrogen-rich material is accreted since some of the nuclear energy emitted by hydrogen burning remains in the accreting model and maintains the temperature in the helium layer higher than when pure helium is accreted at the same effective rate; the mass of the helium layer is larger in the pure helium-accreting case and the helium-burning outburst which eventually occurs is more powerful.

Nomoto and Sugimoto (1977) found that, when the total mass of an initially cold white dwarf accreting helium at an appropriate rate exceeds $\sim$0.65–0.8 $M_\odot$, an explosion of supernova proportions is triggered when the helium layer ignites. Iben & Tutukov (1991) found that, if helium is accreted onto a cold CO white dwarf at the rate $\sim 3 \times 10^{-8}$ $M_\odot$ yr^{-1}, a violent explosion occurs after the accretion of $\Delta M_{\rm He} \sim 0.15\, M_\odot$, nearly independent of the initial mass of the underlying white dwarf. The critical amount of accreted helium increases with decreasing accretion rate, with $\Delta M_{\rm He} > 0.4$ $M_\odot$ for an accretion rate of $\sim 5 \times 10^{-9}\,M_\odot\,yr^{-1}$ (Limongi & Tornambé, 1991). If the white dwarf is initially cold enough and massive enough, helium burning can evolve into a detonation, and an inward moving compression wave can then lead to the detonation of carbon in the core (Nomoto, 1982; Tutukov & Khokhlov, 1992; Woosley & Weaver, 1994). The critical mass ΔM_{He}^{det} of the helium layer for this last eventuality is given roughly by

$$\log\,\Delta M_{He}^{det} \sim -1.2 - 1.79\,(M_{WD} - 0.8).$$

In the first of two sets of experiments, Cassisi, Iben and Tornambé (1998) accrete hydrogen-rich material onto a warm CO white dwarf which initially has a total mass of $0.516\,M_\odot$ and an outer helium layer of mass $4.3 \times 10^{-4}\,M_\odot$. For accretion rates $\dot{M}_{acc} = 4$, 6, and $10 \times 10^{-8}\,M_\odot\,yr^{-1}$, when the mass of the helium layer reaches $\Delta M_{He} \sim 0.09$, 0.08, and $0.06\,M_\odot$, respectively, a powerful helium shell flash develops. In the case of the lowest accretion rate, convection spreads quickly over the entire helium layer. Most of the energy produced by the helium-burning reactions is used up locally in removing electron degeneracy. When $L_{He} \sim 5.4 \times 10^6\,L_\odot$, the outer edge of the helium convective layer touches the inner edge of the hydrogen-rich envelope, and following the evolution further would have necessitated the use of a time-dependent mixing algorithm such as employed by Hollowell, Iben and Fujimoto (1990) and by Iben & MacDonald (1995) who show in similar situations that, as hydrogen penetrates deeply into the convective shell, hydrogen burning forces the formation of a new convective shell which is detached from the first one and then mixes to the surface matter that has experienced partial helium burning followed by hydrogen burning. In the absence of a close companion, the real analog of the model would expand to giant dimensions and presumably lose most of the helium layer in a wind. The presence of a close companion could enhance the extent of mass-loss in consequence of frictional interaction between the compact stellar cores and the wind. For the intermediate accretion rate, the helium-burning luminosity reaches a maximum of $L_{He} \sim 2.28 \times 10^6\,L_\odot$ and thereafter declines. The convective zone maintained by the helium-burning flux continues to grow in mass as L_{He} decreases, and its outer edge eventually touches hydrogen-rich layers, with the same consequences as before. For the highest accretion rate, the helium-burning luminosity reaches a maximum of $L_{He} \sim 1.58 \times 10^5\,L_\odot$, and the convective layer continues to grow in mass until, once again, its outer edge reaches and begins to ingest hydrogen-rich matter.

In models of a CO white dwarf of initial mass $0.8\,M_\odot$ accreting at the rate $\dot{M}_{acc} = 10^{-7}\,M_\odot\,yr^{-1}$, helium burning develops into a thermonuclear runaway after a helium layer of mass $\Delta M_{He} \sim 0.01\,M_\odot$ has been built up. Convection forced by helium burning does not reach the hydrogen-rich layer, but the entire helium envelope expands to red giant dimensions. The maximum helium-burning luminosity attained during the flash is $L_{He} \sim 4.2 \times 10^5\,L_\odot$.

From these experiments, it is evident that, for a constant hydrogen-accretion rate, the larger the initial mass of the white dwarf, the smaller is the mass of the helium layer required to produce a helium shell flash and the smaller is the total power of this flash, even if the peak luminosity is larger. For example, at the accretion rate of $\dot{M}_{acc} \sim 10^{-7}\,M_\odot\,yr^{-1}$, the total energy output of the model of smaller mass is larger ($\sim 0.3 \times 10^{42}$ erg) than that of the model of larger mass ($\sim 0.5 \times 10^{41}$ erg). Nevertheless, because of the smaller mass of the helium layer, the specific energy E_{sp} deposited in the envelope at the peak of the helium flash is larger in the model of larger mass: $E_{sp} \sim 0.5 \times 10^{10}$ erg g^{-1} in the model of smaller mass and $E_{sp} \sim 7.1 \times 10^{10}$ erg g^{-1} in the model of larger mass.

Given the fact that sometimes many hundreds of hydrogen shell flashes must be followed before helium is ignited, it is somewhat impractical to find, by direct computation for a wide range of initial masses and accretion rates, the dependence on white dwarf mass of the mass $M_{\rm He}^{\rm crit}$ of the helium layer required for ignition of helium. It is not only impractical, but it is also not strictly essential, given the fact that most helium-burning outbursts are violent enough to cause the model to expand to giant dimensions; in the real world, this expansion implies, at the very least, extensive mass-loss by a wind and, if the outburst is strong enough, dynamical ejection.

Analytical estimates that have the correct qualitative dependence on the relevant parameters and also give tolerable quantitative accuracy in many mass-accretion regimes can be devised. Assuming that the white dwarf is in a thermal steady state with regard to the accumulation of helium, Iben & Tutukov (1989) construct the approximation

$$M_{\rm He}^{\rm crit} \sim 10^{6.65} R_{\rm WD}^{3.75} M_{\rm WD}^{-0.3} \dot{M}_{-8}^{-0.57}, \tag{3.36}$$

where $R_{\rm WD}$ is the radius of the white dwarf in solar units and $\dot{M}_{-8}$ is the accumulation rate of the helium layer in units of 10^{-8} M$_\odot$ yr^{-1}. The radius of the CO core of a thermally pulsing AGB star (Iben, 1977, 1982) and the radius of a cold white dwarf (Hamada & Salpeter, 1961) can be adequately approximated by (Iben & Tutukov, 1989)

$$R_{\rm WD} \sim 0.0196 \ (1 - 0.59 M_{\rm WD}) \, .$$

The rate at which the mass of the helium layer is increased during the quiescent hydrogen-burning phase of a nova outburst is given approximately by Equation (3.30). Thus,

$$M_{\rm He}^{\rm crit} \sim 0.17 \times (1 - 0.59 M_{\rm WD})^{3.75} M_{\rm WD}^{-0.3} (M_{\rm WD} - 0.26)^{-0.57}. \tag{3.37}$$

When $M_{\rm WD} = 0.6$ M$_\odot$, this equation gives $M_{\rm He}^{\rm crit} \sim 0.07$ M$_\odot$, compared with the values ~0.09, 0.08, and 0.06 M$_\odot$ found by Cassisi, Iben and Tornambé (1998) for $\dot{M}_{\rm acc} = 4$, 6 and 10 $\times 10^{-8}$ M$_\odot$ yr^{-1} and $M_{\rm WD} \sim 0.61$, 0.60, and 0.58, respectively. When $M_{\rm WD} = 0.8$ M$_\odot$, Equation (3.37) gives $M_{\rm He}^{\rm crit} \sim 0.02$ M$_\odot$, compared with the 0.01 M$_\odot$ found by Cassisi, Iben and Tornambé (1998) for $\dot{M}_{\rm acc} = 10^{-7}$ M$_\odot$ yr^{-1}.

3.5.2 *Effect of mixing by particle diffusion and convection*

An understanding of the surface element abundances in classical novae and symbiotic novae requires an exploration of the consequences of mixing between hydrogen-rich accreted matter and matter in the white dwarf during evolution between outbursts. In the absence of rotation, mechanisms which can cause mixing are particle diffusion and convection *prior* to outburst. The study of the effectiveness of particle diffusion has been pioneered by Prialnik and Kovetz (1984); Kovetz and Prialnik (1985); Prialnik (1986); Prialnik and Shara (1986); and Kovetz and Prialnik (1990). Similar results have been obtained by Iben (1990) and by Iben, Fujimoto and MacDonald (1992b). Schwartzman, Kovetz and Prialnik (1994) explore extensively the dependence of these processes on white dwarf mass and interior temperature, the latter being viewed as an adjustable parameter.

Iben, Fujimoto and MacDonald (1991, 1992b) explore the consequences of accretion of hydrogen-rich matter by a 1 M$_\odot$ CO white dwarf in four different approximations: (1) the initial white dwarf is nearly isothermal and cold; (2) the initial white dwarf is in a thermal steady-state (SS) configuration, and pre-outburst convection occurs; (3) the initial white dwarf is in steady state but pre-outburst convection is suppressed; and (4) the initial white

Table 3.3. *M_{crit} (10^{-5} $M_\odot$) for a 1 $M_\odot$ white dwarf. See text for details of the four cases*

$\dot{M}_{acc}$ ($M_\odot$ yr^{-1})	Cold(1)	SS(2)	SS(3)	BL(4)
10^{-8}	2.26	0.60	1.01	
10^{-9}	5.57	1.33	1.51	
10^{-10}	10.2	14.5	14.5	20.9

dwarf is cold but is capped by a helium 'buffer' layer of mass 0.001 $M_\odot$. Accretion rates of $\dot{M}_{acc} = 10^{-10}$, 10^{-9}, and 10^{-8} $M_\odot$ yr^{-1} are adopted.

The motivation for these choices is as follows: (1) in a typical CV, one expects the white dwarf to have cooled before the secondary moves into Roche lobe contact. The cooling age in the experiments is chosen as 10^9 yr and the mean interior temperature is about 10^7 K. (2) After many outbursts, the interior temperature structure between outbursts approaches a steady-state equilibrium wherein the rate at which energy is released locally equals the rate at which energy leaves the local region (Fujimoto & Sugimoto, 1979). The condition for a time-independent solution is that, at every Lagrangian mass point M, the rate at which gravothermal energy is released is

$$\epsilon_{grav} = -(T\,\mathrm{d}S/\mathrm{d}t)_M = +T\,(\partial S/\partial M)(M/M_*)(\mathrm{d}M_*/\mathrm{d}t),$$

where S is the entropy at the mass point M, and M_* is the total mass of the star (Neo *et al.*, 1977; Fujimoto & Sugimoto, 1979, 1982; Kato, 1980). (3) A convective zone forms in the steady-state models, and it is of interest to know the contribution of this convection to surface abundance enhancements. This is accomplished by suppressing convective mixing and allowing only particle diffusion to operate. (4) In general, it is expected that the white dwarf emerges from the pre-CV common envelope phase with a layer of helium above the CO interior. For a 1 $M_\odot$ white dwarf the mass of this layer is expected to be ~ 0.001 $M_\odot$ (Iben & Tutukov, 1985; Iben, Fujimoto and MacDonald, 1992b).

Some results are given in Tables 3.3–3.5. Values of M_{crit}, the amount of hydrogen-rich material accreted prior to an outburst, for the four different cases are given in Table 3.3. For any given case, M_{crit} increases with decreasing $\dot{M}_{acc}$ because the rate of compressional heating decreases with decreasing $\dot{M}_{acc}$ and therefore more matter must be added to bring temperatures in the hydrogen-rich layer up to ignition values.

Table 3.4 gives the dredge-up mass, defined as $M_{dg} = M_{CS} - M_{crit}$, where M_{CS} is the mass of the convective shell when it has reached its maximum size during a thermonuclear flash. The mass M_{dg} is the effective depth in mass to which hydrogen penetrates the underlying heavy-element-rich interior. In practice, the hydrogen mass fraction at the point where the base of the convective shell reaches its maximum inward extent is $X = 0.01$. For any given case, M_{dg} increases with decreasing $\dot{M}_{acc}$ because the time over which diffusion can act increases with decreasing $\dot{M}_{acc}$. In the steady state cases, M_{dg} is larger when pre-outburst convection has not been suppressed.

Table 3.5 gives the values of surface CNO enhancements, in the form [CNO/H] = $\log(N_{CNO}/N_H) - \log(N_{CNO}/N_H)_\odot$, where N_{CNO} is the abundance by number of CNO

Table 3.4. M_{dg} *(*10^{-5} $M_\odot$*) for a 1* $M_\odot$ *white dwarf. See text for details of the four cases*

$\dot{M}_{acc}$ ($M_\odot$ yr^{-1})	Cold(1)	SS(2)	SS(3)	BL(4)
10^{-8}	0.25	0.25	0.10	
10^{-9}	1.3	0.41	0.32	
10^{-10}	2.4	3.5	3.5	2.9

Table 3.5. *Surface enhancements [CNO/H] for a 1* $M_\odot$ *white dwarf. See text for details of the four cases*

$\dot{M}_{acc}$ ($M_\odot$ yr^{-1})	Cold(1)	SS(2)	SS(3)	BL(4)
10^{-8}	1.00	1.34	0.82	
10^{-9}	1.15	1.31	1.21	
10^{-10}	1.28	1.19	1.19	–0.06

elements. Case 4 has been examined for only the smallest $\dot{M}_{acc}$, with the objective of achieving the optimum conditions for interpenetration, through the helium buffer layer, of hydrogen from the accreted layer, and carbon and oxygen from the interior of the white dwarf.

Assuming a knowledge of M_{dg}, M_{crit}, and three additional quantities which can be calculated, it is possible to estimate the element enhancements achieved after enough outbursts have occurred for an 'abundance steady state' to have been achieved (Fujimoto & Iben, 1992; and section 3.6).The three additional quantities are (1) the mass $M_{rm}(H)$ of the hydrogen-rich layer which remains at the surface of the white dwarf after hydrogen burning has ceased following an outburst, (2) the mass $M_{rm}(He)$ of the layer containing fresh helium produced during outburst, and (3) the amount of hydrogen ΔX_{nuc} which must be burned during the convective shell phase of an outburst in order to expand the hydrogen-rich envelope to beyond the Roche lobe. For a 1.01 $M_\odot$ white dwarf with $Z = 0.02$ and $Z_{CNO} \sim 0.01$, Iben (1982) finds a net remnant mass of $M_{rm} = M_{rm}(He) + M_{rm}(H) \sim 1.6 \times 10^{-5}\,M_\odot$ and $\gamma = M_{rm}(H)/M_{rm}(He) \sim 1/3$. For a 1 $M_\odot$ white dwarf, Iben, Fujimoto and MacDonald (1992b) find $M_{rm}(He) \sim 3.7 \times 10^{-6}\,M_\odot$ and $M_{rm}(H) \sim 1.2 \times 10^{-6}\,M_\odot$ when $Z \sim 0.3$. Thus, when the mass of the degenerate CO core $M_{CO} \sim 1\,M_\odot$ and when the Roche-lobe radius is taken to be $R_\odot$, in first approximation,

$$M_{rm}(He) \sim 1.12 \times 10^{-5} \left(\frac{0.01}{Z}\right)^{0.25} \tag{3.38}$$

$$M_{rm}(H) \sim \frac{1}{3} M_{rm}(He),$$

$$\text{and} \quad \Delta X_{nuc} \simeq (G M_{WD}/R_{WD})/E_H \sim 0.04. \tag{3.39}$$

Here $G M_{WD}/R_{WD}$ is a measure of the energy per unit mass required to expand the envelope to a radius much larger than the white dwarf radius, and $E_H = 6 \times 10^{18}$ erg g^{-1} is the energy released per gram by the complete burning of hydrogen.

In each cycle, the mass of the helium layer increases by

$$\Delta M_{\rm He} = M_{\rm rm}({\rm He}) - M_{\rm dg}.$$

Thus, if $M_{\rm dg} < M_{\rm rm}({\rm He})$, the total mass of the helium layer above the CO or ONe white dwarf grows with time until it reaches a critical value somewhere between a minimum value given by Equation (3.36) or Equation (3.37) and a maximum value $\sim 0.14\,{\rm M}_\odot$. For $M_{\rm WD} = 1\,{\rm M}_\odot$, the minimum critical value is $\sim 0.01\,{\rm M}_\odot$.

Let X, Y, and Z be the steady state abundances by mass of hydrogen, helium, and elements heavier than helium at the surface of the model at the end of the convective shell phase of the outburst, and let X_0, Y_0, and Z_0 be the abundances by mass of these same entities in the accreted material. In the presence of a helium buffer layer (within which $Y \sim 1$) and for $M_{\rm dg} < M_{\rm rm}({\rm He})$,

$$M_{\rm CS} Y = (M_{\rm crit} + M_{\rm dg}) Y = M_{\rm acc} Y_0 + M_{\rm rm}({\rm H}) Y + M_{\rm dg} + M_{\rm CS} \Delta X_{\rm nuc}\,,$$

where

$$M_{\rm acc} = M_{\rm crit} - M_{\rm rm}({\rm H}). \tag{3.40}$$

If the Cold (1) case with $\dot{M}_{\rm acc} = 10^{-8}\,{\rm M}_\odot\,{\rm yr}^{-1}$ is applicable, $M_{\rm crit} = 2.26 \times 10^{-5}\,{\rm M}_\odot$ and $M_{\rm dg} = 2.5 \times 10^{-6}\,{\rm M}_\odot$. If $Y_0 = 0.27$ and $Z_0 = 0.02$, the abundance of helium in the nova ejecta is $Y = 0.37$. Adopting, instead, the SS (3) case with $\dot{M}_{\rm acc} = 10^{-8}\,{\rm M}_\odot\,{\rm yr}^{-1}$, one has $M_{\rm crit} = 1.01 \times 10^{-5}\,{\rm M}_\odot$, $M_{\rm dg} = 1.0 \times 10^{-6}\,{\rm M}_\odot$, and $Y = 0.43$. Comparison of the abundance of steady-state models with observed nova abundances suggests that rotation-induced mixing must be operating (Fujimoto & Iben, 1992; see also Sections 3.6, 3.7). In order to understand qualitatively the effect of including rotation-induced mixing, one may double the values of $M_{\rm dg}$ given by the numerical experiments without rotation-induced mixing. Then, the two examples give $Y = 0.49$ (Cold (1) with $\dot{M}_{\rm acc} = 10^{-8}\,{\rm M}_\odot\,{\rm yr}^{-1}$) and $Y = 0.57$ (SS (3) with $\dot{M}_{\rm acc} = 10^{-8}\,{\rm M}_\odot\,{\rm yr}^{-1}$).

We next estimate how long it takes to build up the critical mass of helium for a helium-burning thermonuclear runaway to occur. In the Cold (1) case with $\dot{M}_{\rm acc} = 10^{-8}\,{\rm M}_\odot\,{\rm yr}^{-1}$, $M_{\rm dg} = 2.5 \times 10^{-6}\,{\rm M}_\odot$, $\Delta M_{\rm He} = 0.87 \times 10^{-6}\,{\rm M}_\odot$ and it requires $\sim$1149 cycles to increase the mass of the helium layer to $0.01\,{\rm M}_\odot$. The duration of an outburst cycle is $\tau_{\rm cycle} = (M_{\rm crit} - M_{\rm rm}({\rm H}))/\dot{M}_{\rm acc} = (2.26 - 0.37) \times 10^{-5}/10^{-8} = 1.89 \times 10^3$ years. Hence, it requires at least $1149 \times 1.89 \times 10^6 \sim 2.2 \times 10^6$ years to ignite helium. In case SS (3) with $\dot{M}_{\rm acc} = 10^{-8}\,{\rm M}_\odot\,{\rm yr}^{-1}$, $M_{\rm dg} = 1.0 \times 10^{-6}\,{\rm M}_\odot$, $\Delta M_{\rm He} = 1.02 \times 10^{-6}\,{\rm M}_\odot$, and it requires $\sim$1000 cycles or $1000 \times (1.01 - 0.37) \times 10^{-5}/10^{-8} \sim 6 \times 10^6$ years to increase the mass of the helium layer to $0.01\,{\rm M}_\odot$. These times are longer than those needed to establish thermal steady-state conditions for hydrogen-burning outbursts, but perhaps not long enough for the center of the helium layer to be warm enough for Equations (3.36) and (3.37) to be applicable.

In any case, the time to build the helium layer up to a critical value is large compared with the time which a star can live on the AGB ($\sim 10^5 - 10^6$ years). This, then, may account for the absence of heavy element over-abundances in symbiotic novae with AGB donors (Iben & Tutukov, 1996; see also Sections 3.8, 3.9). In CVs, with typical lifetimes of $\sim 10^8$ years, there is ample time to build up a critical mass of helium, and, once helium is ignited, the explosion will typically be violent enough to remove the entire helium layer. Once the helium buffer layer has been removed, the potential for the mixing of hydrogen-rich material with CO or ONe white dwarf material in successive outburst cycles is improved. The key to this improvement is the fact that the mass $M_{rm}(He)$ decreases with increasing Z (see Equation (3.38)).

3.6 Over-abundances of heavy elements in the cyclical model

Adopting the degree of mixing as a parameter, a steady state cyclical model can be used in conjunction with estimates of abundances in real nova ejecta to gauge the degree of mixing required to achieve the observed abundances (Fujimoto & Iben, 1992).

3.6.1 *The cyclical model*

A full cycle may be broken into four phases. At the beginning of phase (1), when hydrogen burning has just been completed, there are three layers of relevance near the surface of the accretor: the lowest layer consists of unmixed white dwarf material; the uppermost layer, of mass $M_{rm}(H)$, consists of matter which has the same composition as matter in the nova ejecta; the middle layer, of mass $M_{rm}(He)$, has the composition which results from the complete hydrogen burning of matter having the same initial composition as the top layer. During subsequent accretion, hydrogen-rich material is mixed downward into the helium-rich layer, hydrogen and helium are mixed into white dwarf material, and white dwarf material is mixed outward; when the mass of freshly accreted material reaches a critical value M_{acc}, a thermonuclear flash is initiated. At this point three additional layers can be identified: the lowest of these, of mass M_{dg}, defines the region within which significant mixing has taken place; the uppermost layer, of mass M_{acc}, has the composition of accreted matter; the layer between these two is what remains of the layer initially of mass $M_{rm}(He)$. The hydrogen flash begins near the base of the layer within which significant mixing has taken place. Owing to a large flux of energy outward, a convective shell is formed and grows until its outer edge nearly reaches the surface. Because of heating, the position of maximum energy generation and the base of the convective shell move inward in mass to the point where the hydrogen abundance by mass is $X \sim 0.01$. The maximum mass of the convective shell is M_{CS}. We call the phase from the beginning of the outburst to the time when the convective shell has reached its maximum size in mass, phase (2). As the model expands, shell convection disappears. The radius of a model becomes large compared with the dimensions of a host CV system, so that mass is lost from the system until the mass of the hydrogen-rich envelope decreases below a critical value M_{rm}, whereupon the model radius contracts rapidly to less than solar dimensions. Hydrogen continues to burn until the mass of hydrogen-rich matter remaining reaches the value $M_{rm}(H)$; the mass of the matter within which all hydrogen has been freshly converted into helium is $M_{rm}(He) = M_{rm} - M_{rm}(H)$. We call the mass-loss phase, phase (3), and call the subsequent hydrogen-burning phase, phase (4). We have now returned to the beginning of phase (1). Note that if $M_{dg} > M_{rm}(He)$, the mass of the white dwarf is smaller by $\Delta M_{WD} = M_{dg} - M_{rm}(He)$ than the mass of the white dwarf at the end of the previous outburst. Thus, the basic condition for an enhancement of heavy elements to occur is that

$$M'_{dg} = M_{dg} - M_{rm}(He) \tag{3.41}$$

must be larger than zero. An observed enhancement of elements heavier than helium implies that the white dwarf component loses mass on a time-scale long compared with the interburst time-scale.

If $M_{dg} < M_{rm}(\mathrm{He})$, no enhancement of elements heavier than helium will occur, and the white dwarf will increase in mass by $\Delta M_{WD} = M_{rm}(\mathrm{He}) - M_{dg}$ in each cycle. However, in this case, an enhancement of helium still occurs, as discussed in Section 3.5, and the total mass of the helium layer above the CO or ONe white dwarf grows with time.

First approximations to $M_{rm}(\mathrm{H})$ and $M_{rm}(\mathrm{He})$ are given in Section 3.5. Another way of estimating $M_{rm}(\mathrm{H})$ and $M_{rm}(\mathrm{He})$ is actually to remove matter from the surface of an expanding model, once the model radius exceeds a value of the order of the dimensions of a cataclysmic variable, and to do so on a time-scale much shorter than the nuclear-burning time-scale. When the mass of hydrogen remaining decreases below a critical value, the radius of the model shrinks rapidly below the predetermined upper limit. The values of remnant masses obtained in either way are essentially the same.

We are now in a position to construct a quantitative algorithm for estimating abundances in nova ejecta. Let X, Y, and Z be the abundances by mass of hydrogen, helium, and elements heavier than helium at the surface of the model at the end of phase (4), and let X_0, Y_0, and Z_0 be the abundances by mass of these same elements in the accreted material. Then

$$M_{CS} X = M_{acc} X_0 + M_{rm}(\mathrm{H}) X - M_{CS} \Delta X_{nuc}, \tag{3.42}$$

where ΔX_{nuc}, the amount of hydrogen burned in the convective shell, is given by Equation (3.39).

At its maximum extent, the mass M_{CS} of the convective shell is given by

$$M_{CS} = M_{acc} + M_{rm}(\mathrm{H}) + M_{dg}.$$

Thus, Equation (3.42) may be written as

$$X = \frac{M_{acc}}{M_{acc} + M_{dg}} X_0 - \left(1 + \frac{M_{rm}(\mathrm{H})}{M_{acc} + M_{dg}}\right) \Delta X_{nuc}. \tag{3.43}$$

Similarly,

$$Y = \frac{M_{acc}}{M_{acc} + M'_{dg}} Y_0 + \frac{M_{rm}(\mathrm{He})}{M_{acc} + M'_{dg}} X + \frac{M_{acc} + M_{rm}(\mathrm{H}) + M_{dg}}{M_{acc} + M'_{dg}} \Delta Y_{nuc} \tag{3.44}$$

when $M_{dg} > M_{rm}(\mathrm{He})$, and

$$Y = Y_0 + \frac{M_{dg}}{M_{acc}} X + \left(1 + \frac{M_{rm}(\mathrm{H}) + M_{dg}}{M_{acc}}\right) \Delta Y_{nuc} \tag{3.45}$$

when $M_{dg} < M_{rm}(\mathrm{He})$. In these equations, M'_{dg} is defined by Equation (3.41), and ΔY_{nuc} is the amount of helium produced in the convective zone. For CO white dwarfs, ΔY_{nuc} is obtained by subtracting from ΔX_{nuc} the amount of hydrogen required to form $^{14}\mathrm{N}$ from $^{12}\mathrm{C}$. Thus, for CO white dwarfs,

$$\Delta Y_{nuc} \simeq \Delta X_{nuc} - Z_{12}/6$$

if $\Delta X_{\rm nuc} > Z_{12}/6$, and

$$\Delta Y_{\rm nuc} = 0$$

if $\Delta X_{\rm nuc} < Z_{12}/6$, where Z_{12} is the abundance by mass of ^{12}C at the beginning of a flash, averaged over those layers which will ultimately be incorporated into the convective shell at the peak of the flash. For ONe white dwarfs, an adequate approximation is

$$\Delta Y_{\rm nuc} \simeq \Delta X_{\rm nuc}.$$

Finally,

$$Z = \frac{M_{\rm acc} Z_0 + M'_{\rm dg}}{M_{\rm acc} + M'_{\rm dg}} \quad \text{when } M_{\rm dg} > M_{\rm rm}({\rm He}) \tag{3.46}$$

$$\text{and} \quad Z = Z_0 \quad \text{when } M_{\rm dg} < M_{\rm rm}({\rm He}).$$

This last equation is an adequate approximation to the correct expression (see Fujimoto & Iben, 1992) only when $M_{\rm rm}({\rm H}) \ll M_{\rm acc} + M_{\rm dg}$.

3.6.2 Application to 1 $M_\odot$ white dwarf models

As a first illustration, consider the steady-state case for $\dot{M}_{\rm acc} = 10^{-10}\,{\rm M}_\odot\,{\rm yr}^{-1}$. The appropriate choices for $M_{\rm rm}({\rm He})$, $M_{\rm rm}({\rm H})$, and $M_{\rm acc}$ are given by Equations (3.38)–(3.39). Since $M_{\rm dg} > M_{\rm rm}({\rm He})$, Equations (3.44) and (3.46) are applicable, along with Equation (3.43). From Table 3.4, $M_{\rm dg} = 3.5 \times 10^{-5}$, and from Table 3.3, $M_{\rm crit} = 14.5 \times 10^{-5}\,{\rm M}_\odot$. Using Equations (3.38)–(3.39) in Equation (3.46), it follows that

$$Z = \frac{0.21 - 0.067\ (0.02/Z)^{0.43}}{1 - 0.089\ (0.02/Z)^{0.43}}.$$

The solution of this equation is $Z = 0.19$. Hence, $M_{\rm rm}({\rm He}) \sim 0.46 \times 10^{-5}\,{\rm M}_\odot$, $Z_{\rm CNO} = Z - 0.01 = 0.18$, and $Z_{12} = 0.09$. With $\Delta X_{\rm nuc}$ ~0.04 (Fujimoto & Iben, 1992), $\Delta Y_{\rm nuc} \sim 0.025$. Equations (3.43) and (3.45) now give $X = 0.54$ and $Y = 0.27$, respectively. Thus, $Y/X \sim 1.4\ Y_0/X_0$, $Z/X \sim 13\ Z_0/X_0$, and $Z_{\rm CNO}/X \sim 24(Z_{\rm CNO}/X)_0$. Put another way, [CNO/H] $\sim$ 1.38, and [He/H] $= \log(N_{\rm He}/N_{\rm H}) - \log(N_{\rm He}/N_{\rm H})_\odot = \log(Y/X) - \log(Y/X)_\odot \sim 0.15$. Finally, the white dwarf decreases in mass by $\sim 3 \times 10^{-5}{\rm M}_\odot$ in every cycle, compared with the mass $\sim 14 \times 10^{-5}\,{\rm M}_\odot$ accreted between outbursts.

As a next example, consider the steady-state cases when $\dot{M}_{\rm acc} = 10^{-9}\,{\rm M}_\odot\,{\rm yr}^{-1}$. No solution exists for $M_{\rm dg} > M_{\rm rm}({\rm He})$. Therefore, $Z = Z_0 = 0.02$, $M_{\rm rm}({\rm H}) \sim 0.4 \times 10^{-5}\,{\rm M}_\odot$, and $M_{\rm acc} \sim 1.11 \times 10^{-5} (= 1.51 \times 10^{-5} - 0.4 \times 10^{-5})\,{\rm M}_\odot$. Adopting $M_{\rm dg} \sim 0.32 \times 10^{-5}\,{\rm M}_\odot$, Equations (3.43) and (3.45) give $(X, Y) = (0.51, 0.47)$. Thus, $Y/X \sim 2.5\ Y_0/X_0$ and $Z/X \sim 1.4\ Z_0/X_0$. Or, [CNO/H] $\sim$ 0.13 and [He/H] $\sim$ 0.4.

In a similar fashion, the steady-state case (pre-outburst convection neglected) for $\dot{M}_{\rm acc} = 10^{-8}\,{\rm M}_\odot\,{\rm yr}^{-1}$ gives $(X, Y, Z) = (0.56, 0.42, 0.02)$, or $X/Y \sim 2.1\ (X/Y)_0$ and $Z/X \sim 1.3\ Z_0/X_0$. In this case, [CNO/H] $\sim$ 0.13 and [He/H] $\sim$ 0.3.

A useful measure of the degree of mixing is the quantity

$$\zeta = \frac{M_{\rm dg}}{M_{\rm rm}({\rm He})}. \tag{3.47}$$

In order of increasing accretion rate, the three models give $\zeta = 7.6, 0.27$, and 0.08, respectively.

In summary, when only particle diffusion and pre-outburst convection are taken into account, and hydrogen-rich material is allowed to accrete directly onto a 'bare' CO core, large enhancements of elements heavier than helium occur during a thermonuclear outburst. The enhancements are due to the diffusion of hydrogen to a depth $M_{\rm dg}$ below the original surface of the white dwarf model and to the subsequent dredging up of this matter into a convective shell which forms during the outburst. However, after the outburst, when much of the accreted matter is expelled, a layer of helium topped by a less massive layer containing both hydrogen and helium is left behind. If the mass $M_{\rm rm}$(He) of the helium layer is larger than $M_{\rm dg}$, hydrogen will not penetrate below the helium layer during the next accretion phase and no enhancement elements heavier than helium will occur during the next outburst. However, the helium to hydrogen ratio in ejected matter will be larger than in accreted matter.

3.6.3 Comparison with observations

Careful abundance analyses are available for about a dozen novae (see the reviews by Williams, 1985, and by Truran, 1985, 1990). The novae fall into two groups: one for which [CNO/H] (or [ONe]) $\sim 1.7 \pm 0.3$ and [He/H] $\sim 0.12 \pm 0.03$ (case A); and another for which [CNO/H] (or [ONe]) $\sim 1.7 \pm 0.8$ and [He/H] $\sim 0.5 \pm 0.1$ (case B). The last four novae listed in Table 3.6 (Fujimoto & Iben, 1992) belong to group A and the first seven belong to group B.

Using a formalism similar to that described in Section 3.6.2, Fujimoto and Iben (1992) obtain estimates of $M_{\rm acc}$ and $M_{\rm CS}$ for the white dwarf component in 11 systems as a function of assumed white dwarf mass and composition. Results in columns 3 and 4 of Table 3.6 are for a CO white dwarf of mass 1 $M_\odot$ which is accreting matter of solar composition. For case A systems the average value of $M_{\rm acc}$ is $\sim 1.7 \times 10^{-4}$ $M_\odot$, and for case B systems the average value of $M_{\rm acc}$ is $\sim 1.4 \times 10^{-5}$ $M_\odot$. Results do not depend terribly sensitively on the choice of white dwarf mass and similar values of $M_{\rm acc}$ and $M_{\rm CS}$ are obtained on the assumption that the white dwarf is a more massive ONe white dwarf.

Comparing values of $M_{\rm acc}$ in Table 3.6 with entries in Table 3.3, it is clear that the accretion rate in case A systems is on average considerably less than in case B systems. The value of [He/H] ~ 0.15 found in Section 3.5 for an accretion rate of $10^{-10}\,M_\odot\,{\rm yr}^{-1}$ is only slightly larger than the mean for case A systems and the value of [CNO/H] ~ 1.4 is only slightly smaller than the mean for these systems. Further, the value of $M_{\rm acc} \sim 1.4 \times 10^{-4}\,M_\odot$ found in Section 3.5 for $\dot{M}_{\rm acc} = 10^{-10}\,M_\odot\,{\rm yr}^{-1}$ is quite similar to the average estimated for the case A systems. Thus, it is not out of the question that particle diffusion plays an important role in the enhancement of heavy elements in real novae when the mean accretion rate is less than a few times $10^{-10}\,M_\odot\,{\rm yr}^{-1}$.

On the other hand, in case B systems for which estimated values of $M_{\rm acc}$ indicate accretion rates in the range 10^{-9}–$10^{-8}\,M_\odot\,{\rm yr}^{-1}$, not only are the observed enhancements of elements heavier than helium far larger than predicted by non-rotating models with accretion rates in this range ([CNO/H] ~ 0.13), but the enhancements of helium also are larger than the predicted ones ([He/H] ~ 0.3–0.4).

Thus, for case B stars, some mode of mixing in addition to particle diffusion must be operating between outbursts and one might guess at some form of rotationally induced mixing. The extent of such mixing can be described by the 'degree of mixing' parameter ζ defined by Equation (3.47). Values of ζ are given in columns 5–7 of Table 3.6 for three different choices of the mass of the white dwarf: 1.0, 0.8, and 0.6 $M_\odot$ (derived from Fujimoto and

Table 3.6. *Estimated parameters of shell flashes for nova explosions (M_{WD} = $1M_\odot[\Delta X_{nuc} = 0.04]$ and $X_0 = 0.72$)*

Source	Speed[a] class	M_{acc} ($M_\odot$)	M_{CS} ($M_\odot$)	$\zeta(1.0)$	$\zeta(0.8)$	ζ (0.6)
V1370 Aql	VF	2.9×10^{-6}	1.2×10^{-5}	1.90	1.75	1.67
DQ Her	M	6.0×10^{-6}	1.6×10^{-5}	1.90	1.78	1.70
V693 CrA	VF	9.7×10^{-6}	1.9×10^{-5}	1.57	1.51	1.43
GQ Mus[b]	M	1.1×10^{-5}	2.0×10^{-5}	1.31	1.30	1.25
T Aur	M	1.7×10^{-5}	2.8×10^{-5}	1.57	1.50	1.45
RR Pic	S	2.6×10^{-5}	3.7×10^{-5}	1.13	1.14	1.11
HR Del[b]	S	2.2×10^{-5}	3.3×10^{-5}	1.23	1.17	1.15
CP Pup	VF	4.2×10^{-5}	1.0×10^{-4}	14	11	9.6
V1668 Cyg	F	1.0×10^{-4}	1.5×10^{-4}	9.5	5.2	5.0
V1500 Cyg	VF	4.6×10^{-4}	6.4×10^{-4}	39	8.5	6.6
PW Vul[b]	M	8.5×10^{-5}	1.1×10^{-4}	3.9	3.2	2.6

Abundance ratios are taken from Williams (1985)
[a] Designations are according to Bode and Evans (1989)
[b] Abundance ratios are taken from Truran and Livio (1986)

Iben (1992)). Note that the value of ζ for case A systems is not unlike $\zeta = 7.6$ found for the model of smallest accretion rate in Section 3.6.2, whereas the values of ζ for case B systems are considerably larger than the values $\zeta = 0.25$ and 0.08 found for non-rotating models with accretion rates 10^{-9} and 10^{-8} $M_\odot$ yr^{-1}, respectively.

3.7 Mixing due to rotation-induced instabilities

Differential rotation, which is continuously fed by the accretion of angular momentum, gives rise to hydrodynamical instabilities; turbulent motions are generated and lead to the redistribution of angular momentum, and concomitantly, to material mixing and heat transport. In the context of the nova phenomenon, Kippenhahn and Thomas (1978) invoked the Kelvin–Helmholtz instability, and proposed the formation of an accretion belt around the equator of the accretor. MacDonald (1983) subsequently argued that mixing in the horizontal direction is much more efficient than in the vertical direction, and Kutter and Sparks (1989) pointed out that the Kelvin–Helmholtz instability alone is not capable of extracting enough angular momentum from accreted material to cause strong shell flashes.

In studying rotation-induced mixing, the evaluation of the efficiency of turbulent mixing is crucial. Several explorations have invoked combinations of known kinds of instabilities, including the Kelvin–Helmholtz instability and its modification by thermal diffusion (Townsent, 1958; Zahn, 1974), the thermal instability of axisymmetric modes (Goldreich & Schubert, 1967), and even the Eddington–Sweet circulation (e.g. Endal & Sofia, 1976; Pinsonneault *et al.*, 1989). Even though a precise theory of the generating mechanisms and properties of turbulence has yet to be established, the problem can be formulated in a general and self-consistent way (Fujimoto, 1993) which is based on an analysis of the stability of rotating bodies to adiabatic perturbations of non-axisymmetric modes.

3.7.1 Assumptions and method

As described in Fujimoto (1993), turbulent mixing can be approximated by a diffusion process. The viscosity ν due to turbulent motions is taken to be proportional to the upper bound $\omega_{\mathrm{i,max}}$ on the growth rates of the hydrodynamical instabilities of non-axisymmetric, adiabatic modes, and we write

$$\nu = \alpha\, \omega_{\mathrm{i,max}}\, l_{\mathrm{max}}^2, \tag{3.48}$$

where l_{max} is a characteristic length which we set equal to the pressure scale height, and α is a parameter which is presumably much smaller than unity.

We assume that the axisymmetric modes dominate in feeding the turbulence, and contributions of other modes of instability are ignored. The effect of the Eddington–Sweet circulation is also neglected since, in the case at hand, differential rotation is promoted exclusively by the supply of angular momentum at the surface.

When mixing occurs in a stable stratification, work is done against gravity; in the present context, we assume that the energy source is differential rotation. The Richardson number Ri, the flux-Richardson number Rf, which is the fraction of the available kinetic energy that can be converted into the potential energy of mixed material, the turbulent viscosity in the vertical direction $\nu^{(\mathrm{V})}$, and the material diffusivity in the vertical direction $K_\rho^{(\mathrm{V})}$ are related by

$$K_\rho^{(\mathrm{V})} = \nu^{(\mathrm{V})}(Rf/Ri). \tag{3.49}$$

The Richardson number is given by

$$Ri = \mathcal{N}^2/|\mathbf{D}|^2,$$

where $\mathcal{N}$ is the Brunt-Väisälä frequency and $\mathbf{D}$ is the gradient of the rotational velocity. Laboratory experiments show that $Rf \simeq 0.20$ (Fujimoto, 1988). For mixing in the horizontal direction, $K_\rho^{(\mathrm{H})} = \nu^{(\mathrm{H})}$.

Although the interaction between convection and rotation is not completely understood, it seems reaonable to assume that convection tends to smear out shears. So, in the convection zone, we assume that transport can be described as a diffusion process and adopt the diffusion coefficient

$$\nu = K_\rho = v_{\mathrm{conv}}\, l_{\mathrm{mix}},$$

where v_{conv} is the average velocity of convective elements and l_{mix} is a mixing length which we can take equal to the pressure scale height. The convective velocity is calculated in the mixing length approximation.

The equations of stellar structure and evolution are coupled with the equations for angular momentum and element diffusion and, using the prescriptions developed by Kippenhahn and Thomas (1970), all equations can be reduced to a one dimensional form. The boundary condition for angular momentum is replaced by the relationship

$$\dot{M}_{\mathrm{acc}} r^2 \Omega + 4\pi r^3\, \rho\nu\, (\mathrm{d}\Omega/\mathrm{d}r) = \dot{M}_{\mathrm{acc}}\, R^2\, \Omega_{\mathrm{K}}$$

evaluated at the base of the atmosphere, and we have assumed a steady-state distribution for the angular velocity Ω above the base of the atmosphere. Here, R is the surface radius and Ω_{K} is the angular velocity of Keplerian rotation at the surface ($= \surd(GM_{\mathrm{WD}}/R^3)$). Similarly,

$$\dot{M}_{\mathrm{acc}}\, X_i + 4\pi r\, \rho\, K_\rho\, (\mathrm{d}X_i/\mathrm{d}r) = \dot{M}_{\mathrm{acc}}\, X_{i,0},$$

where $X_{i,0}$ is the abundance of the ith element in the accreted material.

Table 3.7. *Characteristics of hydrogen shell flashes*

$\dot{M}_{acc}(M_\odot\ yr^{-1})$	α	$M^a_{acc}(M_\odot)$	$M^b_{cs}(M_\odot)$	M_{cs}/M_{acc}	Z_{CNO}	$L_H(L_\odot)$
10^{-8}	10^{-3}	4.2×10^{-6}	7.4×10^{-6}	1.8	0.47	2.1×10^8
10^{-8}	10^{-5}	3.1×10^{-6}	6.6×10^{-6}	2.1	0.58	1.8×10^8
10^{-9}	10^{-5}	6.9×10^{-6}	1.40×10^{-5}	2.0	0.53	3.1×10^9
10^{-10}	10^{-5}	1.40×10^{-4}	1.84×10^{-4}	1.3	0.27	3.3×10^{12}
10^{-8}	IFM[c]	6.0×10^{-6}	8.4×10^{-6}	1.4	0.30	1.5×10^8
10^{-9}	IFM[c]	1.33×10^{-5}	1.75×10^{-5}	1.3	0.26	1.5×10^9
10^{-10}	IFM[c]	1.45×10^{-4}	1.79×10^{-6}	1.2	0.21	3.3×10^{11}
10^{-8}	no-mix	3.2×10^{-5}	3.2×10^{-5}	1.0	0.02	2.0×10^8

[a] The amount of accreted mass and the amount of mass involved in the convective shell
[b] The amount of mass involved in the convective zone during the flash
[c] Iben, Fujimoto and MacDonald (1992a,b)

3.7.2 *Models of nova explosions*

With these prescriptions, Fujimoto and Iben (1997) have followed the evolution of an accreting CO white dwarf of mass $1\,M_\odot$ through a hydrogen shell flash cycle for several sets of accretion rates and and for several values of α in Equation (3.48). The initial models are the steady-state models of Iben, Fujimoto and MacDonald (1992b), assumed to be rotating uniformly in synchronism with the orbital motion at the rate $\Omega = \sqrt{(GM_\odot/R^3_\odot)}$. The characteristics of resultant hydrogen shell flashes are summarized in Table 3.7. For comparison, the results of model computations with particle diffusion (Iben, Fujimoto & MacDonald, 1992a,b) and results for a model without mixing are also given in Table 3.7.

The calculations show that accreted angular momentum is efficiently transported inwards, and the stellar interior rotates almost uniformly. For the case $\dot{M}_{acc} = 10^{-8}\ M_\odot\ yr^{-1}$ and $\alpha = 10^{-3}$, see Figure 1 in Fujimoto and Iben (1997). The Richardson number, defined by Equation (3.49) with $|\mathbf{D}| = (d\Omega/d\log r)$, remains larger than $\sim 10^4$ over most of the interior. In the outermost layers, the gradient in Ω grows larger as the density drops owing to an increasing temperature brought about by accretion-induced heating; however, Ω is still much smaller than the Keplerian rotation rate at the surface, $\Omega_K \simeq 0.9\ s^{-1}$. When flash convection appears, the angular velocity is effectively averaged in the convection zone, and then, it decreases as the convection zone expands. Because of a large Richardson number, material mixing remains moderate, despite the efficient transfer of angular momentum; near the bottom of accreted layers, $K_\rho \simeq 10^2$, while $\nu \simeq 2 \times 10^6\ cm\ s^{-1}$. Finally, white dwarf material is dredged up to the surface by flash convection, which penetrates into a layer where the hydrogen concentration is $X \sim 0.01$, just as in models with particle diffusion.

The resultant abundances by mass of CNO elements Z_{CNO} in the convective zone are given in the sixth column of Table 3.7. They depend only slightly on α. Smaller α tends to decrease ν, but the increase in the gradient of the rotational velocity counterbalances the change in ν. In contrast, K_ρ increases, though slightly, owing to the reduction in the Richardson number. With increasing accretion rate, the degree of mixing increases, as is intuitively expected; the dependence becomes weaker, however, for higher accretion rates ($\gtrsim 10^{-9}\,\mathrm{M}_\odot\,\mathrm{yr}^{-1}$), since the entropy gradient becomes smaller and the stratification is determined mainly by the gradient in the mean molecular weight.

In summary, rotation-induced mixing effectively redistributes accreted angular momentum and brings the accreting star into nearly uniform rotation. It also produces composition mixing of the right order of magnitude to explain the observed enrichment of heavy elements in nova ejecta, and this provides support for the treatment of rotation-induced mixing which has been adopted. Composition mixing and the attendant heat transfer occur mainly in outer layers and, hence, have little effect on the uniformly rotating interior.

For completeness, we mention that hydrodynamic mixing between white dwarf matter and matter at the base of the convective shell during outburst has been considered by Livne and Arnett (1995), and Livne (1997).

3.8 Symbiotic star evolution

A comparison between theoretical models of thermonuclear outbursts and the properties of symbiotic novae provides a cleaner test of some aspects of the outburst phenomenon than does a comparison between theoretical models and the properties of classical novae. Symbiotic stars are also more attractive than CVs for studying the nature of the wind emitted by the nova. In CVs, when the matter in the outer layers of the erupting accretor expands to encompass both stars, outflowing material may achieve escape velocities in part because of the frictional interaction between the two stars and outflowing material and not entirely because of the conversion of thermonuclear energy into hydrodynamic motions and/or to an optically thick wind during the subsequent quiescent hydrogen-burning phase. This is not a problem for symbiotic stars in which the stellar components are always far enough apart that common envelope activity is unimportant.

3.8.1 Plateau mass–luminosity relationships

Provided it is burning hydrogen in a shell when it is forced to leave the giant branch by the removal of most of its hydrogen-rich envelope either because of a superwind (in the case of single stars) or because of a common-envelope interaction (in close binaries), the remnant of an AGB star evolves to the blue in the H–R diagram along a horizontal track at a plateau luminosity described well (Paczyński, 1970, 1971; Uus, 1970) by:

$$L_{\mathrm{CO}} \sim 60\,000\ (M_{\mathrm{CO}} - 0.52), \tag{3.50}$$

where M_{CO} is the mass of the electron-degenerate CO or ONe core. Similarly, if a low mass red giant loses most of its hydrogen-rich envelope in a common envelope event, its plateau luminosity as it evolves to the blue is related to its mass by Equation (3.14) which, for convenience, we repeat here:

$$L_{\mathrm{He}} \sim 10^{5.6} M_{\mathrm{He}}^{6.5}, \tag{3.29}$$

where M_{He} is the mass of the degenerate helium core. This is a fit by Iben and Tutukov (1984a) to the model results of Mengel *et al.* (1979). The relationships described by Equations (3.50) and (3.14) apply only when the electron-degenerate core is ‘hot’ and has an extended, semi-degenerate mantle separating the hydrogen-rich layers from the electron-degenerate interior.

In the case of CVs and symbiotic stars, the interior of the accreting white dwarf has typically had time to cool to the point that the mantle is so small that the base of the hydrogen-rich layer is essentially at the outer edge of a 'cold' white dwarf. The result is that, in a nova, the base of the hydrogen-burning shell is at a smaller radius (larger gravity and larger pressure) and the hydrogen-burning luminosity is larger than in the case of a planetary nebula nucleus of the same mass and core composition. As discussed by Iben and Tutukov (1996; see their Figure 5), models by a half dozen different sets of authors show that the maximum bolometric luminosity during the plateau phase is related to white dwarf mass by (Iben & Tutukov, 1989; their Figures 3b and 4)

$$L_{\rm nova} = L_{\rm pl} \sim 46\,000\,(M_{\rm WD} - 0.26)\,. \tag{3.111}$$

3.8.2 Lifetimes during red and blue phases of hydrogen burning

During the nuclear burning phase, the rate $\dot{M}_{\rm H}$ at which hydrogen burning processes mass is related to the luminosity of the star by

$$\dot{M}_{\rm H}(\rm M_\odot\,yr^{-1}) \sim 1.33 \times 10^{-11}\, L_{\rm H}, \tag{3.112}$$

where $L_{\rm H}$ is the contribution of hydrogen burning to the luminosity and it has been assumed that $X = 0.7$. Combining the last two equations, we have that, during the plateau phase,

$$\dot{M}^{\rm pl}_{\rm nova}(\rm M_\odot\,yr^{-1}) \sim 6.12 \times 10^{-7}\ (M_{\rm WD} - 0.26)\,. \tag{3.113}$$

From Iben and Tutukov (1989; their Figure 5) the mass $M_{\rm e}$ of the hydrogen-rich envelope during the plateau phase can be approximated by

$$\log M_{\rm e,pl} \sim -4 - 2.69\ (M_{\rm WD} - 0.64) \tag{3.114}$$

and the mass of this envelope at the blue point (the point of maximum $T_{\rm e}$ following the plateau phase) can be approximated by

$$\log M_{\rm e,bp} \sim -4.52 - 2.99\ (M_{\rm WD} - 0.64)\,, \tag{3.115}$$

where 'bp' stands for 'blue point'. The masses $M_{\rm e,pl}$ in Equation (3.114) and $M_{\rm e,bp}$ in Equation (3.115) are in solar units.

It is useful to define a hydrogen-burning time-scale by

$$\tau_{\rm H} = \frac{M_{\rm e,pl}}{\dot{M}_{\rm H,pl}}\,.$$

Using Equations (3.113) and (3.114), we have

$$\tau_{\rm H}\ (\text{in years}) \sim 1600 \times 10^{-2.69(M_{\rm WD}-0.64)}\,. \tag{3.116}$$

If an erupting model star were to reach the plateau luminosity and consume its entire store of hydrogen at this luminosity, $\tau_{\rm H}$ would be its lifetime.

However, the initial mass $M_{\rm e,0}$ of the hydrogen-rich layer of an erupting model depends on the accretion rate, as is illustrated by the quasistatic models in Iben (1982, see Figure 16, and compare Figure 15 with Figure 21). The larger the accretion rate, the smaller is $M_{\rm e,0}$ and the smaller is the maximum radius achieved by the model. If (a) $M_{\rm e,0}$ is substantially larger than $M_{\rm e,pl}$, the model can adopt a giant configuration and spend most of its nuclear burning time at nearly constant radius and at the plateau luminosity; when $M_{\rm e}$ decreases to $M_{\rm e,pl}$, the model evolves rapidly to the blue and burns most of the rest of its nuclear fuel ($= M_{\rm e,pl} - M_{\rm e,bp}$) evolving at small radius from the plateau luminosity to the luminosity at the blue point, where nuclear burning becomes unstable and rapidly goes out. If (b) $M_{\rm e,0} \sim M_{\rm e,pl}$, the duration of the 'red' phase ($L \sim L_{\rm plateau}$ and R large) is brief and the model spends most of its nuclear burning lifetime in the 'blue' phase (L declining from $L_{\rm pl}$ to $L_{\rm bp}$ and R small). If, (c) $M_{\rm e,0}$ is smaller than $M_{\rm e,pl}$, the model fails to adopt a giant configuration and spends all of its nuclear burning time below the blue point.

In case (a), the lifetime of the red phase is

$$t_{\rm red} \sim \frac{M_{\rm e,0} - M_{\rm e,pl}}{\dot{M}_{\rm H,pl}} = \tau_{\rm H}\left(\frac{M_{\rm e,0}}{M_{\rm e,pl}} - 1\right),$$

where $\tau_{\rm H}$ is given by Equation (3.116). The dependence of $M_{\rm e,0}$ on mass and accretion rate can only be obtained by model calculations. During the blue phase for models of both classes (a) and (b), the rate $\dot{M}_{\rm H}$ at which hydrogen burning processes matter is approximately proportional to the mass $M_{\rm e}$ of the hydrogen-rich envelope, so that, if no process other than nuclear burning acts to reduce $M_{\rm e}$, the duration $t_{\rm blue}$ of the blue phase is roughly

$$t_{\rm blue}({\rm yr}) \sim \tau_{\rm H}\ \log_e\left(\frac{M_{\rm e,pl}}{M_{\rm e,bp}}\right). \tag{3.117}$$

Using Equations (3.115)–(3.116) in Equation (3.117), we have

$$t_{\rm blue}({\rm yr}) \sim \frac{196 + 114 \times (M_{\rm WD} - 0.64)}{M_{\rm WD} - 0.26} \times 10^{-2.69(M_{\rm WD}-0.64)}. \tag{3.118}$$

This equation gives $t_{\rm blue} = 32\,000$ yr, 5300 yr, 1800 yr, and 700 yr for $M_{\rm WD} = 0.3$, 0.4, 0.5, and $0.6{\rm M}_\odot$, respectively. For larger masses,

$$t_{\rm blue}({\rm yr}) \sim 700 \times 10^{-3.18(M_{\rm WD}-0.60)}$$

is a reasonable approximation.

The evolution in Iben (1982) of a $1.01{\rm M}_\odot$ quasistatic model star accreting at rates (a) $1.5 \times 10^{-9}\ {\rm M}_\odot\ {\rm yr}^{-1}$ and (b) $1.5 \times 10^{-8}\ {\rm M}_\odot\ {\rm yr}^{-1}$ serves to illustrate the difference between cases (a) and (b). For model (b), $M_{\rm e,0} \sim M_{\rm e,pl} \sim 10^{-4.9}\ {\rm M}_\odot$ and for model (a), $M_{\rm e,0} \sim 2 \times M_{\rm e,pl}$. Model (b) attains a maximum radius of $\sim$50 ${\rm R}_\odot$, and spends essentially all of its $\sim$ 15 year nuclear burning lifetime evolving from a radius of $\sim$ 0.5 ${\rm R}_\odot$ to the blue point at R$\sim$0.015 ${\rm R}_\odot$. Thus, $t_{\rm red}$ $\sim$0 yr and $t_{\rm blue}$ $\sim$15 yr. Model (a) attains a maximum radius of $\sim$ 100 ${\rm R}_\odot$ and spends $\sim$15 yr at this radius and a total of $\sim$21 yr at radii larger than 50 ${\rm R}_\odot$, before evolving in only 2–3 years to a radius of 0.5 ${\rm R}_\odot$. Its evolution thereafter is identical to that of the case (b) model. Thus, in case (a), $t_{\rm red} \sim 24$ yr and $t_{\rm blue} \sim 15$ yr. The quantitative differences between these lifetimes and those derived from static models (Iben & Tutukov, 1989) point out the need for a much more extensive exploration with quasistatic models.

The accretion rate of the case (b) model is approximately ten times smaller than the minimum accretion rate $\dot{M}_{\rm bp} \sim 1.8 \times 10^{-7}\ {\rm M}_\odot\ {\rm yr}^{-1}$ which the static models show is necessary for steady-state hydrogen burning at the base of the hydrogen-rich envelope of a 1.01 ${\rm M}_\odot$ cold white dwarf. Presumably, for values of $\dot{M}_{\rm acc}$ in the range $\sim 1.5\times 10^{-8}\ {\rm M}_\odot\ {\rm yr}^{-1}$ and

Table 3.8. *Observed properties of symbiotic novae*

Star	Distance (kpc)	Period (yr)	$\dot{M}_{rgw}$ (10^{-7} M$_\odot$ yr^{-1})	L_{pl} (L$_\odot$)	R_{max} (R$_\odot$)	τ_{obs}^{red} (yr)	τ_{obs}^{blue} (yr)
AG Peg	0.7	2.26	1.6	4 000	18	60	50
V1329 Cyg	3.7	2.60	8	18 000	26	15	20
RT Ser	9.4	12.0	25	28 000	100	25	40
PU Vul	3.2	13.4	2.5	25 000	50	10	–
V1016 Cyg	3.9	> 15	130	36 000	6	0	> 40
HM Sge	2.9	> 15	100	28 000	20	4	> 20
RR Tel	2.6	> 15	50	17 500	110	7	> 30
RX Pup	1.8	200?	40	16 000	60	4	9

$\dot{M}_{bp}$, a model would experience an outburst when M_e reaches a value smaller than $M_{e,bp}$, and its hydrogen-burning evolution would be confined to luminosities $\leq L_{bp}$. This is a fascinating possibility which remains to be explored numerically, and it has the potential of explaining observed outbursts of moderate luminosity and high surface temperature which might otherwise be attributed to the discharge of an unstable accretion disk.

It is of interest to compare the nova hydrogen-burning time-scale with the time-scale for hydrogen burning by the central stars of planetary nebulae. This latter time-scale is given approximately by (e.g. Iben, 1982)

$$t_{PN}(\text{yr}) \sim 2 \times 10^4 \times \left(\frac{0.6\,\text{M}_\odot}{M_{PN}}\right)^{10},$$

where M_{PN} is the mass of the hot central star. For a typical mass of $M_{WD} = M_{PN} \sim 0.6\,\text{M}_\odot$, $t_{PN} \sim 30 \times t_H$. For $M_{WD} = M_{PN} \sim 0.55\,\text{M}_\odot$, $t_{PN} \sim 50 \times t_H$. The difference in burning times at a given mass arises because (1) the initial reservoir of fuel in a central star of a planetary nebula is several times larger than that in a nova precursor and (2) the burning rate is several times smaller.

For completeness, several quantities which are of interest and which can be explicitly obtained from the information in Iben and Tutukov (1989) are global properties at the blue point for steady-state hydrogen-burning models:

$$\begin{aligned} \log\,(T_{eff})_{bp} &\sim 5.69 + 1.73\,\log M_{WD}, \\ \log L_{bp} &\sim 4.07 + 2.86\,\log M_{WD}, \\ \text{and}\quad \log R_{bp} &\sim -1.82 - 2.04\,\log M_{WD}. \end{aligned} \tag{3.119}$$

3.8.3 *Wind mass-loss versus nuclear processing*

The estimated properties of several symbiotic novae are given in Table 3.8. The entries for RX Pup are from Mikołajewska *et al.* (1999); references to sources of other entries are given in Iben and Tutukov (1996). The wind mass-loss rate $\dot{M}_{rgw}$ of the giant component is given in units of $10^{-7}\,\text{M}_\odot\,\text{yr}^{-1}$, L_{pl} and R_{max} are, respectively, the maximum (plateau) luminosity and maximum radius of the erupting star, and τ_{obs}^{red} and τ_{obs}^{blue} are the observed lifetimes of the red and blue phases, respectively. The estimates of τ_{obs}^{blue} assume

Table 3.9. *Theoretical properties of symbiotic novae*

Star	$M_{\rm WD}$ ($M_\odot$)	$\dot{M}_{\rm acc}$ ($10^{-8}\,{\rm M}_\odot\,{\rm yr}^{-1}$)	$\dot{M}_{\rm bp}$ ($10^{-8}\,{\rm M}_\odot\,{\rm yr}^{-1}$)	$t_{\rm H}$ (yr)	$t_{\rm blue}$ (yr)	$r_{\rm red}$	$T_{\rm red}$ (yr)	$T_{\rm blue}$ (yr)
AG Peg	0.35	0.4	0.9	9600	1300	18	1100	650
V1329 Cyg	0.65	2	5.3	1500	480	16	320	240
RT Ser	0.87	2.5	12	380	100	27	700	50
PU Vul	0.80	0.2	9.5	590	150	20	–	75
V1016 Cyg	1.04	<1	20	130	28	6	0	14
HM Sge	0.87	<10.8	12	380	100	12	50	50
RR Tel	0.64	<0.5	5	1600	510	33	230	250
RX Pup	0.60	0.1	4	2000	680	25	100	340

that the blue phase of hydrogen burning has essentially ended by the time the radius of the erupting star declines below $\sim 0.1{\rm R}_\odot$.

In Table 3.9 are theoretical inferences based on the observed properties (Iben, 2003). The mass estimates make use of Equation (3.30), the hydrogen-burning time-scales $t_{\rm H}$ come from Equation (3.116), and the 'blue' lifetimes $t_{\rm blue}$ come from Equation (3.118). The rate $\dot{M}_{\rm acc}$ at which the hot component is estimated to accrete mass from the wind of the giant component is in units of $10^{-8}\,{\rm M}_\odot\,{\rm yr}^{-1}$.

Even considering all of the uncertainties, it seems clear that theoretical lifetimes are typically an order of magnitude or more larger than observed lifetimes. The only exception is V1016 Cyg, but this may be due to an overestimate of the mass.

Presumably, wind mass-loss from the erupting component is responsible for the discrepancy. Assuming that the wind from this component is a radiative wind of optical depth near unity, one may write

$$\dot{M}_{\rm w} v_{\rm w} = \frac{L_{\rm WD}}{c},$$

where $\dot{M}_{\rm w}$ is the wind mass-loss rate from the erupting star, $v_{\rm w}$ is the velocity of wind matter at large distances from the star, $L_{\rm WD}$ is the luminosity of the star and c is the velocity of light. We write

$$v_{\rm w} = 3\alpha_{\rm w} \left(\frac{2GM_{\rm WD}}{R_{\rm WD}}\right)^{1/2}$$

where $\alpha_{\rm w}$ is an arbitrary parameter. Combining the last two equations gives

$$\dot{M}_{\rm w}({\rm M}_\odot\,{\rm yr}^{-1}) = \frac{3.3\times10^{-11}}{\alpha_{\rm w}} \left(\frac{R_{\rm WD}}{M_{\rm WD}}\right)^{1/2} L_{\rm WD}, \tag{3.120}$$

where now $R_{\rm WD}$, $L_{\rm WD}$, and $M_{\rm WD}$ are in solar units. Assuming that $L_{\rm WD} = L_{\rm H}$, Equations (3.112) and (3.120) give

$$\dot{M}_{\rm w} = \frac{2.5}{\alpha_{\rm w}} \left(\frac{R_{\rm WD}}{M_{\rm WD}}\right)^{1/2} \dot{M}_{\rm H}. \tag{3.121}$$

The ratio $\dot{M}_{\rm w}/\dot{M}_H$ estimated at the maximum radius from Table 3.8 (with $\alpha_{\rm w} = 1$ in Equation (3.121)) is given in the seventh column (labeled $r_{\rm red}$) of Table 3.9, and the lifetime $T_{\rm red}$ obtained by multiplying $\tau_{\rm obs}^{\rm red}$ from Table 3.8 by this ratio is given in the next to last column of Table 3.9.

For the seven stars which attain radii larger than $10\,R_\odot$, the rate at which the star loses mass via a wind is at least an order of magnitude larger than the rate at which it processes mass by hydrogen burning (compare $\tau_{\rm obs}^{\rm red}$ in Table 3.8 with $T_{\rm red}$ in Table 3.9). If anything, the actual wind mass-loss rate during the radially extended phase is larger than estimated in the simple optically thin approximation. The inference that, at high luminosity, the primary accretor emits a strong wind is, of course, also known directly from observations, as shown, for example, by Kenyon *et al.* (1993) for AG Peg.

During the blue phase, with, say, $R_{\rm WD} \sim 0.02$–$1\,R_\odot$, Equation (3.121) predicts that the wind mass-loss rate should be at most comparable to the rate at which hydrogen burning processes matter. In the last column of Table 3.9, $T_{\rm blue} = t_{\rm blue}/2$ is the lifetime of the blue phase predicted by the theory on the assumption that the wind mass-loss rate is on average equal to $\dot{M}_{\rm H}$. The observed blue lifetime $\tau_{\rm blue}$ in Table 3.8 is in some cases substantially smaller than the corresponding $T_{\rm blue}$ in Table 3.9. One may guess that either the blue wind is stronger than given by the simple estimate, or the mass lost during the red phase is such that the mass left in the hydrogen envelope is too small to support hydrogen burning during the blue phase and that this latter phase proceeds on the cooling time-scale of the envelope.

At first sight, it is perplexing that wind mass-loss can so completely control the rate of evolution of symbiotic novae and yet not appear to be similarly dominant in the evolution of central stars of planetary nebulae. How do central stars of planetary nebulae manage to burn hydrogen for typically several times 10^4 years when, for a given luminosity and surface temperature, the wind emitted by a central star should be basically the same as that emitted by a symbiotic nova? The answer is that, for a given stellar mass, the mass of the hydrogen-rich envelope of a central star is initially larger by typically an order of magnitude than the initial envelope mass of an eruping dwarf in a symbiotic system, and the plateau luminosity is typically several times smaller. Neglecting a wind, the theoretical hydrogen-burning lifetime is one to two orders of magnitude larger for the central star of a planetary nebula than for the erupting star in a cataclysmic variable or symbiotic star. Thus, although the rate of evolution during the bolometrically bright hydrogen-burning phases of both central stars and symbiotic novae are controlled by wind mass-loss, the larger initial reservoir of fuel and the smaller rate of decrease in $M_{\rm e}$ (both by nuclear processing and wind mass-loss) act to ensure that the plantary nebula central star has a much longer lifetime as a bolometrically bright star than does a symbiotic nova of the same mass.

3.8.4 Wind accretion and steady state burning

The mass-accretion rates $\dot{M}_{\rm acc}$ listed in column 3 of Table 3.9 are the result of applying Equation (3.22) with the proviso that the accretor can capture at most one-quarter of the mass lost by the giant.

From Iben and Tutukov (1996, see their Figure 5), the mass-accretion rate at the blue point is given approximately by (see also Equations (3.112) and (3.119))

$$\log \dot{M}_{\rm bp} \sim -6.75 + 2.86\ \log M_{\rm WD}.$$

The quantity $\dot{M}_{bp}$ is listed in Table 3.9 in column 4. Its significance is that only if $\dot{M}_{acc} < \dot{M}_{bp}$ can a model accretor experience outbursts. With two exceptions (AG Peg and V1329 Cyg), this criterion is comfortably met. The two exceptions have the shortest orbital periods of systems in the sample, and this reinforces the expectation that Equation (3.22) overestimates the accretion rate for short-period systems, even when one insists that $\dot{M}_{acc} \leq 0.25\dot{M}_{rgw}$.

There is no other apparent correlation between the estimated accretion rate and either the observed characteristics or theoretically inferred characteristics of the symbiotic novae. This may mean only that Equations (3.21) and (3.22) are completely off the mark or that the actual wind from the giant and the corresponding accretion rate are highly variable, so that the secular or long-term average accretion rate which is relevant for determining outburst characteristics may be quite different from the currently estimated rate.

3.9 Remarks on CVs and symbiotic stars as Type Ia supernova precursors

Assuming that a Type Ia supernova is an exploding CO white dwarf, the question is: can the accretor in a CV or symbiotic star retain and convert into carbon and oxygen enough matter to reach the critical CO core mass of $\sim$1.37 $M_\odot$? Since the maximum initial mass of a CO white dwarf is $\sim$1.1 $M_\odot$, the question becomes, can 0.27 $M_\odot$ of hydrogen-rich material be retained?

In the case of CVs, there is direct observational evidence that much of the accreted mass is lost in consequence of a combination of common-envelope mass-loss and wind mass-loss during the outburst. The abundances of heavy elements in the ejecta of perhaps half or more of all classical novae are much larger than are the abundances of these elements in the Solar System distribution, implying that matter dredged up from the white dwarf interior, either during the quiescent phase between outbursts or during outburst, is lost from the system during outburst; this means that more matter is lost during outburst than is accreted between outbursts. Such stars cannot become Type Ia SNe.

In the case of symbiotic stars, the comparison made in Section 3.8 between observed and theoretical lifetimes during the nuclear burning phase provides compelling evidence that the amount of material escaping the erupting star during and after the quiescent hydrogen-burning phase is at least an order of magnitude larger than the amount of matter converted into helium and retained by the star during this phase. Hence, the effective (i.e. secular) accretion rate can be at least an order of magnitude smaller than the rate at which the white dwarf precursor accretes from the wind of the giant component. And, since the efficiency of capture from a wind is typically only a few percent (say $\sim$10% at maximum), the secular accretion efficiency of the white dwarf component must be only several percent or less. This makes it very unlikely that the accreting white dwarf in most symbiotic stars can accrete more than a few hundredths of a solar mass during the lifetime of the system. A corollary is that the likelihood of producing a Chandrasekhar mass white dwarf and a consequent supernova explosion in such a system is quite small.

If a rare system were to accrete of the order of, say, 0.1 $M_\odot$, the helium mantle acquired by repeated burnings of accreted hydrogen-rich matter might be expected to undergo a helium shell flash which would expel the mantle, possibly in an explosion of supernova proportions (e.g. Tutukov & Khokhlov, 1992). In the even rarer system which managed to bring the mass of the accretor up to the effective Chandrasekhar limit, the accretor would probably have to begin as an ONe white dwarf, and electron captures on the magnesium in the core would lead to core collapse to a neutron star. In either case, the explosion would not have the character of a Type Ia supernova, even apart from the fact that the hydrogen-rich material from the mass donor would be present in the spectrum of the outburst. See Chapter 4 for further discussion of this point.

References

Bode, M. F., Duerbeck, H. W., & Evans, A., 1989, in *Classical Novae*, 1st edn, ed. M. F. Bode, & A. Evans. New York and Chichester: Wiley, p. 249.
Burrows, A., & Liebert, J., 1994, *Rev. Mod. Phys.*, **65**, 301.
Burrows, A., Hubbard, W. B., Saumon, D., & Lunine, J. L., 1993, *Ap. J.*, **406**, 158.
Cassisi, S., Iben, I. Jr., & Tornambé, A., 1998, *Ap. J.*, **496**, 376.
D'Antona, F., & Mazzitelli, I., 1994, *Ap. JS*, **90**, 467.
Endal, A. S., & Sofia, S., 1976, *Ap. J.*, **210**, 184.
Fujimoto, M. Y., 1982a, *Ap. J.*, **257**, 752.
Fujimoto, M. Y., 1982b, *Ap. J.*, **257**, 767.
Fujimoto, M. Y., 1988, *A&A*, **198**, 163.
Fujimoto, M. Y., 1993, *Ap. J.*, **419**, 768.
Fujimoto, M. Y., & Iben, I. Jr., 1991, *Ap. J.*, **374**, 631.
Fujimoto, M. Y., & Iben, I. Jr., 1992, *Ap. J.*, **399**, 646.
Fujimoto, M. Y., & Iben, I. Jr., 1997, in *Advances in Stellar Evolution*, ed. R. T. Rood, & A. Renzini. Cambridge: Cambridge University Press, p. 245.
Fujimoto, M. Y., & Sugimoto, D., 1979, *PASJ*, **31**, 1.
Fujimoto, M. Y., & Sugimoto, D., 1982, *Ap. J.*, **257**, 291.
García-Berro, E., & Iben, I. Jr., 1994, *Ap. J.*, **434**, 306.
García-Berro, E., Ritossa, C., & Iben, I. Jr., 1997, *Ap. J.*, **485**, 765.
Giannone, P., & Weigert, A., 1967, *Z. Astrophys*, **67**, 41.
Gil-Pons, P., & García-Berro, E., 2001, *A&A*, **375**, 87.
Gil-Pons, P., & García-Berro, E., 2002, *A&A*, **396**, 589.
Goldreich, P., & Schubert, G., 1967, *Ap. J.*, **150**, 571.
Grossman, A. S., Hays, D., & Graboske, H. C., 1974, *A&A*, **30**, 95.
Hamada, T., & Salpeter, E. E., 1961, *Ap. J.*, **134**, 683.
Hollowell, D., Iben, I. Jr., & Fujimoto, M. Y., 1990, *Ap. J.*, **351**, 245.
Hoyle, F., & Lyttleton, R. A., 1939, *Proc. Camb. Phil. Soc.*, **35**, 405.
Iben, I. Jr., 1977, *Ap. J.*, **217**, 788.
Iben, I. Jr., 1982, *Ap. J.*, **259**, 244.
Iben, I. Jr., 1990, *Ap. J.*, **353**, 215.
Iben, I. Jr., 2003, in *Symbiotic Stars Probing Stellar Evolution*, ed. R. L. M. Corradi, J. Mikołajewska, & T. J. Mahoney. San Francisco: Astronomical Society of the Pacific, p. 177.
Iben, I. Jr. & Tutukov, A. V., 1984a, *Ap. JS*, **54**, 335.
Iben, I. Jr. & Tutukov, A. V., 1984b, *Ap. J.*, **284**, 719.
Iben, I. Jr. & Tutukov, A. V., 1985, *Ap. JS*, **58**, 661.
Iben, I. Jr. & Tutukov, A. V., 1989, *Ap. J.*, **342**, 430.
Iben, I. Jr. & Tutukov, A. V., 1991, *Ap. J.*, **370**, 615.
Iben, I. Jr. & Tutukov, A. V., 1993, *Ap. J.*, **418**, 343.
Iben, I. Jr. & Tutukov, A. V., 1995, in *Cataclysmic Variables*, ed. A. Bianchini, M. Della Valle, & M. Orio. Dordrecht: Kluwer, p. 201.
Iben, I. Jr. & Tutukov, A. V., 1996, *Ap. JS*, **105**, 145.
Iben, I. Jr., Nomoto, K., Tornambé, A., & Tutukov, A. V., 1987, *Ap. J.*, **317**, 717.

Iben, I. Jr., Fujimoto, M.Y., & MacDonald, J., 1991, *Ap. J.*, **375**, L27.
Iben, I. Jr., Fujimoto, M.Y., & MacDonald, J., 1992a, *Ap. J.*, **384**, 580.
Iben, I. Jr., Fujimoto, M.Y., & MacDonald, J., 1992b, *Ap. J.*, **388**, 521.
Iben, I. Jr., & MacDonald, J., 1995, in *White Dwarfs*, ed. D. Koester, & K. Werner. Berlin: Springer Verlag, p. 45.
Iben, I. Jr., Ritossa, C., & García-Berro, E., 1997, *Ap. J.*, **489**, 772.
José, J., Hernanz, M., & Isern, J., 1993, *A&A*, **269**, 291.
Kato, M., 1980, *Prog. Theor. Phys.*, **64**, 847.
Kato, M., & Hachisu, I., 2004, *Ap. J.*, **613**, L132.
Kenyon, S. J., Mikołajewska, J., Mikołajewski, M., Polidan, R. S., & Slovak, M. H., 1993, *AJ*, **106**, 1573.
Kippenhahn, R., & Thomas, H.-C., 1970, in *Stellar Rotation*, IAU Colloquium 4, ed. A. Slettebak. New York: Gordon & Breach, p. 20.
Kippenhahn, R., & Thomas, H.-C., 1978, *A&A*, **63**, 625.
Kovetz, A., & Prialnik, D., 1985, *Ap. J.*, **291**, 812.
Kovetz, A., & Prialnik, D., 1990, in *Physics of Classical Novae*, ed. A. Cassatella, & R. Viotti. Berlin: Springer, p. 394.
Kumar, S. S., 1963, *Ap. J.*, **340**, 985.
Kutter, G. S., & Sparks, W. M., 1989, *Ap. J.*, **340**, 985.
Landau, L. D., & Lifshitz, E., 1951, *The Classical Theory of Fields*. Cambridge: Addison-Wesley.
Laughlin, G., & Bodenheimer, P., 1993, *Ap. J.*, **403**, 303.
Laughlin, G., Bodenheimer, P., & Adams, F. C., 1997, *Ap. J.*, **482**, 420.
Limongi, M., & Tornambé, A., 1991, *Ap. J.*, **371**, 317.
Livio, M., Prialnik, D., & Regev, O., 1989, *Ap. J.*, **341**, 299.
Livne, E., 1997, in *Thermonuclear Supernovae*, ed. P. Ruiz-Lapuente, R. Canal, & J. Isern. Dordrecht: Kluwer, p. 425.
Livne, E., & Arnett, D., 1995, *Ap. J.*, **452**, 62.
MacDonald, J., 1983, *Ap. J.*, **273**, 289.
Mengel, J. G., Sweigart, A. V., Demarque, P., & Goss, P. G., 1979, *Ap. JS*, **40**, 733.
Mikołajewska, J., 2003, in *Symbiotic Stars Probing Stellar Evolution*, ed. R. L. M. Corradi, J. Mikołajewska, & T. J. Mahoney. San Francisco: Astronomical Society of the Pacific, p. 9.
Mikołajewska, J., Brandi, E., Hack, W. *et al.*, 1999, *MNRAS*, **305**, 190.
Neo, S., Miyaji, S., Nomoto, K., & Sugimoto, D., 1977, *PASJ*, **29**, 249.
Nomoto, K., 1982, *Ap. J.*, **257**, 780.
Nomoto, K., & Sugimoto, D., 1977, *PASJ*, **29**, 765.
Paczyński, B., 1970, *Acta Astron.*, **20**, 47.
Paczyński, B., 1971, *Acta Astron.*, **21**, 417.
Paczyński, B., 1976, *Structure and Evolution of Close Binary Systems*, ed P. P. Eggleton, S. Mitton, & J. Whelan. Dordrecht: Reidel, p. 75.
Paczyński, B., & Sienkievicz, R., 1983, *Ap. J.*, **268**, 825.
Paczyński, B., & Żytkow, A. N., 1978, *Ap. J.*, **222**, 604.
Patterson, J., 1984, *Ap. JS*, **54**, 443.
Piersanti, L., Cassisi, S., Iben, I. Jr., & Tornambé, A., 1999, *Ap. J.*, **521**, L62.
Piersanti, L., Cassisi, S., Iben, I. Jr., & Tornambé, A., 2000, *Ap. J.*, **535**, 932.
Pinsonneault, M. H., Kawaler, S. D., Sofia, S, & Demarque, P., 1989, *Ap. J.*, **338**, 424.
Prialnik, D., 1986, *Ap. J.*, **310**, 222.
Prialnik, D., & Kovetz, A., 1984, *Ap. J.*, **281**, 367.
Prialnik, D., & Shara, M. M., 1986, *Ap. J.*, **311**, 172.
Ritossa, C., García-Berro, E., & Iben, I. Jr., 1996, *Ap. J.*, **460**, 489.
Ritossa, C., García-Berro, E., & Iben, I. Jr., 1999, *Ap. J.*, **515**, 381.
Savonije, G. J., deKool, M., & van den Heuvel, E. P. J., 1986, *A&A*, **155**, 51.
Schwartzman, E., Kovetz, A., & Prialnik, D., 1994, *MNRAS*, **269**, 323.
Sion, E. M., & Starrfield, S., 1986, *Ap. J.*, **303**, 130.
Sion, E. M., Acierno, M. J., & Tomczyk, S., 1979, *Ap. J.*, **230**, 832.
Skumanich, A., 1972, *Ap. J.*, **171**, 565.
Sparks, W. M., & Kutter, G. S., 1987, *Ap. J.*, **321**, 394.
Starrfield, S., Sparks, W. M., & Truran, J. W., 1974a, *Ap. JS*, **28**, 247.
Starrfield, S., Sparks, W. M., & Truran, J. W., 1974b, *Ap. J.*, **192**, 647.
Townsent, A. A., 1958, *J. Fluid Mech.*, **4**, 361.

Truran, J. W., 1985, in *Proceedings of the ESO Workshop on Production and Destruction of C, N, O Elements*, ed. J. Danzinger, F. Matteucci, & K. Kät. Garching: European Southern Observatory, p. 221.
Truran, J. W., 1990, in *Physics of Classical Novae*, IAU Colloquium 122, eds. A. Cassatella, & R. Viotti. Berlin: Springer, p. 373.
Truran, J. W., & Livio, M., 1986, *Ap. J.*, *308*, 721.
Tutukov, A. V., 1983, University of Illinois preprint IAP 83–25.
Tutukov, A. V., & Khokhlov, A. M., 1992, *Astron. Zh.*, **69**, 784 = *Soviet Astron.*, **36**, 401.
Tutukov, A. V., & Yungelson, L. R., 1976, *Astrofizika*, **12**, 521 = *Astrophysics*, **12**, 342.
Uus, U. H., 1970, *Nauch. Inf.*, **17**, 48.
Webbink, R. F., 1990, in *Accretion Powered Compact Binaries*, ed. C. W. Mauche. Cambridge: Cambridge University Press, p. 177.
Williams, R. E., 1985, in *Production and Distribution of CNO Elements*, ed. I. J. Danziger, European Southern Observatory, Garching, p. 225.
Woosley, S. E., & Weaver, T. A., 1994, *Ap. J.*, **423**, 371.
Zahn, J. P., 1974, in *Stellar Instability and Evolution*, ed. P. Ledoux, A. Noels, & A. W. Rogers. Dordrecht: Reidel, p. 185.

4

Thermonuclear processes

Sumner Starrfield, Christian Iliadis and W. Raphael Hix

4.1 Introduction

The classical nova outburst is one consequence of the accretion of hydrogen-rich material onto a white dwarf (WD) in a close binary system. Over long periods of time the accreting material gradually forms a layer of fuel on the WD and the bottom of this layer is gradually compressed and heated by the strong surface gravity of the WD. Ultimately, the bottom of the layer becomes electron-degenerate. The degeneracy of the material then contains the explosion so that, once nuclear burning in the layer bottom reaches thermonuclear runaway (TNR) conditions, the temperatures in the nuclear burning region will exceed 10^8 K under almost all circumstances. As a direct result, a major fraction of the nuclei in the envelope capable of capturing a proton (C, N, O, Ne, Mg . . .) are transformed into β^+-unstable nuclei, which limits nuclear energy generation on the dynamical time-scale of the runaway and yields distinctively non-solar CNO isotopic abundance ratios in the ejected gases.

Observations of the outburst show that a classical nova explosively ejects metal enriched gas and grains and this material is a source of heavy elements for the interstellar medium (ISM). The observed amount of metal enrichment demands that mixing of the accreted material with core material occur at some time during the evolution of the outburst. The ejection velocities measured for classical nova ejecta exceed, in most cases, 10^3 km s^{-1} so that this material is rapidly mixed into the diffuse gas and then incorporated into molecular clouds before being formed into young stars and planetary systems during star formation. Therefore, classical novae must be included in studies of Galactic chemical evolution (see also Chapters 3, 6 and 8). They are predicted to be the major source of ^{13}C, ^{15}N, and ^{17}O in the Galaxy and contribute to the abundances of other isotopes in this atomic mass range.

Infrared observations have confirmed the formation of carbon, SiC, hydrocarbons, and oxygen-rich silicate grains in classical nova ejecta, suggesting that some fraction of the pre-solar grains identified in meteoritic material (Zinner, 1998; José *et al.*, 2004) may come from novae (Gehrz *et al.*, 1998; see also Chapters 6, 8, and 13). The mean mass ejected during a classical nova outburst is $\sim 2 \times 10^{-4}$ M$_\odot$ (Gehrz *et al.*, 1998). Using the observed classical nova rate of 35 ± 11 per year in our Galaxy (Shafter, 1997; see also Chapter 14), it follows that they introduce $\sim 7 \times 10^{-3}$ M$_\odot$ yr^{-1} of processed matter into the ISM. Observations suggest

Classical Novae, 2nd edition, ed. Michael Bode and Aneurin Evans. Published by Cambridge University Press.

that more material is ejected by a classical nova than currently believed and, therefore, this value is a lower limit (Saizar & Ferland, 1994; Gehrz *et al.*, 1998).

This review is an updated version of a chapter on thermonuclear processes in the first edition of *Classical Novae* (Bode & Evans, 1989; Starrfield, 1989). That chapter was written in 1983 and updated in 1987. A great deal has been learned about the thermonuclear evolution of the classical nova outburst since that time. We will describe the advances that have been made in our understanding of the progress of the outburst, and outline some of the puzzles that are still outstanding. In addition to the other relevant chapters in this book, recent reviews of the classical nova phenomenon can be found in Gehrz *et al.* (1998), Hernanz and José (2002) and Warner (1995). In the next section, we will present the basic physics of the TNR and the importance of the β^+-unstable nuclei. We follow that with sections on (a) the initial conditions, (b) the effects of new reaction rates, (c) multidimensional studies of the TNR, (d) nucleosynthesis in CO and ONe novae, (e) the ejecta mass discrepancy, and (f) the proposed relationship of classical novae to the progenitors of supernovae of Type Ia. We end with a summary and an outline of future work.

4.2 The thermonuclear runaway

Hydrodynamic studies have shown that the consequence of accretion from the secondary is a growing layer of hydrogen-rich gas on the WD. When both the WD luminosity and the rate of mass accretion onto the WD are sufficiently low, such that the deepest layers of the accreted material have become both hot and electron-degenerate, a TNR occurs at or near the base of the accreted layers depending on the degree of degeneracy. For the physical conditions of temperature and density that occur in this environment, nuclear processing proceeds by hydrogen burning, first from the proton–proton chain (which must include the pep reaction: $\mathrm{p} + \mathrm{e}^- + \mathrm{p} \rightarrow \mathrm{d} + \nu$; Schatzman, 1958; Bahcall & May, 1969) and, subsequently, via the carbon, nitrogen, and oxygen (CNO) cycles. If there are heavier nuclei present in the nuclear burning shell (for example, ^{20}Ne and ^{24}Mg), then they will contribute significantly to the nucleosynthesis. Moreover, for the range of temperatures typically sampled in classical nova outbursts, these heavier nuclei could give rise to significant energy production and this is now being investigated in current hydrodynamic simulations. Both compressional heating and the energy released by the nuclear reactions heat the accreted material ($\sim 10^{-6}\,\mathrm{M}_\odot$ to $\sim 10^{-4}\,\mathrm{M}_\odot$, depending upon WD mass) until an explosion occurs.

While the proton–proton chain is important during the accretion phase of the outburst during which time the amount of accreted mass is determined, it is the CNO-cycle reactions and, ultimately, the hot CNO sequences that power the final stages of the TNR and the evolution to the peak of the explosion. Energy production and nucleosynthesis associated with the CNO hydrogen burning reaction sequences impose interesting constraints on the energetics of the runaway. In particular, the rate of nuclear energy generation at high temperatures ($T > 10^8$ K) is limited by the time-scales of the slower and temperature-insensitive β^+-decays, particularly ^{13}N ($\tau_{1/2} = 598$ s), ^{14}O ($\tau_{1/2} = 71$ s), and ^{15}O ($\tau_{1/2} = 122$ s). The behavior of the β^+-decay nuclei holds important implications for the nature and consequences of classical nova outbursts. For example, significant enrichment of CNO nuclei in the nuclear burning regime is required to ensure high levels of energy release on a hydrodynamic time-scale (seconds for WDs) and thus produce a violent outburst (Starrfield, 1989; Starrfield *et al.*, 1998b).

The large abundances of the β^+-decay nuclei, at the peak of the outburst, have important and exciting consequences for the evolution.

(1) Since the energy production in the CNO cycle comes from proton captures, interspersed by β^+-decays, the rate at which energy is produced, at temperatures exceeding 10^8 K, depends only on the half-lives of the β^+-decay nuclei and the numbers of C, N, O, Ne, and Mg nuclei initially present in the envelope.
(2) When temperatures in the nuclear burning region significantly exceed 10^8 K, a major fraction of those nuclei capable of capturing a proton do so, and most of the CNO nuclei become β^+-unstable nuclei.
(3) Since convection operates throughout the entire accreted envelope, it brings unburned C, N, O, Ne, and Mg nuclei into the shell source, when the temperature is rising extremely rapidly, and keeps the nuclear reactions operating far from equilibrium.
(4) Since the convective turn-over time-scale is $\sim 10^2$ s near the peak of the TNR, a significant fraction of the β^+-decay nuclei are able to reach the surface before decaying and the rate of energy generation at the surface can exceed 10^{13} to 10^{15} erg g^{-1} s^{-1} (depending upon the enrichment).
(5) Their half-lives are longer than the hydrodynamic expansion time in the outer layers and these same nuclei decay when the temperatures in the envelope have declined to values that are too low for any further proton captures to occur, yielding isotopic ratios in the ejected material that are distinctly different from the ratios predicted from studies of equilibrium C, N, O, Ne, Mg burning.
(6) The decays of these nuclei provide an intense heat source throughout the envelope that flattens the temperature gradient and ultimately shuts off convection.
(7) Finally, the energy release from the β^+-decays helps eject the material off the WD.

Theoretical studies of this mechanism show that sufficient energy is produced, during the evolution described above, to eject material with expansion velocities that agree with observed values and that the predicted bolometric light curves for the early stages are in reasonable agreement with the observations (Starrfield, 1989; Starrfield *et al.*, 1998b). The hydrodynamic studies also show that at least three of the observational behaviors of the classical nova outburst are strongly dependent upon the complicated interaction between the nuclear reactions and convection that occurs during the final minutes of the TNR. Those are:

(1) The early evolution of the observed light curves of classical novae on which their use as 'standard candles' is based.
(2) The observed peak luminosity of fast novae which is typically super-Eddington; in some cases for as long as two weeks (Schwarz *et al.*, 2001).
(3) The composition of matter ejected by a classical nova which depends on the amount and composition of the material dredged up from the underlying CO or ONe WD core. (The existence of mixing is demanded by observations of classical nova ejecta; Gehrz *et al.*, 1998.) Predicting the ejecta composition is also critical to questions concerning the possibility of observing γ-rays from nearby classical novae (^{7}Be and/or ^{22}Na; see Chapter 11), and the contributions of classical novae both to Galactic chemical evolution and to some of the isotopic anomalies observed in pre-solar grains (José *et al.*, 2004). Moreover, the abundances of the core nuclei in the ejected material imply that the WD in a nova system is losing mass as a result of continued outbursts, and it cannot be a SN Ia progenitor (MacDonald, 1984; Starrfield *et al.*, 2000).

As already mentioned, the β^+-decay heating of the outermost regions of the nova envelope reduces the temperature gradient and, in turn, curtails convection in the surface layers around the time of peak temperature in the TNR. The growth of convection from the burning region to the surface and its subsequent retreat in mass, as the envelope relaxes from the peak of the runaway on a thermal time-scale, implies that there should exist considerable variations in the elemental and isotopic abundances in the ejected gases. Observations that provide data on abundance gradients in nova ejecta can critically constrain our knowledge of the history of convection during the TNR.

4.3 Initial conditions for the outburst

A description of the history of the TNR hypothesis for the classical nova outburst is given in Starrfield (1989) and will not be repeated here. One of the important developments since that review has been the calculations of the amount of hydrogen-rich material that can be accreted before the TNR is triggered. In the 1980s there were both analytic (Fujimoto, 1982a,b) and semi-analytic (MacDonald, 1983) calculations to determine the amount of material that could be accreted onto a WD before the TNR occurred. Since that time, there have been a number of studies of accretion onto WDs using Lagrangian hydrostatic or hydrodynamic computer codes which follow the evolution of the material as it gradually accretes onto the WD. These calculations show that the amount of material accreted is a function of the WD mass, WD luminosity, the composition of the accreted matter, and the evolutionary history of the WD (Townsley & Bildsten, 2004). If mixing of accreted material with core material occurs during the accretion process, then the opacity in the nuclear burning region will increase and trap more heat in this region than if no mixing had occurred. As a result, the temperature in the nuclear burning region will increase rapidly, reduce the time to TNR and, therefore, the total amount of accreted (and ejected) material (Starrfield *et al.*, 1998b).

Given that the evolution begins with a 'bare' WD core, which has a surface layer rich in helium remaining from the previous outburst (Shara, 1989; Krautter *et al.*, 1996; Starrfield *et al.*, 1998b), most of the time is spent and most of the mass is accreted during the phase when the principal energy production mechanism is the proton–proton chain (Starrfield *et al.*, 1998b, 2000). In this phase of evolution there is a competition between the energy production, which has a T^4 dependence, degenerate electron conduction into the interior, and radiative diffusion to the surface. Since the thickness of the surface layers is small and convection is not yet important, most of the energy produced at the bottom of the accreted layers is transported to the surface and radiated while only a small fraction is transported into the interior. Therefore, the temperature in the nuclear burning region increases extremely slowly as a function of the mass accretion rate (Townsley & Bildsten, 2004).

The amount of mass accreted during the proton–proton phase also depends on the metallicity of the material. Independent of when during the evolution of the TNR mixing occurs between core matter and the accreted layers, increasing the metallicity of the accreted material results in an increase in the opacity. The increased opacity results in more of the heat (produced by compression and nuclear burning in the deeper layers of the accreted material) being trapped in the region where it is being produced and the temperature increasing faster per unit accreted mass (Starrfield *et al.*, 1998b) than for a sequence with a lower opacity (metallicity). In contrast, lowering the metallicity by accreting material representative of the LMC (one-third solar metallicity or less) reduces the opacity and increases the rate of radiative heat transport out of the nuclear burning layers. As a result, the temperatures increase more slowly than for higher metallicity material and more material is accreted. A more massive accreted layer implies a higher density at the bottom and a more violent explosion (Starrfield *et al.*, 1999). This result is in agreement with the observations of classical novae in the LMC (Della Valle *et al.*, 1992, 1994).

If, however, the accreted material mixes with core material during this phase, either by shear mixing (Kutter & Sparks, 1987; Sparks & Kutter, 1987; Rosner *et al.*, 2001; Alexakis *et al.*, 2004) or by elemental diffusion (Prialnik & Kovetz, 1984; Kovetz & Prialnik, 1985), this will enrich the heavy nuclei in the accreted layers and, in turn, increase the opacity in the nuclear burning layers. This enrichment will reduce the amount of material accreted before the onset of the TNR and, thereby, the amount of material ejected during the outburst. Given that the theoretical predictions of the amount of material ejected during the outburst are far lower than the observations, increasing the metals in the accreted layers by early mixing exacerbates this disagreement. There is an interesting corollary to this discussion. As the opacities have been improved (more levels, better line profiles, more elements included, better equations-of-state) by the various groups working in this area (for example, Rogers & Iglesias, 1994; Rogers, Swenson & Iglesias, 1996), they have also increased for a given temperature and density. We have also found that the amount of accreted material has decreased as we have included more modern opacities. Therefore, even without mixing core material with accreted material or changing the metallicity of the accreting material, our latest simulations have increased the discrepancy between theory and observations (Starrfield *et al.*, 2000).

In order to study this effect, we updated and improved our 1D, hydrodynamic, evolution code by including the latest OPAL opacities (Iglesias & Rogers, 1993; Rogers, Swenson & Iglesias, 1996). We also improved the nuclear reaction rate libraries but will discuss those results in the next section. We calculated new evolutionary sequences for 1.25 $M_\odot$ WDs, in an attempt to simulate the outburst of V1974 Cyg (Starrfield, Truran & Gehrz, 1997; Starrfield *et al.*, 1998b). The revised opacities had profound effects on the simulations. Because the modern opacities were larger than those we had been using (the Iben (1982) fit to the Cox and Stewart (1971) and Cox and Tabor (1976) opacities), we found that the heat from the nuclear reactions was trapped more efficiently in the layers where it was produced. Our new simulations ejected a factor of 10 less mass than was inferred from observations for the outburst of V1974 Cyg (Starrfield, Truran & Gehrz, 1997; Starrfield *et al.*, 1998b). This discrepancy was also found in a study of accretion onto ONe WDs (José, Hernanz & Coc, 1997; José & Hernanz, 1998).

In Starrfield *et al.* (1998b), we proposed a possible solution to this problem. The WD spends a major fraction of the time during the accretion phase generating energy from the proton–proton reaction sequences which have (approximately) a T^4 to T^6 temperature dependence. Any change in the physical conditions that lengthens the time spent in this phase will increase the accreted mass. One way to accomplish this is to reduce the opacity, which allows a greater fraction of the nuclear energy to be transported away from the region where it is being produced. Mixing of the accreted hydrogen-rich material into a residual helium-rich shell (the remnant of previous outbursts) will have such an effect. This will also reduce the hydrogen mass fraction and, thereby, the rate of energy generation from the proton-proton sequence ($\epsilon_{\rm nuc} \sim X^2$).

Prialnik *et al.* (1982) were the first to show that there was a strong effect of the rate of mass accretion on the ignition mass. They reported that increasing the rate of mass accretion increased compressional heating and, thereby, caused the temperatures in the accreted layers to rise more rapidly (per given amount of accreted mass) than for lower mass accretion rates. The observed mass accretion rates of $\sim 10^{-9}\,M_\odot\,yr^{-1}$ (Townsley & Bildsten, 2003, 2004) result in much smaller amounts of material being accreted when compared with simulations where the rate of mass accretion has been reduced by a factor of 10 to 100.

It has also been found that increasing the mass accretion rate much above $10^{-8}\,M_\odot\,yr^{-1}$, on *low* luminosity WDs, causes extremely weak flashes. In this case, the large amount of heat released by compression keeps the degeneracy low and the outburst does not reach high temperatures; if there is a TNR at all. Even higher rates cause the accreted envelope to grow to large radii in less than 100 yr, which rules out large mass accretion rates in classical nova systems.

All other parameters held constant, the internal luminosity (or internal temperature) of the underlying WD also affects the amount of mass accreted prior to the TNR. It has been found that as the luminosity of the WD declines, the amount of accreted material increases. There are two reasons that the luminosity is important. First, in addition to compressional heating from accreted, the heat from the underlying WD also heats the nuclear burning layers. As the WD evolves and cools, this heat source becomes less important for the accreted layers. Second, a cooler interior implies (prior to accretion) cooler surface layers so that nuclear reactions begin later in the evolution of the WD. Once the WD has undergone a series of outbursts, the luminosity is determined by the long-term evolution of the WD (Townsley & Bildsten, 2004).

However, the discussion up to this point is most relevant for *low* luminosity WDs. As the luminosity of the WD is increased because either the WD is 'younger' or it has not yet reached quiescence after a classical nova explosion, then the temperatures in the surface layers are high enough for nuclear burning to occur as soon as the accreted material arrives on the surface. As a result, as the surface luminosity (temperature) increases, the amount of material that can be accreted, for the same rate of mass accretion, becomes smaller and the outburst occurs earlier with less accreted mass. More important, at some value of the WD surface luminosity, as yet undetermined, the infalling material burns at the same rate that it is accreted and no TNR occurs. Paczyński and Zytkow (1978), Sion, Acierno and Tomczyk (1979), Fujimoto (1982a,b) and Iben (1982) introduced the idea of 'Steady Burning' which is accretion at a high rate onto WDs. The Steady Burning mass accretion rate is a few $\times 10^{-7}\,M_\odot$ yr^{-1} and there is only one mass accretion rate for a given WD mass. These authors assumed that Steady Burning would occur at all WD luminosities but Starrfield *et al.* (2004b) have recently shown that quiescent hydrogen burning in the surface layers, a phenomenon that resembles steady burning but is not the Steady Burning defined in the literature, occurs only for *hot, luminous* WDs. They call it Surface Hydrogen Burning (SHB). SHB is important in trying to understand the evolution of the Super Soft X-ray Binaries (SSS) since the surface conditions of the evolutionary sequences of Starrfield *et al.* resemble those observed for the SSS (van den Heuvel *et al.*, 1992; Kahabka & van den Heuvel 1997).

The SSS were discovered by the Einstein satellite (for example, two members are CAL83 and CAL87: Long, Helfand & Grabelsky, 1981) but they were not identified as a stellar class until the ROSAT survey of the LMC (Trümper *et al.*, 1991). SSS are luminous, $L \sim 10^{37}\,erg\,s^{-1}$, with surface temperatures ranging from 30 to 50 eV (or higher). Optical studies show that they are close binaries containing a WD (Cowley *et al.*, 1998). Van den Heuvel *et al.* (1992) proposed that steady burning of the hydrogen-rich material accreted from the secondary was occurring on the surface of the WD component of the SSS binary. As a result, no TNR would occur, no mass would be ejected, and the mass of the WD could grow to the Chandrasekhar limit. Starrfield *et al.* (2004) have now shown that their conjecture is correct but it is SHB that is occurring, and it only occurs on luminous WDs. Their simulations, the first study of accretion at high rates onto luminous, massive WDs for long times, demonstrate that no TNR occurs, no mass is ejected, and the mass of the WD grows to the Chandrasekhar limit. Thus, they also predict that the SSS are one class of Supernova Ia progenitors.

The final parameter that affects the amount of material that is accreted prior to the TNR is the mass of the WD. It is a well-known result that the amount of material accreted, all other parameters held constant, is inversely proportional to the mass of the WD (see, for example, MacDonald, Fujimoto and Truran (1985) and references therein). Specific values of the ignition mass as a function of WD mass are given in Starrfield (1989).

Equation (4.1):

$$P_{\rm crit} = \frac{G M_{\rm WD} M_{\rm crit}}{4\pi R_{\rm WD}^4}, \tag{4.1}$$

illustrates how the ignition mass, $M_{\rm crit}$, can be estimated. The critical pressure $P_{\rm crit}$ is assumed to be $\sim 10^{20}$ dyne cm^{-2} and a mass–radius relation for WDs gives the ignition mass, $M_{\rm crit}$. Equation (4.1) is obtained by realizing that a critical pressure must be achieved at the bottom of the accreted layers before a TNR can occur (Fujimoto, 1982a,b; Gehrz *et al.*, 1998). Note, however, that the actual value of the critical pressure is also a function of WD composition and rate of accretion (Starrfield, 1989). If one assumes the above value for the pressure, then the amount of accreted mass can range from less than 10^{-5} M$_\odot$ for WDs near the Chandrasekhar limit to values exceeding 10^{-2} M$_\odot$ for 0.5 M$_\odot$ WDs. In addition, because the surface gravity of a low mass WD is smaller than that of a massive WD, the bottom of the accreted layers is considerably less degenerate at the time the TNR occurs. Therefore, the peak temperature, for a TNR on a low mass WD, may not even reach 10^8 K and no interesting nucleosynthesis occurs.

4.4 The effects of the nuclear reactions and reaction rates on the outburst

In the previous section, we discussed the effects of the opacities on the outburst. Here, we present the results of the effects of changes in the nuclear reaction rates (see also Chapter 6). As has been shown in numerous papers on the classical nova outburst, the behavior of the TNR depends critically both on the nuclear reactions considered and on the values of the reaction rates used in the simulations. It is the operation of the CNO reactions at high temperatures and densities that imposes severe constraints on the energetics and nucleosynthesis of the outburst. In order to demonstrate this, we report on new studies (done, in part, for this chapter) in which all initial conditions are held constant and only the nuclear reaction rate libraries are changed. Here, we report on the results of accretion of hydrogen-rich matter onto both 1.25 M$_\odot$ and 1.35 M$_\odot$ WDs using three different nuclear reaction rate libraries and holding all other parameters constant. Two of those libraries have been used and cited in previous publications (Politano *et al.*, 1995; Starrfield *et al.*, 1998b). In this chapter, we combine the libraries used in those papers with a current library of one of us (C.I.). We distinguish the libraries by the paper in which each was used.

We use the one-dimensional, fully implicit, Lagrangian, hydrodynamic computer code, NOVA, described in Starrfield *et al.* (1998b) and (2000) with one major change and some minor changes. The major change is that we no longer use the nuclear reaction network of

Weiss and Truran (1990) but now use that of Hix and Thielemann (1999). As part of the effort in updating the network, we performed a number of tests with NOVA in order to understand the source of any differences between the various versions of the code and the two networks. We discovered during these tests that the Weiss and Truran (1990) network does not include the pep reaction ($p + e^- + p \rightarrow d + \nu$; see Schatzman, 1958; Bahcall & May, 1969). While energy generation from the pep reaction is unimportant for solar models (but not the neutrino losses: Rolfs & Rodney, 1988), the density can reach values exceeding 10^4 g cm^{-3} in a WD envelope and thus the initial rate of energy generation is increased by about 40% (Starrfield *et al.*, in preparation). This difference is caused by the density dependence of the pep reaction (Bahcall & May, 1969).

With the energy generation higher in the early stages of accretion, the temperature rises faster per unit accreted material so that the additional energy generation from the pep reaction has the direct effect of reducing the amount of accreted material (Starrfield *et al.* 1998b, 2000). As a result, peak temperatures are lower for models evolved with the same nuclear reaction rate library used in our previous studies. Another difference between this work and our previous studies is that we do not initiate nuclear burning until the temperatures have reached 9 million degrees. Previously, nuclear burning was switched on at 4 million degrees. We made this change because the reaction rates in the latest libraries are not fit at temperatures below about 10 million degrees so that at about 8 million degrees some of the rate fits diverge rapidly to large values. We ran a number of tests and found that 9 million degrees was a good cut-off value. However, this also means that the very outermost layers, which have temperatures below 9 million degrees until near the peak of the outburst (when the energy and products of nuclear burning are brought to the surface by convection), do not experience nuclear burning during the accretion phase of the outburst. As a consequence, when the outer layers are finally linked to the nuclear burning regions near the peak of the TNR (when convection reaches the surface from below), fresh unburned material is brought down to high temperatures increasing the energy generation.

The reaction library used in Politano *et al.* (1995) obtained its rates from Caughlan and Fowler (1988) and Thielemann, Arnould and Truran (1987, 1988). The library was provided by Thielemann and made available to Truran and Starrfield and was included in the Weiss and Truran (1990) network. Starrfield *et al.* (1998b) used an updated reaction rate library which contained new rates calculated, measured, and compiled by Thielemann and Wiescher. A discussion of the improvements is provided in that paper. Our latest reaction library is the latest compilation of one of us (C.I) and is current at the time of writing. This library is an update of the library described in Iliadis *et al.* (2001).

A discussion of the improvements in the library since Iliadis *et al.* (2001) will appear in a future publication (Starrfield *et al.*, in preparation) so we only provide a few details here. In total, the rates of 11 and 33 proton-induced reactions were adopted from Angulo *et al.* (1999) and Iliadis *et al.* (2001), respectively. For 17 proton-induced reactions, new rates have been evaluated based on more recent experimental information. Those include, for example, the (p, γ) and (p, α) reactions on ^{17}O, ^{18}F, and ^{23}Na. A number of rates for α-particle-induced reactions, including those for ^{14}O(α, p), ^{18}Ne(α, p), and ^{15}O(α, γ) that may be important for the breakout from the hot CNO-cycles, have also been updated. The ground and isomeric state of ^{26}Al are treated as separate nuclei (Ward & Fowler, 1980) and the communication between those states through thermal excitations involving higher-lying excited ^{26}Al levels is taken explicitly into account. The required γ-ray transition probabilities are adopted from Runkle, Champagne and Engel (2001).

We evolved three new evolutionary sequences for each WD mass. In all cases, we assumed an initial WD luminosity of $\sim 3 \times 10^{-3}\, L_{\odot}$ and a mass accretion rate of $10^{16}\, g\, s^{-1}$. This mass accretion rate is 5 times lower than the lowest rate used in Starrfield *et al.* (1998b) and was chosen to maximize the amount of accreted material given the increased energy generation from the pep reaction. We use the same composition for the accreting material as used and described in Politano *et al.* (1995), Starrfield *et al.* (1998b, 2000) and Starrfield *et al.* (2001): a mixture of half-solar and half-ONeMg. By using this composition, we assume that core material has mixed with accreted material from the beginning of the evolution. This composition also affects the amount of accreted mass at the peak of the TNR since it has a higher opacity than if no mixing were assumed. The results of the evolution are given in Table 4.1. The rows are the WD mass (in solar masses), the particular reaction library that we used (the references are given below the table), and the conditions at the peak of the TNR (temperature, energy generation, surface luminosity). We also provide the total accreted mass, the amount of mass ejected in the explosive phase of the outburst, and the maximum expansion velocity for the ejected matter. Below the horizontal line we provide the abundances (mass fraction) in the ejecta.

We provide two different columns for Politano *et al.* (1995). The first, with the superscript 1, is taken from a recent calculation using the Politano *et al.* library and the Weiss and Truran (1990) network in NOVA. The second, with the superscript 2, uses the same reaction library as Politano *et al.* but with the Hix and Thielemann (1999) network. These two columns show the effects of including the pep reaction on the TNR simulations. The difference between these two calculations is large at $1.35\, M_{\odot}$ and small at $1.25\, M_{\odot}$.

Clearly, the gross features of the simulations do not depend significantly on the choice of reaction library. The major changes, as expected, come from including the pep reaction. The simulations with the pep reaction included accrete less mass, reach a lower peak temperature and a lower peak energy generation, and eject less mass moving at lower velocities. For the simulations with the Hix and Thielemann (1999) network, the accreted mass, M_{acc}, is reasonably constant for each WD mass but is larger for the smaller mass WD. We note that the peak temperature, tabulated at the core–envelope interface which is typically the zone where the highest temperature occurs during the TNR, is reasonably constant at $1.25\, M_{\odot}$ but declines slightly for the sequence at $1.35\, M_{\odot}$ with the newest reaction rate library. The results of the evolutionary sequences for $1.35\, M_{\odot}$ show that because the WD mass is larger and the radius is smaller, they reach higher densities and higher peak temperatures than the sequences at lower WD mass (Starrfield, 1989).

Figure 4.1 shows the variation of temperature with time for the deepest hydrogen-rich zone for the $1.25\, M_{\odot}$ evolutionary sequences. In this figure and all other figures in this chapter we only plot the simulations done with the pep reaction included (i.e., with the Hix and Thielemann (1999) network). The specific evolutionary sequence is identified on the plot and the time coordinate is arbitrary and chosen to show each curve clearly. Interestingly enough, the rise time and peak temperature are nearly the same for the three sequences. Figure 4.2 shows the same plot for the sequences at $1.35\, M_{\odot}$ but here we can see differences between the three simulations. Peak temperature drops from about 413 million degrees to 392 million degrees and peak nuclear energy generation drops by about a factor of 2.5 from the oldest library to the newest library (8.4×10^{17} erg $g^{-1}\, s^{-1}$ to 4.4×10^{17} erg $g^{-1}\, s^{-1}$). The temperature

Table 4.1. *Comparison of the results from three reaction libraries*

WD mass	$1.25M_\odot$	$1.25M_\odot$	$1.25M_\odot$	$1.25M_\odot$	$1.35M_\odot$	$1.35M_\odot$	$1.35M_\odot$	$1.35M_\odot$
Reaction library	P1995[a]	P1995[b]	S1998[c]	I2005[d]	P1995[a]	P1995[b]	S1998[c]	I2005[d]
$T_{peak}\,(T_6)$	331	321	321	320	454	413	414	392
ϵ_{nuc}(peak) (10^{16} erg g^{-1} s^{-1})	23	21	21	13	204	84	86	44
$L_{peak}\,(10^4 L_\odot)$	27	26	23	26	78	96	80	59
$M_{acc}\,(10^{-5} M_\odot)$	6.7	6.0	6.0	6.1	3.8	3.3	3.3	2.8
$M_{ej}\,(10^{-5} M_\odot)$	2	1.8	1.5	1.5	3.2	2.3	2.3	1.7
V_{max} (km s^{-1})	3900	3076	2860	3143	6184	5239	4755	4513
^{12}C	7.0×10^{-3}	8.0×10^{-3}	8.6×10^{-3}	4.4×10^{-3}	4.0×10^{-3}	8.0×10^{-3}	1.2×10^{-2}	6.2×10^{-3}
^{13}C	4.5×10^{-3}	5.6×10^{-3}	7.7×10^{-3}	2.6×10^{-3}	1.4×10^{-3}	2.8×10^{-3}	4.0×10^{-3}	2.4×10^{-3}
^{14}N	4.8×10^{-3}	5.4×10^{-3}	7.4×10^{-3}	9.6×10^{-3}	3.1×10^{-3}	4.3×10^{-3}	4.8×10^{-3}	8.4×10^{-3}
^{15}N	8.3×10^{-2}	7.5×10^{-2}	7.5×10^{-2}	4.6×10^{-2}	1.1×10^{-1}	1.1×10^{-1}	1.1×10^{-1}	6.0×10^{-2}
^{16}O	1.2×10^{-2}	1.3×10^{-2}	1.2×10^{-2}	9.4×10^{-3}	1.8×10^{-3}	1.2×10^{-3}	1.1×10^{-3}	2.4×10^{-3}
^{17}O	2.7×10^{-2}	3.3×10^{-2}	3.0×10^{-2}	7.7×10^{-2}	1.1×10^{-3}	1.1×10^{-3}	1.0×10^{-3}	6.7×10^{-2}
^{22}Na	7.8×10^{-3}	4.8×10^{-3}	7.3×10^{-3}	4.5×10^{-3}	6.4×10^{-2}	3.5×10^{-2}	5.1×10^{-2}	2.3×10^{-2}
^{26}Al	4.0×10^{-3}	5.2×10^{-3}	1.9×10^{-3}	2.1×10^{-3}	1.4×10^{-3}	2.8×10^{-3}	2.1×10^{-2}	3.0×10^{-3}
^{27}Al	2.0×10^{-2}	1.9×10^{-2}	1.9×10^{-2}	1.0×10^{-2}	3.2×10^{-2}	2.8×10^{-2}	3.4×10^{-2}	1.4×10^{-2}
^{32}S	6.5×10^{-3}	5.7×10^{-3}	1.1×10^{-2}	1.3×10^{-2}	1.6×10^{-2}	2.1×10^{-2}	2.8×10^{-2}	4.0×10^{-2}

[a] Politano *et al.* (1995) library: pep reaction not included (Weiss & Truran (1990) network)
[b] Politano *et al.* (1995) library: pep reaction included (Hix & Thielemann (1999) network)
[c] Starrfield *et al.* (1998b) library: pep reaction included (Hix & Thielemann (1999) network)
[d] Iliadis (unpublished reaction library used in this chapter) library: pep reaction included (Hix & Thielemann (1999) network)

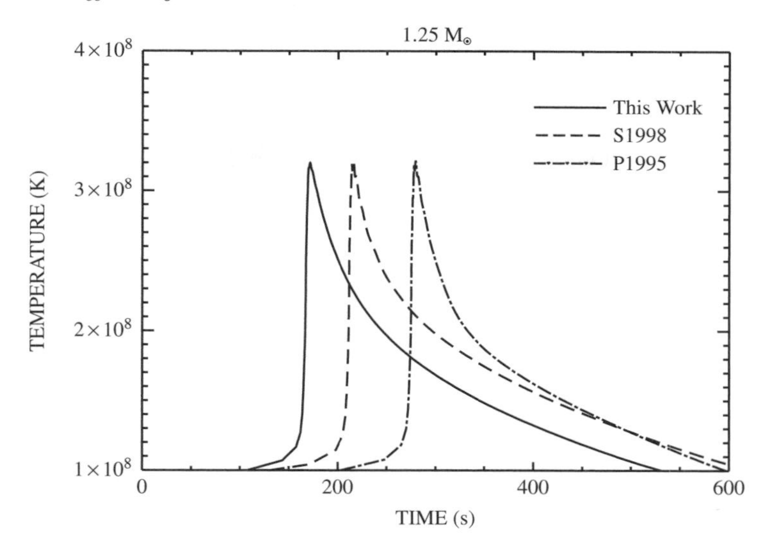

Fig. 4.1. The variation with time of the temperature in the deepest hydrogen-rich zone around the time when peak temperature occurs. We have plotted the results for three different simulations on a 1.25 $M_{\odot}$ WD. The identification with a specific library is given on the plot. In this plot and all following plots, S1998 refers to Starrfield *et al.* (1998b) and P1995 refers to Politano *et al.* (1995). The curve for each sequence has been shifted slightly in time to improve its visibility.

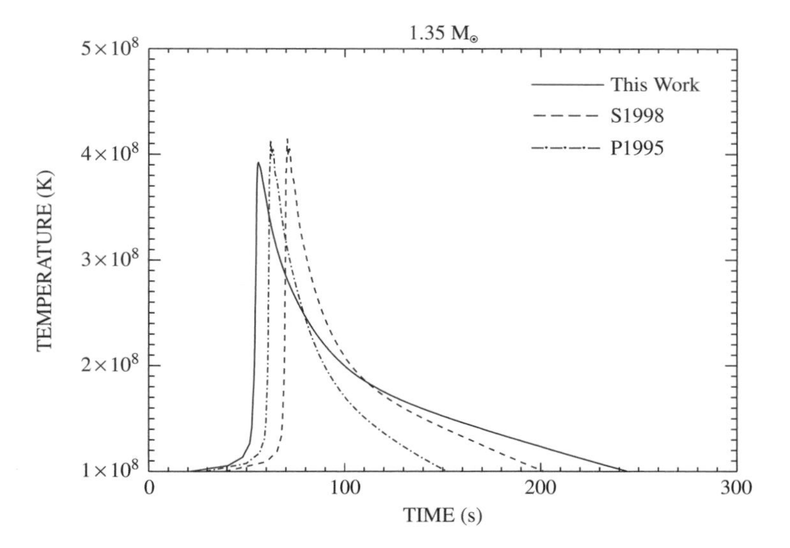

Fig. 4.2. Same as for Figure 4.1 but for a WD mass of 1.35 $M_{\odot}$.

declines more rapidly for the sequence computed with the oldest reaction library (Politano *et al.*, 1995) because it exhibited a larger release of nuclear energy throughout the evolution which caused the overlying zones to expand more rapidly and the nuclear burning layers to cool more rapidly. In contrast, the newest library, with the smallest expansion velocities, cools slowly. Note the difference of a factor of two in the time coordinate used for these two plots.

The differences in total nuclear energy generation (erg s^{-1}) as a function of time for each mass are shown in Figure 4.3 (1.25 $M_\odot$) and Figure 4.4 (1.35 $M_\odot$). The time coordinate is consistent with that used for Figures 4.1 and 4.2. Again, there is hardly any difference at the lower mass but at 1.35 $M_\odot$ the peak for the latest library is definitely lower than seen in the earlier libraries. Apparently, the changes in the libraries are more important for the more massive isotopes and become more important as higher temperatures are reached.

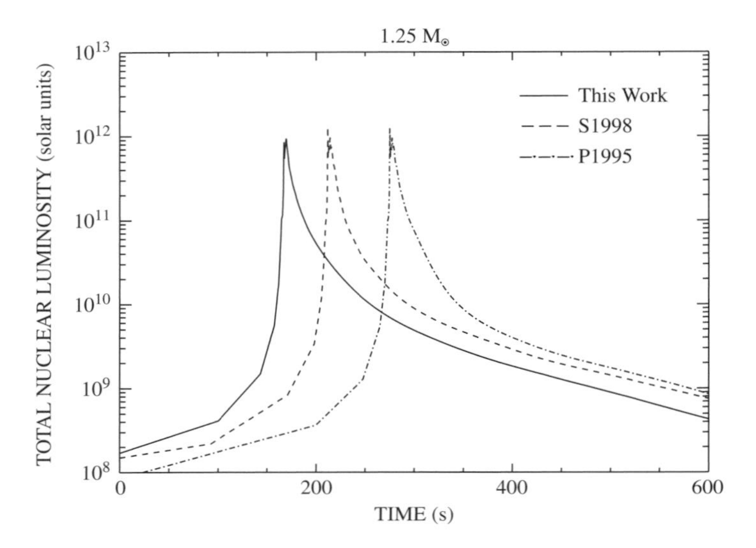

Fig. 4.3. The variation with time of the total nuclear luminosity (erg s^{-1}) in solar units ($L_\odot$) around the time of peak temperature during the TNR on a 1.25$M_\odot$ WD. We integrated over all zones taking part in the explosion. The identification with each library is given on the plot. The time coordinate is chosen to improve visibility.

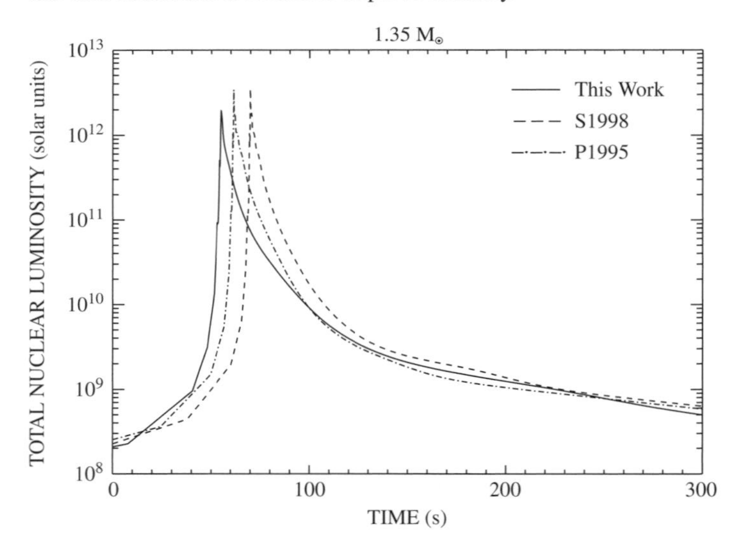

Fig. 4.4. Same as for Figure 4.3 but for a 1.35 $M_\odot$ WD.

Figures 4.5 and 4.6 show the variation of the effective temperature with time as the layers begin their expansion. We have plotted the results on the same time-scale as in Figure 4.1 and Figure 4.2, and the plots show how rapidly the energy and β^+-unstable nuclei reach and heat the surface. The large amplitude oscillations seen in the sequences in the older libraries

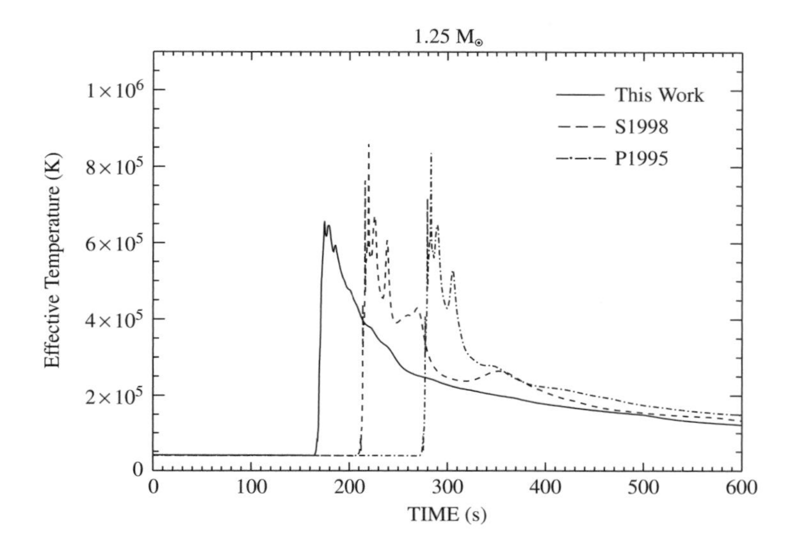

Fig. 4.5. The variation with time of the effective temperature around the time when peak temperature is achieved in the TNR for the sequence on the 1.25 $M_\odot$ WD. The time-scale is identical to that shown in Figure 4.1 and shows how rapidly the nuclear burning products are transported from the depths of the hydrogen burning shell source to the surface. The different evolutionary sequences are labeled on the plot.

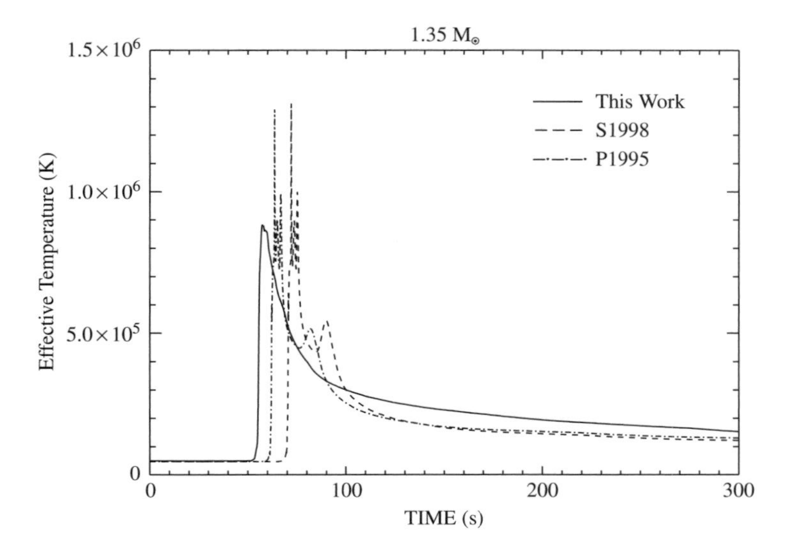

Fig. 4.6. Same as for Figure 4.5 but for a 1.35 $M_\odot$ WD.

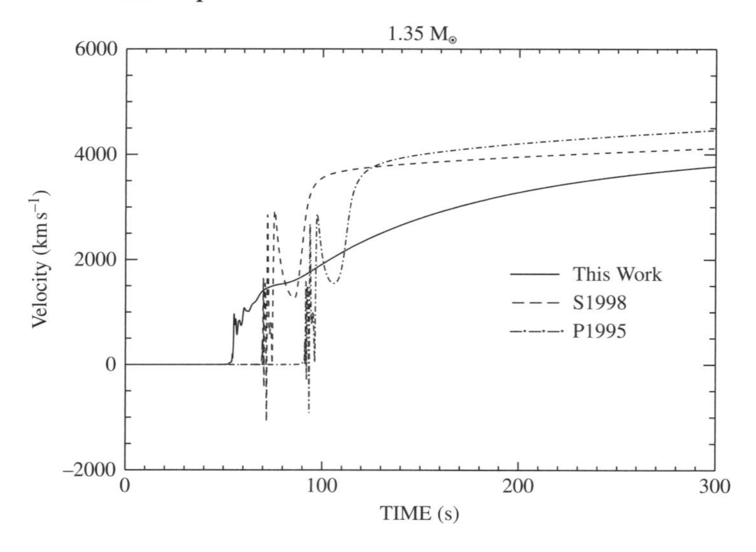

Fig. 4.7. The variation with time, over the first 300 s of the outburst, for the velocity of the surface zone using the three different reaction libraries which are labeled on the plot. We have offset the time on the Politano *et al.* (1995) sequence to make the curves more visible.

and not in that from the current library are caused by intense heating at the surface causing the layers to expand rapidly, cool and collapse back onto the surface, and then expand again. Hardly any expansion has occurred at the time of the oscillation and so the 'quasi'-period is that of the free-fall time for the underlying WD. After a few seconds the outer layers are expanding sufficiently rapidly so the oscillations cease. They are not present in the latest sequence because the surface heating is less important. The outburst is also more gradual and the star has started to expand when the β^+-unstable nuclei reach the surface. This can also be seen in Figure 4.7 which shows the variation of the surface velocity as a function of time at the time of peak temperature in the nuclear burning region. The intense heat from the β^+-unstable nuclei quickly causes the luminosity to become super-Eddington and the layers to begin expanding. However, they are still deep within the potential well of the WD and oscillate for a few seconds before reaching and then exceeding escape velocity. With the latest library, the sequence does not become as luminous and the initial oscillations exhibit a much smaller amplitude. (We have slightly offset the time on the Politano *et al.* (1995) sequence from the earlier plots to make the curves in this plot more visible.)

Figures 4.8 and 4.9 show the variation with time of the surface luminosity (for the first 11 hours of the TNR) for the three libraries at each of the two WD masses. These figures demonstrate that if we could observe a classical nova sufficiently early in the outburst then it should be super-Eddington. The initial spike (at a time of about 100 s) is caused by a slowing of the expansion as the energy from the β^+-decays decreases. After this time, expansion and cooling of the outer layers causes the opacity to increase and radiation pressure then accelerates the layers outward. The continuous flow of heat from the interior, combined with the increase in opacity, causes another increase in luminosity until the peak is reached.

Not including the sequence done without the pep reaction, if we examine the abundance predictions for the 1.25 $M_\odot$ sequences, we see that the differences caused by improving the

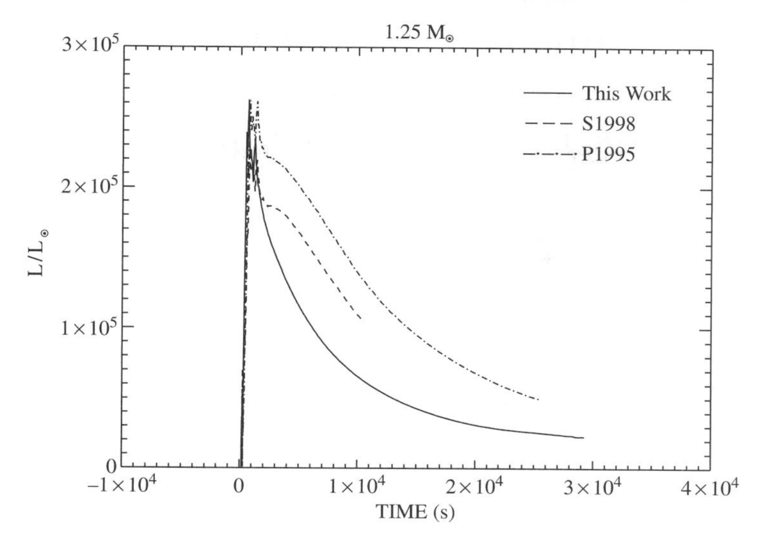

Fig. 4.8. The variation in time, over the first 11 hours of the outburst, for the surface luminosity of the evolutionary sequences using the three different reaction libraries; the WD mass is 1.25 $M_{\odot}$. The label which identifies each different sequence is given on the plot. Note that as the nuclear physics has improved, the peak luminosity and the luminosity at later times have decreased.

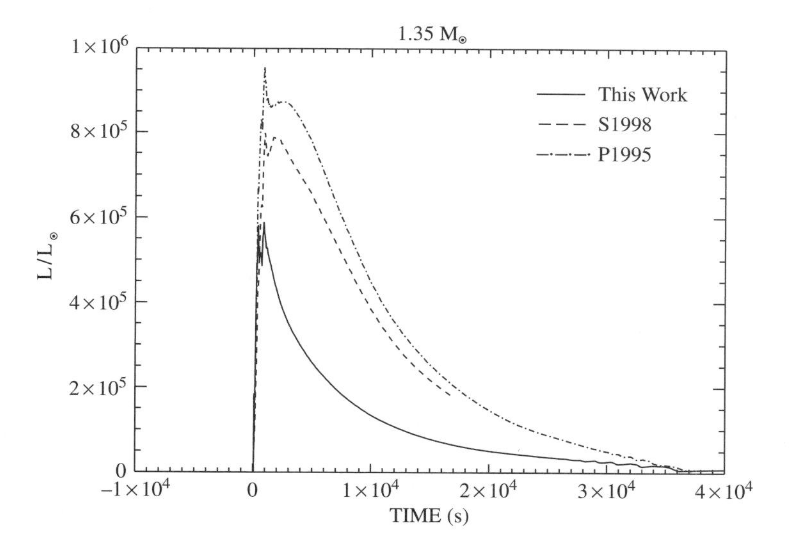

Fig. 4.9. Same as Figure 4.8 but for a WD mass of 1.35 $M_{\odot}$.

reaction library are small except for a few nuclei. For example, ^{12}C varies by about a factor of two, ^{14}N by less than a factor of two, and ^{16}O by about a factor of 1.5. The low mass odd isotopes (^{13}C, ^{15}N, and ^{17}O) all vary by about a factor of two. However, both ^{12}C and ^{13}C are depleted in the latest sequence, as is ^{15}N, while ^{17}O is enriched by the latest reaction rate library. In addition, in all three sequences the ejected oxygen exceeds carbon as found in our earlier studies. This result continues to be puzzling in light of the production of carbon-rich dust in classical nova ejecta. It is likely that the carbon-dust-forming classical novae occur on lower mass WDs which never develop sufficiently hot nuclear burning temperatures to deplete the carbon as compared with oxygen. As we examine the more massive nuclei at 1.25 $M_\odot$, we see that ^{22}Na, ^{26}Al, and ^{27}Al are depleted as we go to the latest library while ^{32}S is enhanced. A larger tabulation of the ejecta abundances will appear in Starrfield *et al.* (in preparation).

The effects of changing the nuclear reaction library are also apparent for the sequences at 1.35 $M_\odot$. Both ^{12}C and ^{13}C drop in abundance in the 2005 library while ^{14}N, ^{16}O, and ^{17}O increase in abundance. In addition, there is hardly any difference in the abundances as a function of WD mass except for ^{14}N which is lower by about a factor of 3 at 1.35 $M_\odot$. We also find that while the abundance of ^{22}Na is lowest for the 2005 library, it is still a factor of about 5 more abundant at 1.35 $M_\odot$ than at 1.25 $M_\odot$. However, the abundances of both ^{26}Al and ^{27}Al are nearly unchanged from one sequence to another and do not depend on WD mass. This is not what we reported for the studies done without the pep reaction and older libraries where we found that the abundance of ^{26}Al declined as the WD mass increased. Finally, we note that the abundance of ^{32}S is largest for the 2005 library at 1.35 $M_\odot$. In fact, it reaches 4% of the ejected material.

4.5 Multi-dimensional studies of the thermonuclear runaway

This section is the first in a series devoted to the more interesting problems remaining in our understanding of the thermonuclear processes in the classical nova outburst. Two of those problems are:

(1) the growth and development of the convective region in response to the TNR; and
(2) how and when the WD core nuclei are mixed into the accreted matter.

As already shown in the one-dimensional hydrodynamic calculations, the transport of heat and β^+-decay nuclei to the surface by convection as the TNR rises to its peak is extremely rapid and may influence a number of observable features of the classical nova outburst that can be used both to guide and constrain new simulations. Two of these features are:

(1) the early evolution of the visual light curves of fast classical novae on which their use as 'standard candles' is based. During this phase the bolometric luminosity of a nova can remain more than an order of magnitude above the Eddington luminosity for several days (confirmed for LMC 1991 by Schwarz *et al.*, 2001).
(2) the composition of matter ejected by a nova as a function of time. It is possible that material ejected early in the outburst may not have the same composition (isotopic or elemental) as material ejected later. Both of these features depend on the amount and composition of the material dredged up from the underlying CO or ONe WD core and when during the TNR they are dredged up.

The close binary nature of CVs (see Chapter 2) and the observed asymmetry of accretion in these systems make it clear that realistic modeling of classical nova outbursts must be done in three dimensions. Nevertheless, to date, virtually all hydrodynamic studies of the nova problem have been carried out in one dimension, with convection and mixing treated by mixing-length theory. Fryxell and Woosley (1982) first discussed the importance of multidimensional effects for TNRs that occurred in thin stellar shells. For classical novae, they assumed initiation at a point and calculated the lateral burning velocity of the deflagration front that spread the burning along the surface. Subsequently, Shankar, Arnett and Fryxell (1992) and Shankar and Arnett (1994) carried out two-dimensional hydrodynamic calculations of this problem. They restricted their survey to strong, instantaneous, temperature fluctuations that developed on a dynamical time-scale. However, they found that the initially intense burning at a point extinguished on a short time-scale, as the perturbed region rapidly rose, expanded, and cooled.

Owing to advances in computing power, it has become possible to treat convection at or near the peak of the TNR in both two and three dimensions. This is possible for classical nova studies since the relevant time-scales are all on the order of seconds. For example, the dynamical time-scale $\tau_{\rm hyd}$, at a density $\sim 10^4$ g cm^{-3} is of the order of seconds. The nuclear burning time-scale decreases from years to seconds, once the temperature rises above 10^8 K, until constrained by the β^+-decay lifetimes. Finally, the convective turn-over time-scale is of the order of seconds near the peak of the runaway (Starrfield *et al.*, 1998b).

Glasner, Livne and Truran (1997) explored the consequences of thermonuclear ignition and explosive hydrogen burning in classical novae with a two-dimensional fully implicit hydrodynamic code. They followed the evolution of a convectively unstable hydrogen-rich envelope accreted onto a CO WD at a time close to the peak of the TNR and found a flow pattern that effectively dredged up sufficient material from the core to explain the observed levels of heavy element enrichment in classical nova ejecta ($\sim$30% to 40% by mass: Gehrz *et al.*, 1998). The redistribution of nuclear energy generation over the envelope, caused by the outward transport of short-lived β^+-decay nuclei, was also found to play a significant role in the outburst. In a complementary study, Kercek, Hillebrandt and Truran (1998) examined the early stages of the evolution, using the same initial model. While their simulations confirmed the finding of Glasner, Livne and Truran, mixing was not as strong and occurred over a longer time-scale.

Kercek, Hillebrandt and Truran (1999) then performed two- and three-dimensional studies, using the same input model and physics as before, but with improved resolution. Their results displayed less mixing with core material and a completely different flow structure which cast doubts on this mixing mechanism. Therefore, Rosner *et al.* (2001) re-examined shear mixing (Kippenhahn & Thomas, 1978; Sparks & Kutter, 1987; Kutter & Sparks, 1987) and suggested, based on semi-analytical and time-scale arguments plus two-dimensional calculations, that this mechanism could also be responsible for the mixing. Alexakis *et al.* (2004) studied the development of shear mixing on a 1.0 $M_\odot$ WD in two dimensions. Their initial model consisted of a completely convective layer moving at a large velocity tangential to the surface of the WD. They did not include nuclear reactions. They found core material was mixed into the envelope but they failed to address how such a thick layer could have formed on the surface of the WD since the accreting material must have mixed far earlier in the evolution. In addition, since they mixed CO material into a H-rich layer with a peak temperature of 10^8 K, they should have obtained an explosion (Starrfield *et al.*, 1999).

Given the differences in these calculations, the general inferences that can be drawn from the existing multi-dimensional calculations are that:

(1) the amount of convective mixing occurring prior to the TNR is negligible;
(2) the amount of convective mixing occurring during the early stages of the TNR is a sensitive function of the degree of degeneracy; and

(3) convective overshoot mixing, which dredges up CO- or ONe-rich matter from the underlying WD core (Woosley, 1986), may be sufficient to produce the observed enrichments of nova ejecta.

Since the heavy element enrichment of the envelope via the latter mechanism does not occur until late in the evolution to the TNR, the envelope composition during accretion is that of the material being transferred by the secondary. This keeps core material out of the accreted layers until the peak of the TNR and, in turn, the opacity stays low and the amount of accreted material is increased (Starrfield *et al.*, 1998b). In summary, while the multi-dimensional studies have provided some insight into the progress of convection during the TNR, they are still in their infancy and further work is needed.

4.6 Nucleosynthesis in CO and ONe novae

The measured abundances for classical nova ejecta confirm the levels of enrichment required by the theoretical studies to reproduce the dynamical features of classical nova outbursts and, in addition, establish that both CO and ONe WDs occur in cataclysmic variable binary systems (Gehrz *et al.*, 1998; Starrfield *et al.*, 1998b). Further, the significant enhancements of heavy elements in classical nova ejecta, taken together with the observational determinations of the masses of their ejecta, confirm that classical novae contribute to the abundances of some of the CNO isotopes in the Galaxy. Finally, the elemental and isotopic compositions of classical nova ejecta are predicted to be both distinctive and incorporated into grains. Recently a signature of nova processing was identified in the pre-solar grains found in meteorites (José *et al.*, 2004; see also Chapter 6).

The extensive database of new atmospheric and nebular elemental abundances for classical nova ejecta (Gehrz *et al.*, 1998; Starrfield *et al.*, 1998b) constitutes an important and powerful tool both for constraining the modeling of their outbursts and for determining their contributions to Galactic chemical evolution. The degree to which elements such as silicon, sulfur, and argon are enriched, according to previous nucleosynthesis studies, is a sensitive function of the temperature history of the burning shell, as are the abundances of ^{22}Na and ^{26}Al (José, Hernanz & Coc, 1997; José, Coc & Hernanz, 1999, 2001; Starrfield *et al.*, 2001; Iliadis *et al.*, 2002; Hix *et al.*, 2003; Parete-Koon *et al.*, 2003). Moreover, the abundances seen in the ejecta of V838 Her suggest that breakout (alpha captures on ^{14}O and ^{15}O) from the CNO cycle may have occurred during the TNR which, in turn, implies peak temperatures exceeding 5×10^8 K (Schatz, 2005, private communication). Finally, the abundance of ^{7}Be in classical nova ejecta is sensitive to the rate at which it is transported to the surface regions prior to its decay to ^{7}Li. The extent of classical nova contributions to the abundance of ^{7}Li in the Galaxy and expectations for the detection of γ-rays from ^{7}Be decay in classical nova ejecta remain open questions (Romano *et al.*, 1999; Romano & Matteucci, 2003; see also Chapter 11).

Since both CO and ONe WDs are found in classical nova systems, it is crucial to calculate evolutionary sequences for consistent choices of WD mass, envelope mass, thermal structure, and composition (CO or ONe) that can be compared directly to data for individual classical nova systems. While we have evolved one-dimensional sequences designed to fit the observed properties of the ONe nova V1974 Cyg (Starrfield *et al.*, 2000), these simulations predicted sufficient ^{22}Na production by this classical nova that its γ-ray emission should have been detected by the Compton Gamma-Ray Observatory (CGRO), but it was not seen (Shrader *et al.*, 1994; Iyudin *et al.*, 1995; Leising, 1998). Another discrepancy was that comparison of the abundance predictions with observations suggested that the simulations in Starrfield *et al.* (1998b) were over-producing nuclei in the mass region past magnesium. One source of these discrepancies appears to be that Starrfield *et al.* used the post-carbon burning abundances of Arnett and Truran (1969) for their core abundances. In contrast, the study of carbon burning nucleosynthesis by Ritossa, Garcia-Berro and Iben (1996) predicted lower abundances for Mg and Si. If their distribution is implemented, however, a lower level of ^{26}Al production is obtained (José & Hernanz, 1998; Starrfield *et al.*, 2000). Therefore, nucleosynthesis calculations for ONe classical novae using updated initial abundances and updated nuclear reaction libraries should be done and compared to the large amount of observational data now available.

CO WDs produce the largest amounts of dust (Starrfield, Gehrz & Truran, 1997; Gehrz *et al.*, 1998; see also Chapters 8 and 13) and infrared observations indicate that the same CO nova can form dust with radically different compositions (graphite, SiC, hydrocarbons, and silicates) at different times during its outburst (Gehrz *et al.*, 1998; again see Chapters 8 and 13). As already noted, the isotopic abundance patterns predicted for CO novae ejecta show similarities to some of those found in pre-solar grains in meteorites (José *et al.*, 2004). Laboratory studies of these grains are providing significant constraints on stellar evolution calculations, convective mixing, and Galactic chemical evolution (Zinner, 1998). Because of the importance of identifying pre-solar contributors, it is necessary to redo the abundance predictions for CO novae with new reaction rates and updated initial models.

One problem with our understanding, however, is that the compositions determined for the ejected shells of classical novae from optical and ultraviolet studies typically show O/C $>$ 1 (Gehrz *et al.*, 1998; Starrfield *et al.*, 1998b; see Chapter 9 for a full discussion). However, grains of C, SiC, and hydrocarbons were identified early in the outburst and O-rich silicate grains later in the outburst for both QV Vul and V705 Cas (Gehrz *et al.*, 1998), which suggests that distinct regions with O/C $>$ 1 and O/C $<$ 1 can occur in the ejecta of the same classical nova (Rawlings & Evans, 2002; see also Chapter 13).

A major challenge, therefore, is to identify the conditions that produce carbon-rich nova ejecta. Most of the early studies of nucleosynthesis associated with CO novae assumed that the composition of the material dredged up from the WD core was 50% ^{12}C and 50% ^{16}O. Since the rate of the ^{12}C(p,γ)^{13}N reaction exceeds that of the ^{16}O(p,γ)^{17}F reaction, this ensures that even the earliest matter ejected by a classical nova will have O/C $>$ 1. More recent stellar evolution studies of the evolution to WDs (Althaus *et al.*, 2002; Gil-Pons *et al.*, 2003) imply that the outer regions of WD CO cores will have C/O ratios that differ strongly from unity. Therefore, if carbon is more abundant than oxygen in the core, then when core material is dredged up into the envelope, the surface composition of the envelope might exhibit both C/O $>$ 1 and extremely non-solar isotopic abundances for the CNO elements.

4.7 Mass of the ejecta

Another extremely important problem is the discrepancy between observations and predictions of the amount of mass ejected in the outburst (Warner, 1995; Gehrz *et al.*, 1998; Starrfield *et al.*, 1998b). Infrared and radio analyses, combined with optical and UV studies of nebular emission lines, appear to provide reliable and independent estimates of the ejected mass (Warner, 1995; Gehrz *et al.*, 1998; see also Chapters 7 and 8, while noting the caveats pointed out in Chapter 6). In contrast to the observationally determined masses, numerical simulations of TNRs on both CO and ONe WDs predict ejecta masses that can be smaller by up to a factor ~10 (Prialnik & Kovetz, 1995; José, Coc & Hernanz, 1999; Gehrz *et al.*, 1998; Starrfield *et al.*, 1998b). Moreover, calculations of the envelope mass required to initiate a TNR are in approximate agreement with the theoretical calculations and not with the observations (MacDonald, 1983; Townsley & Bildsten, 2004).

There are two reasons that the cause of the ejected mass discrepancy must be determined. First, a solution should provide an improved understanding of the development of the classical nova outburst; and second, most estimates of the contributions of classical novae to Galactic chemical evolution use ejecta masses determined from the theoretical predictions. If the masses inferred from observations are used in the chemical evolution studies, then classical novae become even more important for production of the odd isotopes of the light elements in the Galaxy (particularly ^{13}C, ^{15}N, and ^{17}O) than previously believed.

In addition to the nucleosynthesis implications of increased ejecta mass, the nuclear reaction sequences that proceed in larger mass envelopes differ from those occurring in lower mass envelopes for otherwise similar sequences (Starrfield *et al.*, 1999). One-dimensional calculations show that a TNR initiated with a larger envelope mass has a higher density and degeneracy at the time of the TNR. In turn, this increases the peak temperature and energy generation rate during the TNR, changes the character of convective mixing, modifies the predictions of element and isotopic production during the TNR, and ejects material at higher velocities (Starrfield *et al.*, 1998b, 1999). For example, in a preliminary study of this problem, the envelope mass was increased to about five times the values reported in Starrfield *et al.* (1998b) and it was found that the peak temperature, in a 1.25 $M_\odot$ sequence, exceeded 6×10^8 K (and breakout from the CNO cycle occurred as suggested for V838 Her: Vanlandingham *et al.*, 1996). In another sequence on a 1.0 $M_\odot$ CO WD, where the envelope mass was increased to agree with *observed* values, the resulting evolution ejected $\sim 10^{-4}$ $M_\odot$ at speeds exceeding 6000 km s^{-1} (in good agreement with the observations of LMC 1991: Schwarz *et al.*, 2001). Further work on this problem is warranted.

4.8 Relationship of classical novae to supernovae of Type Ia

Supernovae of Type Ia (SNe Ia) are those supernovae in which neither hydrogen nor helium is seen in any spectra obtained during the outburst. They have similar light curves, which can be further calibrated (Phillips, 1993), making them excellent standard candles to redshift $z > 1$ (Filippenko, 1997). Thus, they have become extremely important because they can be used to determine the structure and evolution of the Universe (Leibundgut, 2000, 2001, and references therein). Supernovae Ia are also important because they contribute a major fraction of the iron group elements to the Galaxy. In the past few years, tremendous effort has gone into studies of their light curves and the variations in observed properties (cf. Hillebrandt & Leibundgut, 2003). Nevertheless, the progenitor(s) of SN Ia explosions are, as yet, unknown. Whelan and Iben (1973) proposed that the explosion occurred on a CO WD which accreted material from a binary companion until its mass reached the Chandrasekhar limit and a carbon deflagration/detonation occurred (Nomoto, Thielemann & Yokoi, 1984; Branch *et al.*, 1995; Hillebrandt & Niemeyer, 2000). Of relevance to this chapter, typical classical nova systems can be ruled out because the WD is decreasing in mass as a result of repeated nova explosions and cannot be growing to the Chandrasekhar limit (MacDonald, 1984; Gehrz *et al.*, 1998; Starrfield *et al.*, 1998c; see also Chapter 3). In addition, the absence of hydrogen and helium in the spectrum of a SN Ia rules out most other CV systems since the WDs are accreting hydrogen- and helium-rich material. If the WD were to explode, then the accreted envelope would be carried along with the supernova ejecta and be seen in the spectrum (Marietta, Burrows & Fryxell, 2000; Starrfield, 2003).

The systems that most recently have been proposed as SN Ia progenitors are the Super Soft X-ray Sources (Branch *et al.*, 1995; Starrfield *et al.*, 2004b; van den Heuvel *et al.*, 1992; Kahabka & van den Heuvel, 1997). These systems are close binaries which contain an extremely hot and luminous WD. Photometric studies show that there are variations both in their X-ray and optical emission (Lanz *et al.*, 2004; Kahabka & van den Heuvel, 1997). There is observational evidence for accretion disks, jets, and winds which implies that material is both being transferred from the secondary and also lost from the system (Cowley *et al.*, 1998). The current explanation of their properties is that they are WDs accreting material from their companions and the infalling material burns at the Steady Burning rate (van den Heuvel *et al.*, 1992). Thus, no TNR occurs and the mass of the WD grows to the Chandrasekhar limit (van den Heuvel *et al.*, 1992; Kahabka & van den Heuvel, 1997). However, the Steady Burning hypothesis requires that this behavior occur only for a narrow range of mass accretion rates for a given WD mass.

Starrfield *et al.* (2004b) have now used their one-dimensional hydrodynamic code to study accretion onto *hot* WDs and, in contrast, find that hydrogen burns to helium (and helium to carbon and oxygen) in the surface layers for a broad range of mass accretion rates (Surface Hydrogen Burning [SHB]: Starrfield *et al.*, 2004b). In order to begin the sequences, they evolve a WD through a classical nova outburst. Once all the ejected material is expanding faster than the escape speed, has reached radii exceeding 10^{13} cm, and is optically thin, it is then removed from the calculations. They then take the remnant WD and either wait until it has cooled to minimum and restart accretion (to evolve a classical nova explosion) or begin accretion when the WD is still hot to study the evolution of SSS. They report that accretion, from $1.6 \times 10^{-9}\,\mathrm{M_{\odot}\,yr^{-1}} < \dot{M}_{\mathrm{acc}} < 8 \times 10^{-7}\,\mathrm{M_{\odot}\,yr^{-1}}$, onto hot ($2.3 \times 10^5$ K), luminous ($30\,\mathrm{L_{\odot}}$), massive ($1.25\,\mathrm{M_{\odot}}$, $1.35\,\mathrm{M_{\odot}}$) CO WDs results in the accreted material burning to helium in the surface zones. This evolution differs from Steady Burning since Starrfield *et al.* (2004b) find that the material can burn at rates that are both lower and higher than the single value assumed for the Steady Burning rate. Steady Burning also assumes that only helium is produced by this mechanism so that a helium layer accumulates below the surface zones. However, because of the energy release from hydrogen burning near the surface, Starrfield *et al.* (2004b) find that the helium layer remains hot and helium then steadily burns to carbon, oxygen, and more massive nuclei without experiencing a TNR. No mass is ejected and the WD grows in mass to the Chandrasekhar limit. Some sequences have been evolved for more than 10^6 yr and no TNR has occurred. Energy release from both nuclear burning and compression caused by the increasing mass (and decreasing radius) of the WD over this time is sufficient to prevent the WD from cooling.

Since most of the hydrogen and helium that is accreted from the companion star burns to carbon and oxygen, if the WDs that we are evolving explode in a SN Ia outburst, then there will be almost no hydrogen or helium present in the ejecta (and spectrum) and the event will be classified as a SN Ia. In addition, the luminosities and effective temperatures of our evolutionary sequences fit the observations of SSS such as CAL83 and CAL87. The results of Starrfield *et al.* (2004b), therefore, strengthen the proposed connection between SSS and SN Ia progenitors. While these results are encouraging, a great deal more work is necessary to demonstrate that the SSS are indeed progenitors of SN Ia explosions.

4.9 Summary and discussion

In this chapter, we have reviewed our current knowledge about the thermonuclear processing that occurs during the evolution of the classical nova outburst. A TNR in the accreted hydrogen-rich layers on the *low* luminosity WDs in cataclysmic variable binary systems is the outburst mechanism for classical, recurrent, and symbiotic novae. The interaction between the hydrodynamic evolution and nuclear physics lies at the basis of our understanding of how the TNR is initiated, evolves, and grows to the peak of the explosion. The *observed* high levels of enrichment of classical nova ejecta in elements ranging from carbon to sulfur confirm that there is dredge-up of matter from the core of the WD and enable classical novae to contribute to the chemical enrichment of the interstellar medium. Therefore, studies of classical novae are leading to an improved understanding of Galactic nucleosynthesis, the sources of pre-solar grains, the extragalactic distance scale, and the nature of the progenitors of SN Ia.

It is now recognized that the characteristics of the classical nova explosion depend on the complex interaction between nuclear physics (the β^+-limited CNO reactions) and convection during the final stages of the TNR. The light curves, the peak luminosities (which can exceed the Eddington luminosity), the levels of envelope enrichment, and the composition of classical nova ejecta are all strongly dependent upon the extent and time-scale of convective mixing during the explosion. The characteristics of the outburst depend upon the white dwarf mass and luminosity, the mass accretion rate, the chemical composition of both the accreting material and white dwarf core material, the evolutionary history of the white dwarf, and when and how the accreted layers are mixed with the white dwarf core. The importance of nuclear physics to our understanding of the progress of the outburst can be seen when we compare a series of evolutionary sequences in which the only change has been the underlying nuclear reaction rate library. Although there are only small changes at lower WD masses, there are large changes at the highest white dwarf mass that we studied: 1.35 $M_\odot$. In order to make meaningful comparisons of theory with observations, we need to use the best nuclear physics and opacities that are available to us. We also discussed a number of problems with our understanding of the outburst. For example, theoretical predictions of the amount of material ejected during the outburst do not agree with the observations. This problem has important implications and must be solved before we can claim a better understanding of the outburst.

Acknowledgements

We are grateful to a number of collaborators who over the years have helped us to better understand the nova outburst. Many of them are the authors of other chapters in this book. We have benefited from discussions with A. Champagne, A. Evans, R. D. Gehrz, P. H. Hauschildt, M. Hernanz, J. José, J. Krautter, J.-U. Ness, M. Orio, G. Schwarz, H. Schatz, S. N. Shore, G. Shaviv, E. M. Sion, W. M. Sparks, P. Szkody, J. Truran, K. Vanlandingham, M. Wiescher, and C. E. Woodward. S. S. acknowledges partial support from NASA and NSF grants to ASU. W.R.H. acknowledges partial support from DoE and NSF. Oak Ridge National Laboratory is managed by UT-Battelle, LLC, for the US Department of Energy under contract DE-AC05-00OR22725. C.I. acknowledges partial support from DoE.

References

Alexakis, A. C., Calder, A. C., Heger, A. *et al.*, 2004, *Ap. J.*, **602**, 931.
Althaus, L. G., Serenelli, A. M., Corsico, A. H., & Benvenuto, O. G., 2002, *MNRAS*, **330**, 685.
Angulo, C., Arnould, M., Rayet, P. *et al.*, 1999, *Nucl. Phys. A*, **656**, 3.
Arnett, W. D., & Truran, J. W., 1969, *Ap. J.*, **157**, 339.
Bahcall, J. N., & May, R. M., 1969, *Ap. J.*, **155**, 501.
Bode, M., & Evans, A., 1989, *Classical Novae*, 1st edn. New York and Chichestes: Wiley.
Branch, D., Livio, M., Yungelson, L. R., Boffi, F. R., & Baron, E., 1995, *PASP*, **107**, 1019.
Caughlan, G., & Fowler, W. A., 1988, *Atom Data Nucl. Data* **40**, 291.
Cowley, A., Schmidtke, P., Crampton, D., & Hutchings, J., 1998, *Ap. J.*, **504**, 854.
Cox, A. N., & Stewart, J. N., 1971, *Ap. JS*, **19**, 243.
Cox, A. N., & Tabor, J. E., 1976, *Ap. JS*, **31**, 271.
Della Valle, M., Bianchini, A., Livio, M., & Orio, M., 1992, *A&A*, **266**, 232.
Della Valle, M., Rosino, L., Bianchini, A., Livio, M., 1994, *A&A*, **287**, 403.
Filippenko, A. V., 1997, *ARA&A*, **35**, 309.
Fryxell, B., & Woosley, S., 1982, *Ap. J.*, **261**, 332.
Fujimoto, M. Y., 1982a, *Ap. J.*, **257**, 752.
Fujimoto, M. Y., 1982b, *Ap. J.*, **257**, 767.
Gehrz, R. D., Truran, J. W., Williams, R. E., & Starrfield, S., 1998, *PASP*, **110**, 3.
Gil-Pons, P., Garcia-Berro, E., José, J., Hernanz, M., & Truran, J. W., 2003, *A&A*, **407**, 1021.
Glasner, S. A., Livne, E., & Truran, J. W., 1997, *Ap. J.*, **475**, 754.
Hernanz, M., & José, J., 2002, *Classical Nova Explosions*. New York: American Institute of Physics.
Hillebrandt, W., & Leibundgut, B., 2003, in *From Twilight to Highlight: The Physics of Supernovae*. Heidelberg: Springer.
Hillebrandt, W., & Niemeyer, J., 2000, *ARA&A*, **38**, 191.
Hix, W. R. & Thielemann, F.-K., 1999, *J. Comp. Appl. Maths*, **109**, 321.
Hix, W. R., Smith, M. S., Mezzacappa, A., Starrfield, S., & Smith, D. L., 2003, in *Proceedings of the 7th International Symposium on Nuclei in the Cosmos*, ed. S. Kubono. *Nuclear Physics A*, **718**, 620.
Iben, I., 1982, *Ap. J.*, **259**, 244.
Iglesias, C. A., & Rogers, F. J., 1993, *Ap. J.*, **412**, 752.
Iliadis, C., D'Auria, J. M., Starrfield, S., Thompson, W. J., & Wiescher, M., 2001, *Ap. JS*, **134**, 151.
Iliadis, C., Champagne, A., José, J., Starrfield, S., & Tupper, P., 2002, *Ap. JS*, **142**, 105.
Iyudin, A. F., Bennett, K., Bloemen, H. *et al.*, 1995, *A&A*, **300**, 422.
José, J., & Hernanz, M., 1998, *Ap. J.*, **494**, 680.
José, J., Hernanz, M., & Coc, A., 1997, *Ap. J.*, **479**, L55.
José, J., Coc, A., & Hernanz, M., 1999, *Ap. J.*, **520**, 347.
José, J., Coc, A., & Hernanz, M., 2001, *Ap. J.*, **560**, 897.
José, J., Hernanz, M., Amari, S., Lodders, K., & Zinner, E., 2004, *Ap. J.*, **612**, 614.
Kahabka, P., & van den Heuvel, E. P. J., 1997, *ARA&A*, **35**, 69.
Kercek, A., Hillebrandt, W., & Truran, J. W., 1998, *A&A*, **337**, 379.
Kercek, A., Hillebrandt, W., & Truran, J. W., 1999, *A&A*, **345**, 831.
Kippenhahn, R., & Thomas, H.-C., 1978, *A&A*, **63**, 265.
Kovetz, A., & Prialnik, D., 1985, *Ap. J.*, **291**, 812.

Krautter, J., Ögelman, H., Starrfield, S., Wichmann, R., & Pfeffermann, E., 1996, *Ap. J.*, **456**, 788.
Kutter, G. S., & Sparks, W. M., 1987, *Ap. J.*, **321**, 386.
Lanz, T., Telis, G. A., Audard, M. *et al.* 2004, *Ap. J.*, **602**, 342.
Leibundgut, B., 2000, *A&A Reviews*, **10**, 179.
Leibundgut, B., 2001, *ARA&A*, **39**, 67.
Leising, M., 1998, in *Proceedings of 4th Compton Symposium*, ed. C. D. Dermer, M. S. Strickman, & J. D. Kurfess. New York: American Institute of Physics, p. 163.
Long, K. S., Helfand, D. J., & Grabelsky, D. A., 1981, *Ap. J.*, **248**, 925.
MacDonald, J., 1983, *Ap. J.*, **267**, 732.
MacDonald, J., 1984, *Ap. J.*, **283**, 241.
MacDonald, J., Fujimoto, M. Y., & Truran, J. W., 1985, *Ap. J.*, **294**, 263.
Marietta, E., Burrows, A., & Fryxell, B., 2000, *Ap. JS*, **128**, 615.
Nomoto, K., Thielemann, F.-K., & Yokoi, K., 1984, *Ap. J.*, **286**, 644.
Paczyński, B., & Zytkow, A. N., 1978, *Ap. J.*, **222**, 604.
Parete-Koon, S., Hix, W. R., Smith, M. S. *et al.*, 2003, *Ap. J.*, **598**, 1239.
Phillips, M., 1993, *Ap. J.*, **413**, L105.
Politano, M., Starrfield, S., Truran, J. W., Sparks, W. M., & Weiss, A., 1995, *Ap. J.*, **448**, 807.
Prialnik, D., & Kovetz, A., 1984, *Ap. J.*, **281**, 367.
Prialnik, D., & Kovetz, A., 1995, *Ap. J.*, **445**, 789.
Prialnik, D., Livio, M., Shaviv, G., & Kovetz, A., 1982, *Ap. J.*, **257**, 312.
Rawlings, J. M. C., & Evans, A., 2002, in *Classical Nova Explosions*, ed. M. Hernanz, & J. José. New York: American Institute of Physics, p. 270.
Ritossa, C., Garcia-Berro, E., & Iben, I. Jr., 1996, *Ap. J.*, **460**, 489.
Rogers, F. J., & Iglesias, C. A., 1994, *Science*, **263**, 50.
Rogers, F. J., Swenson, F., & Iglesias, C. A., 1996, *Ap. J.*, **456**, 902.
Rolfs, C. E., & Rodney, W. S., 1988, *Cauldrons in the Cosmos: Nuclear Astrophysics*. Chicago: University of Chicago Press.
Romano, D., Matteucci, F., Molaro, P., & Bonifacie, P., 1999, *A&A*, **352**, 117.
Romano, D., & Matteucci, F., 2003, *MNRAS*, **342**, 185.
Rosner, R., Alexakis, A., Young, Y.-N., Truran, J. W., & Hillebrandt, W. 2001, *Ap. J.*, **562**, L177.
Runkle, R. C., Champagne, A. E., & Engel, J., 2001, *Ap. J.*, **556**, 970.
Saizar, P., & Ferland, G. J., 1994, *Ap. J.*, **425**, 755.
Schatzman, E., 1958, *White Dwarfs*. Amsterdam: North Holland Press.
Schwarz, G. J., Shore, S. N., Starrfield, S. *et al.*, 2001, *MNRAS*, **320**, 103.
Shankar, A., & Arnett, W. D., 1994, *Ap. J.*, **433**, 216.
Shankar, A., Arnett W. D., & Fryxell, B. A., 1992, *Ap. J.*, **394**, L13.
Shafter, A., 1997, *Ap. J.*, **487**, 226.
Shara, M. M., 1989, *PASP*, **101**, 5.
Shrader, C., Starrfield, S., Truran, J. W., Leising, M. D., & Shore, S. N., 1994, *BAAS*, **26**, 1325.
Sion, E. M., Acierno, M. J., & Tomczyk, S., 1979, *Ap. J.*, **230**, 832.
Sion, E. M., & Starrfield, S., 1994, *Ap. J.*, **421**, 261.
Sparks, W. M., & Kutter, G. S., 1987, *Ap. J.*, **321**, 394.
Starrfield, S., 1989, in *Classical Novae*, 1st edn, ed. M. F. Bode, & A. Evans. New York and Chichester: Wiley, p. 39.
Starrfield, S., 2003, in *From Twilight to Highlight: the Physics of Supernovae*, ed. W. Hillebrandt, & B. Liebundgut. Heidelberg: Springer, p. 128.
Starrfield, S., Gehrz, R. D., & Truran, J. W., 1997, in *Astrophysical Implications of the Laboratory Study of Pre-solar Grains*, ed. T. Bernatowicz, & E. Zinner. New York: American Institute of Physics, p. 203.
Starrfield, S., Schwarz, G., Truran, J. W., & Sparks, W. M., 1998a, *BAAS*, **30**, 1400.
Starrfield, S., Truran, J. W., Wiescher, M. C., & Sparks, W. M., 1998b, *MNRAS*, **296**, 502.
Starrfield, S., Schwarz, G., Truran, J. W., & Sparks, W. M., 1999, *BAAS*, **31**, 977.
Starrfield, S., Sparks, W. M., Truran, J. W., & Wiescher, M. C., 2000, *Ap. JS*, **127**, 485.
Starrfield, S., Iliadis, C., Truran, J. W., Wiescher, M., & Sparks, W. M., 2001, in *Nuclei in the Cosmos 2000*, ed. J. Christensen-Dalsgaard, & K. Langange. Amsterdam: Elsevier, p. 110c (see also *Nuclear Physics A*, **688**, 110c).
Starrfield, S., Dwyer, S., Timmes, F. X., *et al.* 2004a, in *Cosmic Explosions in Three Dimensions*, ed. P. Höflich, P. Kumar, & J. C. Wheeler. Cambridge: Cambridge University Press, p. 87.
Starrfield, S., Timmes, F. X., Hix, W. R., Sion, E. M., Sparks, W. M., & Dwyer, S. J., 2004b, *Ap. J.*, **612**, L53.

Thielemann, F.-K., Arnould, M., & Truran, J. W., 1987,
in *Advances in Nuclear Astrophysics*, ed. E. Vangioni-Flam *et al.* Gif sur Yvette: Éditions Frontiéres, p. 525.
Thielemann, F.-K., Arnould, M., & Truran, J. W., 1988,
in *Capture Gamma-Ray Spectroscopy*, ed. K. Abrahams, & P. van Assche. Bristol: Institute of Physics, p. 730.
Townsley, D., & Bildsten, L., 2003, *Ap. J.*, **596**, L227.
Townsley, D., & Bildsten, L., 2004, *Ap. J.*, **600**, 390.
Trümper, J., Hasinger, G., Aschenbach, B., Braeuninger, H. & Briel, U. G., 1991, *Nature*, **349**, 579.
van den Heuvel, E. P. J., Bhattacharya, D., Nomoto, K., & Rappaport, S. A., 1992, *A&A*, **262**, 97.
Vanlandingham, K., Starrfield, S., Wagner, R. M., Shore, S. N., & Sonneborn, G., 1996, *MNRAS*, **282**, 563.
Ward, R. A., & Fowler, W. A., 1980, *Ap. J.*, **238**, 266.
Warner, B., 1995, *Cataclysmic Variable Stars*. Cambridge: Cambridge University Press.
Weiss, A., & Truran, J. W., 1990, *A&A*, **238**, 178.
Wiescher, M., Görres, J., Graff, S., Buchmann, L., & Thielemann, F. K., 1989, *Ap. J.*, **343**, 352.
Whelan, J., & Iben, I., 1973, *Ap. J.*, **186**, 1007.
Woosley, S. E., 1986, in *Nucleosynthesis and Chemical Evolution*, ed. B. Houck, A. Maeder, & G. Meynet. Sauverny: Geneva Observatory, p. 1.
Zinner, E., 1998, *ARE&PS*, **26**, 147.

5

Nova atmospheres and winds

Peter H. Hauschildt

5.1 Introduction

The modeling and analysis of early nova spectra have made significant progress since the first edition of this book. The main culprit is the author, via the construction of detailed model atmospheres and synthetic spectra for novae (Hauschildt *et al.*, 1992, 1994a,b, 1995, 1996, 1997; Pistinner *et al.*, 1995; Schwarz *et al.*, 1997; Short *et al.*, 1999; Schwarz *et al.*, 2001; Short *et al.*, 2001; Shore *et al.*, 2003).

In the early stages of the nova outburst, the spectrum is formed in an optically *very* thick (in both lines and continua) shell with a flat density profile, leading to very extended continuum and line-forming regions (hereafter, CFR and LFR respectively). The large variation of the physical conditions inside the spectrum-forming region makes the classical term 'photosphere' not very useful for novae. The large geometrical extension leads to a very large electron temperature gradient within the CFR and LFR, allowing for the observed simultaneous presence of several ionization stages of many elements. Typically, the relative geometrical extension $R_{\rm out}/R_{\rm in}$ of a nova atmosphere is ~ 100–1000, which is much larger than the geometrical extension of hydrostatic stellar atmospheres (even in giants $R_{\rm out}/R_{\rm in}$ is typically less than 2) or supernovae (SNe).

The electron temperatures and gas pressures typically found in nova photospheres lead to the presence of a large number of spectral lines, predominantly Fe-group elements, in the LFR and a corresponding influence of line blanketing on the emergent spectrum. The situation is complicated significantly by the velocity field of the expanding shell which leads to an enhancement of the overlapping of the individual lines. This in turn makes simplified approximate treatments of the radiative transfer by, e.g., the Sobolev approximation, very inaccurate, and more sophisticated radiative transfer methods, which treat the overlapping lines and continua simultaneously, must be used in order to obtain reliable models. The line blanketing also leads to strong wavelength redistribution of the radiative energy; therefore, the temperature structure of the shell must be calculated including the effects of the line blanketing.

The situation is further complicated by the fact that the (electron) densities of the CFR and LFR in a nova atmosphere are very low compared with classical stellar atmospheres. This leads to an overwhelming dominance of the radiative rates over the collisional rates. In

Classical Novae, 2nd edition, ed. Michael Bode and Aneurin Evans. Published by Cambridge University Press.

addition, the radiation field is very non-grey. These two effects lead to large departures from local thermodynamic equilibrium (LTE) in the CFR and LFR. Therefore, non-LTE effects must be included self-consistently in the model construction, in particular in the calculation of the temperature structure and the synthetic spectra. As discussed in the previous paragraph, the effects of the line blanketing on the radiative rates require a careful treatment of the radiative transfer in the NLTE calculations, and simple approximations can lead to incorrect results.

5.2 Modeling nova atmospheres and spectra

5.2.1 Model assumptions

Nova specific

Following the discussion of Bath and Shaviv (1976), we approximate nova photospheres to be spherical and expanding but stationary configurations. Therefore, we assume that all time-dependent terms both in the hydrodynamics and in the radiative transfer equation can be neglected and all quantities depend only upon the radial coordinate (except for the specific intensity of the radiation field which, in addition, depends on the angle to the radial direction). These assumptions are justified since the hydrodynamic time-scale is much longer than the time-scales for photon thermalization and for the establishment of excitation and ionization equilibria. The assumption of spherical symmetry is only an approximation: all novae eject non-spherical and clumpy shells. However, in the *early* stages of the nova outburst the clumps are still very close together and the gas between the individual clumps will be optically thick. A three-dimensional treatment of the full problem is currently not feasible (but may be standard procedure in time for the future editions of this book!) and simple approximations, e.g. neglecting NLTE effects and/or line blanketing, will lead to wrong conclusions.

Using the results from hydrodynamic calculations of thermonuclear runaways in the accreted envelopes on white dwarf stars (see Chapter 4), we assume that the density varies according to a power law, $\rho \propto r^{-p}$. For the earliest, fireball, phase of the nova outburst, the parameter p is of the order of 10, a value typical for supernova envelopes. In later phases, the value of p is much smaller, around 2–3, leading to very large geometrical extensions of nova atmospheres. We treat the velocity field as a free parameter, to be derived from the observed spectrum. In general, the velocity field can be either ballistic $V \propto r$ (in the fireball phase), or corresponding to $\dot{M}(r) = \text{const.}$ (in the wind phase). The velocity field may also follow a standard 'wind' profile, i.e. $V(r) = V_\infty (1 - a/r)^b$ (and the corresponding density profile for $\dot{M}(r) = \text{const.}$) with three free parameters V_∞, a, and b, to be derived from the spectrum. In our model construction, we can allow for all three of the above velocity profiles.

General

The complicated interaction of overlapping lines and continua as well as the velocity field present in the LFR require a sophisticated treatment of the radiative transfer in the nova atmosphere. We solve the full special relativistic non-grey equation of radiative transfer for both lines and continua. The maximum velocities typically found in novae would, in principle, allow us to restrict the radiative transfer to complete first order in $\beta = v/c$, but this restriction would not simplify the calculations (Hauschildt *et al.*, 1994c). Neglecting the advection and aberration terms, which are of order β, would lead to errors of $\sim 5\beta$ in the mean intensity (Mihalas et al., 1976; Hauschildt & Wehrse, 1991). At maximum observed velocities of $\sim 5000\,\mathrm{km\,s^{-1}}$, this corresponds to errors of $\sim 8\%$ in the mean intensity J, which could lead to measurable changes in the radiative rates and observable changes in the emergent spectra. In addition, we solve the radiative energy equation in the Lagrangian frame including all velocity terms so that the temperature structure is fully self-consistent with the radiation field. The temperature structure is computed using the condition of radiative equilibrium in the co-moving frame. Radioactive decays do not have a significant influence on the temperature structure for atmospheres with $p \geq 3$ (Pistinner *et al.*, 1995) and are thus neglected in the models shown in this chapter.

The low densities and complicated radiation fields in the atmosphere require that the species that are most important for the opacities and the spectra are treated using multi-level NLTE. *PHOENIX* uses a rate-operator splitting method to solve the non-linear NLTE radiation transport and statistical equation problem (see Hauschildt and Baron, 1999, for details). The iteration scheme for the solution of the multi-level non-LTE problem can be summarized as follows: (1) for given n_i and n_e, solve the radiative transfer equation at each wavelength point and update the radiative rates and the approximate rate operator, (2) solve the linear system of equations given in Hauschildt and Baron (1999) for each group for a given electron density, (3) compute new electron densities (by either fixed point iteration or the generalized partition function method), (4) if the electron density has not converged to the prescribed accuracy, go back to step 2, otherwise go to step 1. The iterations are repeated until a prescribed accuracy for the n_e and the n_i is reached. It is important to account for coherent scattering processes during the solution of the wavelength-dependent radiative transfer equation, as this explicitly removes a global coupling from the iterations.

5.2.2 Model parameters

We choose as a numerically convenient set of basic parameters the effective temperature T_eff, the luminosity L, and the density exponent p. The effective temperature is not well defined for extended atmospheres; our definition is based on $L = 4\pi R^2_{\tau_\mathrm{std}=1} \sigma\, T_\mathrm{eff}{}^4$. Here, $R_{\tau_\mathrm{std}=1}$ is the reference radius at $\tau_\mathrm{std} \equiv \tau_{5000,\mathrm{cont}} = 1$; see Pistinner *et al.* (1995) for a discussion about this set of model parameters.

The radial form of the velocity field is given by either $\dot{M}(r) = \mathrm{const.}$ or by assuming a velocity law. We report here calculations made for either a ballistic velocity law ($V \propto r$) or for 'radiatively driven wind' velocity laws of the general form $V \propto V_\infty\,(1 - a/r)^b$, as noted above.

For nova atmosphere models with $p \leq 3$ we also need to choose the total mass of the envelope by using a cut-off outer density ρ_out. This is required simply because the integral $\int_0^\infty \rho(r)\,\mathrm{d}r$ diverges for $p \leq 3$. Additional parameters are, of course, the elemental abundances.

5.2.3 Model computation

We compute the nova atmosphere models and synthetic spectra with our general-purpose NLTE atmosphere code *PHOENIX*, version 16. The details of the numerical methods and algorithms used in *PHOENIX* are described in Hauschildt and Baron (1999), so we give only a brief summary here. *PHOENIX* solves the non-grey radiation transport and NLTE rate equations with an operator splitting method using nested iterations. The temperature correction procedure is based on an extension of the Unsöld–Lucy scheme. Elements and ions that are not treated in NLTE are treated as LTE background species. Their line opacities are computed with a direct opacity sampling method which allows for depth-dependent individual line profiles for each line. All NLTE lines also have individual depth-dependent profiles.

For this chapter, we have calculated a set of new nova atmosphere models that span a $T_{\rm eff}$ range of 5000 to 60 000 K in increments of 1000–5000 K. The most important model parameters are the maximum expansion velocity of atmosphere, $V_{\max} = 2000\,\mathrm{km\,s^{-1}}$, the exponent of the ρ power law, $p = 3$, the micro-turbulent velocity broadening width, $\xi = 50\,\mathrm{km\,s^{-1}}$, and the abundances, which are solar if not noted otherwise. A linear velocity law and a value of p equal to 3 are consistent with a constant mass-loss rate ($\dot{M}$ = constant). The radius $R = r(\tau_{5000} = 1)$, where τ_{5000} is the continuum optical depth at 5000 Å, is adjusted for each value of $T_{\rm eff}$ to keep the bolometric luminosity, $L_{\rm bol}$, equal to $50\,000\mathrm{L}_\odot$ during the early phases of the outburst, i.e. before the nebular lines emerge. These models cover the typical range of observed novae. The spectra are available at ftp://phoenix.hs.uni-hamburg.de/.

5.3 Results: theory

5.3.1 The structure of nova atmospheres

The structure of nova atmospheres is strongly affected by both the large temperature gradient and the low densities in the LFR. In Figure 5.1 we show the concentration of Fe I – Fe IX in a nova model atmosphere with an effective temperature of 25 000 K. The plot

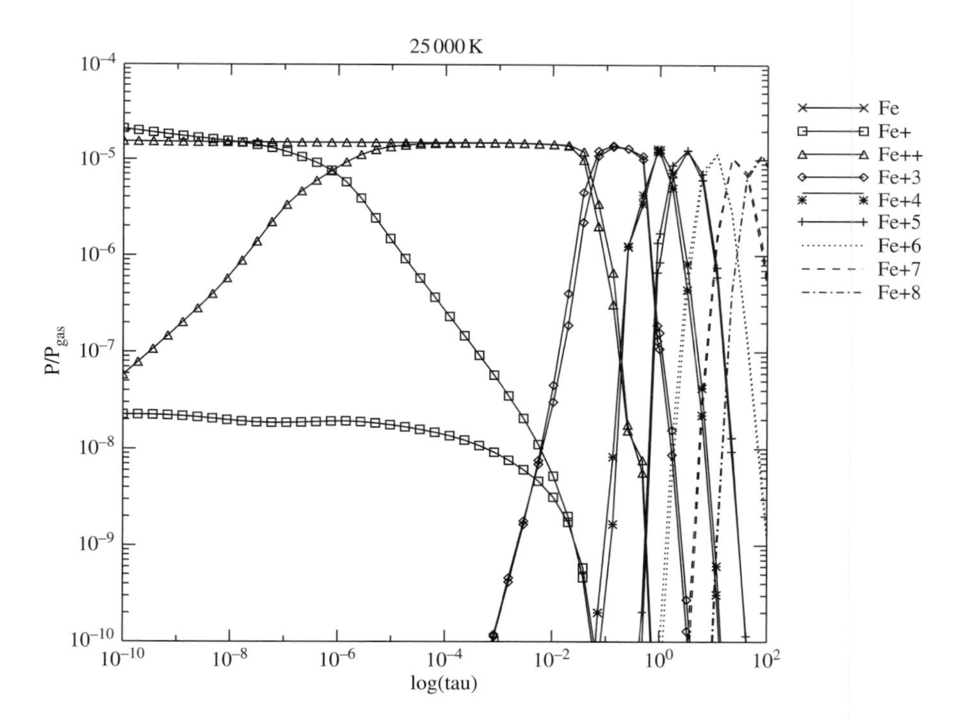

Fig. 5.1. Relative concentration of iron atoms and ions in a nova atmosphere model with $T_{\rm eff} = 25\,000$ K, $p = 3$, and solar abundances. The plot gives the partial pressures of the indicated species relative to the gas pressure as functions of standard optical depth $\tau_{\rm std}$. The thick lines and symbols are for a NLTE model (which includes Fe I– VI in NLTE) whereas the thin lines and symbols are for a LTE model.

demonstrates that multiple ionization stages of an element can be present simultaneously in the atmosphere. In fact, the Fe ionization sequence forms a number of nested 'Strömgren spheres' in the atmosphere. Furthermore, the figure shows the fundamental importance of NLTE effects: NLTE effects prevent the recombination of Fe III to Fe II in the outer atmosphere, thus dramatically changing the appearance of the spectrum compared with the LTE case. It is also very important to include a sufficient number of ionization stages for each element in NLTE, cf. Hauschildt *et al.* (1997). Otherwise, the ionization balance may be skewed because of artificial enforcement of LTE ionization/recombination processes for important ions. The other elements show similar behaviors.

5.3.2 The effects of the line blanketing

Line blanketing in the ultraviolet and optical spectral ranges is the dominant source of opacity in nova atmospheres. In Figures 5.2 and 5.3 we display synthetic spectra for models with $T_{\rm eff} = 15\,000$ K and 25 000 K with a varying degree of line blanketing included. In each panel, the continuous spectrum is compared to a spectrum with the indicated number of spectral lines included. Line blanketing has an enormous impact on the spectrum emitted by the model with $T_{\rm eff} = 15\,000$ K, in particular in the ultraviolet. Practically all lines in the ultraviolet are *absorption* lines; only a few regions of relative transparency in the 'line haze' exist, e.g. the broad features at about 1350 Å, 1500 Å, 1600 Å, 2000 Å, and 2640 Å. These appear as 'emission' features but are merely holes in the 'iron curtain', a term coined by

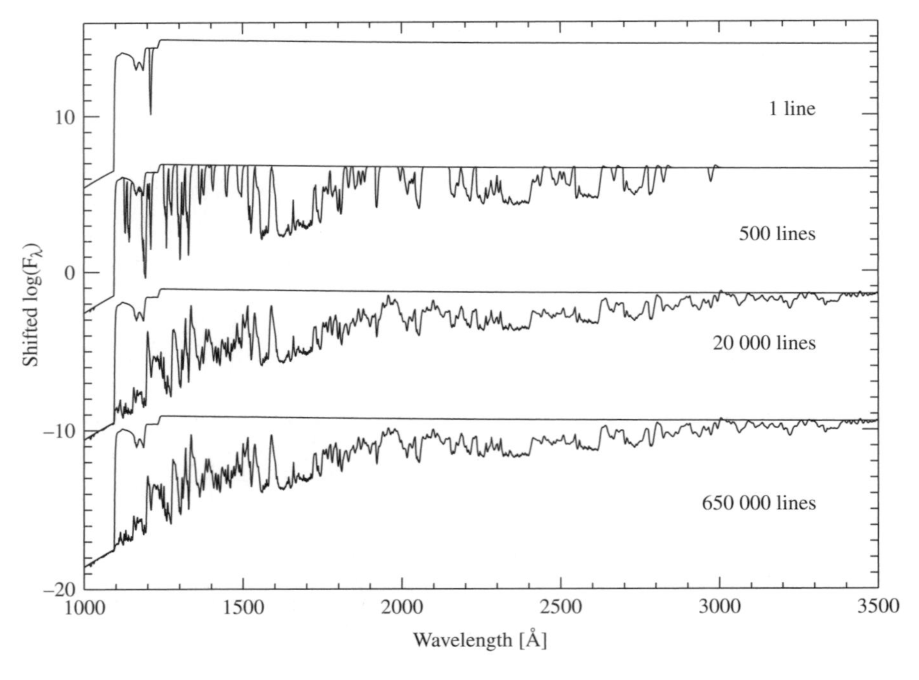

Fig. 5.2. The effect of steadily increasing the line blanketing on synthetic LTE (for simplicity) nova spectra. The nova model atmospheres have the common parameters $p = 3$, $v_{\rm max} = 2000\,{\rm km\,s^{-1}}$, solar abundances and $T_{\rm eff} = 15\,000$ K. Each sequence shows the effects of the indicated number of spectral lines, selected according to their opacities relative to the surrounding continuum, compared with the bound–free and free–free continuum.

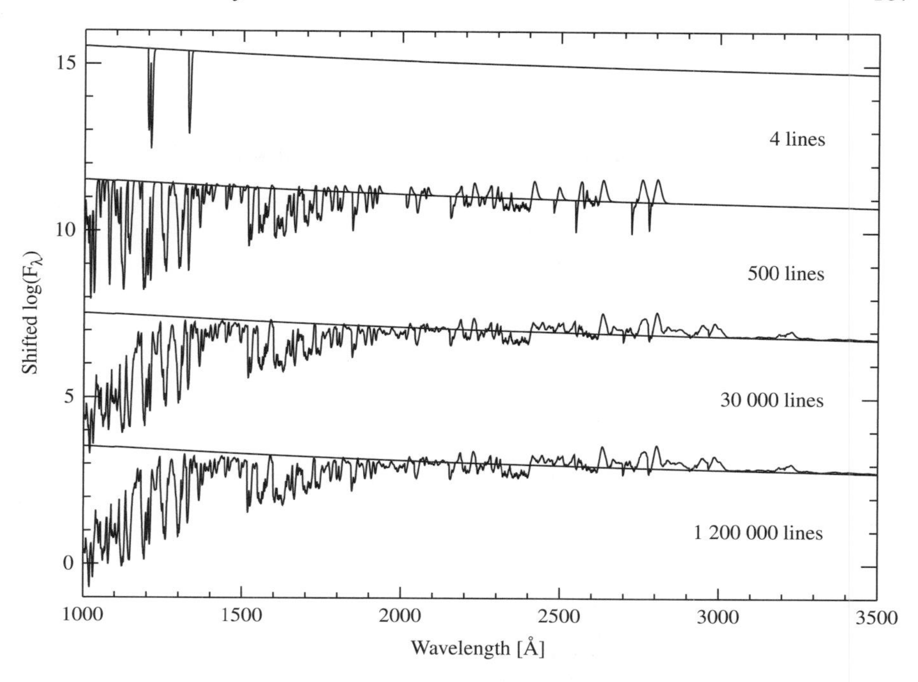

Fig. 5.3. As Figure 5.2 but for $T_{\rm eff} = 25\,000$ K.

Steven Shore to highlight that most of the ultraviolet line opacity is delivered by Fe lines. In the optical spectral range the line blanketing is more localized in overlapping absorption and P Cygni lines, but the effect is nowhere near as strong as in the ultraviolet. At $T_{\rm eff} = 25\,000$ K, the optical lines show marginally overlapping P Cygni profiles; this is also the case for the ultraviolet lines above ~1500 Å. Below 1500 Å the lines are mostly in absorption and the spectrum is formed in a very similar way to the ultraviolet spectrum of the 10 000 K model. Note that some features change their character when the effective temperature increases. For example, the 2640 Å feature is, for $T_{\rm eff} \approx 15\,000$ K, formed by a gap between strong Fe II lines, but at $T_{\rm eff} \approx 25\,000$ K it is formed by the overlapping emission parts of Fe II lines just short of 2640 Å. In most models, about 100 000 atomic lines are required to 'saturate' the spectrum, i.e. the synthetic spectrum does not change significantly if more lines are included but would change considerably if fewer lines were used. We demonstrate this in Figures 5.2 and 5.3 which show the effects of increasing the number of lines on the synthetic spectrum. The model structures were taken from a fully self-consistent LTE model including ~400 000 lines.

Figure 5.4 shows the effect of line blending for the Mg II $h + k$ doublet. The dotted line gives the pure Mg II profile, i.e. every line other than the $h + k$ lines is neglected, whereas the full curve gives the line profile calculated including other lines. The changes in the line profile are obvious and significant. Therefore, one must be very careful in analyzing observed nova line profiles and include blending in the analysis. We find that about 500 000 lines are stronger than about 10^{-4} times the local absorptive continuum for $\xi = 50\,{\rm km\,s^{-1}}$. The vast majority of these lines have wavelengths smaller than 4000 Å, so that on the average 125 lines lie within 1 Å. This corresponds to an average spacing of 2.4 km s^{-1} at 1000 Å, which is much smaller than the width of the individual lines, leading to strong overlap. For $\xi = 2\,{\rm km\,s^{-1}}$, we

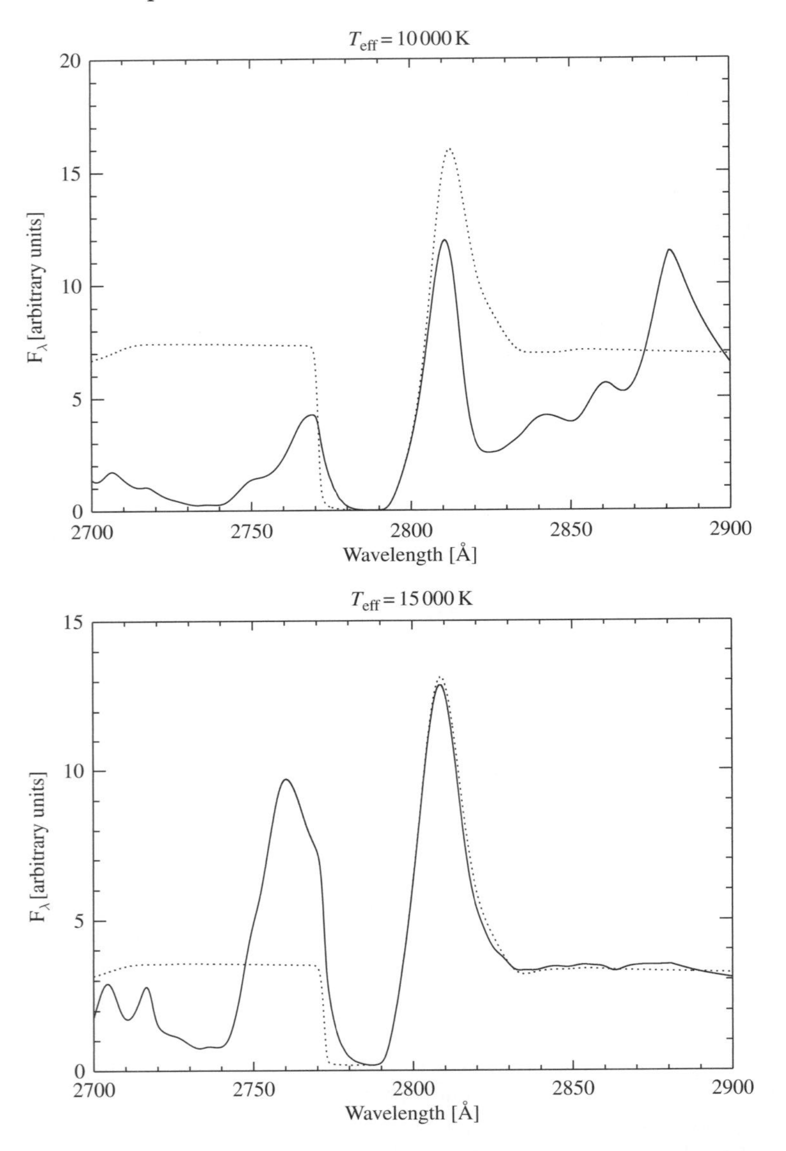

Fig. 5.4. The region around the Mg II $h+k$ lines for two models with $T_{eff} = 10\,000$ K (upper panel) and $T_{eff} = 15\,000$ K (lower panel). The dotted curves give the spectrum obtained if line blanketing by Fe II lines is neglected; the full curves give the synthetic spectra including Fe II line blanketing in addition to the Mg II lines. The significant change in the Mg II $h+k$ line profiles due to overlapping Fe II lines is very apparent in this plot and demonstrates the importance of a proper treatment of overlapping lines in the model construction and the computation of synthetic spectra of novae. From Hauschildt *et al.* (1995).

find that about 4.8×10^6 lines are stronger than 10^{-4} of the local continuum, i.e. their average spacing is about $0.25\,\mathrm{km\,s^{-1}}$ so that they are also completely overlapping. The vast majority of these lines are relatively weak and form a slowly varying 'pseudo-continuum' which serves as a background on which the few very strong lines form individual P Cygni profiles.

5.3.3 Departures from LTE

In Figure 5.5 we show the departure coefficients b_i of a few important species for a nova model with an effective temperature of 25 000 K. It can be seen that the departures from LTE are large throughout both the LFR and the CFR. In addition, the b_i can be both larger and smaller than unity, depending on the species, the level under consideration, and the optical depth. Line blending has an *enormous* influence on the results of NLTE calculations through the radiative rates due to the high line density in the ionizing ultraviolet. Neglecting line blanketing leads to much larger departures from LTE (the radiation field is super-Planckian in the outer atmosphere) which, in turn, leads to huge errors in abundance determinations. This means that a fully self-consistent solution of the NLTE radiative transfer, rate and energy equations, including line blanketing, is *necessary* in order to obtain reliable physical models of nova atmospheres.

5.3.4 The effects of the form of the velocity field

The form of the velocity field in the atmosphere also affects the emitted line profiles. We can use this additional information to derive the form of the velocity field from observed

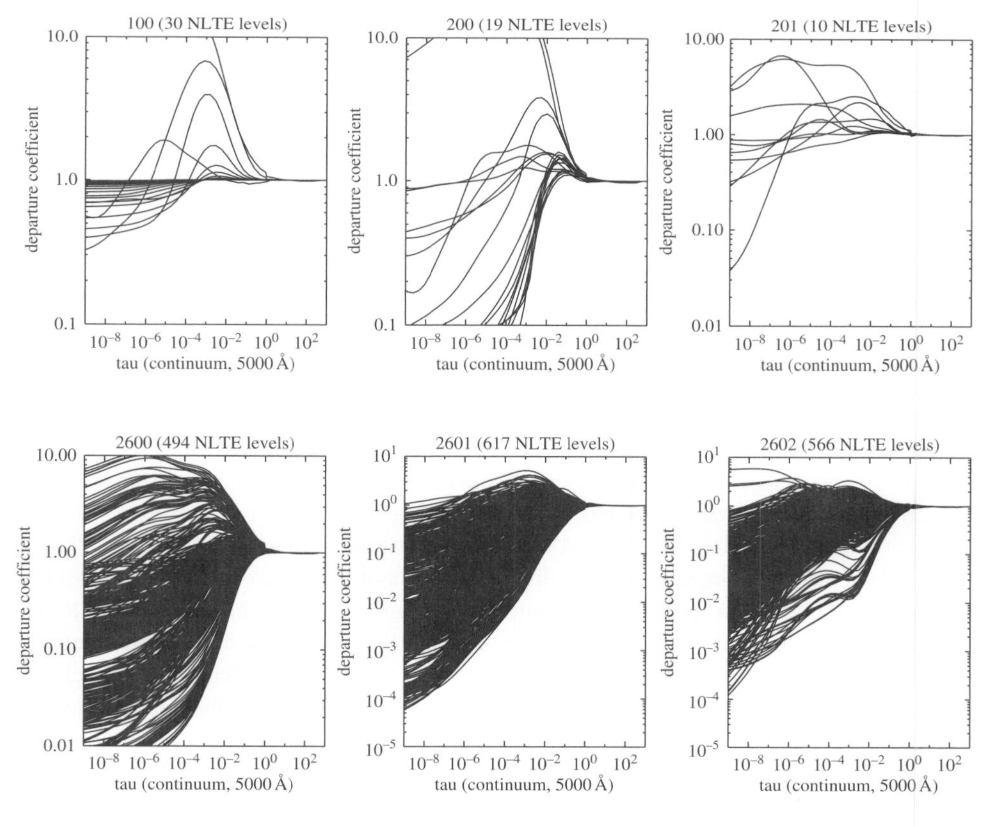

Fig. 5.5. Departure coefficients b_i as functions of standard optical depth $\tau_{\rm std}$ (continuum at 5000 Å) for a nova atmosphere model with $T_{\rm eff} = 25\,000$ K. The species shown are H I (code: 100), He I (200), He II (201), Fe I (2600), Fe II (2601) and Fe III (2602). The number of levels in each model atom is indicated.

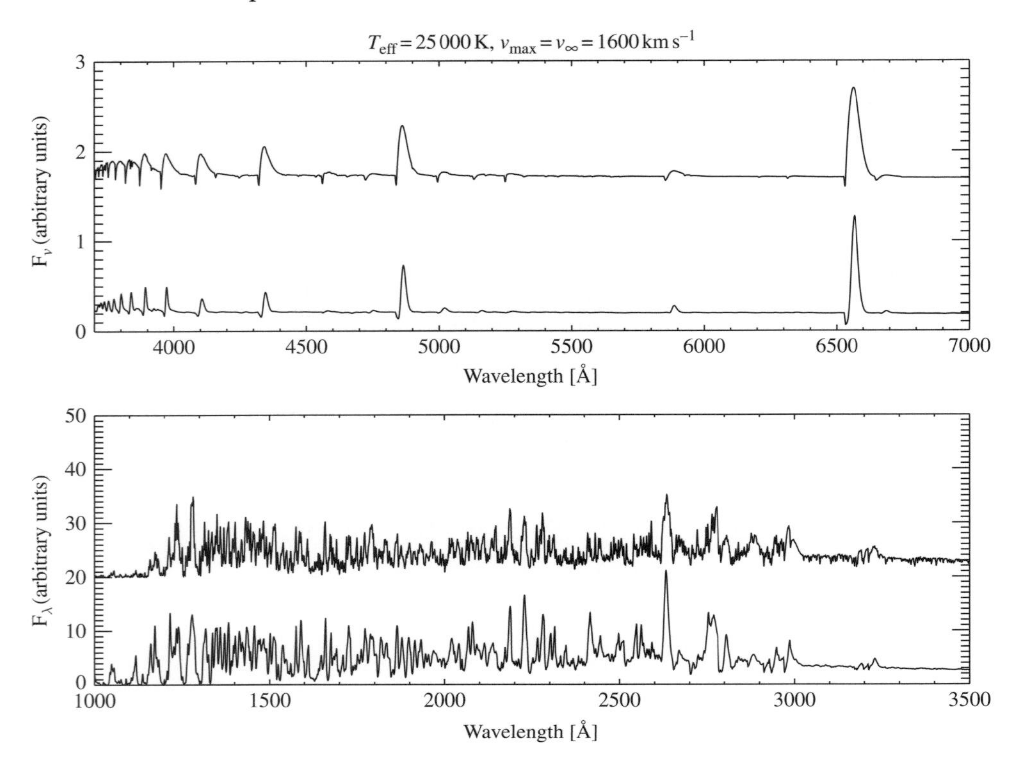

Fig. 5.6. Effect of two different velocity laws on the spectrum for a nova model atmosphere with $T_{\rm eff} = 25\,000\,{\rm K}$ and $v_{\rm max} = v_\infty = 1600\,{\rm km\,s^{-1}}$. The top curve is the spectrum calculated using a 'wind' velocity field of the general form $V(r) \propto V_\infty\,(1 - a/r)^b$ whereas the bottom curve represents the results for the ballistic (linear) velocity law. Both models have the common parameter $p = 3$ and twice the solar metal abundances and are self-consistently calculated with their respective velocity fields. From Hauschildt *et al.* (1995).

line profiles, which in turn indicates the ejection mechanism of the shell. In Figure 5.6 we compare synthetic spectra calculated using a ballistic (linear) velocity field (bottom spectrum in each plot) with spectra computed using a wind-type velocity law with $b = 0.1$ (top spectrum). The model has $T_{\rm eff} = 25\,000\,{\rm K}$ and a maximum velocity of $1600\,{\rm km\,s^{-1}}$. The figure shows that there are significantly different effects of the form of the velocity field on the optical and ultraviolet spectra. In the optical, the emission parts of the P Cygni profiles are broader and the absorption parts are narrower for the 'wind' velocity field when compared to the ballistic velocity field. This is caused by the fact that for the optical lines the velocity gradient in the LFR of the 'wind' velocity field is much smaller than in the ballistic case.

In the ultraviolet the effects are more pronounced than in the optical and different for cool ($T_{\rm eff} \approx 15\,000\,{\rm K}$) and hot ($T_{\rm eff} \approx 25\,000\,{\rm K}$) models (Hauschildt *et al.*, 1995). In cool models, the appearance of the ultraviolet spectrum changes significantly; the shapes of the different features are altered. This is caused by the different velocity gradients in the LFRs of cool and hot models, which directly influences the amount of the overlap between lines. In the hot model, the features remain essentially the same for both velocity laws, but the 'wind' velocity field model shows much more 'fine structure' superimposed on the broader features. This general behavior makes the ultraviolet less attractive as a velocity field indicator for hotter nova atmospheres because the 'fine structure' can easily be confused with noise in the observed spectrum if the signal-to-noise and/or the resolution of the observations are not high enough.

5.3.5 *The effects of the luminosity on the spectrum*

The influence of different parameters chosen for the model construction on the synthetic spectra has been discussed in detail by Pistinner *et al.* (1995). They found both from analytical and simplified numerical test calculations that some of the 'classical' parameters may have only a marginal influence on the emitted spectrum and, therefore, cannot be determined from the spectrum. We demonstrate that this behavior is also true for full NLTE model atmosphere calculations of novae. As an example, we calculate the influence of the luminosity L, formally an input parameter, on the emergent spectrum (Figure 5.7). All remaining parameters, in particular the outer density and the maximum expansion velocity, are identical in both models (therefore, model atmospheres with larger luminosity also have larger masses and mass-loss rates). The figure shows that the luminosity does *not* have a

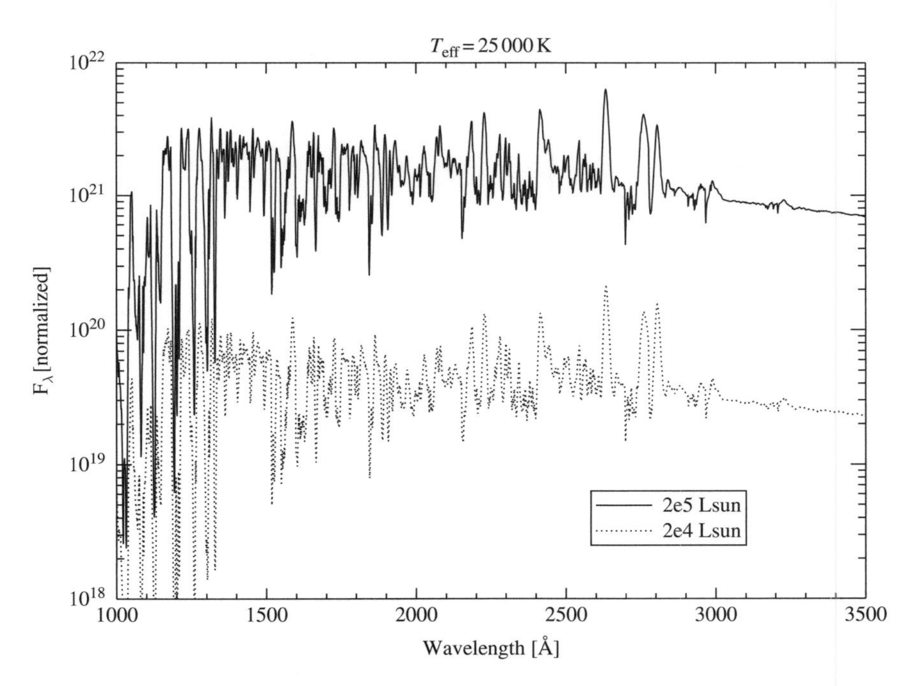

Fig. 5.7. Effect of different luminosities on the spectra of nova atmospheres for a model atmosphere with $T_{\rm eff} = 25\,000$ K and $v_{\rm max} = v_\infty = 1600\,{\rm km\,s^{-1}}$. The top curve is the spectrum calculated for $L = 200\,000\,L_\odot$ whereas the bottom curve represents the results for $L = 20\,000\,L_\odot$. The curves have been displaced relative to each other in order to make the plot clearer. Both (simplified) models have the common parameter $p = 3$ and solar abundances, and are self-consistently calculated with their respective luminosity. From Hauschildt *et al.* (1995).

significant effect on the emergent spectrum, since the spectra shown in Figure 5.7 are nearly identical. The reason is that the radiative transfer and the temperature structure are sensitive only to the *relative* extension of the atmosphere. For p as low as 3, the geometrical extension of the atmosphere does not change significantly with the absolute radius (and therefore with the luminosity through $L \propto R^2$). The only effect is due to changes in the density of the CFR and LFR owing to spherical geometry. Very large changes in density are required to produce discernible changes in the emergent spectrum. Thus, there is nearly total insensitivity of the emitted spectrum to luminosity changes.

5.3.6 The radiation pressure

Although the details of the spectra are insensitive to the luminosity of the nova atmosphere, both the radiation pressure and the radiative acceleration *are* sensitive to the luminosity. In order to study the ejection mechanism of a nova shell we plot in Figure 5.8 the ratio of radiative acceleration $a_{\rm rad}$ to gravity $g = a_{\rm grav}$ for a 1.25 $M_\odot$ central object at a luminosity of 70 000 $L_\odot$ (which is slightly above the classical Eddington luminosity for a 1.25 $M_\odot$ CO white dwarf) for a number of different effective temperatures and for both

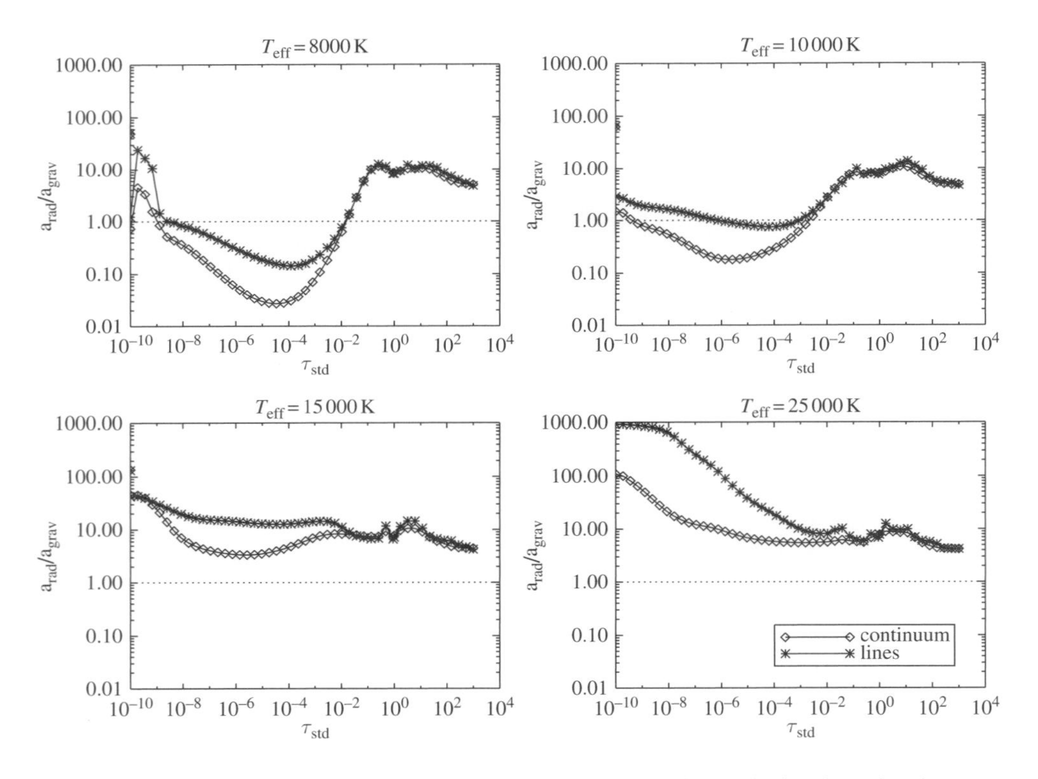

Fig. 5.8. The ratio of the radiation force per mass $a_{\rm rad}$ to the gravitational acceleration $a_{\rm grav}$ (assuming $L = 70\,000\,L_\odot$ and a 1.25 $M_\odot$ white dwarf) for four NLTE nova atmosphere models with a ballistic (linear) velocity law and $p = 3$. The curves annotated with diamond-shaped symbols give the results obtained by neglecting lines whereas the curves annotated with asterisks give the results obtained including both continua and lines. $\tau_{\rm std}$ is the absorption optical depth in the continuum at $\lambda = 5000$ Å. From Hauschildt *et al.* (1995).

continua and lines + continua. If the ratio is larger than unity, then the radiative acceleration is larger than the gravity and the material can be ejected by radiation alone. Generally, $a_{rad}/a_{grav} > 1$ for optical depths $\tau_{std} > 0.1$ at all T_{eff} shown in the figure. Only for $T_{eff} \leq 8000$ K do we find an outer zone for which the ratio is less than unity. In these zones, however, the material has already been accelerated to speeds larger than the escape velocity and has, in fact, been ejected. In the hotter models, the radiative acceleration is always larger than gravity even for a massive white dwarf. This means that (a) radiative acceleration alone can eject the nova shell for the conditions shown here, and (b) the velocity field inside the nova atmosphere may be changed with time owing to continuous radiative acceleration. The latter effect would be able to transform a ballistic velocity field to a 'wind' velocity profile.

5.4 Results: comparison with observations

5.4.1 The ONe nova V1974 Cyg

Because of the rapid response of observers and the density of observational coverage in time, the classical ONe nova V1974 Cyg is one of the best-observed novae in history (see Chapters 8, 9, and 10 in particular). The existence of a frequent, well-spaced set of low

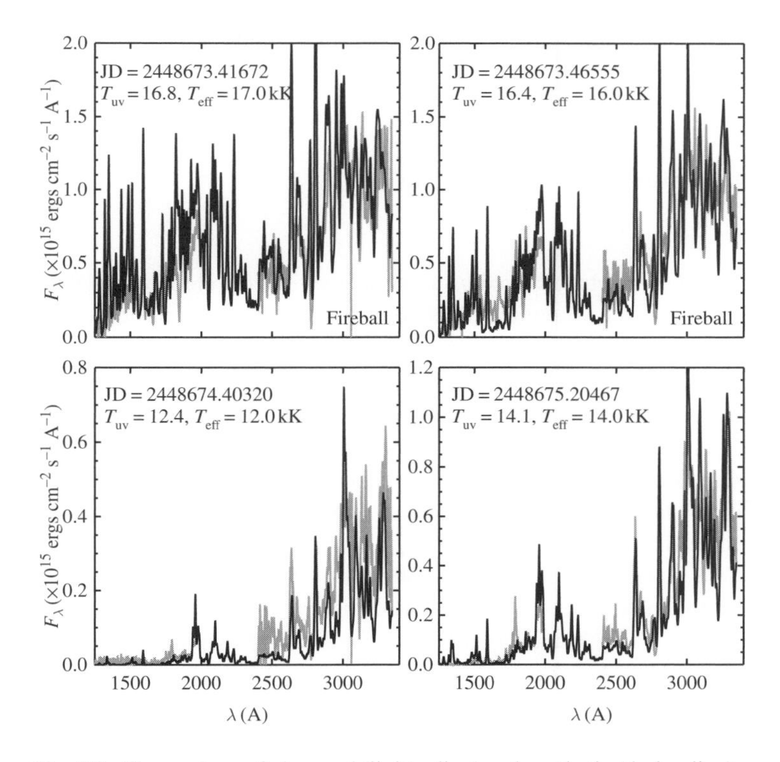

Fig. 5.9. Comparison of observed (lighter line) and synthetic (darker line) spectra at the first four phases of nearly simultaneous IUE low resolution observations of V1974 Cyg. The gap in λ for the observed spectra occurs because the reduced spectra in the 'SWP' and 'LWP' cameras are disjointed in λ. Each panel is labeled with the JD of the observed spectrum, the derived UV color temperature T_{UV} from the $(SWP - LWP)$ color, and the value of T_{eff} of the model used to synthesize the spectrum. From Short *et al.* (2001).

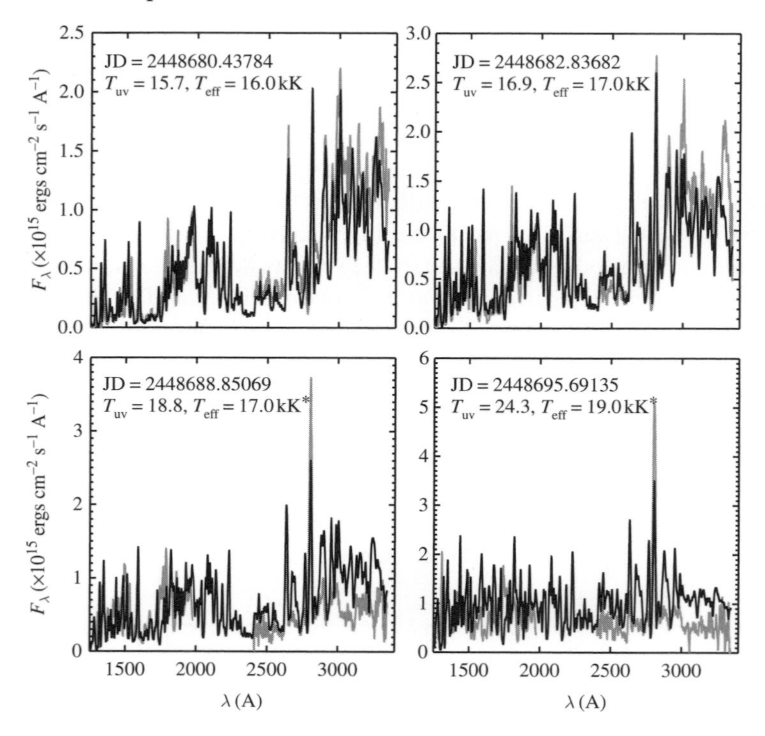

Fig. 5.9. *continued*

and high resolution IUE spectra allowed us to model the time development of the expanding gas shell in a spectral region where the object emits most of its flux during the fireball and optically thick wind phases of the explosion (Short *et al.*, 2001).

Figure 5.9 shows, for some of the available epochs, the comparison between the observed spectra and the synthetic spectrum for the model whose $T_{\rm eff}$ value is closest to the $T_{\rm UV}$ value of the observed spectrum. Because we do not know the angular diameter of V1974 Cyg, we arbitrarily adjust the flux level of the observed spectra to approximately match the absolute flux level of the synthetic spectra. However, note that, in keeping with the condition of constant $L_{\rm bol}$, the *same* scale factor has been used in all plots (i.e. individual pairs of observed and synthetic spectra have not had their flux calibration 'tuned'). The observed spectrum in each panel has a gap in the middle because the reduced spectra in the IUE SWP and LWP (Short Wavelength Prime and Long Wavelength Prime) cameras are disjoint in λ. For times later than JD 2448683, the Mg II $h + k$ resonance doublet at 2800 Å becomes increasingly strong in emission with respect to the pseudo-continuum, which indicates that the nova is becoming increasingly nebular at that time.

Figure 5.10 shows a comparison of the observed spectrum and synthetic spectra from the best-fit solar abundance model ($T_{\rm eff} = 14\,000$ K) and the best-fit hydrogen-deficient model (with a 1000 K higher effective temperature) at two times during the optically thick wind phase. From a visual inspection, both synthetic spectra provide approximately the same goodness of fit at both times. We conclude that the result of fitting hydrogen-deficient

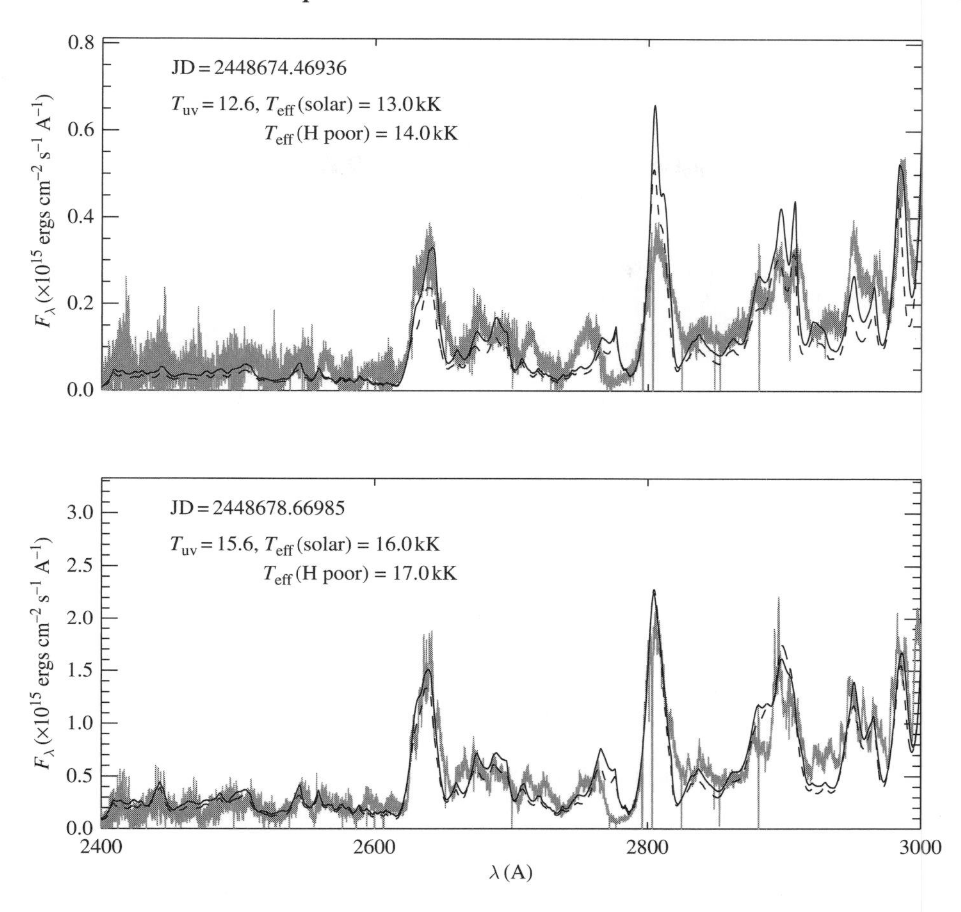

Fig. 5.10. Comparison of observed (darker line) and synthetic (lighter lines) high resolution IUE spectra at two representative times. Lighter solid line: [Fe/H] = 0 (solar abundance) models; lighter dashed line: [Fe/H] = 0.3 models. From Short *et al.* (2001).

models to the observed spectral sequence is to shift the derived $T_{\rm eff}$ evolution upward by approximately 1000 K. There are no spectral features in this range that distinguish between two such models. Therefore, there is a degeneracy in $T_{\rm eff}$ and $n_{\rm H}$, at least within the range of parameters explored here. Superficially, the degeneracy arises because a larger value of $T_{\rm eff}$ enhances the ultraviolet flux, whereas the increased value of [Fe/H] suppresses the ultraviolet flux. The net result is that increasing both in a particular proportion will lead to an approximately similar ultraviolet flux spectrum (see also Pistinner *et al.*, 1995, for a discussion of this effect).

5.4.2 CO novae in the LMC

Nova LMC 1988 #1 was a moderately fast, CO type, dust-forming classical nova and was the first extragalactic nova to be observed with the IUE satellite. The combined ultraviolet and optical spectra during the CML phase were best fitted by a synthetic spectrum with $T_{\rm eff}$ = 14 000 K (Schwarz, 1999). In Figure 5.11, we present the best-fit synthetic

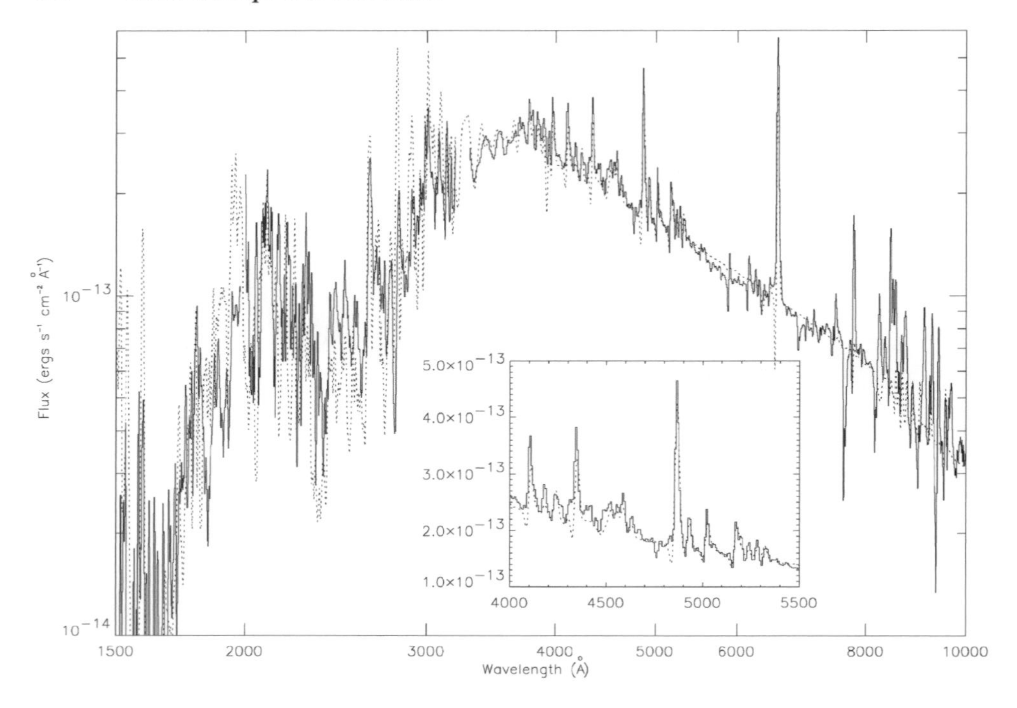

Fig. 5.11. The spectrum of nova LMC 1988 #1 on 1988 March 25 from the UV (SWP33158 and LWP12922) to the optical (Williams *et al.*, 1991) (scaled to match the flux level of the IUE spectra) compared to a nova model spectrum for $T_{\rm eff} = 14\,000$ K. From Schwarz (1999).

spectrum (dotted line) to the 25 March IUE and optical spectra (solid line). The optical spectrum was taken approximately 17 hours earlier than the IUE spectra. In order to account for any differences in flux between the two observations or any inaccuracies in the absolute flux calibration of the optical, we scaled the optical spectrum by 1.45 so that the spectra match in the region where they overlap (3200–3300 Å). The resolution of the synthetic spectra was degraded to the low resolution of IUE below 3200 Å and to the resolution of the optical observations above 3200 Å by convolving with a Gaussian kernel with a half-width corresponding to the IUE and optical resolutions, respectively.

Nova LMC 1991 was a very fast, classical nova and the brightest nova ever observed in the Large Magellanic Cloud. It was extensively observed during both its early optically thick and its nebular evolution in the optical and ultraviolet wavelength regions. The early spectra of nova LMC 1991 were analyzed by Schwarz *et al.* (2001). The first observation was obtained by Williams, Phillips and Hauy (1994), 4 days before visual maximum and 19 hours before the first IUE spectrum, still within the fireball phase of the nova outburst. The fit to the IUE and optical spectra of this phase is shown in Figure 5.12. This spectrum was taken just after visual maximum, and shows some dramatic changes when compared to the previous optical spectrum. The most obvious one is an increase in the emission line widths, which roughly double. The fits to these spectra are shown in Figure 5.13. The terminal component of the P Cygni profiles now indicates an expansion velocity exceeding $4500\,{\rm km\,s^{-1}}$. An increase in the expansion velocity with time has been observed in many novae. For example,

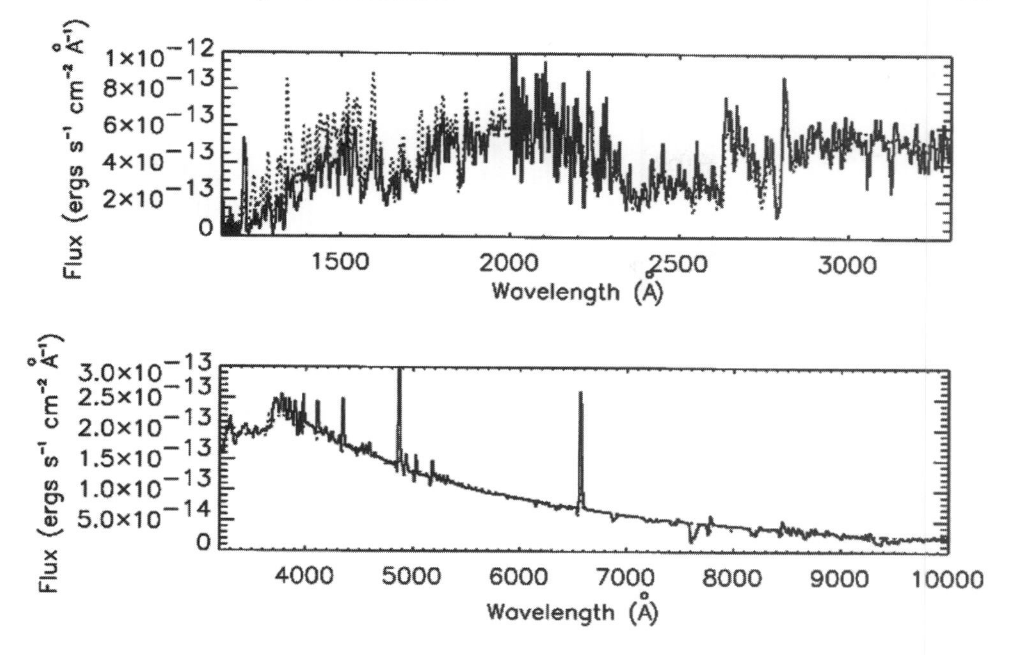

Fig. 5.12. Nova LMC 1991 during the fireball phase. Top panel: fit to the IUE spectrum of 1991 April 20.8 (solid line) with $T_{\rm eff} = 11\,000$ K; bottom panel: optical spectrum on 1991 April 20 compared to a synthetic spectrum for $T_{\rm eff} = 12\,000$ K. The models have the common parameters $p = 7$, $V_{\rm max} = 2000\,{\rm km\,s^{-1}}$, $Z = 0.1$ and CNO abundances increased by a factor of 10. From Schwarz *et al.* (2001).

the 'very fast' novae V603 Aql, CP Lac and GK Per showed a factor of 2 to 4 increase in velocity between pre-maximum and the t_2 time (Payne-Gaposchkin, 1957). The 1991 LMC nova showed a very low (even for the LMC) metallicity of about 1/10 solar, and a very large ejected mass; see Schwarz *et al.* (2001) for details.

5.5 Summary and conclusions

NLTE model atmospheres for novae show that the early evolution of a nova shell, the 'optically thick' phase, can be divided into at least three very different epochs. The first, and very short-lived, stage is the 'fireball' stage, first named by Gehrz (1988) and analyzed in the ultraviolet by Hauschildt *et al.* (1994a). In this stage, the density gradient in the nova atmosphere is high, $p \approx 15$ and the effective temperatures are dropping from $\sim 15\,000$ to $< 10\,000$ K (lower $T_{\rm eff}$ are probable but have not yet been observed). In this stage, the nova spectrum resembles that of a supernova with low velocities ($v_{\rm max} \approx 4000\,{\rm km\,s^{-1}}$ for V1974 Cyg).

As the density and temperatures of the expanding fireball drop, the material becomes optically thin and deeper layers become visible. In this stage, the 'optically thick wind phase', the atmosphere evolves to a very flat density profile, $p \approx 3$, and the LFR, as well as the CFR, has a very large geometrical extension. Values of $\Delta R/R \approx 100$ in the LFR are common. The large geometrical extension causes a very large temperature gradient in the LFR; typically the electron temperatures range between 3300 K and 35 000 K for a model with $T_{\rm eff} = 10\,000$ K

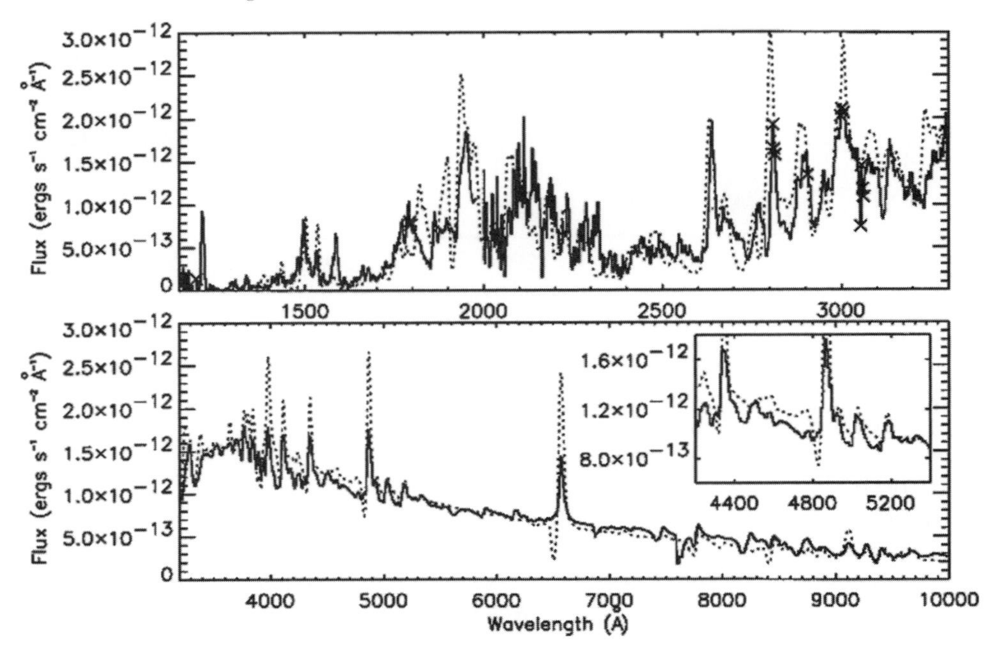

Fig. 5.13. Nova LMC 1991 during the continuous mass-loss phase. Top panel: fit to the IUE spectrum of 1991 April 25.7 (solid line) with $T_{\rm eff}$ = 14 000 K; bottom panel: optical spectrum on 1991 April 24.9 compared to synthetic spectrum for $T_{\rm eff}$ = 13 000 K. The models have the common parameters $p = 3$, $V_{\rm max} = 4000\,{\rm km\,s^{-1}}$, $Z = 0.1$ and CNO abundances increased by a factor of 10. From Schwarz *et al.* (2001).

and from 6700 K to 70 000 K for a model with $T_{\rm eff}$ = 25 000 K (we emphasize that all of the regions may be visible simultaneously in the emitted spectrum). This explains the observed fact that nova spectra can show, simultaneously, lines from different ions of the same element (see Chapter 2). The structure of the atmosphere and the calculated optical spectra are sensitive to changes of the abundances of iron, carbon, nitrogen, and oxygen, which makes abundance determinations of these elements possible. For reliable abundance determinations, NLTE effects need to be included self-consistently for all important opacity sources as well as the elements under consideration. Consistent with the results of Pistinner *et al.* (1995), we also find that the synthetic spectra are insensitive to large changes in the luminosity. However, the spectra are very sensitive to changes in the form of the velocity profile inside the atmosphere. In terms of the classical scheme of nova spectrum classification (McLaughlin, 1960), the increase in the effective temperature during the wind stage corresponds to the transition from the 'pre-maximum spectrum' to the 'diffuse enhanced spectrum' (see also Chapter 2).

The ultraviolet spectra of novae are totally dominated by the line blanketing of many thousands of overlapping spectral lines, mostly due to Fe ions. The line blanketing changes the structure of the atmosphere, and the emitted spectrum, so radically that it must always be included in the models in order to be able to compare the results to observed spectra. Furthermore, the strong coupling between continua and lines, as well as the strong overlap of lines, requires a unified treatment of the radiative transfer to (at least) full first order in V/c (i.e., including advection and aberration terms). Methods that separate continua and lines (e.g. the Sobolev approximation) give unreliable results. The 'density' of lines in the ultraviolet is so high that practically the complete range is blocked by lines. In fact, most of the observed 'emission lines' during the early phases of the outburst are merely gaps between 'clusters' of lines: thus they are regions of transparency which have less opacity than the surrounding wavelength regions. In addition, the radiation fields inside the nova continuum and line-forming regions nowhere resemble blackbody or grey distributions. This condition, combined with the low densities, causes very large departures from LTE. The NLTE radiative transfer and rate equations *must* be solved self-consistently including the effects of line blanketing in the ultraviolet and optical spectral regions. The effects of neglecting the line blanketing on the departure coefficients are enormous. In addition, NLTE has to be considered a zeroth-order effect in nova atmospheres.

In nova models with low effective temperatures, $T_{\rm eff} \leq 10\,000$ K, we find that molecular lines and bands are important. The synthetic spectra for the coolest model predict the presence of molecular lines in the optical and near-infrared spectra (Hauschildt *et al.*, 1995), as has been observed in a number of historical (e.g. DQ Her) and modern (e.g. V705 Cas) novae (see also Chapter 13).

The agreement between our synthetic spectra and observed nova spectra is very good from the ultraviolet to the optical. We are able to reproduce the high-resolution ultraviolet and low-resolution optical spectra of very different types of classical novae (cf. Hauschildt *et al.*, 1994a,b; Schwarz *et al.*, 2001 for quantitative comparisons between synthetic and observed nova spectra).

The basic physics and the modeling of early nova atmospheres is now well understood and we are currently working on further improvements of the models (e.g. extending the modeling into the X-ray regime) and on the analysis of those novae which have good spectral coverage in the ultraviolet and the optical. Another important step will be to investigate the effects of density inversions (clumps) on the emitted spectrum, and a more systematic investigation of the effects of different velocity fields on nova spectra is also required. Furthermore, we are currently working on a detailed treatment of the 'pre-nebula' phase, i.e. the phase where allowed, semi-forbidden and sometimes forbidden lines are simultaneously present in the observed nova spectra. This will not only require very detailed model atoms for the species showing these lines but will also extend the model atmospheres into later stages of the outburst.

Acknowledgements

It is a pleasure to thank S. Starrfield, S. Shore, E. Baron, and F. Allard for stimulating discussions. During this work I was supported in part by the Pôle Scientifique de Modélisation Numérique at ENS-Lyon. Some of the calculations presented here were performed at the Höchstleistungs Rechenzentrum Nord (HLRN), at the National Energy Research Supercomputer Center (NERSC), supported by the US DoE, and at the San Diego Supercomputer Center (SDSC), supported by the NSF. I thank all these institutions for a generous allocation of computer time.

References

Bath, G. T., & Shaviv, G., 1976, *MNRAS*, **197**, 305.
Gehrz, R. D., 1988, *ARA&A*, **26**, 377.
Hauschildt, P. H., & Baron, E., 1999, *J. Comp. Appl. Math.*, **102**, 41.
Hauschildt, P. H., & Wehrse, R., 1991, *JQSRT*, **46**, 81.
Hauschildt, P. H., Wehrse, R., Starrfield, S., & Shaviv, G., 1992, *Ap. J.*, **393**, 307.
Hauschildt, P. H., Starrfield, S., Austin, S. J. *et al.*, 1994a, *Ap. J.*, **422**, 831.
Hauschildt, P. H., Starrfield, S., Shore, S. N., Gonzales-Riestra, R., Sonneborn, G., & Allard, F., 1994b, *AJ*, **108**, 1008.
Hauschildt, P. H., Störzer, H., & Baron, E., 1994c, *JQSRT*, **51**, 875.
Hauschildt, P. H., Starrfield, S., Shore, S. N., Allard, F., & Baron, E., 1995, *Ap. J.*, **447**, 829.
Hauschildt, P., Baron, E., Starrfield, S., & Allard, F., 1996, *Ap. J.*, **462**, 386.
Hauschildt, P. H., Schwarz, G. J., Baron, E. *et al.*, 1997, *Ap. J.*, **490**, 803.
McLaughlin, D., 1960, in *Stars and Stellar Systems*, Vol. VI, *Stellar Atmospheres*, ed. J. L. Greenstein. Chicago: University of Chicago Press, p. 585.
Mihalas, D., Kunasz, P. B., & Hummer, D. G., 1976, *Ap. J.*, **206**, 515.
Payne-Gaposchkin, C., 1957, *The Galactic Novae*. New York: Dover.
Pistinner, S., Shaviv, G., Hauschildt, P. H., & Starrfield, S., 1995, *Ap. J.*, **451**, 451.
Schwarz, G. J., 1999, Multiwavelength analyses of classical carbon–oxygen novae. Unpublished Ph. D. thesis, Arizona State University.
Schwarz, G. J., Hauschildt, P. H., Starrfield, S. *et al.*, 1997, *MNRAS*, **284**, 669.
Schwarz, G. J., Shore, S. N., Starrfield, S. *et al.*, 2001, *MNRAS*, **320**, 103.
Shore, S. N., Schwarz, G., Bond, H. E. *et al.*, 2003, *AJ*, **125**, 1507.
Short, C. I., Hauschildt, P. H., & Baron, E., 1999, *Ap. J.*, **525**, 375.
Short, C. I., Hauschildt, P. H., Starrfield, S., & Baron, E., 2001, *Ap. J.*, **547**, 1057.
Williams, R. E., Hamuy, M., Phillips, M. M. *et al.*, 1991, *Ap. J.*, **376**, 721.
Williams, R. E., Phillips, M. M., & Hamuy, M., 1994, *Ap. JS*, **90**, 297.

6

Observational mysteries and theoretical challenges for abundance studies

Jordi José and Steven N. Shore

6.1 Introduction

Unlike core collapse supernovae, we have no observation of the initial moments of a classical nova outburst. But, as described elsewhere in this volume, we can reconstruct – at least in a broad sense – the explosive and mixing phenomena because of the comparative simplicity of the ejection process. The thermonuclear runaway is the foundation of our understanding. Yet many details of the dynamics remain to be explored and some of the signposts for change are observational since at the moment of ejection, the nucleosynthetic products are contained in – and expelled with – the ejecta. They provide an interpretable record of the stages of the explosion that are otherwise inaccessible to direct observation. The ambiguities of this record, understood only through modeling, will be the subject of our discussion in this chapter.

Decoding the abundance patterns in various subclasses of novae and individual objects has proven a great challenge over the decades since the first edition of this book appeared. Considerable progress has been made in obtaining consistent pictures from a variety of analyses, as we will discuss. Here we will highlight a few of these methods, examine their various limitations and systematic uncertainties, and indicate some directions that new methods of spectroscopic analysis and theoretical modeling can take to improve the picture in the near and long term.

6.2 Photo-ionization analysis of the ejecta

6.2.1 Masses and structures

The need for new approaches to the analysis of ejecta and the calculations of the details of the explosions is highlighted by the structures observed during the earliest stages of the outburst. As discussed in other chapters, because the velocity distribution in the ejecta is simple, a line profile contains a surprising amount of information that can be relatively easily interpreted. If we look at any collection of spectra taken with even modest resolution, the data unambiguously indicate that the ejecta are frequently axially symmetric. Whether this is a signature of jets, annuli, or some combination, is not clear. But the treatment for many systems has employed spherical symmetry and that is clearly not the case in general. What

Classical Novae, 2nd edition, ed. Michael Bode and Aneurin Evans. Published by Cambridge University Press.

this means for the mass determinations is also unclear, but it poses a serious challenge for theoretical models of the explosion dynamics. A moment's reflection on the images of HR Del, obtained with WFPC-2 on the Hubble Space Telescope (see Figure 12.2 in Chapter 12), should prove convincing that this is indeed a very difficult problem (see also the discussion in Chapters 7 and 12 of structures in the resolved radio and optical remnants). Individual regions, all of which leave their imprint at specific projected velocities on the line profile, may have very different ionization depending on the local density variations. For example, note that the [O III] lines are typically formed on the convex surfaces of the knots facing the central star, but these are seen in projection and may not be at unique radial velocities, depending on the geometry of the ejecta.

What imposes this symmetry and when? From optical Balmer line profiles, it is likely that some of the structure is due to changes in the optical depth of the transition since even when no P Cygni trough is observed the profiles are often weaker on the negative velocity side at the first observations. Some profiles then switch their shape from stronger red to stronger blue peaks (e.g. V382 Vel) but others become more symmetric and then stay stable (e.g. V4160 Sgr). Almost all have double peaks, which for an annulus with a constant velocity gradient indicates shell thicknesses $\Delta R/R < 1$. A further complication comes from the changes in the optically thin transitions, which may show spherical symmetry at the beginning of the outburst followed by more axisymmetric profiles during the transition to the nebular stage (e.g. V1187 Sco). That this structure persists throughout the ejection is shown by the persistence of the behavior across frequency; the infrared profiles show the same structure earlier than the optical as would be expected if the only contaminating effect is the opacity. As the profiles show this development from the earliest observed stages of the outburst, it must be imposed at the source and during the ejection since, once the ejecta are freely expanding, subsequent structural changes should not occur in the highly supersonic gas. A variety of instabilities recommend themselves, including those driven by strong density gradients during the earliest stages of the expansion (e.g. Richtmyer–Meshkov), buoyancy-driven (Rayleigh–Taylor), or a combination of these present in detonations (e.g. Landau–Darieus), but none is certainly responsible and their elucidation in the multi-dimensional models is still in its infancy (see below).

The profile fine structures present another important puzzle and clue. In most abundance studies to date, the filling factor – the fraction of the shell volume occupied by gas – is taken as a free parameter. In photo-ionization codes, there is no gas between the hypothesized clumps so the only effect is to reduce the absorption efficiency and thus the luminosity ratio of the lines to the central star. In some analyses this has been extended to include a covering factor that combines the emission from two regions of very different density but, in general, the analyses ignore the detailed information in the profiles and attempt to match the integrated line fluxes with two components whose properties are not well constrained. If the structure is imposed at the explosion, and the evidence is strong that this is the case, then there may also be differences in the composition that reflect the mixing processes active during the ejection. It is also possible that the symmetries of the different components may differ. At least for V1974 Cyg the lowest density gas seems to be more nearly spherical while the later, narrower profiles that show the previously mentioned axisymmetry also show much more fine structure that repeats among the profiles at the same velocities. Different densities also mean the recombination times may be drastically different (since this depends on the density) and this should be included in both the abundance and mass estimates. Only V1974 Cyg was followed at high resolution in the resonance lines for long enough to show the effect, but it did occur after the cessation of nuclear burning on the WD.

Is it possible to discriminate between jets, annuli, and more complex structures in the ejecta from the line profiles alone (perhaps using something like Doppler tomography)? How can photo-ionization codes be extended to include the effects described? It is not merely propaganda for a new space-based effort to lament the lacuna in the foreseeable future of high resolution ultraviolet spectroscopy, but for novae, and many other astrophysically important problems, this intermediate energy region is the seat of essential information that now can only be reconstructed by analogy with previously observed objects.

6.2.2 Techniques

Novae are dynamical from the start and, although their spectra appear to be similar to H II regions and planetary nebulae at various stages of the outburst, the formation of the spectrum occurs under very different conditions. The central WD continues to irradiate the ejecta while it gradually relaxes toward equilibrium. At the earliest phase of the ejection, when the material is completely optically thick, the spectrum is formed similarly to a stellar wind and, indeed, the closest analog to the *iron curtain* phase is that of luminous blue variables (LBVs, also called S Dor stars) in outburst; these are massive supergiants whose winds form a pseudo-photosphere. Unlike the LBVs, however, the optically thick surface progresses inward within the ejecta and because of the finite duration of the ejection event and fixed mass of the shell, the material eventually becomes transparent, marking the start of the *nebular* phase. This latter stage is the one generally used in emission line studies based on analogies to either static H II region or coronal conditions (e.g. Williams *et al.*, 1981; Andreä, Drechsel and Starrfield, 1994) in which there is either continual heating by ultraviolet irradiation or distributed non-thermal (magnetic) sources such as MHD turbulence, or simple static cooling (e.g. electron conduction and/or radiation). Novae, in contrast, freeze in their ionization at the peak of the hard radiation just before the central source turns off but, because of the expansion, the density drops sufficiently rapidly to slow the recombination and maintain a state of ionization far higher than the contemporary radiation warrants. For this reason, analyses of the first epoch require an active ultraviolet source, whatever its input spectrum, that is different from the later stages. In practice, the actual ionization state of the gas is difficult to assess for much of the expansion because the census of available ionic lines depends on the wavelength region and the species. For instance, for most outbursts, before the coronal line phase the gas may be four or five times ionized but intensities can be measured for only lower ionization states. Therefore it is often necessary to correct for the unobserved/unobservable fraction of the element, a technique called the ionization correction factor (ICF) (see Andreä, Drechsel and Starrfield (1994) for an extensive application to nova spectra). Classical photo-ionization analyses have been carried out using *MAPPINGS*, *CLOUDY* and *NEBU* (see Ferland (2003), and Osterbrock and Ferland (2006), for general reviews of existing photo-ionization codes).

To solve for the structure and abundances simultaneously, an alternative method is to use an iterative least squares minimization for all lines together rather than individually, with the input parameters being the reddening and the ejecta structure. This has been used successfully for a number of studies with the algorithm *MINUIT*, adapted from the CERN library, being used for the parameter search, but other aggregate analytical methods can also be used (e.g. those already included in *CLOUDY*).

Recent progress has been made in three dimensional radiative transfer which, at this stage, must still be performed using Monte Carlo methods (e.g. Och, Lucy & Rosa, 1998;

Lucy, 1999, 2001, 2003; Ercolano, Barlow & Storey, 2005; Morisset, Stasinska & Peña, 2005). Such codes, while accurate and able to reproduce much of the observed structure, cannot yet cope with the known chemical inhomogeneities. While not yet applied to novae, a recent advance has been made in the analysis of complex H II regions in which chemical inhomogeneities are certainly present (Tsamis & Péquignot, 2005). Pertinent to the ejecta problem, they identify a fundamental bias in the He abundances of H II regions when the composition is not uniform. The same technique should be used as well for novae, as highlighted by the spatially resolved studies of V1974 Cyg (for which the Ne/C ratio varies by a factor >2 between knots, while the He/H ratio appears to be roughly constant; Shore *et al.* (1997)). A fundamental problem in the current models is still the need, mainly because of the limitations of available computational facilities, to restrict the degree of inhomogeneities present in the ejecta. This is not likely to be overcome soon because, although the geometry can be treated self-consistently within a Monte Carlo environment, the introduction of both small-scale structure and abundance variations introduces too many free parameters for the current models. This can be treated statistically only by computing separately models for a variety of conditions – varying the density and abundance set – for plane parallel structures embedded in a diffuse background from scattering and then folding them into a set of line profiles that reproduce the observed structures.

6.2.3 Abundances

Elemental abundances are most often quoted only relative to some standard *solar* values, usually the compilation by Grevesse and Noels (1993) or Grevesse, Noels and Sauval (1996) but in the pre-1994 literature are referenced to Anders and Grevesse (1989). However, the solar values have recently undergone significant revisions (see Asplund, 2005; Asplund, Grevesse and Sauval, 2005, especially for CNO and NeMg), thus changing the *overabundances* quoted in the literature. For instance, the range in $(O/H)/(O/H)_\odot$ is about a factor of 1.5 but since the solar value has now been reduced by about 0.3 dex, this increases the oxygen abundances accordingly. Carbon and nitrogen have also been revised downward, although to a lesser extent for nitrogen. Because the O/Ne ratio has been fixed independently of the oxygen abundances,[1] the solar neon abundance has been decreased by the same factor as oxygen. The summary effect is a change of a factor of about 0.1 to 0.3 dex in the C N O Ne Mg overabundances. Thus, *correcting for the new solar values*, it is instructive to compare the range in a single sample of novae – the Galactic ONe class – done homogeneously using *CLOUDY* (see Vanlandingham *et al.* (2005) and references therein). For V1974 Cyg, Vanlandingham *et al.* (2005) find overabundance factors of $He/H = 1.2 \pm 0.2$, $C/H = 1.0 \pm 0.2$, $N/H = 45 \pm 11$, $O/H = 22 \pm 7$, $Ne/H = 71 \pm 17$, $Mg/H = 5.2 \pm 3$, and $Al/H \geq 1$; the study also reports an Fe/H enhancement of 4.9 ± 4 relative to solar (the value of which is not changed in the latest compilations), indicating a significant hydrogen deficiency in these ejecta. In a phenomenologically almost identical nova, V382 Vel, Shore *et al.* (2003) found enhancement factors of $He/H = 1.0 \pm 0.1$, $C/H = 0.9 \pm 0.3$, $N/H = 17 \pm 4$, $O/H = 5.8 \pm 0.3$,

[1] There is a problem for neon that has yet to be resolved at the time of writing. In the Asplund, Grevesse and Sauval (2005) compilation the O/Ne ratio was fixed at the older solar value, 6.6, so it is assumed that the neon abundance should decrease by the same factor as oxygen. It is possible, however, that this ratio is different for Galactic B stars (Cunha, Hubeny & Lanz, 2006): their *derived* Ne abundance is about 1.9 times the assumed solar Ne, with an attendant smaller O/Ne ratio ~ 3.5. Given the uncertainty, we have chosen to use the published solar ratio but the reader should be warned this may change.

Table 6.1. *Sample overabundances: photo-ionization analyses*

Nova	He	C, N, O	Ne, Mg, Al, S
V1974 Cyg	1.2±0.2	1.0±0.2, 45±11, 22±7	71±17, 5.2±3, ≥1, ...
V382 Vel	1.0±0.1	0.9±0.3, 17±4, 5.8±0.3	29±3, 2.9±0.1, 7±2, ...
V693 CrA	1	4, 13, 24	420±200, 9, 76, ...
QU Vul	1.2±0.1	0.3±0.1, 12±0.6, 4.3±1.3	36.9±1.7, 11±5, 67±15, ...
V838 Her	1.4±0.1	7.5±0.5, 24.5±3.4, 2.0±0.3	182±2, 1.5±1, 29±17, 27.9±2.7
CP Cru	1.6	..., 1.7, 9	41, ..., ..., 1.4
LMC 1991	0.8±0.2	7±2, 85±20, 11±2	..., ..., ..., ...
GQ Mus	2.7	7.4, 316, 36	4.5, 2.2, ..., 3.2

Ne/H = 29 ± 3, Mg/H = 2.9 ± 0.1, Al/H = 7 ± 2, and Si/H = 0.6 ± 0.3, again correcting for the new solar values. For V693 CrA, Vanlandingham *et al.* (1997) found He/H = 0.1, C/H ≈ 4, N/H ≈ 135, O/H ≈ 24, Ne/H ≈ 420 (exceptionally large but with a considerable uncertainty of 50%), Mg/H ≈ 9, Al/H ≈ 76, and Si/H ≈ 24. For QU Vul, an unusually prolific dust-forming ONe nova, Schwarz (2002) shows the results of intercomparison of a number of methods. The importance is that, like few others, QU Vul was well observed throughout the outburst at almost all portions of the spectrum (lacking only a far ultraviolet constraint). A maximum likelihood set of abundances were derived using *CLOUDY* as with many of the novae we have described here, with the results being enhancements of He/H = 1.2 ± 0.1, C/H = 0.3 ± 0.1, N/H = 12 ± 0.6, O/H = 4.3 ± 1.3, Ne/H = 36.9 ± 1.7, Mg/H = 11 ± 5, Al/H = 67 ± 15, Si/H = 2.2 ± 1.6, and Fe/H = 0.53 ± 0.03. This last value is exceptional for a Galactic nova and suggests that at least some of the refractory elements were actually depleted by the dust formation. However, the results show possible systematic biases introduced by the various corrections for incomplete ionic samples. While He agrees between the ICF and full analyses (at least within a factor of 20%), C/H changes by a factor of 5 between analyses, N/H by a factor of 4, and Ne/H by almost a factor of 10. Under such conditions, it is difficult to assess – other than a biased choice – which analysis to accept since inherently any derived metallicities for the ejecta are purely model-dependent. A particularly useful constraint again comes from the Fe/H ratio, which shows that a large part of the variance comes from a variance of more than a factor of 5 in Fe/H. Finally, for the most extreme ONe nova, V838 Her, the enhancements were greatest for the elements above the CNO group: He/H = 1.4 ± 0.1, C/H = 7.5 ± 0.5, N/H = 24.5 ± 3.4, O/H = 2.0 ± 0.3, Ne/H = 182 ± 2, Mg/H = 1.5 ± 1, Al/H = 29 ± 17, Si/H = 7 ± 2, and S/H = 27.9 ± 2.7. The Fe/H enhancement was 1.3 ± 0.4, suggesting that again the hydrogen is depleted. Notably, this nova, the most rapidly expanding Galactic ONe nova (based on the maximum velocity, 7000 $\mathrm{km\,s^{-1}}$, and t_2 = 4 days), had an exceptionally pronounced spectrum of S II and S III and appears to have processed the ejecta well above the usual group seen in the ONe novae. In Table 6.1 we collect these results for convenience.

Unfortunately, there are few independent abundance analyses for most novae, and in general the intercomparisons are hampered by the partial data sets or incompatible methods of analysis. The largest compilation for CO novae remains that of Andreä, Drechsel and Starrfield (1994) using the two methods of ionization correction factors and ‘differential analysis' using species with similar ionization potentials, but the ranges in the quoted abundances remain large. This comparison should serve as a guide to the remaining range in uncertainties among the abundance studies. For V1974 Cyg, for example, the only independent analyses are by Paresce *et al.* (1995) using *CLOUDY* for one epoch and finding $Ne/Ne_{\odot} \geq 25$, $3 < O/O_{\odot} < 6$, and He enrichment by perhaps a factor of two. Hayward *et al.* (1996) and Moro-Martín, Garnavich and Noriega-Crespo (2001) provide only partial analyses but a comparison of Hayward *et al.* (using infrared lines) and Vanlandingham *et al.* (2005) is particularly important. The derived Ne and Mg enhancements agree to within 20%, the only species for which independent line measurements were made in different spectral ranges.

The most comprehensive study of a single CO nova to date with photo-ionization modeling is GQ Mus, for which an extensive set of ultraviolet and optical observations spanned the entire outburst well into the nebular and coronal stage. Morisset and Péquignot (1996) find elemental enhancements by number, correcting for the new solar abundances, of He/H = 2.7, C/H = 7.4, N/H = 316, O/H = 36, Ne/H = 4.5, Mg/H = 2.2, S/H = 3.2, and Fe/H$\approx$He/H, supporting the contention that the ejecta were helium-enriched. In the optical, the absence of carbon lines hampers many comparisons with the ultraviolet-based studies but recent results have been consistent with the general enhancement of CNO by an order of magnitude relative to solar composition.

A few objects stand apart from the usual classification. CP Cru, called a transition object between CO and ONe novae by Lyke *et al.* (2003), displayed enhancement factors relative to solar of He/H = 1.6, N/H = 1.7, O/H = 9, Ne/H = 41 and S/H = 1.4. Other than individual nebular stage studies or the Andreä, Drechsel and Starrfield (1994) collation, the CO group has been little studied in a systematic way. In general the abundances of CNO are significantly enhanced (>5 times) relative to solar. For the LMC nova LMC 1991 (one of the most intrinsically luminous novae ever observed), He/H $= 0.8 \pm 0.2$, C/H $= 7 \pm 2$, N/H $= 85 \pm 20$, and O/H $= 11 \pm 2$ relative to solar values; the other elements were approximately consistent with the LMC (Schwarz, 2002). Both the Galactic recurrent nova U Sco (Iijima, 2002) and the LMC nova LMC 1990 no. 2 (Shore *et al.*, 2003) had He/H ratios above solar by about 50% but more complete analyses are hampered by the differences among the circumstellar environments and the speed of the events: those with red giant companions are dominated in the nebular phase by emission arising from the ionized wind, while there are few ultraviolet observations of the compact systems in outburst.

An important constraint is available for novae based on our understanding of the nucleosynthesis that occurs during the TNR. The temperatures and durations are not sufficient to reach into the iron peak ($Z \geq 20$) and, therefore, the iron abundance provides a normalization for any photo-ionization analysis, except when dust is important and some elemental depletion can occur. The two novae for which we report examples of the number density abundances, V382 Vel and V1974 Cyg, were nearly identical ONe novae that were observed with similar resolution over the entire spectral range, from the far ultraviolet through the visible and near infrared. Although lacking the time coverage of V1974 Cyg, the fact that high resolution *FUSE* spectra were available for V382 Vel makes this comparison especially significant. When the Fe is overabundant relative to hydrogen, this probably indicates a hydrogen deficiency from which we can conclude that many of the ejecta are actually helium rich, and that the mass-losing companion has either been stripped or is more evolved than main sequence, or perhaps both. For understanding the dynamics of the ejection and the possible importance of a non-solar composition for the accreted mass, this is especially significant and is a frequently seen feature of the spectra.

Shell structures, as we have seen from the previous discussion of line profiles, can be extraordinarily complicated. The resolved ejecta illustrate this dramatically, particularly HR Del for which [N II], [O III], and Hα images are available with $0.1''$ resolution. The knots inferred from line profiles, possibly due to abundance variations but certainly indicating variations in the local densities and geometry of the shells, have been well studied in unresolved cases. Returning now to the question of structure of the ejecta at early times, a remaining spectroscopic 'mystery' is the persistence of O I 1304 Å during the early outburst. This line, a resonance transition, remains one of the strongest in the near ultraviolet throughout the iron curtain and early nebular stages; in fact, because of interstellar absorption, the strength of the line is probably somewhat underestimated. Williams (1992) proposed that the ejecta display large density contrasts with the neutral species surviving in the shadows created by denser clumps. It is now clear that this suggestion is correct and must be accounted for in any photo-ionization analysis. The detection of such shadows is rendered easier by the simplicity of the velocity structure of the ejecta in the sense that individual knots can be identified in each ionization stage and the relative ratios computed as a function of their velocity. This procedure, used frequently for the analysis of interstellar absorption lines, could be adapted to the nova case by using the thermal line width as the standard or separating the portions of the ejecta that are clearly moving differently. Abundance inhomogeneities are, however, very difficult to detect and separate from simple ionization effects of this sort without either very high resolution time sequences or spatially resolved spectra (e.g. Shore *et al.*, 1997).

One last point should be mentioned, that there is now an indication that even well into the nebular stage the ejecta are not yet completely frozen in their ionization state. Infrared observations of the ONe nova V382 Vel about 5000 days after optical maximum indicate the infrared neon ions are still being powered by recombination and that this occurs very inhomogeneously within the ejecta. For V838 Her and other fast novae, not observable at this late stage, this is also true. Thus any late stage analysis will have to take into account, at least statistically, this effect of structure on the line intensities. A proper treatment, still some time away, should reduce considerably the observational dispersion resulting from the incomplete treatment of ejecta filamentation and dynamics.[2]

6.3 Dust in novae: some open questions

6.3.1 Formation mechanisms: nucleation and kinetic processes

As discussed in Chapters 8 and 13, many (although not all) types of classical novae form dust at some time during the expansion. The same can be said of supernovae and the same problems arise in understanding the process. Chemically, the phase most favored for formation of dust is one in which the carbon and/or silicon is supersaturated. In homogeneous

[2] A difficulty is that the velocity field, while essentially monotonic, is not unique. This is unambiguous only under limited circumstances, e.g. when during the optically thick stage individual filaments are seen in absorption that, later, coincide with isolated, narrow emission peaks on the nebular phase line profiles. Some improvement is afforded by analyzing the lines using two components separated on the basis of positive or negative radial velocity relative to line center and then performing a complete photo-ionization analysis of each individually, but because different transitions have different density and temperature dependences, even this method will not mine all the information of the thermodynamic and abundance diagnostics.

nucleation theory, the dependence of the nucleation rate on this parameter is extremely sensitive to minor changes, regardless of the potential used to model the interactions among the atomic species. Interstellar grains are known to grow by kinetic processes, charged particle interactions with the grains that are themselves charged by the diffuse interstellar radiation field. The metallic depletions seen in this diffuse environment indicates the relative efficiency of the process of adsorption.

Novae present an ideal medium in which to condense solids, but it is not clear if the process is in equilibrium or essentially kinetic. Several models have been proposed, beginning with the analysis by Clayton and Hoyle (1976), that are mainly chemical and involve some form of nucleation. How the dust forms is an especially important issue for understanding and studying nucleosynthetic yields by comparing pre-solar grain abundances with explosion models. The dust thus formed may be extremely large, as in the case of the luminous blue variables for which the reflection spectra of the stellar photosphere have been detected in some of the dusty rings surrounding the star (e.g. AG Car, R127 in the Large Magellanic Cloud), while there is still some uncertainty about the size distribution of grains in the nova case. X-ray scattering from the ejecta and infrared emission indicate that the grains may be normal size for interstellar dust, less than 0.1 μm (Evans *et al.*, 2005) even though observations of the dust formed during the V705 Cas outburst show the near ultraviolet to have been nearly grey. The presence of amorphous phases, and mixtures of silicates and carbonaceous grains, complicates the picture (Joiner, 1999; Pontefract and Rawlings, 2004, and references therein). An important constraint is the observation of CO overtone emission in V705 Cas (Evans *et al.*, 1997); there are even earlier reports of molecular absorption, the optical CN band (λ3883, 4216 Å) in DQ Her during what would now be called the iron curtain stage (e.g. Wilson & Merrill, 1935; Barbier & Chalonge, 1940; Antipova, 1969; see also Payne-Gaposchkin 1957, and references therein). Dust formation is especially important for understanding and studying nucleosynthetic yields by comparing pre-solar grain abundances with models. An important diagnostic, to date little applied to novae in early stages of outburst, will be spectropolarimetry – especially in the longer wavelength parts of the optical and near-infrared spectrum and during the early stages of the outburst where the pseudo-photosphere is seen through the part of the ejecta where the dust is most likely to form, to study the grain size distribution and symmetries.

6.3.2 Pre-solar grains: where do we stand?

Pre-solar grains, found in primitive meteorites, are characterized by huge isotopic anomalies that are usually explained in terms of nucleosynthetic processes in their stellar sources. So far, silicon carbide (SiC), graphite (C), diamond (C), silicon nitride (Si_3N_4), and oxides have been identified as pre-solar grains. Isotopic abundance ratios obtained by ion microprobe analyses have helped to attribute a likely paternity for individual grains, often related to AGB stars and supernovae progenitors (Zinner, 1998).

Infrared (Evans, 1990; Gehrz *et al.*, 1998; Gehrz, 1999) and ultraviolet observations (Shore *et al.*, 1994) of the evolution of nova light curves have shown that novae form grains in their expanding ejected shells. Apparently, ONe novae are not so prolific dust producers as CO novae. The reason for this is not well known but it has been suggested that the lower-mass, high-velocity ejecta in ONe novae could hinder the condensation of appreciable amounts of dust. Observations of the condensation of dust containing different species, such as silicates, SiC, carbon, and hydrocarbons, have been reported for a number of novae (Gehrz *et al.*,

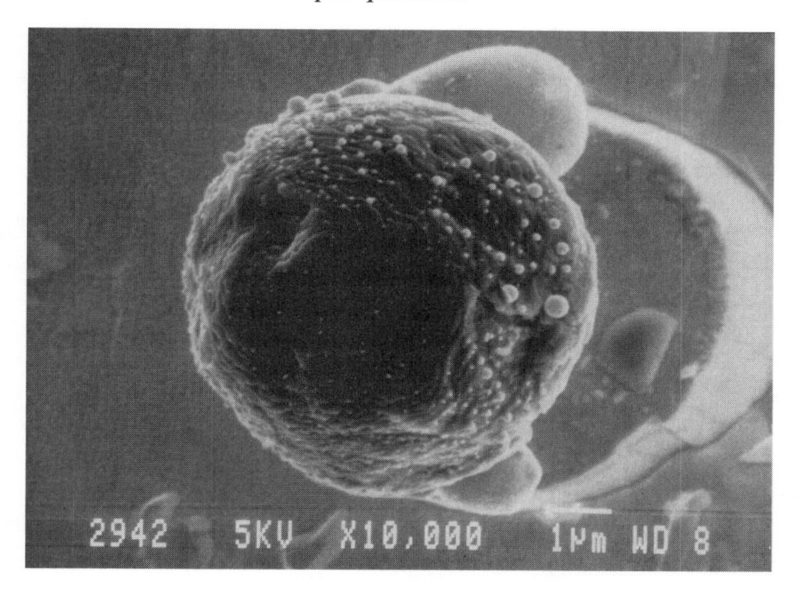

Fig. 6.1. Image of the nova candidate graphite grain KFC1a-511 after secondary ion mass spectrometry (SIMS) analysis. Image courtesy of S. Amari.

1998). Until recently, it was generally believed that C > O was needed for the formation of SiC and/or graphite grains: if oxygen is more abundant than carbon, essentially all C is locked up in the very stable CO molecule, and the excess O leads to formation of oxides and silicates as condensates. On the other hand, if carbon is more abundant than oxygen, essentially all O is tied up in CO and the excess C can form reduced condensates such as SiC or graphite. However, it seems that there are ways to overcome such constraints: first, recent studies have shown that the presence of Al, Ca, Mg, and Si in a C<O environment (such as in ONe novae) can dramatically affect the chemistry of C and O, leading to the formation of C-rich (SiC) and O-rich dust (corundum (Al_2O_3), enstatite ($MgSiO_3$)) simultaneously (José *et al.*, 2004); and second, the process of condensation in a nova environment is likely to take place kinetically (rather than at equilibrium) through induced dipole reactions (Shore & Gehrz, 2004), hence reducing the role of the CO molecule in the process.

Since the first studies of dust formation in classical novae (Clayton & Hoyle, 1976), the identification of pre-solar nova grains from meteorites has only relied on low $^{20}Ne/^{22}Ne$ ratios (with ^{22}Ne attributed to ^{22}Na decay; see Amari, Lewis and Anders (1995)). Recently, five SiC and two graphite grains that exhibit several isotopic signatures characteristic of nova nucleosynthesis have been isolated from the Murchison and Acfer meteorites (Amari *et al.*, 2001; Amari, 2002): the SiC grains have very low $^{12}C/^{13}C$ and $^{14}N/^{15}N$ ratios, while the graphite grains have low $^{12}C/^{13}C$ but normal $^{14}N/^{15}N$ ratios. However, the original $^{14}N/^{15}N$ ratios of these two graphite grains could have been much lower, because there is evidence that indigenous N in pre-solar graphites has been isotopically equilibrated with terrestrial nitrogen; $^{26}Al/^{27}Al$ ratios have been determined only for two SiC grains (KJGM4C-100-3 and KJGM4C-311-6) and are very high ($>10^{-2}$); $^{20}Ne/^{22}Ne$ is only available for the graphite grain KFB1a-161 (<0.01). The low ratio suggests that ^{22}Ne most likely originated from *in*

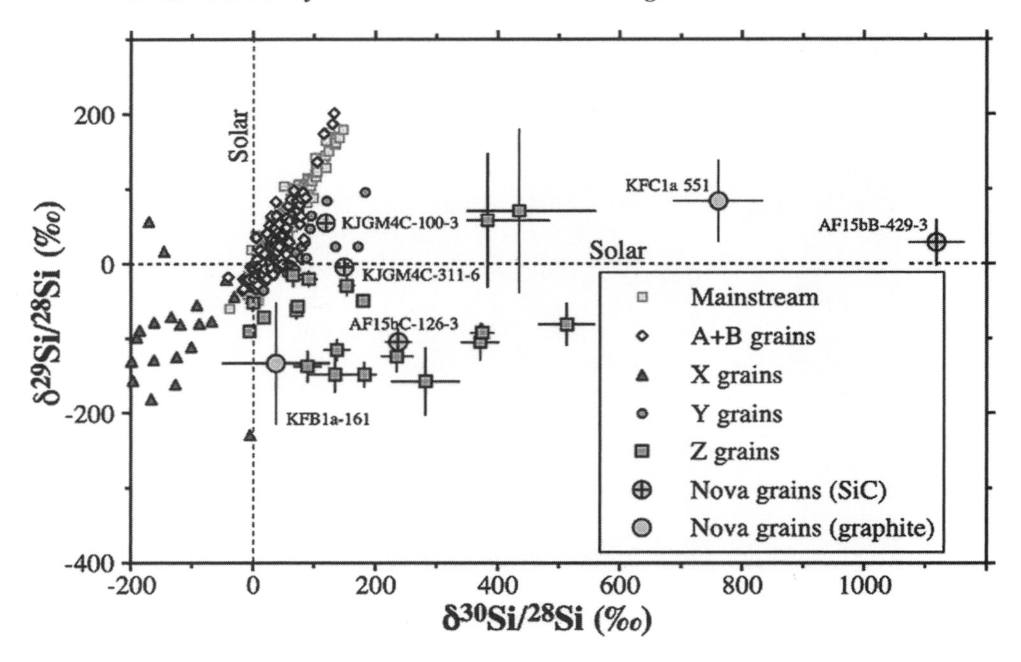

Fig. 6.2. Silicon isotopic ratios of pre-solar silicon carbide (SiC) grains, including nova candidate grains (SiC and graphite). Ratios are plotted as δ-values, deviations from Solar System values ($^{29}Si/^{28}Si = 0.0506$; $^{30}Si/^{28}Si = 0.0334$) in per mil. Error bars are smaller than the symbols. From José *et al.* (2004).

situ decay of ^{22}Na. Silicon isotopic ratios of the five SiC grains are characterized by ^{30}Si excesses and close-to- or slightly lower-than-solar $^{29}Si/^{28}Si$ ratios (see Figure 6.2). All in all, the isotopic signatures of these grains qualitatively agree with current predictions from hydrodynamic models of nova outbursts. In fact, a detailed comparison between grain data and nova models suggests that these grains probably formed in an ONe nova with a white dwarf mass of at least 1.25 $M_\odot$ (Amari *et al.*, 2001). However, two main problems, related to the paternity of these grains, remain to be solved: first, the connection with an ONe nova is challenging because, as stated before, these are not prolific dust producers, and second, in order to match the grain data quantitatively, one has to assume a mixing process between newly synthesized material in the nova outburst and more than ten times as much unprocessed, isotopically close-to-solar, material before or during grain formation.

Concerns about the possible nova paternity of these grains have been raised by Nittler and Hoppe (2005) on the basis of new measurements of three micrometer-sized SiC grains isolated from the Murchison meteorite. These three grains have very low $^{12}C/^{13}C$ and $^{14}N/^{15}N$ ratios, similar to the previous pre-solar nova grain candidates, but also additional imprints that point instead toward a Type II supernova origin: grains 347-4 and 151-4 are, indeed, highly enriched in ^{30}Si as well as overabundant in one or more Ti isotopes. It is not clear if measurements in these grains can be used to infer the origin of other grains for which no Ti determinations were done. Indeed, even a potential supernova origin for these grains is puzzling: no single supernova zone is predicted to have ^{30}Si enhancement and ^{29}Si depletion as observed in some of these grains; and furthermore, such very low $^{12}C/^{13}C$ and $^{14}N/^{15}N$ ratios are not reproduced by current supernova models. It remains to be analyzed whether models with very massive ONe white dwarfs, evolved at lower mass-accretion rates, initial metallicities and/or luminosities, and for which a more powerful event is expected, may account for these isotopic anomalies.

It is clear that, statistically, the number of pre-solar nova grain candidates is still too scarce. But one would expect more nova grain candidates to be identified soon. This will benefit from more sophisticated equipments, such as the NanoSIMS, recently operational. Specifically, it would be important to isolate SiC grains but also oxide grains. Apparently, there is no unambiguous evidence of novae among the oxide grain population isolated so far. However, as has also been pointed out by Nittler and Hoppe (2005), the highly ^{17}O-rich and slightly ^{18}O-depleted corundum grain T54 (Nittler *et al.*, 1997) is consistent with recent models of CO novae (José *et al.*, 2004).

There are a number of critical studies and measurements to tighten the links between nova nucleosynthesis and pre-solar grains: first, systematic infrared observations of novae are required to better establish the asymmetry between CO and ONe novae as dust-forming environments; second, it has to be understood whether the nova ejecta can efficiently mix with the solar-like material from the companion to account for the dilution process suggested; third, a better knowledge of the $^{30}P(p,\gamma)$ rate (which competes with $^{30}P(\beta^{+})^{30}Si$) is required to better understand the origin and extent of the ^{30}Si excesses; and fourth, models of explosions on massive ONe white dwarfs yield large overproduction of ^{33}S (José, Coc & Hernanz, 2001), which could provide additional imprints of a nova origin. However, the chances to measure such excesses in pre-solar grains are scarce: so far, no sulfur isotopic measurements have been made on pre-solar SiC grains. Although calculations predict that sulfides might be incorporated into such grains (Lodders & Fegley, 1995), it is likely that the S abundances may be dominated by contamination during the chemical separation procedure.

6.4 Multi-dimensional modeling of a nova outburst: progress and prospects

The departures from ideal geometries and structures so evident in the observational data, even at the earliest moments of the outburst, raise the question of multi-dimensionality in the approach to nova hydrodynamics and detonation. The usual assumption of spherical symmetry in nova models (and in general, in stellar explosions) excludes an entire sequence of features associated with the way a TNR initiates (presumably as a point-like ignition) and propagates. As first suggested by Shara (1982), on the basis of semi-analytical models, it is likely that TNRs on white dwarfs begin as localized events that may spread to the rest of the star's surface or give rise to some sort of *volcanic-like* TNRs, depending on the efficiency of heat transport. Indeed, this study suggested that localized, volcanic-like TNRs were likely to occur (mainly in $M_{\rm WD} \geq 1.2\,{\rm M}_\odot$ white dwarfs), as the diffusively propagated burning wave may require tens of years to extend throughout the stellar surface. However, this analysis, based only on radiative and conductive transport, ignored the major (and critical) role played by convection on the lateral thermalization of a TNR.

The importance of multi-dimensional effects for TNRs in thin stellar shells (including classical nova outbursts) was later revisited by Fryxell and Woosley (1982), who concluded that the most likely situation in nova outbursts involves TNRs propagated by small-scale turbulence. From dimensional analysis and flame theory, the authors derived a relation for the velocity of the deflagration front spreading around the stellar surface: $V_{\rm def} \sim (h_{\rm p} V_{\rm conv}/\tau_{\rm burn})^{1/2}$, where $h_{\rm p}$ is the pressure scale height, $\tau_{\rm burn}$ is the characteristic time-scale for burning the fuel, and $V_{\rm conv}$ the characteristic convective velocity. Typical values for nova outbursts yield $V_{\rm def} \sim 10^4\,{\rm cm\,s^{-1}}$ (that is, the flame propagates halfway through the stellar surface in about 1.3 days).

The first, pioneering studies that addressed this problem in the framework of multi-dimensional codes were performed by Shankar, Arnett and Fryxell (1992), and Shankar and Arnett (1994). An accreting 1.25 $M_\odot$ white dwarf was evolved with a spherically symmetric (1-D) hydro code and mapped into a 2-D domain (a spherical-polar grid of 25×60 km). The explosive event was then followed with a 2-D version of the Eulerian code *PROMETHEUS*. A 12-isotope network, ranging from H to ^{17}F, was included to account for the energetics of the explosion. Unfortunately, the subsonic nature of the problem, coupled with the use of an explicit code (with a time step limited by the Courant–Friedrichs–Levy condition), posed severe limitations on the study, which was forced to very extreme cases, characterized by an artificial increase of the temperature in small, local regions of the envelope's base of about 100–600%, to mimic the onset of the TNR. The overall computed time was also extremely small: about 1 second. The calculations revealed that instantaneous, local temperature fluctuations cause Rayleigh–Taylor instabilities. The rapid rise and subsequent expansion, in a dynamical time-scale, cools down the hot material and halts the lateral spread of the burning front, suggesting that such local temperature fluctuations do not play a relevant role in the initiation of a TNR (in particular at early stages). The study, therefore, favored the occurrence of local volcanic-like TNRs, as suggested by Shara.

At the turn of the century, Glasner and Livne (1995), and Glasner, Livne and Truran (1997) extended these early attempts. New 2-D simulations were performed with another tool, *VULCAN*, an arbitrarily Lagrangian Eulerian (ALE) hydro code that can handle both explicit and implicit time steps. As in previous work (cf. Shankar *et al.*), a slice of the star (0.1π, in radians) was used in spherical-polar coordinates with reflecting boundary conditions and with a resolution near the envelope's base of about 5 km. As before, the evolution of an accreting 1 $M_\odot$ CO white dwarf was initially followed by means of a 1-D hydro code (to overcome the early, computationally challenging phases of the TNR), and then mapped into a 2-D domain as soon as the temperature at the envelope's base reached $T_{\rm b} \sim 10^8$ K. As in Shankar, Arnett and Fryxell (1992), and Shankar and Arnett (1994), the 2-D runs included a 12-isotope network. The simulations revealed a good agreement with the *gross* picture obtained with 1-D models (for instance, the critical role played by the β^+-unstable nuclei ^{13}N, ^{14,15}O, and ^{17}F, in the ejection stage, and consequently, the presence of large amounts of ^{13}C, ^{15}N, and ^{17}O in the ejecta). However, some remarkable differences were also found. The TNR was initiated by a myriad of irregular, localized eruptions at the envelope's base caused by convection-driven temperature fluctuations. Hence, combustion proceeds as a chain of many localized flames (not as a thin front), each surviving only a few seconds. Nevertheless, they concluded that turbulent diffusion efficiently dissipates any local burning around the core. As a result, the fast stage of the TNR cannot be localized and therefore the runaway must spread through the entire envelope. Contrary to 1-D models, the core–envelope interface is now convectively unstable, providing a source for the envelope's metallicity enhancement through a Kelvin–Helmholtz instability (a mechanism that bears some resemblance with the convective overshooting proposed by Woosley (1986)). The efficient dredge-up of CO material from the outermost white dwarf layers accounts for a ~30% metal enrichment in the envelope (the accreted envelope was assumed to be solar-like, without any arbitrary pre-enrichment prescription), in agreement with the inferred metallicities in the nova shells ejected from CO novae. And finally, larger convective eddies, extending up to 2/3 of the envelope's height, with characteristic velocities of $V_{\rm conv} \sim 10^7$ cm s^{-1}, were found in these simulations. Nevertheless, despite these differences, the expansion and progress of the TNR toward the outer envelope was almost spherically symmetric (although the initial burning process was not).

Models by Kercek, Hillebrandt and Truran (1998) used a version of the Eulerian *PROMETHEUS* code; a similar domain (a box of about 1800 km $\times$ 1100 km) was adopted despite a Cartesian, plane-parallel geometry to allow the use of periodic boundary conditions. Two runs, a coarser 5 km $\times$ 5 km grid and a finer 1 km $\times$ 1 km, were performed using the same initial model as Glasner, Livne and Truran (1997). The results were qualitatively similar to Glasner, Livne and Truran, but the outbursts were somewhat less violent (i.e. longer TNRs with lower $T_{\rm peak}$ and $V_{\rm ej}$). This was due to major differences in the convective flow patterns: whereas in Glasner *et al.* a few large convective eddies dominated the flow, most of the early TNR in the Kercek *et al.* models were governed by small, very stable eddies (with $l_{\rm max} \sim 200$ km) that limited dredge-up and mixing episodes.

The only 3-D nova simulation to date has also been performed by Kercek, Hillebrandt and Truran (1999). The run, using a computational domain of $1800 \times 1800 \times 1000$ km^3 with a resolution of 8 km in each dimension, displayed flow patterns dramatically different from those found in the 2-D simulations (much more erratic in the 3-D case): mixing by turbulent motions took place on very small scales (not fully resolved with the adopted resolution); the peak temperatures achieved were slightly lower than in the 2-D case (a consequence of the slower and more limited dredge-up of core material). Moreover, the envelope attained a maximum velocity that was a factor $\sim$100 times smaller than the escape velocity and, presumably, no mass ejection would result. In view of these results, the authors concluded that CO mixing *must* take place prior to the TNR, in contrast with the main results reported by Glasner *et al.* In view of the differences reported by Kercek *et al.*, Glasner, Livne and Truran (2005) performed a sensitivity study of the role played by the outer boundary conditions – hence, the geometry assumed in the simulation – on multi-dimensional studies of the nova outburst, a problem originally posed by Nariai, Nomoto and Sugimoto (1980), in the context of Lagrangian models. A suite of different 2-D tests was adopted, including an Eulerian scheme with a free outflow at the outer boundary (in which high-velocity material can be lost from the computational grid), an Eulerian scheme with modified mass flux at the outer boundary (a model that assumes a 'ghost shell' allowing material in a convective shell, initially leaving the grid, to re-enter the computational domain), an Eulerian scheme with no mass flux at the outer boundary (that is, matter is forced to stay within the computational domain), and finally, a sophisticated arbitrarily ALE scheme (a method that includes remapping in a way that conserves mass). They concluded that calculations where material can freely escape the computational domain lead to an artificial quenching of the runaway, and thus no nova outburst results. On the contrary, calculations performed with the ALE scheme are compatible with the main features of a nova explosion. It is therefore likely that the different boundary conditions used by Glasner *et al.* (reflecting boundaries) and Kercek *et al.* (periodic boundaries), partially imposed by the different geometry assumed for their computational grids, are at the origin of the discrepancies between these studies.

We conclude that more efforts in multi-dimensional modeling are needed, at this stage, particularly to clarify the role played by the outer boundary conditions. This can take advantage of state-of-the-art, massive parallel architectures. Indeed, it would be critical to investigate the early stages of the TNR. Actually, the studies performed by Glasner *et al.* and Kercek *et al.* begin with a 1-D, spherically symmetric simulation that is currently mapped into a 2- or 3-D domain when the temperature at the envelope's base is already 10^8 K. For that, the use of an implicit code that may overcome the limitations posed by the Courant–Friedrichs–Levy condition seems mandatory. Moreover, the use of gridless codes, such as the *Smoothed-Particle Hydrodynamics* (SPH), may be useful for an analysis of a nova outburst from a Lagrangian perspective, free of the problems posed by an Eulerian formulation on the outer boundary conditions. Such SPH simulations are currently being carried out within the Barcelona group.

6.5 The nuclear perspective

6.5.1 *The puzzling synthesis of ^{7}Li*

One of the long-standing predictions of the TNR model for nova outbursts is the synthesis of significant amounts of ^{7}Li, but this is a suggestion that has raised a vivid controversy and still constitutes a matter of debate. The first claim of a substantial synthesis of ^{7}Li during nova outbursts was based on pioneering calculations performed in the framework of crude parametric models (Arnould & Nørgaard, 1975). Following the so-called *beryllium-transport mechanism* (Cameron, 1955), it is initiated by ^{3}He(α, γ)^{7}Be (both the accreted ^{3}He plus the contribution coming from ^{1}H(p,e$^+$ ν_e)^{2}H(p,γ)^{3}He), which is ultimately transformed into ^{7}Li ($\tau \sim 77$ days), in its first excited state, by an electron capture. This state de-excites by emission of a single 478 keV γ-ray photon that might be detectable by a sensitive enough instrument (see Chapter 11 for further discussion).

These preliminary results were confirmed by hydrodynamic simulations (Starrfield *et al.*, 1978) that, however, assumed envelopes in place, thus neglecting the accretion phase and the possible influence of the initial stages of the TNR on the evolution. Later, parametric one/two-zone models (Boffin *et al.*, 1993) claimed that the inclusion of the ^{8}B(p,γ) reaction, not considered in all previous works, throttled the synthesis of ^{7}Li in novae at the typical densities and explosion time-scales reported in hydrodynamic simulations. The scenario was revisited by Hernanz *et al.* (1996) and José and Hernanz (1998), who performed further hydrodynamic calculations taking into account both the accretion and explosion stages using a large nuclear reaction network (including ^{8}B(p,γ)). These studies confirmed that nova outbursts are likely sites for a large overproduction of ^{7}Li (showing in turn the risky overstatements that can be made on the basis of parameterized models), and also pointed out that one of the most critical properties affecting the synthesis of this fragile isotope is the final amount of ^{3}He that survives the early TNR. In particular, the different time-scales of the TNRs achieved in CO and ONe novae, which strongly depend on the initial ^{12}C content in the envelope, lead to larger amounts of ^{7}Be in CO novae. It is worth noting that ^{7}Be avoids destruction from (p,γ) reactions because of the very efficient inverse photo-disintegration reaction on ^{8}B.

Directly observing ^{7}Li in the ejecta is extraordinarily challenging. Unsuccessful attempts carried out by Friedjung (1979), on the basis of optical spectra of three different novae, led only to upper limit estimates of Li/Na $<$ 3.8–4.5(Li/Na)$_\odot$. More recently, the presence of this elusive isotope in a nova shell has been tentatively claimed: an observed feature compatible with the doublet at 6708 Å of Li I has been reported in the spectra of V382 Vel (Della Valle *et al.*, 2002). Although the inferred lithium abundances are compatible with theoretical estimates for a fast nova, it has been argued (Shore *et al.*, 2003) that the observed feature may instead be another low-ionization emission centered at around 6705 Å, e.g. a N I doublet or some other weak neutral or singly ionized species.

Finally, it is worth noting that the potential contribution of classical novae to the Galactic ^{7}Li content is rather small (less than 15%, according to Hernanz *et al.*, 1996, and José and Hernanz, 1998). However, a nova contribution is required to match the ^{7}Li content in some calculations of Galactic chemical evolution (Alibés, Labay & Canal, 2002; Romano *et al.*, 1999; Romano & Matteucci, 2002), although Della Valle *et al.* (2002) do not require a nova contribution. The importance of this isotope for cosmological studies is notable since lithium is a relic tracer of primordial nucleosynthesis, along with ^{4}He and ^{3}He, and serves as a probe of the thermodynamic conditions – hence indirectly as a measure of the expansion rate – at a critical epoch before the formation of the cosmic background radiation. Any later stellar contributors to its abundance must therefore be properly deleted from the observed values before an assessment of the primordial contribution can be made. Unfortunately, these estimates usually rely on a mean contribution of novae, directly averaged from existing models, ignoring that the ^{7}Li yield depend critically on the mass of the white dwarf where the nova explosion will take place. Moreover, there are reasons to believe that novae have not contaminated the interstellar medium in the same way during the 10^{10} yr of Galactic history. A better knowledge of the distribution of novae in the Galaxy as well as the evolution of the nova frequency during the overall Galaxy's age is required to better assess these numbers.

A final question associated with the challenging observation of ^{7}Li is the intrinsic interest of such observation. What can we learn from an unambiguous detection of ^{7}Li in a nova shell? Actually, there are enough arguments favoring the thermonuclear runaway model, so that the detection of ^{7}Li (optically or through γ-ray emission at 478 keV) is not a crucial issue. However, important insights into the physics of the explosion could be learned, such as the efficiency of convection and the overall timescales of the early TNR.

6.5.2 *Nuclear inputs and uncertainties*

The most important reactions in nova outbursts are initiated by protons since these result in the smallest Coulomb barriers and, thus, the largest reaction cross-sections. Consequently, the predominant nuclear reactions considered in nova nucleosynthesis calculations are proton-capture reactions. Current nova model calculations predict peak temperatures $<4 \times 10^{8}$ K. Such temperatures seem insufficient to initiate a break-out (see, for instance, Davids *et al.* (2003a) for the role of ^{15}O(α,γ) in stellar explosions). Indeed, because of these limited peak temperatures achieved, α-capture reactions are not feasible during nova outbursts (with the exception of the ^{3}He(α,γ)^{7}Be reaction). As a result, the nuclear reaction path runs close to the stability valley, between the line of stable nuclei and the proton drip line, for $A < 40$ mass nuclei. Therefore, novae involve a restricted number of nuclear processes (a few hundred reactions). It is worth noting that a significant number of nuclear reactions of interest for nova nucleosynthesis have been measured in the laboratory, including stable and radioactive ion beams. Although certain key nuclear reactions have not been measured yet (and may not be measured in the foreseeable future), we can conclude that novae represent unique events, likely the only stellar explosions for which the nuclear physics input is (or will soon be) primarily based on experimental information (José, Hernanz & Iliadis, 2006).

In the past, the identification of important reaction-rate uncertainties was frequently based on intuitive guesses (often ignoring the complex interplay between the suite of nuclear reactions, convection and hydrodynamical processes). For this reason, substantial efforts have been invested recently in identifying important reaction-rate uncertainties through numerical calculations. Different strategies, somewhat complementary, have been adopted. A first category includes *one-zone nucleosynthesis calculations*, in which T–ρ profiles from either hydrodynamic simulations or from analytic models are coupled to a detailed nuclear reaction network. The most detailed and exhaustive example of this approach is the recent sensitivity study carried out by Iliadis *et al.* (2002): it is striking that for the vast majority of nuclear processes included in the network calculations (i.e. several hundred), reaction-rate variations have no significant effect on the final abundances in current nova models. Instead, only a reduced number of key reaction rates play any remarkable role. Another important result of that investigation is that nova model predictions of final Li, Be, C, and N abundances are not affected by present reaction-rate uncertainties.

To overcome the limitations posed by the oversimplifications of this approach, another, more direct, strategy, involving state-of-the-art hydrodynamic simulations, has been pursued. In order to check the impact of a nuclear uncertainty affecting a specific reaction on the final nova yields, a series of hydro tests has to be performed: one, for instance, assuming the recommended rate, while for subsequent calculations the rate is changed by specific factors within the quoted rate uncertainty. A decision regarding the importance of the particular reaction-rate uncertainty can then be based on the extent of final isotopic abundance variations predicted by these calculations. The procedure would then be repeated for other nuclear reactions that are of potential interest for novae. Of course, such an approach involves a hydrodynamic code coupled directly to a nuclear reaction network and is computationally very intensive; this procedure has been applied so far only to a small number of reaction-rate changes (cf. José, Coc & Hernanz, 1999, 2001; Coc *et al.*, 2000; Fox *et al.*, 2004; Rowland *et al.*, 2004).

Recent experimental efforts have focused on a number of nuclear reactions of critical importance for nova nucleosynthesis studies. These include ^{17}O (p, γ) and ^{17}O (p, α) combining direct and indirect studies (Fox *et al.*, 2004, 2005; Chafa *et al.*, 2005); ^{21}Na(p, γ)^{22} Mg, measured at KVI using the indirect transfer reaction ^{24}Mg$(p,t)^{22}$Mg (Davids *et al.*, 2003b), and mainly through the first direct experiment carried out at TRIUMF (Bishop *et al.*, 2003; D'Auria *et al.*, 2004); ^{22}Na$(p,\gamma)^{23}$Mg, measuring ^{12}C$(^{12}$C$,n)^{23}$Mg with Gammasphere at ANL; Jenkins *et al.* (2004); ^{23}Na(p, γ) and ^{23}Na(p, α), through a direct ^{23}Na(p, γ) measurement; Rowland *et al.* (2004); and ^{30}P(p, γ)^{31}S, measuring ^{12}C(^{20}Ne, n)^{31}S and ^{12}C(^{20}Ne, p)^{31}P with Gammasphere at ANL; Jenkins *et al.* (2006). Special efforts have been devoted to the challenging ^{18}F(p, α) (and ^{18}F(p, γ)) reaction, including elastic (p, p$'$) scattering studies (Bardayan *et al.*, 2000, 2001; Graulich *et al.*, 2001), direct (p, α) measurements on ^{18}F (Bardayan *et al.*, 2001, 2002; Graulich *et al.*, 2001), and a suite of indirect transfer reactions to populate levels in the compound nucleus ^{19}Ne (Utku *et al.*, 1998; de Séréville *et al.*, 2003, 2006; Kozub *et al.*, 2005). New measurements are available for the ^{14}N(p, α) reaction, both at LENA (North Carolina; Runkle *et al.*, 2005) and, underground, at LUNA (Grand Sasso, Italy; Formicola *et al.*, 2004). In summary, most of the key reactions relevant for nova nucleosynthesis that involve stable target nuclei have now been directly measured in the laboratory. The resulting reaction rate uncertainties, typically $\leq$30%, are small enough for quantitative nova model predictions. On the other hand, direct measurements of reactions involving *unstable* target nuclei have only recently begun. In fact, only two of these reactions, $^{18}F(p, \alpha)^{15}O$ and $^{21}Na(p, \gamma)^{22}Mg$, have so far been measured directly in the nova Gamow window.

Current uncertainties affecting nova nucleosynthesis studies involve only a handful of nuclear reaction rates, particularly $^{18}F(p, \alpha)$, $^{25}Al(p, \gamma)$, $^{26}Al^{g}(p, \gamma)$, and $^{30}P(p, \gamma)$, for which several experiments have been proposed at different facilities around the world (TRIUMF-ISAC, Oak Ridge National Laboratory, Louvain-la-Neuve, or Argonne National Laboratory, among others).

6.6 Closing remarks

Although throughout this book the emphasis has been on Galactic novae, it will not be long before similar spatial and spectral resolutions will be available for nova studies out to almost the Virgo cluster. The new generation of extremely large telescopes, and the growth of optical and millimeter interferometry – including polarimetry – are making more detailed studies possible. With the experience of panchromatic observations of Galactic and Large Magellanic Cloud novae, the extension of this experience should permit the detailed modeling of novae occurring in very different stellar populations in, for instance, ellipticals and spirals beyond the Local Group. It has been the purpose of this chapter to highlight some of the clouds remaining within the standard model that has developed over the last half century and indicate how some may be blown away.

Acknowledgements

We would like to thank S. Amari, J. Aufdenberg, A. Calder, A. E. Champagne, A. Coc, J. D'Auria, M. Della Valle, J. Dursi, D. Galli, R. D. Gehrz, M. Hernanz, C. Iliadis, G. Schwarz, S. Starrfield, J. Truran, K. Vanlandingham, M. Wiescher, R. E. Williams, and E. Zinner for the stimulating collaborations and feedback on several topics addressed in this manuscript. This work has been partially supported by the MCYT grant AYA2004-06290-C02-02 and by MUIR and INFN-Pisa.

References

Alibés, A., Labay, J., & Canal, R., 2002, *Ap. J.*, **571**, 336.
Amari, S., 2002, *New Astron. Rev.*, **46**, 519.
Amari, S., Lewis, R. S., & Anders, E., 1995, *Geochim. Cosmochim. Acta.*, **59**, 1411.
Amari, S., Gao, X., Nittler, L. R. *et al.*, 2001, *Ap. J.*, **551**, 1065.
Anders, E., & Grevesse, N., 1989, *Geochim. Cosmochim. Acta.*, **53**, 197.
Andreä, J., Drechsel, H., & Starrfield, S., 1994, *A&A*, **291**, 869.
Arnould, M., & Nørgaard, H., 1975, *A&A*, **42**, 55.
Antipova, L. I., 1969, *A.Zh.*, **46**, 366 (translation in *Soviet Astron.*, **13**, 288).
Asplund, M., 2005, *ARA&A*, **43**, 481.
Asplund, M., Grevesse, N., & Sauval, A. J., 2005, in *Cosmic Abundances as Records of Stellar Evolution and Nucleosynthesis*, ed. T. G. Barnes, & F. N. Bash. Astronomical Society of the Pacific Conference Series, **336**, p. 25.
Barbier, D., & Chalonge, D., 1940, *Ann. Astrophys.*, **3**, 41.
Bardayan, D. W., Blackmon, J. C., Bradfield-Smith, W. *et al.*, 2000, *Phys. Rev. C*, **62**, 42802.
Bardayan, D. W., Blackmon, J. C., Bradfield-Smith, W. *et al.*, 2001, *Phys. Rev. C*, **63**, 065802.
Bardayan, D. W., Batchelder, J. C., Blackmon, J. C. *et al.*, 2002, *Phys. Rev. Lett.*, **89**, 262501.
Bishop, S., 2003, *Phys. Rev. Lett.* **90**, 162501; Erratum: *Phys. Rev. Lett.*, **90**, 229902.
Boffin, H. M. J., Paulus, G., Arnould, M., & Mowlavi, N., 1993, *A&A*, **279**, 173.
Cameron, A. G. W., 1955, *Ap. J.*, **121**, 144.
Chafa, A., Tatischeff, V., Aguer, P. *et al.*, 2005, *Phys. Rev. Lett.*, **95**, 031101.
Clayton, D. D., & Hoyle, F., 1976, *Ap. J.*, **203**, 490.
Coc, A., Hernanz, M., José, J., & Thibaud, J.-P., 2000, *A&A*, **357**, 561.
Cunha, N., Hubeny, I., & Lanz, T., 2006, *Ap. J.*, **647**, L 143.
D'Auria, J. M., Azuma, R. E., Bishop, S., *et al.*, 2004, *Phys. Rev. C*, **69**, 065803.
Davids, B., 2003a, *Phys. Rev. C*, **67**, 065808.
Davids, B., 2003b, *Phys. Rev. C*, **68**, 055805.
de Freitas Pacheco, J. A., & Codina, S. J., 1985, *MNRAS*, **214**, 481.
de Séréville, N., Coc, A., Angulo, C. *et al.*, 2003, *Phys. Rev. C*, **67**, 052801-R.
de Séréville, N., Coc, A., Angulo, C. *et al.*, 2007, Nucl. Phys. A 791, 251.
Della Valle, M., Pasquini, L., Daou, D., & Williams, R. E., 2002, *A&A*, **390**, 155.
Ercolano, B., Barlow, M. J., & Storey, P. J., 2005, *MNRAS*, **362**, 1038.
Evans, A., 1990, in *Physics of Classical Novae*, ed. A. Cassatella, & R. Viotti. Berlin: Springer, p. 253.
Evans, A., Geballe, T. R., Rawlings, J. M. C., & Scott, A. D., 1997, *MNRAS*, **292**, 1049.
Evans, A., Tyne, V. H., Smith, O. *et al.*, 2005, *MNRAS*, **360**, 1483.
Ferland, G. J., 2003, *ARA&A*, **41**, 517.
Formicola, A., Imbriani, G., Costantini, H. *et al.*, 2004, *Phys. Lett. B*, **591**, 61.
Fox, C., 2004, *Phys. Rev. Lett.*, **93**, 081102.
Fox., C. *et al.*, 2005, *Phys. Rev. C*, **71**, 55801.
Friedjung, M., 1979, *A&A*, **77**, 357.
Fryxell, B. A., & Woosley, S. E., 1982, *Ap. J.*, **261**, 332.
Gehrz, R. D., 1999, *Phys. Rep.*, **311**, 405.
Gehrz, R. D., Truran J. W., Williams, R. E., & Starrfield, S. G., 1998, *PASP*, **110**, 3.

Glasner, S. A., & Livne, E., 1995, *Ap. J.*, **445**, L149.
Glasner, S. A., Livne, E., & Truran, J. W., 1997, *Ap. J.*, **475**, 754.
Glasner, S. A., Livne, E., & Truran, J. W., 2005, *Ap. J.*, **625**, 347.
Graulich, J. S., Cherubini, S., Coszach, R. *et al.*, 2001, *Phys. Rev. C*, **63**, 11302.
Grevesse, N., & Noels, A., 1993, in *Origin and Evolution of the Elements*, ed. N. Prantzos, E. Vangioni, & M. Cassé. Cambridge: Cambridge University Press. p. 15.
Grevesse, N., Noels, A., & Sauval, A. J., 1993, in *Proceedings of the 6th Annual October Astrophysics Conference*, ed. S. S. Holt, & G. Sonneborn. San Francisco: Astronomical Society of the Pacific, p. 117.
Hayward, T. L., Saizar, P., Gehrz, R. D. *et al.*, 1996, *Ap. J.*, **469**, 854.
Hernanz, M., José, J., Coc, A., & Isern, J., 1996, *Ap. J.*, **465**, L27.
Iijima, T., 2002, *A&A*, **387**, 1013.
Iliadis, C., Champagne, A., José, J., Starrfield, S., & Tupper, P., 2002, *Ap. JS*, **142**, 105.
Jenkins, D. G., Lister, C. J., Janssens, R. V. *et al.*, 2004, *Phys. Rev. Lett.*, **92**, 031101.
Jenkins, D. G., Lister, C. J., Carpenter, M. P. *et al.*, 2005, *Phys. Rev. C*, **72**, 31303.
Jenkins, D. G., Meadowcroft, A., Lister, C. J. *et al.*, 2006, *Phys. Rev. C*, **73**, 65802.
Joiner, D. A., 1999, The nucleation and growth of dust grains in nova shells. Unpublished Ph.D. thesis, Rensselaer Polytechnic Institute.
José, J., & Hernanz, M., 1998, *Ap. J.*, **494**, 680.
José, J., Coc, A., & Hernanz, M., 1999, *Ap. J.*, **520**, 347.
José, J., Coc, A., & Hernanz, M., 2001, *Ap. J.*, **560**, 897.
José, J., Hernanz, M., Amari, S., Lodders, K., & Zinner, E., 2004, *Ap. J.*, **612**, 414.
José, J., Hernanz, M., & Iliadis, C., 2006, *Nucl. Phys. A*, in press.
Kercek, A. Hillebrandt, W., & Truran, J. W., 1998, *A&A*, **337**, 379.
Kercek, A., Hillebrandt, W., & Truran, J. W., 1999, *A&A*, **345**, 831.
Kingdon, J. B., & Ferland, G. J., 1995, *Ap. J.*, **450**, 691.
Kozub, R. L., Bardayan, D. W., Batchelder, J. C. *et al.*, 2005, *Phys. Rev. C*, **71**, 032801.
Lodders, K., & Fegley, B., 1995, *Meteoritics*, **30**, 661.
Lucy, L. B., 1999, *A&A*, **344**, 282.
Lucy, L. B., 2001, *MNRAS*, **326**, 95.
Lucy, L. B., 2003, *A&A*, **403**, 261.
Lyke, J., Koenig, X. P., Barlow, M. J. *et al.*, 2003, *AJ*, **126**, 993.
Moro-Martín, A., Garnavich, P. M., & Noriega-Crespo, A., 2001, *AJ*, **121**, 1636.
Morisset, C., & Pequignot, D., 1996, *A&A*, **312**, 135.
Morisset, C., Stasinska, G., & Peña, M., 2005, *MNRAS*, **360**, 449.
Nariai, K., Nomoto, K., & Sugimoto, D., 1980, *PASJ*, **32**, 473.
Nittler, L. R., & Hoppe, P., 2005, *Ap. J.*, **631**, L89.
Nittler, L. R., Alexander, C. M. O'd., Gao, X., Walker, R. M., & Zinner, E., 1997, *Ap. J.*, **483**, 475.
Och, S. R., Lucy, L. B., & Rosa, M. R., 1998, *A&A*, **336**, 301.
Osterbrock, D. E., & Ferland, G. J., 2006, *Astrophysics of Gaseous Nebulae and Active Galactic Nuclei*, 2nd edn. New York: University Science Books.
Paresce, F., Livio, M., Hack, W., & Korista, K., 1995, *A&A*, **299**, 823.
Payne-Gaposchkin, C., 1957, *The Galactic Novae*. New York: Dover.
Pontefract, M., & Rawlings, J. M. C., 2004, *MNRAS*, **347**, 1294.
Romano, D., Matteucci, F., Molaro, P., & Bonifacio, P., 1999, *A&A*, **352**, 117.
Romano, D., & Matteucci, F., 2002, in *Classical Nova Explosions*, ed. M. Hernanz, & J. José. New York: American Institute of Physics, p. 144.
Rowland, C., Iliadis, C., Champagne, A. E. *et al.* 2004, *Ap. J.*, **615**, L37.
Runkle, R. C., Champagne, A. E., Angulo, C. *et al.*, 2005, *Phys. Rev. Lett*, **94**, 2503.
Schwarz, G. J., 2002, *Ap. J.*, **577**, 940.
Shankar, A., & Arnett, D., 1994, *Ap. J.*, **433**, 216.
Shankar, A., Arnett, D., & Fryxell, B. A., 1992, *Ap. J.*, **394**, L13.
Shara, M. M., 1982, *Ap. J.*, **261**, 649.
Shore, S. N., Starrfield, S., Ake, T. B., & Hauschildt, P. H., 1997, *Ap. J.*, **490**, 393.
Shore, S. N. Schwarz, G., Bond, H. E. *et al.*, 2003, *AJ*, **125**, 1507.
Shore, S. N., & Gehrz, R. D., 2004, *A&A*, **417**, 695.
Shore, S. N. Starrfield, S., Gonzalez-Riestra, R., Hauschildt, P. H., & Sonneborn, G. *et al.*, 1994, *Nature*, **369**, 539.

Starrfield, S., Truran, J. W., Sparks, W. M., & Arnould, M., 1978, *Ap. J.*, **222**, 600.
Tsamis, Y. G., & Péquignot, D., 2005, *MNRAS*, **364**, 687.
Utku, S., Ross, J. G., Bateman, N. P. T. *et al.*, 1998, *Phys. Rev. C*, **57**, 2731.
Vanlandingham, K. M., Starrfield, S., & Shore, S. N., 1997, *MNRAS*, **290**, 395.
Vanlandingham, K. M., Schwarz, G. J., Shore, S. N., Starrfield, S., & Wagner, R. M., 2005, *Ap. J.*, **624**, 914.
Williams, R. E., 1992, *AJ*, **104**, 725.
Williams, R. E., Sparks, W. M., Gallagher, J. S. *et al.*, 1981, *MNRAS*, **251**, 221.
Wilson, O. C., & Merrill, P. W., 1935, *PASP*, **47**, 53.
Woosley, S. E., 1986, in *Nucleosynthesis and Chemical Evolution*, ed. B. Hauck, A. Maeder, & G. Meynet. Sauverny: Geneva Observatory, p. 1.
Zinner, E., 1998, *ARE&PS*, **26**, 147.

7

Radio emission from novae

E. R. Seaquist and M. F. Bode

7.1 Introduction

In this chapter, we review the observation of radio emission from classical novae. In the majority of cases such emission is undoubtedly dominated by that from the thermal bremsstrahlung process and we begin by giving an overview of such emission arising from relatively simple geometries. We then move on to describe various models of the kinematics in novae and show how these have been combined with the thermal emission process described earlier to fit the radio light curves of several well-observed novae. However, it has become apparent that relatively simple spherically symmetric models fail to describe some of the details in the observations of more recent objects and more complex modelling is introduced. This is reinforced when we consider the evolution of the resolved radio remnants as illustrated in Chapter 12.

In the first edition of this book in 1989, radio emission was reported to have been detected in nine classical novae and, of these, two were spatially resolved (Seaquist, 1989). Since then, we have seen the advent of improvements to the sensitivity of radio arrays. Here we consider observations of radio emission from a total of eighteen classical novae, and as reported here and in Chapter 12, at least six of these have published radio imagery. Of these, GK Per shows an unusual, long-lived, non-thermal radio remnant which is discussed in detail in Chapter 12 (see also Chapter 13). In addition to the classical novae considered here, we briefly discuss emission from related objects (symbiotic stars and recurrent novae).

7.2 Nature and measurement of radio emission from novae

The first classical novae to be detected at radio wavelengths were HR Del and FH Ser by Hjellming and Wade (1970). The early work is summarized by Hjellming (1974). The observations were made with the three-element interferometer of the National Radio Astronomy Observatory at 2.7 and 8.1 GHz. The characteristic time-scale of the radio variation for these stars is several years, substantially greater than that for the optical light curve, and the maximum radio flux density occurs much later than optical maximum (see Figure 12.1). During the rising part of the 'radio light curve' the radio spectral index $\alpha > 0$ ($f_\nu \propto \nu^\alpha$), whereas during the decay $\alpha \sim -0.1$. These characteristics are consistent with

Classical Novae, 2nd edition, ed. Michael Bode and Aneurin Evans. Published by Cambridge University Press.

thermal bremsstrahlung emitted by an ionized gas shell ejected at the time of the outburst. This interpretation is supported by a simple estimate of the brightness temperature

$$T_{\mathrm{b}} \sim \frac{f_\nu \lambda^2 D^2}{2\pi k V_{\mathrm{ej}}^2 t^2}$$

during the optically thick phase of the radio light curve. Here f_ν is the flux density at time t, V_{ej} is the expansion velocity of the envelope or ejecta and D is the distance. For FH Ser at $\lambda = 3.7$ cm (8.1 GHz), $f_\nu = 35$ mJy[1] at $t = 0.5$ yr, $V_{\mathrm{ej}} = 1100\,\mathrm{km\,s^{-1}}$ (Hutchings, 1972) and $D = 710$ pc (Hutchings & Fisher, 1973) which yields $T_{\mathrm{b}} \approx 10^4$ K. This figure is comparable with the electron temperature of a photo-ionized plasma. The prolonged time-scale of the radio emission occurs because the emitting gas remains optically thick at radio wavelengths long after optical maximum. Other radio novae generally follow the same qualitative behaviour as HR Del and FH Ser, and are therefore thermal radio emitters, although there are hints at some contributions from non-thermally emitting regions at some phases of development in a few objects (see below). A truly exceptional case, however, is GK Per whose spatially resolved remnant is dominated by non-thermal emission, as noted above.

Even the brightest radio novae observed have peak flux densities less than 100 mJy at centimetre wavelengths. Therefore, unlike the case for optical radiation, large telescopes are essential to measure the radio light curves accurately. In addition, high resolution is necessary to avoid the effects of confusion produced by background sources at these faint levels. Most of the radio observations to date have been made with the three-element interferometer at Green Bank, West Virginia, the Westerbork Synthesis Radio Telescope (WSRT), the Very Large Array (VLA) in New Mexico, or the *MERLIN* network of telescopes, centred around Jodrell Bank.

Bode, Seaquist and Evans (1987) carried out the first systematic search for radio emission from classical novae, using the *VLA* at 1.5 to 4.9 GHz. The search included two dozen novae of which two were detected as weak (thermal) sources. No counterparts to the non-thermal (synchrotron emission) remnant of GK Per were found, and from this Bode *et al.* determined that classical novae supply $<1\%$ of the cosmic ray electron energy to the Galactic disk compared with supernovae. Table 7.1 contains all classical novae for which there are reported measurements of radio flux densities. Upper limits for other classical novae may be found in Bode, Seaquist and Evans (1987), Haddock *et al.* (1963) and Turner (1985). Since this chapter is concerned primarily with classical novae, related objects are not included in this list, but are considered separately in Section 7.8.

7.3 Radio thermal bremsstrahlung from an expanding cloud

7.3.1 *Uniform-slab model*

The radio characteristics of a nova may be anticipated by first considering the expansion of a uniform cube of ionized gas. The flux density f_ν at a frequency ν is given by

$$f_\nu = B_\nu \, (\ell/D)^2 \, (1 - \mathrm{e}^{-\tau_\nu}), \tag{7.1}$$

[1] 1 jansky = $10^{-26}\,\mathrm{W\,m^{-2}\,Hz^{-1}}$

Table 7.1. *Classical novae detected at radio wavelengths (in chronological order)*

Name	Refs. to radio measurements	Comments
GK Per	Reynolds & Chevelier (1984); Seaquist *et al.* (1989)	Non-thermal emission detected from resolved remnant in 1983
HR Del	Hjellming & Wade (1970); Wade & Hjellming (1971)	Multi-frequency data
FH Ser	Hjellming & Wade (1970); Wade & Hjellming (1971)	Multi-frequency data
V368 Sct	Herrero, Hjellming & Wade (1971)	Detection, but no radio light curve
V1500 Cyg	Hjellming (1975); Hjellming *et al.* (1979); Seaquist *et al.* (1980);	Very detailed multi-frequency light curves
NQ Vul	Bode, Seaquist & Evans (1987)	Weak emission detected 1984
V1370 Aql	Turner (1985); Snijders *et al.* (1987)	Sparsely sampled declining radio light curve
V4077 Sgr	Bode, Seaquist & Evans (1987)	Weak emission detected in 1984
PW Vul	Hjellming (1990)	Only observed during optically thin decay
QU Vul	Taylor *et al.* (1987)	Intense, very steeply rising radio spectrum initially. Radio image
V1819 Cyg	Hjellming (1996b)	Unpublished data
V827 Her	Hjellming (1996b)	Unpublished data
V838 Her	Hjellming (1991, 1996b)	Multi-frequency detections at one epoch published; more data unpublished
V351 Pup	Hjellming (1992, 1996b)	Multi-frequency detections at one epoch published; more data unpublished
V1974 Cyg	Ivison *et al.* (1993); Pavelin (1993); Hjellming (1995, 1996a,b); Eyres, Davis & Bode (1996)	Detailed multi-frequency images and light curves
V705 Cas	Eyres *et al.* (2000)	Multi-frequency images and light curves
V723 Cas	Heywood *et al.* (2002, 2005)	5 GHz MERLIN images and light curve
V1494 Aql	Eyres *et al.* (2005)	Multi-frequency light curves and MERLIN image

where $B_\nu = 2kT_e\nu^2/c^2$ is the Planck function (in the Rayleigh–Jeans approximation) in terms of the electron temperature T_e, ℓ is the size of the cubic emitting region and τ_ν is the free–free optical depth at frequency ν. At radio frequencies the optical depth is given by Lang (1980) as

$$\tau_\nu = 8.235 \times 10^{-2} \left(\frac{T_e}{\mathrm{K}}\right)^{-1.35} \left(\frac{\nu}{\mathrm{GHz}}\right)^{-2.1} \left(\frac{\mathrm{EM}}{\mathrm{cm^{-6}\,pc}}\right) \tag{7.2}$$

where $\mathrm{EM} = \sum_i Z_i^2 n_i n_e \, d\ell$ is the emission measure, the sum being taken over all ionic species of nuclear charge Z_i.

Equations (7.1) and (7.2) may be combined to show that, if the cloud expands uniformly conserving mass M_{ej}, then the flux density reaches a maximum f_{max} (in mJy) when the angular size is θ_{max} (in arcsec) at time t_{max} (in years), given by

$$f_{max} \approx 6.7 \left(\frac{\nu}{\text{GHz}}\right)^{1.16} \left(\frac{T_e}{10^4\ \text{K}}\right)^{0.46} \left(\frac{M_{ej}}{10^{-4}\ \text{M}_\odot}\right)^{0.80} \left(\frac{D}{\text{kpc}}\right)^{-2} \tag{7.3}$$

$$\theta_{max} \approx 1.1 \left(\frac{\nu}{\text{GHz}}\right)^{-0.42} \left(\frac{T_e}{10^4\ \text{K}}\right)^{-0.27} \left(\frac{M_{ej}}{10^{-4}\ \text{M}_\odot}\right)^{0.40} \left(\frac{D}{\text{kpc}}\right)^{-1} \tag{7.4}$$

$$t_{max} \approx 2.6 \left(\frac{\nu}{\text{GHz}}\right)^{-0.42} \left(\frac{T_e}{10^4\ \text{K}}\right)^{-0.27} \left(\frac{M_{ej}}{10^{-4}\ \text{M}_\odot}\right)^{0.40} \times \left(\frac{V_{ej}}{10^3\ \text{km s}^{-1}}\right)^{-1} \tag{7.5}$$

where V_{ej} is the ejecta velocity. The time dependence of the flux density during the optically thick and thin phases is as follows:

$$t \ll t_{max} \quad (\tau_\nu \gg 1) \qquad f_\nu \propto \nu^2 t^2 \tag{7.6}$$

$$t \gg t_{max} \quad (\tau_\nu \ll 1) \qquad f_\nu \propto \nu^{-0.1} t^{-3}. \tag{7.7}$$

If it is assumed for illustrative purposes that useful radio light curves can nowadays be obtained when f_{max} is more than a few mJy at 5 GHz, then Equation (7.3) with $M_{ej} = 10^{-4}\ \text{M}_\odot$ shows that useful observations can still be made when $D \sim 1$–2 kpc. According to Warner (1989) 12 novae were discovered within 1 kpc in 70 yr (i.e. about 0.2 per year within 1 kpc). This would correspond to about 1 nova per year within 2 kpc (assuming a disk-like distribution). Reports of radio discoveries are 0.5 per year, which suggests that not all detectable novae are being observed or reported. This rather lower figure is not surprising as (a) not all novae are accessible to the VLA and MERLIN and (b) systematic Target of Opportunity campaigns have not been in operation at these facilities continuously for this period.

The equations for the simple cube model also serve to highlight the fact that information may be obtained over a period extending several years after the outburst when the gas density is very low. In addition, the angular resolution of, for example, the VLA ($0.3''$ at 5 GHz) is comparable with the angular extent at maximum (about $0.5''$ at 5 GHz if $D = 1$ kpc according to Equation (7.4)), so that the envelopes of bright novae can be resolved and the evolution of their morphology followed (see also Chapter 12).

The limitations of the simple cube model may be investigated by direct comparison with the radio data. For example, in the early days of attempting to fit models to the radio data, Hjellming (1974) noted that the behaviour of FH Ser during the pre-radio maximum phase was $f_\nu \propto t^{0.74}$, in disagreement with Equation (7.6). He noted that this behaviour indicates a possible deceleration in the expansion of the nova envelope. In addition, according to Seaquist and Palimaka (1977), the pre-radio maximum emission possessed a spectral index $\alpha \sim +1$, whereas Equation (7.6) predicts $\alpha = +2$. Seaquist and Palimaka suggested that these results together may be explained by an inhomogeneity in the shell without the need to invoke a non-uniform rate of expansion. Specifically, a thick or wide spherical shell in which the density decreases outward would reproduce both results. As the expanding shell becomes less dense, it becomes less optically thick and the radio photosphere (where $\tau \sim 1$) lags behind the expansion of the outer shell boundary. If the flow becomes steady, the radius of this photosphere and the flux density become constant with respect to time. Another departure from the uniform cube model is that the peak of the radio light curve is broader than that expected for a homogenous source. This effect is consistent with that expected from a density gradient in the shell. Some portions of the shell remain optically thick well past maximum, thus broadening the peak in the radio light curve.

Other complications may arise as well. At very high radio frequencies, the light curve peaks very early, and the effects of incomplete ionization in the early stages of the shell development might be reflected in the data. An instrumental effect which may occur when an interferometer is used is that extended structure in the nova envelope can be resolved out. There are other additional details of the temporal and spectral development that are addressed by recourse to other plausible physical causes described later. First we consider the next step on the road to developing more complex and physically realistic models.

7.3.2 *Spherically symmetric shell with a density gradient*

Most of the models used to fit the radio data to date consist of a spherically symmetric fully ionized shell with a density gradient. The density gradient is incorporated by including a density profile of the form

$$n = A/r^p \tag{7.8}$$

where n is the ion number density. Figure 7.1 illustrates the shell geometry and the definition of the quantities used. The solution to the equation of radiative transfer is

$$f_\nu = \frac{2\pi B_\nu(T_\mathrm{e})}{D^2}\left(\int_0^{R_\mathrm{i}} q\ \{1-\exp[-\tau_1(q)]\}\,\mathrm{d}q + \int_{R_\mathrm{i}}^{R_\mathrm{o}} q\,\{1-\exp[-\tau_2(q)]\}\right) \tag{7.9}$$

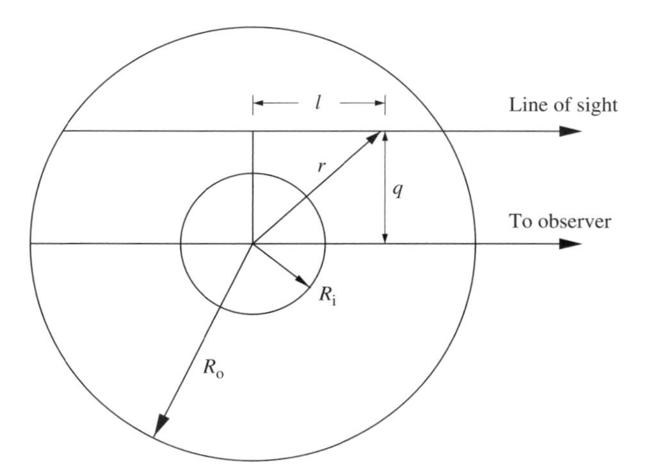

Fig. 7.1. Geometry used to derive the equations for the flux density of an expanding spherical cloud. See Section 7.3.2 for details.

where the optical depths $\tau_1(q)$ and $\tau_2(q)$ are given by

$$\tau_1(q) = 2\kappa(\nu, T_e)\gamma A^2 \int_{(R_i^2-q^2)^{1/2}}^{(R_o^2-q^2)^{1/2}} \frac{dl}{(l^2+q^2)^p} \qquad 0 \le q \le R_i$$

$$\text{and} \quad \tau_2(q) = 2\kappa(\nu, T_e)\gamma A^2 \int_0^{(R_o^2-q^2)^{1/2}} \frac{dl}{(l^2+q^2)^p} \qquad R_i < q \le R_o\,. \tag{7.10}$$

Here $\kappa(\nu, T_e)$ is the free–free absorption coefficient for unit emission measure, and γ is the electron-to-ion number ratio. The first and second terms in Equation (7.9) are the contributions to the flux density from the regions $0 \le q \le R_i$ and $R_i < q \le R_o$ respectively. The equations include neither occultation of a portion of the shell by the star nor the contribution to the flux density of the star itself. Since the stellar diameter is far less than the shell diameter at $t = t_{max}$, these latter effects are negligible. The equation may be conveniently parameterized in the following way:

$$\begin{aligned} R_i &= x\, R_o \\ R_o &= y\, D\, T_e^{-1/2} \\ A &= z\, \gamma^{-1/2} D^{3/2} T_e^{-0.075}. \end{aligned} \tag{7.11}$$

In the case where $p = 2$, which provides a good fit to much of the data on novae so far, Equation (7.9) may be rewritten, after carrying out the integrals in Equation (7.10):

$$f_\nu = 2\pi \frac{2k\nu^2}{c^2} y^2 \left(\int_0^x u\{1 - \exp[-\tau_1(u)]\}du + \int_x^1 u\{1 - \exp[-\tau_2(u)]\}du \right) \tag{7.12}$$

where

$$\tau_1(u) = \frac{2\kappa'(\nu)z^2}{y^3}\left[\frac{1}{2u^2}(1-u^2)^{1/2} + \frac{1}{2u^3}\arccos u - \frac{1}{2x^2u^2}(x^2-u^2)^{1/2} - \frac{1}{2u^3}\arccos\frac{u}{x}\right] \qquad 0 \le u \le x \tag{7.13}$$

$$\tau_2(u) = \frac{2\kappa'(\nu)z^2}{y^3}\left[\frac{1}{2u^2}(1-u^2)^{1/2} + \frac{1}{2u^3}\arccos u\right] \qquad x < u \le 1\,. \tag{7.14}$$

In Equations (7.13) and (7.14) $\kappa(\nu, T_e)$ has been written as

$$\kappa(\nu, T_e) = \kappa'(\nu) T_e^{-1.35}.$$

The integration of Equation (7.12) must be carried out numerically, yielding f_ν as a function of the three parameters (x, y, z) which characterize the shell. Analytical forms in the optically thick ($\tau_{1,2} \gg 1$), partially opaque ($\tau_1 \gg 1$, $\tau_2 \sim 1$) and optically thin ($\tau_{1,2} \ll 1$) regimes are:

$$\begin{aligned} \tau_{1,2} \gg 1 \qquad & f_\nu = \pi\left(\frac{2k\nu^2}{c^2}\right)y^2 \\ \tau_1 \gg 1,\ \tau_2 \sim 1 \qquad & f_\nu = \left(\frac{8\pi}{3}\right)\left(\frac{\pi}{2}\right)^{2/3}\left(\frac{2k\nu^2}{c^2}\right)[\kappa'(\nu)]^{2/3} z^{4/3} \\ \tau_{1,2} \ll 1 \qquad & f_\nu = 4\pi\left(\frac{2k\nu^2}{c^2}\right)\kappa'(\nu)\frac{z^2}{y}\left(\frac{1}{x} - 1\right). \end{aligned} \tag{7.15}$$

An explicit kinematic model may be used to specify the temporal behaviour of the parameters (x,y,z), and Equation (7.12) may then be used to compute the flux density as a function of time and frequency. Thus a set of radio light curves for different frequencies is produced. At a given frequency, the regimes represented by Equation (7.15) reflect three distinct regimes in the temporal behaviour of the radio light curve, as seen later.

7.4 Shell kinematics

The first applications of inhomogenous spherical shell models to the radio data were made by Seaquist and Palimaka (1977). Since then, similar analyses have been carried out by Hjellming *et al.* (1979), Seaquist *et al.* (1980) and Kwok (1983). The types of models employed by these investigators may be conveniently divided according to their kinematic behaviour. These are *self-similar flows* and *variable stellar winds*. Neither of these approaches is firmly rooted in a detailed understanding of the mass-loss mechanism, but they do constitute at least an initial framework for interpreting the radio data, and for obtaining insight into this mechanism.

Self-similar flows are flows with scale-free internal structure arising when there are no forces acting on a given element of the flow, and consequently its momentum is conserved. Hjellming *et al.* (1979) provided the basis for applying such ideas to nova envelopes. We are led to consider (a) envelopes with outflow speed proportional to the radial distance, i.e. a 'Hubble flow', and (b) variable winds with constant speed. In addition, a third variant called the unified model was introduced by Hjellming (1990, 1995, 1996a,b) essentially combining the properties of models (a) and (b) together. Figure 7.2 shows the schematic behaviour of all three models. We discuss these in turn.

7.4.1 Hubble flow model

In the Hubble flow model, all the gas is ejected at one instant (impulsive ejection) with a spread in the expansion velocity. A variant of this model, the linear gradient model, introduced by Hjellming *et al.* (1979), allows for finite initial inner and outer radii. The density distribution depends on how the mass is distributed in velocity. In the case of a uniform distribution, the density ρ as a function of radius r is given by (e.g. Hjellming, 1996b)

$$\rho = \left(\frac{1}{4\pi r^2}\right)\frac{M}{[R_o(t) - R_i(t)]} \tag{7.16}$$

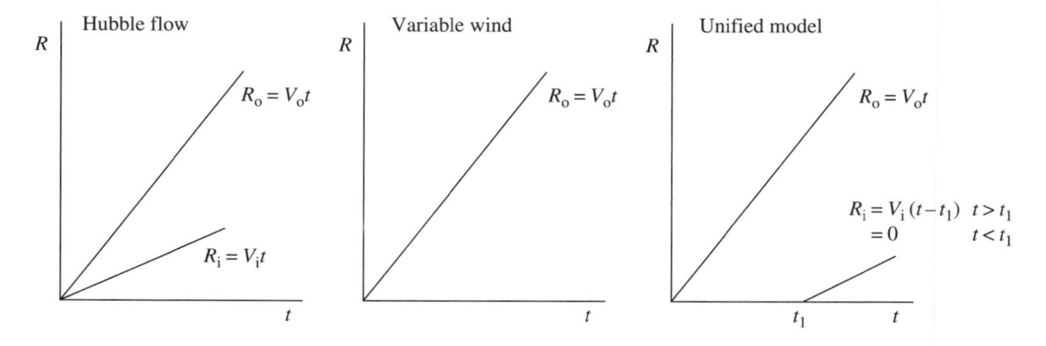

Fig. 7.2. Schematic diagrams of the shell kinematics for (a) Hubble flow, (b) variable wind and (c) unified models (see text for details). Adapted from Hjellming (1990).

where M is the shell mass, and R_i and R_o are the inner and outer radii increasing uniformly with velocities V_i and V_o respectively, and $V_i < V_o$.

The corresponding time-dependent gas number density is of the form

$$n = \frac{A_0}{r^2}\left(\frac{t_0}{t}\right), \tag{7.17}$$

where A_0 is the density parameter at $t = t_0$. Combined with Equations (7.11) and (7.15), this leads to the following time dependences for the flux densities for the three regimes in Equation (7.15):

$$\begin{aligned} \tau_{1,2} \gg 1 \qquad & f_\nu \propto \nu^2 t^2 \\ \tau_1 \gg 1,\ \tau_2 \sim 1 \qquad & f_\nu \propto \nu^{0.6} t^{-4/3} \\ \tau_{1,2} \ll 1 \qquad & f_\nu \propto \nu^{-0.1} t^{-3}\,. \end{aligned} \tag{7.18}$$

The dependence on ν in the middle expression of (7.18) is based on the $\nu^{-2.1}$ dependence of κ', reflected also in Equation (7.2) for the optical depth.

This behaviour is identical with the uniform slab model with the exception of the intermediate behaviour $\nu^{0.6}t^{-4/3}$. In this phase, the radio emission emerges primarily from a 'radio photosphere' between the inner and outer radii, whose location depends on both frequency and time.

With $p = 2$ in Equation (7.8), the mass M and kinetic energy KE in this model are given by

$$M = 4\pi\ \mu m_{\rm H} A_0 t_0\,(V_o - V_i) \tag{7.19}$$

where μ is the mean atomic weight and

$$\mathrm{KE} = \frac{1}{6} M V_o^2 \left[\frac{(1 - x^3)}{(1 - x)}\right] \tag{7.20}$$

where $x = R_i/R_o = V_i/V_o$.

Thus, the Hubble flow model is essentially a three-parameter model, in which the M, V_i, V_o are derivable from the parameters (x, y, z) once values are adopted for (D, T_e). Note that the parameter x is fixed, whereas the parameters y and z vary directly and inversely with time respectively.

This model has been remarkably successful in accounting for the radio light curves at various frequencies. The essential reason is that the different material velocities lead to a thickening shell with a significant density gradient. These conditions broaden the light curves compared with the homogeneous model described by Equation (7.1) because different parts of the shell become optically thick at different times. This broadening is present in the observed light curve of FH Ser, as pointed out by Seaquist and Palimaka (1977).

7.4.2 *Variable wind model*

There is considerable evidence that at least some of the mass-loss in nova eruptions takes the form of prolonged stellar winds (Gallagher & Starrfield, 1978; see also Chapters 5 and 9). The mass-loss rate is high at the epoch of eruption, and declines slowly with time. The mass-loss is probably sustained by the large and prolonged bolometric luminosity near the Eddington limit, exhibited by some novae after visual maximum. In this case the solution to the continuity equation for constant flow speed V_{ej} is

$$\rho = \frac{1}{4\pi r^2} \frac{\dot{M}(t - r/V_{ej})}{V_{ej}} \tag{7.21}$$

where $\dot{M}(t)$ is the time-dependent mass-loss rate. This model is distinguished from 'impulsive injection' models in the radio context by the condition that the duration of detectable mass-loss from the star is comparable to (or greater than) the duration of the radio light curve. Since envelopes remain detectable at radio wavelengths for long periods after visual maximum, radio observations may offer the possibility of distinguishing between 'wind' and 'impulsive ejection' models.

If we neglect the transit time effects, i.e. the time-scale for variation in $\dot{M}$ is much greater than r/V_{ej}, then Equation (7.21) may be written

$$\rho = \frac{1}{4\pi r^2} \frac{\dot{M}}{V_{ej}} .$$

If

$$\dot{M} = (\dot{M})_0 \left(\frac{t_0}{t}\right)^p \qquad t \geq t_0$$

then

$$\rho = \frac{1}{4\pi r^2} \left(\frac{(\dot{M})_0}{V_{ej}}\right) \left(\frac{t_0}{t}\right)^p .$$

Hjellming (1990) considers $p = 1$, and writes the density in the form

$$\rho = \frac{1}{4\pi r^2} \left(\frac{M}{R_o}\right).$$

In this case, $M = \dot{M} t_0$ represents the total mass ejected between $t = t_0$ and $t = 2.7 t_0$.

We can also write the number density,

$$n = \frac{A_0}{r^2} \left(\frac{t_0}{t}\right)^p$$

and the flux density varies as

$$f_\nu \propto \nu^{0.6} t^{-4p/3} . \tag{7.22}$$

Equation (7.22) is essentially applicable to a nearly steady flow in which the mass flow rate is taken to be a slowly varying function of time, i.e. $t_0 \gg r/V_{ej}$.

Such models are generally not applicable to the whole range of behaviour exhibited by the radio light curves because the latter require an inner boundary to produce the t^{-3} at large t (Hjellming, 1996b). However, they may be applicable to modelling the behaviour at optical or infrared wavelengths since these light curves reach their peaks at much earlier times, prior to the detachment of the shell from the star.

7.4.3 *Unified model*

The unified model is characterized by a more complex geometry required to fit the light curves and images of V1974 Cyg. This model essentially combines the Hubble flow and variable wind models to give a representation of the temporal history of the ejection over both early and late phases of the evolution of the envelope. A variable wind is ejected for a period of time t_1 after the initiation of the nova event, after which the shell becomes detached (see Figure 7.2). Furthermore, the velocity of ejection varies linearly with time according to

$$V_{ej} = V_o - (V_o - V_i)\left(\frac{t}{t_1}\right) \qquad 0 \leq t \leq t_1 \tag{7.23}$$

where V_i and V_o are the final and initial velocities respectively. Note that Equation (7.23) corrects a typographical error in this formula in Hjellming (1996b).

The essential features of this model are a variable wind at $t < t_1$ and a Hubble flow behaviour when $t \gg t_1$. The Hubble flow behaviour occurs because a linear velocity gradient is introduced by the velocity variation in Equation (7.23).

7.5 Discussion of selected classical novae: first class data

The classical novae that have been studied most extensively at radio wavelengths are the slow nova HR Del; the moderate-speed nova FH Ser; the very fast nova V1500 Cyg; and the fast nova V1974 Cyg. There is also more limited published light curve coverage for the fast nova V1370 Aql; the moderate-speed nova QU Vul; the slow nova V723 Cas; and the very fast nova V1494 Aql. We will discuss these latter objects in the next section. Sporadic data on several other novae are as referenced in Table 7.1.

V1500 Cyg has been well observed in both the radio and the infrared, and there is strong evidence that the infrared emission, like the radio, is thermal bremsstrahlung (see also Chapter 8). By contrast, V1974 Cyg was the first nova to be observed in any detail in the millimetre/submillimetre part of the spectrum, and deviations from the expected behaviour here and from spatial mapping have led to the formulation of at least two variants on the hitherto accepted model of radio emission. We therefore discuss these two objects separately since the comparison of the radio and shorter wavelength emission yields considerable further insight into the behaviour of fast nova envelopes. Radio outbursts from related objects, including the recurrent nova RS Oph, have also been observed extensively at radio wavelengths, and are discussed briefly in Section 7.8.

7.5.1 *FH Ser and HR Del*

The data for FH Ser have been analysed by various investigators using the two different types of basic model. Seaquist and Palimaka (1977) explored a variable wind model and a Hubble flow model, including both spherically symmetric and polar shell geometries. Hjellming *et al.* (1979) later considered only the Hubble flow model using more extensive data, while Kwok (1983) examined a variable wind model.

In the variable wind model considered by Seaquist and Palimaka (1977), a stellar wind at a fixed outflow rate and speed occurs for a specific period of time. There are four parameters, namely the inner and outer shell radii, the density parameter A and the flow speed V_{ej}. In this case, $p = 2$ in Equation (7.8) and the parameter A is related to the mass-loss rate $\dot{M}$ by the relation

$$A = \frac{\dot{M}}{4\pi \mu m_H V_{ej}}. \tag{7.24}$$

Seaquist and Palimaka used this model to fit data published by Hjellming (1974) at 2.7 and 8.1 GHz covering a period of approximately 3 years. Data over a more extended frequency range were also available for epoch $t = 0.62$ yr. The model was found to be inadequate since no set of parameters provides a satisfactory fit to both the radio spectrum at $t = 0.62$ yr and the temporal data. For example, if a suitable fit is made to the radio light curves at 2.7 and 8.1 GHz, then the model shell considerably under-predicts the 90 GHz flux density at $t = 0.62$ yr.

In order to remove this problem, a velocity gradient was found to be necessary. The fast-moving outer part of the envelope accounts for the rapid increase in flux density at earlier epochs, and the slower-moving inner part of the envelope produces a higher 90 GHz flux density at $t = 0.62$ yr. The latter effect is due to the maintenance of a high optical depth at 90 GHz in the inner region. Therefore, the self-similar Hubble flow model is preferred. Only models with $p = 2$ were considered.

Hjellming *et al.* (1979) subsequently used a linear gradient model, also with $p = 2$, to obtain satisfactory fits to the data for HR Del and FH Ser (see Figure 7.3). Similar results for V1500 Cyg are described in Section 7.5.2. The fit of the linear gradient models is excellent, considering their simplicity, involving only three parameters.

Kwok (1983) used stellar wind models with constant outflow speed and variable mass flow rate given by

$$\dot{M}(t) = \dot{M}_0(t_0/t)^{\delta}. \tag{7.25}$$

These models have been fitted to the radio data for HR Del, FH Ser and V1500 Cyg. The motivation for this approach is that the post-maximum decline in visible light can be reproduced by assuming an optically thick wind. His analysis leads to the conclusion that

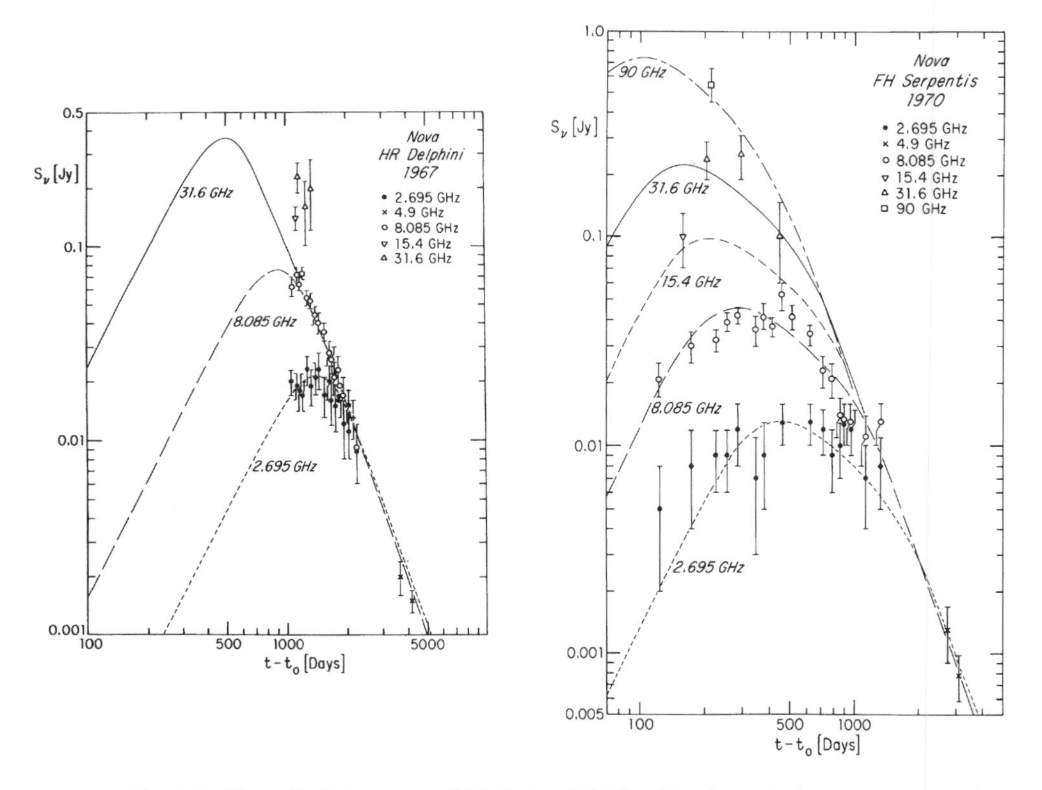

Fig. 7.3. The radio light curves of HR Del and FH Ser. The theoretical curves correspond to spherical shell linear gradient models with density index $p = 2$. From Hjellming *et al.* (1979).

the *optical* data for the first two novae require $\delta = 2$ and $t_0 = 0.1$ yr, but that these parameters do not produce the wide shell necessary to account for the radio data. Instead, a more slowly varying stellar wind with $t_0 \sim 1$ yr is required. Thus a second phase of mass-loss is envisaged with a substantially different time-scale (see Chapters 5 and 12 for further discussion of different phases of ejection). This picture is reasonably satisfactory, but the curves considerably under-predict the flux densities at $\nu = 90$ GHz, a problem encountered in the constant-outflow-speed model considered by Seaquist and Palimaka (1977), as noted above.

A second difficulty with this 'two-phase mass-loss' picture is that the radio flux densities of the mass ejected in the first phase (necessary to explain the optical light curve) cannot be ignored even though it provides an unsatisfactory fit to the radio data. With $V_{ej} \sim 10^3$ km s^{-1}, $\dot{M} = 10^{-4}$ M$_\odot$ yr^{-1}, $\delta = 2$ and $t_0 = 0.1$ yr, the total mass ejected in this phase would be approximately 10^{-5} M$_\odot$ confined to a shell of thickness 3×10^{14} cm. Such a shell would remain optically thick at centimetre wavelengths for a period of more than one year. It would consequently dominate the radio spectrum during this period, and would remain a visible influence thereafter as well. Therefore, models involving a second phase of mass ejection must include the observable effects of the first phase in order to reproduce the radio data.

It is worth noting that the first phase with $t_0 = 0.1$ yr is short compared with the time-scale of the radio variation so that a stellar wind with this time-scale is compatible with the Hubble flow model, but a large variation in the wind outflow speed would be necessary to produce a velocity gradient.

7.5.2 V1500 Cyg

Despite undergoing outburst over 30 years ago, V1500 Cyg is still one of the best-studied novae to date. In addition to the radio observations, there is a wealth of data in the ultraviolet, optical and infrared. As in the case of HR Del and FH Ser, the radio light curves are well accounted for by thermal bremsstrahlung emitted by an ionized shell with a linear velocity gradient.

Ennis *et al.* (1977) published extensive infrared data for V1500 Cyg covering a total period of about one year (see also Chapter 8). Since the radio light curve spans a period of more than two years, the two data sets for this nova together provide a means for investigating the shell structure in more detail. It is noteworthy that in other novae, such as FH Ser, significant infrared emission is produced by either dust formation as the shell cools (e.g. Gallagher (1977) and Chapters 8 and 13). Ennis *et al.* concluded, however, that their data could be explained by bremsstrahlung alone from a narrow ionized shell expanding at $V_{ej} = 2000$ km s^{-1}. The proposed shell has nearly constant width during the first 60 days in order to account for the shallow slope ($f \propto t^{-2}$) of the decay of the infrared light curve during this period. The shell width increases at the sound speed ($\sim$10 km s^{-1}) and the flux density variation develops a steeper slope ($f \propto t^{-3}$) at $t \sim 60$ days when the total increasing width is comparable to the initial width (10^{13} cm). An analysis by Seaquist *et al.* (1980) showed that this model does not fit the radio data, however. Nor does the thick shell model, required to fit the radio data, provide an adequate fit to the infrared data. Hjellming *et al.* (1979) made reference to a similar discrepancy. The picture suggested by the infrared behaviour seems to be supported however by the optical emission-line strengths in the nebular phase since they also have a time dependence expected for a narrow shell (Ferland & Shields, 1978).

Hjellming *et al.* (1979) proposed an alternative explanation for this transition. If the inner shell boundary forms at $t = 60$ days, then the behaviour prior to this epoch would reflect that of a variable stellar wind, whereas the development after this epoch would reflect a t^{-3}

behaviour if the envelope is optically thin in the infrared. Essentially this corresponds to the use of Hjellming's unified model to account for both the radio and infrared behaviour.

In the context of the Hubble flow model, the discrepancy between the radio and infrared behaviour is also resolved to a significant degree if the change at $t = 60$ days is produced by the transition from an ionization-bounded to a density-bounded shell (Seaquist *et al.*, 1980). The bremsstrahlung radiation is emitted only by the ionized component in the shell. Thus different parts of the expanding shell contribute to the radiation at different stages of the expansion. Seaquist *et al.* (1980) considered linear gradient models of the type described in the previous section, but included the effect of incomplete ionization to reconcile the radio and infrared data.

Following Seaquist *et al.* (1980), Figures 7.4 and 7.5 show the behaviour of the radio and infrared light curves for Hubble flow models with $p = 2$ and $p = 3$. The temporal behaviour of the infrared light curve prior to $t = 60$ days is controlled by the ultraviolet luminosity which ionizes a thin zone at the inner boundary of the thick expanding shell. This thin zone is

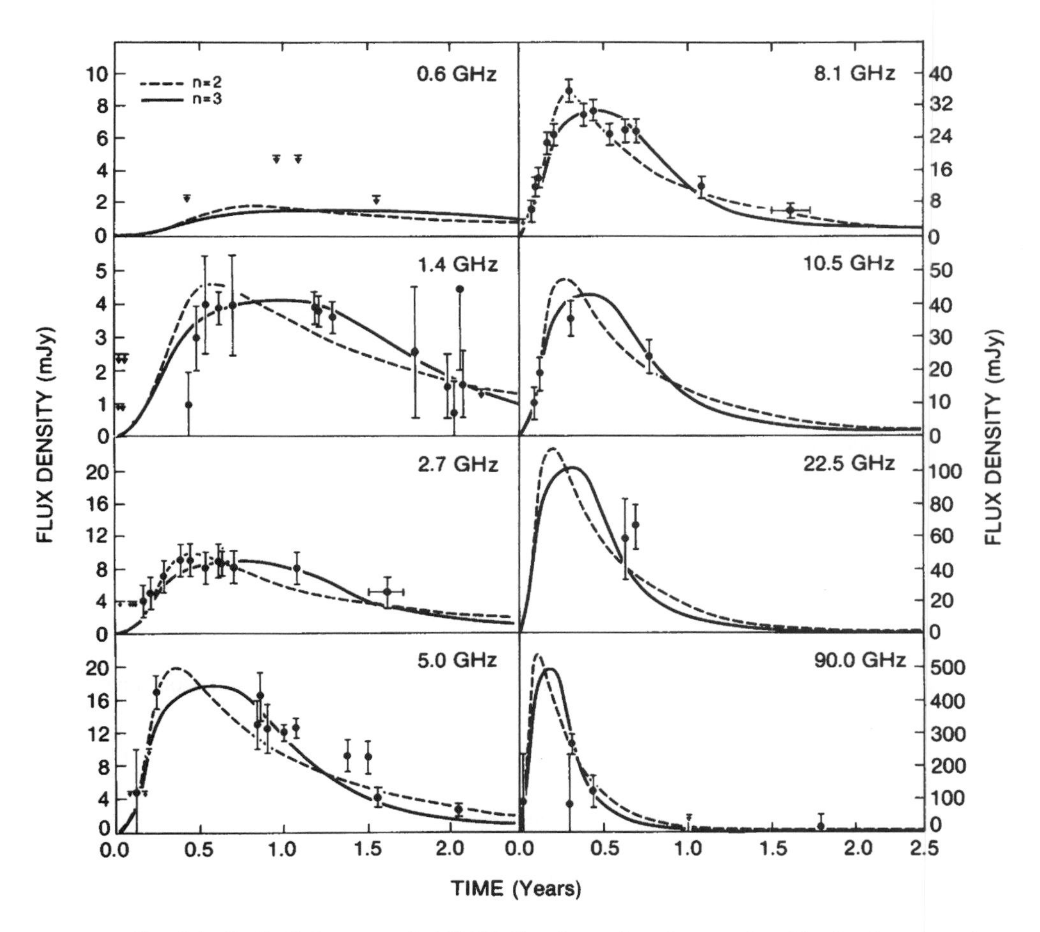

Fig. 7.4. Radio light curves for V1500 Cyg for various frequencies. The fitted curves relate to spherical shell Hubble flow models with $p = 2$ and $p = 3$. From Seaquist *et al.* (1980).

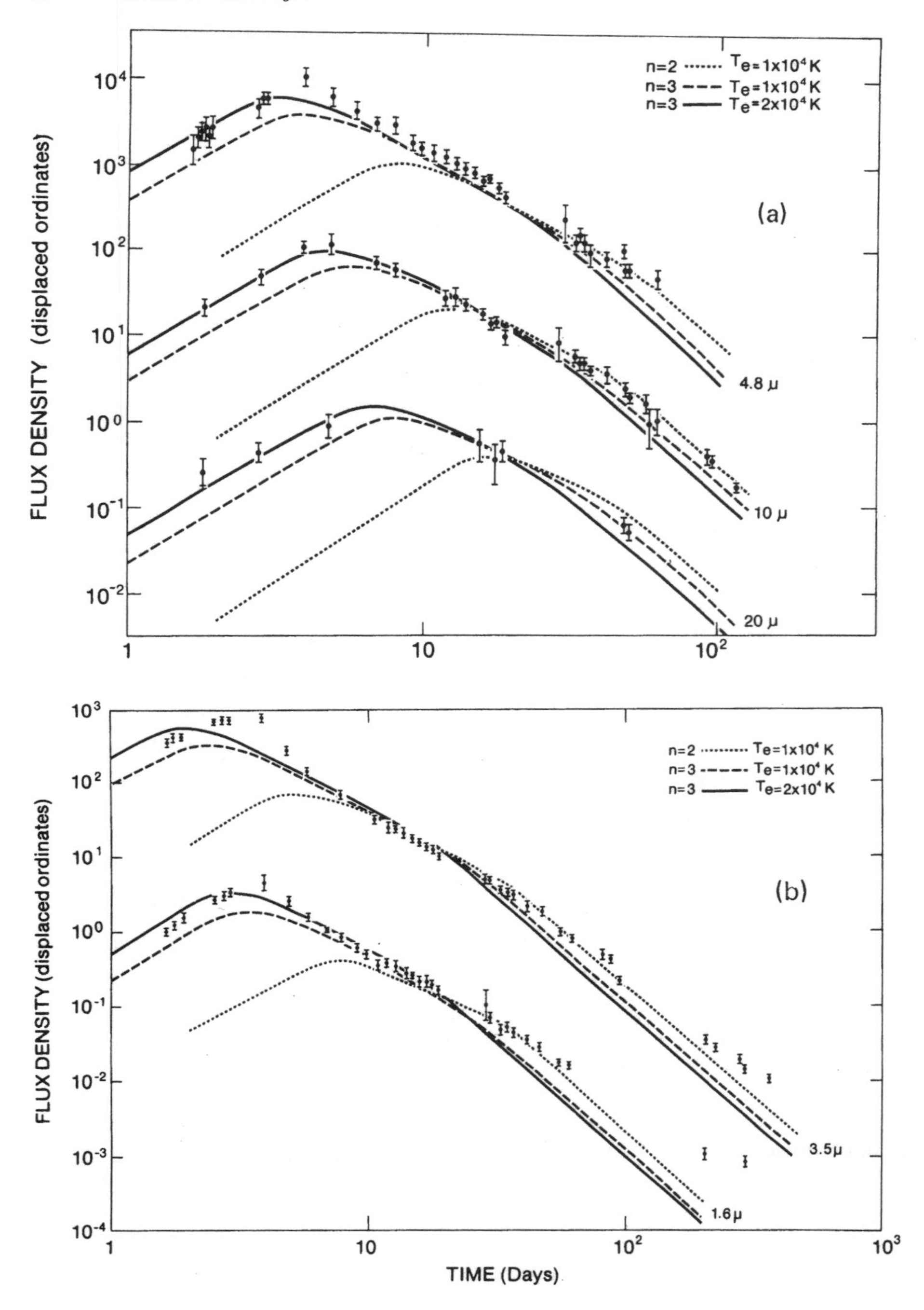

Fig. 7.5. Infrared light curves for V1500 Cyg for various wavelengths. The fitted curves relate to ionization bounded models using Hubble flow shell structures which fit the radio data. The curves are from Seaquist *et al.* (1980) and the data from Ennis *et al.* (1977).

optically thin in the infrared and consequently the infrared light curve mimics the ultraviolet light curve. The Lyman continuum luminosity L_u required to account for the infrared light curve shown by Ennis *et al.* (1977) must be of the form $L_u \sim t^{-2}$, and the epoch of complete photo-ionization is at $t = 46$ days if $p = 2$ and at $t = 22$ days if $p = 3$.

Both Gallagher (1977) and Strittmatter *et al.* (1977) estimate that the time-scale for complete ionization of the shell of V1500 Cyg is about one month, roughly in accord with these model values. Ferland and Shields (1978) also note that the existence of neutral oxygen lines throughout the optical decline indicates that portions of the ejecta were optically thick to ionizing photons.

As in the case of HR Del and FH Ser, there is excellent agreement with the Hubble flow model in the radio, and pretty good agreement in the infrared, particularly for $p = 3$. A temperature of 2×10^4 K provides a better fit at $t < 30$ days and 1×10^4 K thereafter. This may indicate that the shell cooled as it expanded. Shell cooling has not, however, generally been incorporated in the radio models for nova envelopes.

7.5.3 V1974 Cyg

V1974 Cyg was observed by both the VLA and MERLIN. Radio light curves with the VLA at 1.49, 4.9, 8.4, 14.9 and 22.5 GHz span the period from 21 to 1185 days after outburst (Hjellming, 1995, 1996a,b; unfortunately, the data and details of their analysis appear only in these brief conference proceedings). In addition, MERLIN imagery at 1.7 and 5 GHz shows an apparently rapidly evolving remnant, first resolved only 80 days after outburst (Pavelin *et al.*, 1993; Eyres, Davis & Bode, 1996). It was also the first classical nova to be observed in detail with the Hubble Space Telescope (see also Chapters 8, 9 and 12). In addition, Ivison *et al.* (1993) obtained 0.45–2.0 mm continuum observations with the James Clerk Maxwell Telescope (JCMT) at epochs from 66 to 358 days after outburst.

Ivison *et al.* (1993) found that neither of the basic models (Hubble flow, variable wind) provided good fits to the evolution of the millimeter/submillimeter data (see Figure 7.6). They noted that from the imagery of Pavelin *et al.* (1993), subsequently confirmed by Eyres, Davis and Bode (1996), there appeared to be deviations from spherical symmetry and there was evidence for temperature gradients not accounted for in these simple models.

Figure 7.7 shows a fit to the radio light curves by a modified version of the unified model. Hjellming (1995, 1996a,b) used a unified model with an initial wind whose velocity decreases linearly with time, terminating at some finite time after outburst (in this case, 180 days) providing the initial state for a Hubble flow model. In addition, he noted that the MERLIN images appeared to show marked evolution of the source such that the position angle of the major axis of the radio remnant showed dramatic changes over the epochs of observation (a note of caution must, however, be sounded here when interpreting the evolution of such relatively low surface brightness features derived from a relatively sparse array such as MERLIN–see Chapter 12).

In order to fit the early light curve behaviour, Hjellming included an increase in ionized gas temperature with time. The modified unified model also includes inner and outer ellipsoidal boundaries (included at least in part to explain the evolution of the radio images – see Chapter 12) with the major axes of the ellipsoids normal to each other. The inner ellipsoid develops when mass ejection effectively stops at $t = 180$ days (note that Hjellming comments that a similar change in decay light curves in V1500 Cyg indicates a delay of around 60 days in the formation of a detached shell in that object).

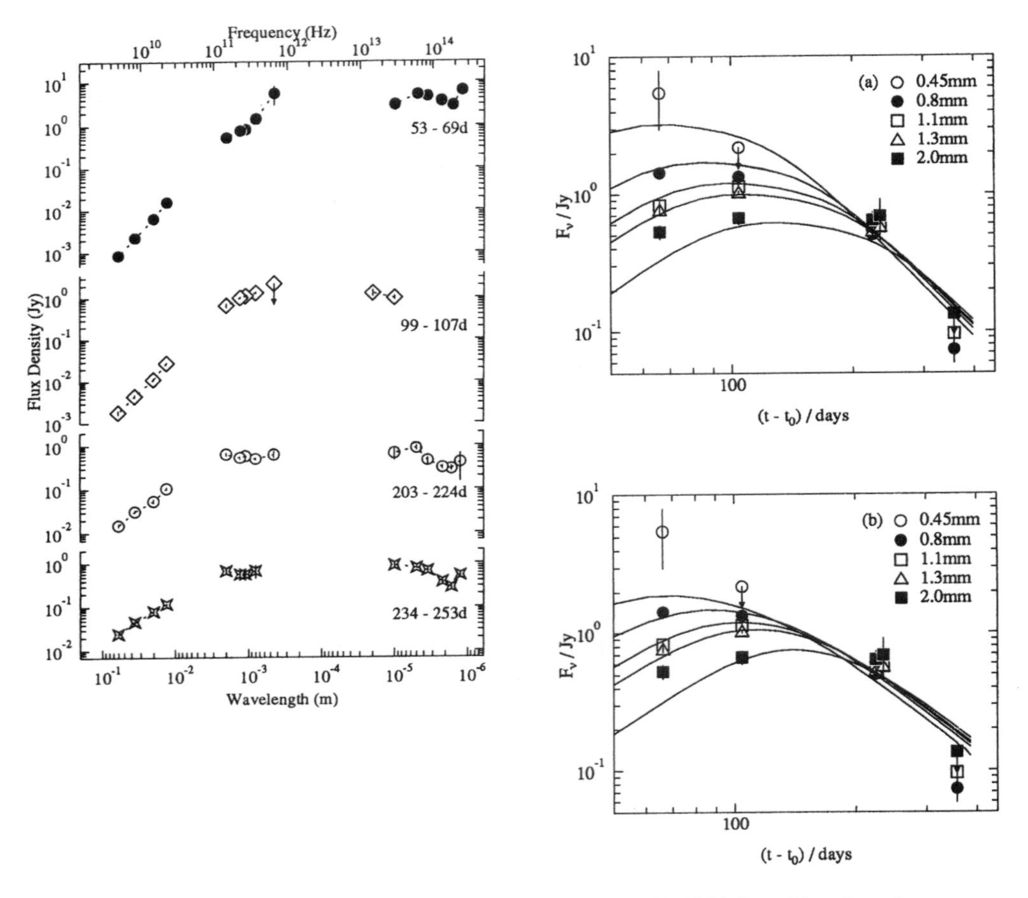

Fig. 7.6. (Left panel) The radio-infrared continuum of V1974 Cyg. The plots, from top to bottom, represent fluxes obtained in 1992 April, June, September and October respectively. They reveal the changing shape of the spectrum as the expanding ejecta become progressively more transparent at longer wavelengths. (Right panel) Data points and best fits to mm data for (a) Hubble flow and (b) variable wind models. The curves correspond to, in ascending order, $\lambda = 2.0$, 1.3, 1.1, 0.8 and 0.45 mm. See Ivison *et al.* (1993), from which these figures were taken, for further details.

In the case of V1974 Cyg, Hjellming furthermore noted that a spherically symmetric version of the unified model would produce flat-topped emission lines (Hjellming, 1995, 1996a,b). At later times at least, these are a very poor representation even of the gross features of the emission line profiles in this (and other) novae (see also Chapter 9). Not only are the radio light curves of V1974 Cyg well fitted by his adapted model (i.e.with inner and outer ellipsoids), but also the gross profile of the emission lines can be better replicated (though obviously not the multiple peaks).

The Hjellming model of the evolution of V1974 Cyg is admittedly somewhat ad hoc, however. A more physically justified model is that developed by Lloyd, O'Brien and Bode (1996), who describe the evolution of the radio flux arising from their models of the shaping of nova remnants (see also Chapter 12). Briefly, their model comprises a wind from the central

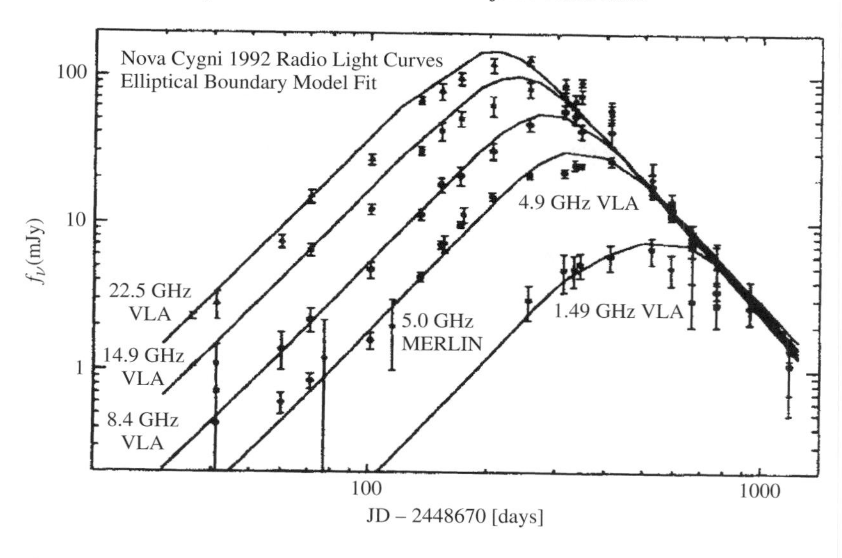

Fig. 7.7. The evolution of radio fluxes of V1974 Cyg at 1.49 (VLA), 4.9 (VLA), 5 (MERLIN), 8.4 (VLA), 14.9 (VLA) and 22.5 (VLA) GHz. Also shown are model light curves obtained from fitting a linear velocity gradient, ellipsoidal boundary model. From Hjellming (1996b).

nova system with secularly increasing velocity which gives rise to a shell of hot, shocked gas. An additional source of hot gas is located near the centre of the flow where the secondary star interacts with the outflowing envelope.

Models were run for four different speed classes of nova (the time for which the slow wind blows being assumed proportional to the decay time of the optical light curve for example). Although the gross behaviour may be similar, the detailed behaviour of the radio flux then differs from the canonical model. For example, initially (over the order of a few tens of days), the radio flux at any given frequency increases more slowly than t^2, although the ejecta are completely optically thick and the radio photosphere coincides with the outer boundary of the (assumed fully ionized) ejecta. This is due to the innermost very hot gas, seen at early times, being obscured by the overlying cooler wind. This gives rise to a decrease in average T_b during this phase. Later in the evolution, regions of high brightness temperature begin to appear in the synthetic maps as the optical depth in the cool overlying wind decreases (for example at around 100 days) revealing the emission from the hot shocked shell beneath. At this time, the radio flux density increases very rapidly and the spectral index is greater than the value of +2 expected for an optically thick isothermal source. Figure 7.8 illustrates the behaviour of light curves at three different frequencies for four models of differing speed class. There is a qualitative similarity particularly between the curves shown in Run 3 for a moderately fast nova and the observations of V1974 Cygni.

There is no doubt that the detailed observations of V1974 Cyg have encouraged some of the refinements that are needed to models of the radio emission. Indeed, there is obviously still work to be done in this regard. This is also the case, although to a lesser degree, for some of the novae with less comprehensive data sets.

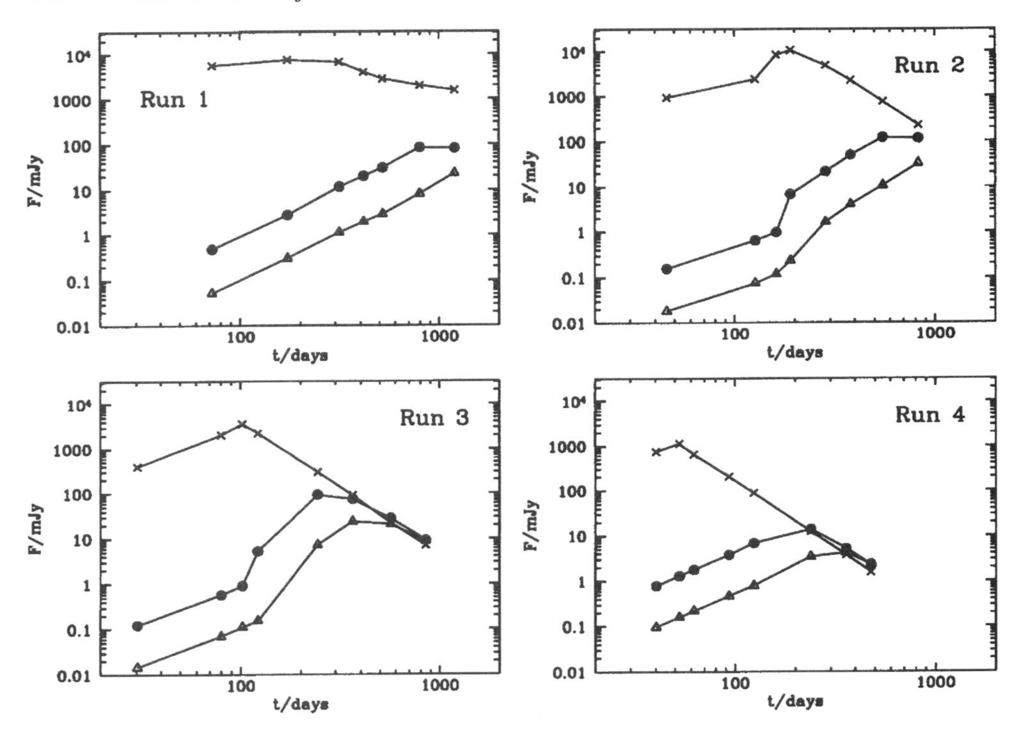

Fig. 7.8. Radio light curves resulting from four nova models of Lloyd, O'Brien and Bode (1996). The frequencies are 1.7 GHz (triangles), 5 GHz (squares), and 273 GHz (crosses). The runs are for four different speed classes of nova with time for the visual light curve to decline 3 magnitudes, t_3, equal to 230, 100, 50, and 30 days respectively – see text for further details.

7.5.4 *The significance of the inferred velocity gradients*

The relative success of the Hubble flow and unified models implies that the matter in all four novae (HR Del, FH Ser, V1500 Cyg and V1974 Cyg) was ejected with a large velocity dispersion, the total range corresponding to the separation in velocity between the inner and outer shell boundaries. This velocity dispersion should be detectable in the optical *absorption*-line spectra which contain information on the dispersion in the radial velocities. Novae, indeed, generally exhibit multiple absorption components and frequently temporal variations in the absorption-line velocities as well (see e.g. Chapters 2, 9 and 12). For example, Hutchings, Smolinski and Grygar (1972) note that the principal shell of FH Ser exhibits a complex spectrum with velocity components ranging at least from -600 to -1100 km s^{-1}. A dispersion in absorption-line widths may also be attributed to dispersion in the outflow *direction* of gas adjacent to and seen projected against the distended photosphere produced by the outflowing wind. However, Ferland (1980) inferred from the difference in widths of the Balmer and Fe II lines that large velocity gradients exist in the expanding envelope of V1500 Cyg. There does appear therefore to be optical evidence for velocity dispersions in nova shells.

7.6 Discussion of selected classical novae: second class data

7.6.1 *V1370 Aql*

Radio emission at 0.6, 1.4 and 5 GHz from this very fast nova was observed using the WSRT. Results are presented and analysed by Snijders *et al.* (1987). The maximum flux density recorded at 1.4 GHz was approximately 14 mJy. Most of the data for this nova cover only the decay portion of the radio light curve, and are rather fragmentary, thereby hampering the analysis.

The derived fluxes appear too large to be explained as emission from the medium-velocity gas ($V_{ej} \sim 4000$ km s^{-1}). A thermal, Hubble flow model for emission from the high-velocity gas ($V_{ej} \sim 10\,000$ km s^{-1}) requires unacceptably high ejecta kinetic energy and gas temperature. On the other hand, the radio spectral index poses problems for a purely non-thermal (synchrotron) mechanism.

7.6.2 *QU Vul*

Multi-frequency monitoring of this moderate-speed nova using the VLA is reported by Taylor *et al.* (1987). The results reveal a unique radio light curve in which a strong outburst with flux density 63.4 mJy at 14.9 GHz, 206 days after the optical maximum, precedes the appearance of normal radio emission from the principal ejecta of the nova by at least 100 days.

The early emission had an extremely high spectral index ($\alpha = 2.4$) which cannot of course be explained by any geometry of an isothermal bremsstrahlung source. The inferred brightness temperature is in excess of 10^5 K. Synchrotron emission was thus thought to be a major contributor to the early radio flux in this object. A model involving thermal emission from shocked gas resulting from the impact of a high-velocity wind with the principal ejecta was invoked. The mass-loss rate in this wind was estimated to be of order $10^{-5}\,M_{\odot}\,yr^{-1}$, lasting a period of 200–300 days after optical outburst.

Taylor *et al.* were able to make the first radio map of the ejecta of a classical nova shortly after outburst (see also Chapter 12). Fits to the last two epochs of images using the unified model with $T_e = 17\,000$ K yielded the values of parameters given in Table 7.2 (Taylor *et al.*, 1988).

7.6.3 *V705 Cas*

Radio emission from this moderately fast nova was detected in the radio 221 days after visual maximum and data were obtained at several frequencies between 1.4 GHz and 22 GHz (Eyres *et al.*, 2000). At this stage, the radio emission was still optically thick and the remnant was imaged from this time through to the optically thin decline phase. From a combination of optical spectroscopy and the radio imagery, the distance to the nova was determined to be 570 ± 40 pc. Further details of the imaging and spectroscopy of this nova remnant are given in Chapter 12.

7.6.4 *V723 Cas*

Heywood *et al.* (2002, 2005) observed this unusually slow nova with MERLIN on ten occasions from 477 to 2255 days after discovery, producing radio maps at nine epochs (note that the parameters given in Table 7.2 arise from the more detailed treatment given in Heywood *et al.*, (2005)). Again, the radio images revealed a complex structure which appeared to change radically over time (see also Chapter 12). These authors fitted the radio

Table 7.2. *Parameters derived from model fits (see text for detailed discussion)*

Object	V_o (km s^{-1})	V_i/V_o	M_{ej} ($M_\odot$)	KE (erg)	D (kpc)	Reference
HR Del	460	0.44	8.6×10^{-5}	9.9×10^{43}	0.8	Hjellming (1996b)
FH Ser	990	0.05	4.5×10^{-5}	1.5×10^{44}	0.65	*ibid.*
V1500 Cyg	5600	0.036	2.4×10^{-4}	2.6×10^{46}	1.4	*ibid.*
QU Vul	1010	0.87	3.6×10^{-4}	3.2×10^{45}	3.6	Taylor *et al.* (1988)
V1819 Cyg	690	0.2	9.0×10^{-5}	1.8×10^{44}	5.0	Hjellming (1996b)
V827 Her	1400	0.25	4.6×10^{-5}	3.9×10^{44}	2.6	*ibid.*
V838 Her	17600	0.042	1.0×10^{-4}	1.1×10^{47}	3.1	*ibid.*
V351 Pup	2500	0.069	1.0×10^{-3}	2.2×10^{46}	4.1	*ibid.*
V1974 Cyg	1940	0.46	3.1×10^{-4}	6.5×10^{45}	2.0	*ibid.*
V723 Cas	414	0.24	1.1×10^{-4}	8.1×10^{43}	2.4	Heywood *et al.* (2005)

data, which encompassed the initial optically thick rise, through a transition, and into an optically thin decline phase, with a Hubble flow incorporating a gas temperature of 17 000 K.

7.6.5 V1494 Aql

This very fast nova was optically one of the brightest of recent years, reaching $V \approx 4$ at maximum. Eyres *et al.* (2005) report a combination of optical spectroscopy, sub-millimetre (JCMT) and radio (VLA plus MERLIN) observations from 4 to 284 days post-outburst. A single epoch MERLIN image was obtained at 5 GHz on day 136 which again showed a structured remnant (see also Chapter 12). As was the case for V1974 Cyg, these authors find that the simple spherical shell models could not simultaneously fit the early sub-millimetre and later centimetre data. Their optical spectroscopy indicates cessation of mass-loss at $t \leq 15$ days post-maximum. A very simple geometrical model yielded $M_{ej} = 1.8 \times 10^{-5}\,M_\odot$ for a derived distance of 1.6 kpc. This is around an order of magnitude less than the ejected mass calculated from a Hubble flow model for this nova.

7.7 Estimates of distance, shell mass and kinetic energy

7.7.1 Distances using radio data

Radio observations allow estimation of the distances to novae by comparing the angular expansion rates derived from the radio with the absorption line velocities. In the case of older data where the shells are not resolved, but excellent multi-frequency data were obtained, the radio light curves were fitted to a Hubble flow model to get the angular expansion rates. The kinetic temperature must be assumed since the optically thick flux densities depend linearly on both the temperature and the angular area. The three novae yielding distances by this method were HR Del, FH Ser, and V1500 Cyg. Hubble flow angular velocities for the outer shells of these novae are given by Seaquist and Palimaka (1977); Hjellming *et al.* (1979) and Seaquist *et al.* (1980). Comparisons with maximum absorption line velocities yield respective radio distances for HR Del, FH Ser and V1500 Cyg of 0.91 kpc, 0.71 kpc, and 1.4 kpc if we assume $T_e = 10^4$ K. These are in good agreement with the respective optical distances of 0.85 kpc, 0.71 kpc, and 1.35 kpc (Kohoutek, 1981; Hutchings & Fisher,

1973; Becker & Duerbeck, 1980) and those used by Hjellming (1996b) (see Table 7.2). More recent determinations have relied on direct radio imaging to detect the angular expansion rates, which are then compared with maximum velocities (e.g. V1974 Cyg, QU Vul, V705 Cas). Distances using this method have, for example, been found for V1974 Cyg, QU Vul, and V705 Cas, yielding respectively 2.0 kpc (Hjellming, 1995), 3.6 kpc (Taylor *et al.*, 1987) and 0.57 kpc (Eyres *et al.*, 2000).

7.7.2 Shell masses and kinetic energies

Table 7.2 gives the masses and kinetic energies for a number of novae for which the values are well determined from the radio data, with the masses and velocity information compiled for the most part in Hjellming (1996b). The table includes V1819 Cyg, V827 Her, V838 Her, and V351 Pup, for which the observational data are otherwise unpublished. In all cases tabulated, a Hubble flow or unified model was used (the latter being the case for V1500 Cyg and V1974 Cyg, with t_1 approximately 60 days and 180 days respectively). The kinetic energies were derived from the masses and velocity information using Equation (7.20) in Section 7.4.1. Derived masses using Hubble flow models tend to be larger than those based on optical emission from the shell (e.g. Hjellming, 1996a,b), and the kinetic energies are also significantly larger. The reason is that the radio light curves are sensitive to the outer regions of the shell which contain significant amounts of mass and most of the kinetic energy. On the other hand, any method relying solely on the optically thin emission gives most weight to the inner dense parts of the shell since the emissivity is dependent on the square of the density. Thus optical methods very likely underestimate the mass and kinetic energy, whereas the Hubble flow method will overestimate these quantities if the actual density profile is steeper than r^{-2}.

7.7.3 Benefits and caveats regarding the radio analyses

Radio methods of determining the parameters of nova shells have both benefits and shortcomings. On the beneficial side, external and/or internal extinction by dust is negligible, and since the emission mechanism is almost universally thermal bremsstrahlung, a LTE analysis is exact. This is especially useful for resolved images which permit an estimate of the optically thick radio surface brightness and hence the electron temperature. As noted earlier, the radio light curves are sensitive to the outer portions of the nova shells which contain significant amounts of mass and kinetic energy. This is not the case for optical emission which is optically thin and weights the inner less massive and less energetic regions.

There are also a number of caveats. It is important to bear in mind that Hubble flow or unified models yield estimates of mass and kinetic energy which are model-dependent. The density gradient in these models is usually assumed to be an inverse square law (corresponding to uniformly distributed mass with expansion velocity) and any departure from this law will affect the derived shell parameters. In principle, the radio light curves contain information on the density gradient as well, but require data of exceptionally high quality at many frequencies, and the results based on such an analysis would still be model-dependent. Other assumptions implicit in most applications include spherical symmetry and a geometric filling factor ϕ of unity. Complex geometries (e.g. rings and polar caps) may lead to errors in the derived mass of several tens of percent (see e.g. Seaquist and Palimaka, 1977), and since thermal bremsstrahlung varies as the square of the local density, any clumping of the gas will lead to an overestimate of the mass by $\phi^{-1/2}$. The availability of resolved images of novae in the recent work can help enormously to remove geometrical ambiguities, and can lead to improved results, as shown by the ellipsoidal shell models fit to images of V1974 Cyg by Hjellming (1995).

It is clear that millimetre, sub-millimetre and infrared observations are revealing the inadequacy of the simplest Hubble flow and variable wind models. Bremsstrahlung emission at these wavelengths reaches its peak at much earlier times than at centimetre wavelengths and is thus sensitive to conditions earlier in the envelope evolution. Unified models involving a delayed detachment of the inner shell boundary appear to help here but other complex effects, such as variation of the kinetic temperature and incomplete ionization of the shell in the early phases of development, need also to be quantitatively considered.

7.8 Radio emission from related objects

7.8.1 Symbiotic stars

Several symbiotic stars have been classed as 'very slow' novae by various authors. These are objects characterized optically by eruptive events followed by a very slow decline lasting many decades or more. The origin of the outbursts may lie in thermonuclear events similar to those in classical novae (although for some symbiotics, gross changes in optical brightness may be more reasonably linked to accretion events). However, in the case of the very slow novae, there is a continuous or very prolonged mass-loss (Kwok, 1992). Their central systems also exhibit spectral characteristics of a red giant or Mira variable accompanied by highly excited emission lines. The presence of these evolved mass-donating stars, in what then must be a widely separated binary system, makes this class distinct from that of the classical novae (see Seaquist, Krogulec and Taylor (1993) for a radio survey of these objects and Bode (2004) for a review of radio imagery).

7.8.2 Recurrent novae

Radio detections have been claimed for three recurrent novae: T CrB (Turner, 1985), V745 Sco (Hjellming, 1989) and RS Oph (Padin, Davis & Bode, 1985; Hjellming *et al.*, 1986; O'Brien *et al.*, 2006). Of these three, only RS Oph has been studied extensively. We may note also that the detection of T CrB is open to question because of confusion problems (K. C. Turner, private communication).

Although RS Oph had had four previously recorded outbursts, that in January 1985 was the first to be studied in the radio. Emission at the level of 15 mJy at 5 GHz was detected by Padin, Davis & Bode (1985) only 18 days after the optical outburst. The flux density in 1982 February was less than 0.35 mJy (Seaquist, Taylor & Button, 1984), and this is presumably the quiescent level. The 18 day interval for the appearance of detectable radio emission is remarkably short compared to that for classical novae. Multi-frequency observations using the VLA (Hjellming *et al.*, 1986) showed that the radio emission peaked at around 1 month after outburst, and that the spectrum contained at least two components whose evolution was complex, and quite unlike that of classical novae. The high flux levels (over 60 mJy peak at 5 GHz), and implied brightness temperatures in excess of 10^7 K, meant that VLBI experiments were feasible (Porcas, Davis & Graham, 1987). These were carried out at 40 and 77 days after optical outburst at 5 and 1.7 GHz respectively. The results suggest that the ejection of material in the outburst was highly non-spherical, and that the radio emission comprised a thermal high-frequency component and a non-thermal low-frequency component (Taylor *et al.*, 1989).

Bode and Kahn (1985) developed a model closely resembling that of extragalactic Type II supernovae, some of which become non-thermal radio sources shortly after outburst (Weiler *et al.*, 1982; Sramek, Panagia & Weiler, 1984; Sramek & Weiler, 2003). Here, the high-velocity supernova ejecta interact with a slow-moving pre-outburst stellar wind. If a similar picture is invoked for RS Oph, then the total energy content of the relativistic electrons produced by the interaction is 0.02% of the total energy of the remnant. If this picture is correct, then RS Oph could be an important source of information for the study of high-energy particle acceleration in shocks.

Lloyd *et al.* (1993) considered the origin of the bipolar structure evident in the VLBI (EVN) map shown in Taylor *et al.* (1989). Two possibilities were investigated using a 2-D hydrodynamic code. The first of these was that the ejecta ran into an equatorially enhanced density distribution of the pre-existing wind of the red giant. However, it was found to be impossible to transport sufficient energy to the distances of the observed lobes of radio emission to account for the equipartition energies derived by Taylor *et al.* (1989). The second possibility that the ejecta leave the binary system in a highly collimated flow appeared more satisfactory. Lloyd and Bode (1996) show preliminary fits to the radio light curves on the basis of the hydrodynamic models of a bipolar explosion and assuming that the electrons responsible for the synchrotron emission are accelerated in the forward shock driven into the wind by the ejecta and the reverse shock travelling into the ejecta itself. Although the declines in radio flux could be relatively well fitted, the rise could not. Particularly difficult to account for is the fact that the radio flux peaks at all frequencies at almost the same time. It is suggested that the rise is due to the finite time required for diffusive acceleration of the relativistic electrons, which for reasonable values of the diffusion coefficient can be tens of days.

A further outburst of RS Oph occurred on 2006 Febraury 12. On this occasion, radio detection was made even earlier (4.4 days) and observations were made with the VLA, MERLIN, VLBA and EVN which promise to give unprecedented detail on the progress of the outburst in the radio (O'Brien *et al.*, 2006).

7.9 Concluding remarks

Radio observations of classical novae are an important probe of the physical conditions within, and parameters of, the expanding ejecta from the earliest times until several years after outburst. The increased sensitivity of radio arrays means that nowadays most optically discovered Galactic classical novae are potentially detectable at centimetre wavelengths. However, systematic and frequent radio monitoring campaigns are somewhat lacking at present.

Over the next few years the advent of e-MERLIN and EVLA, with their order of magnitude increases in sensitivity, should see all Galactic classical novae become readily observable radio sources. Many of these will be resolvable only weeks after outburst and of course, even at kiloparsec distances, such observations will be unaffected by interstellar extinction. In addition, the Atacama Large mm/sub-mm Array (ALMA) will give us complementary advances in imagery in the millimetre/sub-millimetre range and in the more distant future the Square Kilometre Array (SKA) promises to revolutionize the whole field of radio astronomy, including studies of Galactic novae.

Models of the radio evolution of novae have been relatively successful in fitting the data presently available. However, the aforementioned increases in sensitivity and resolution, plus extensive frequency coverage in instruments such as e-MERLIN, EVLA, SKA and ALMA, will place severe demands on theoretical models to account for the complexities that are likely to be detected. The models discussed at present are simplistic and somewhat ad hoc since they do not convey any physical insight into the origins of the observed shell behaviour. What are needed are physical models of the nova explosions predicting the morphology and velocity structure of the ejecta to fit to the data from these new telescopes.

Acknowledgements

The authors are grateful to Elizabeth Bode, Robert King, Matthew Ley and Andrew Newsam for help in preparation of the manuscript and figures. ‘E.R.S. acknowledges the support of a Discovery Grant from the Canadian University Funding Agency NSERC’. M.F.B. thanks the UK PPARC for the provision of a Senior Fellowship, during which this work was undertaken. We also note the untimely and tragic death of Robert M. Hjellming on 29 July 2000. Dr Hjellming made seminal contributions to the subject of radio emission from classical novae. We take this opportunity to pay tribute to his deep insights and communication skills which transformed this and other fields of the study of stellar radio emission.

References

Becker, R. H., & Duerbeck, H. W., 1980, *PASP*, **92**, 792.

Bode, M. F., 2004, in *Symbiotic Stars Probing Stellar Evolution*, ed. R. L. M. Corradi, J. Mikolajewska, & T. J. Mahoney. San Francisco: Astronomical Society of the Pacific, p. 359.

Bode, M. F., & Kahn, F. D., 1985, *MNRAS*, **217**, 205.

Bode, M. F., Seaquist, E. R., & Evans, A., 1987, *MNRAS*, **228**, 217.

Ennis, D., Becklin, E. E., Beckwith, J. *et al.*, 1977, *Ap. J.*, **214**, 478.

Eyres, S. P. S., Davis, R. J., & Bode, M. F., 1996, *MNRAS*, **279**, 249.

Eyres, S. P. S., Bode, M. F., O'Brien, T. J., Watson, S. K., & Davis, R. J., 2000, *MNRAS*, **318**, 1086.

Eyres, S. P. S., Heywood, I., O'Brien, T. J. *et al.*, 2005, *MNRAS*, **358**, 1019.

Ferland, G. J., 1980, *Ap. J.*, **236**, 847.

Ferland, G. J., & Shields, G. A., 1978, *Ap. J.*, **226**, 172.

Gallagher, J. S., 1977, *Ap. J.*, **82**, 209.

Gallagher, J. S., & Starrfield, S., 1978, *ARA&A*, **16**, 171.

Haddock, E. T., Howard, W. E. III, Malville, J. M., & Seling, T. V., 1963, *PASP*, **75**, 456.

Herrero, V., Hjellming, R. M., & Wade, C. M., 1971, *Ap. J.*, **166**, 149.

Heywood, I., O'Brien, T. J., Eyres, S. P. S., Bode, M. F., & Davis, R. J., 2002, in *Classical Nova Explosions*, ed. M. Hernanz, & J. Josó. New York: American Institute of Physics, p. 242.

Heywood, I., O'Brien, T. J., Eyres, S. P. S., Bode, M. F., & Davis, R. J., 2005, *MNRAS*, **362**, 469.

Hjellming, R. M., 1974, in *Galactic and Extragalactic Radio Astronomy*, ed. G. L. Vershuur, & K. I. Kellermann. New York: Springer, p. 159.

Hjellming, R. M., 1975, *IAUC*, 2853.

Hjellming, R. M., 1989, *IAUC*, 4853.

Hjellming, R. A., 1990, in *The Physics of Classical Novae*, ed. A. Cassatella, & R. Viotti. Berlin: Springer, p. 169.

Hjellming, R. M., 1991, *IAUC*, 5234.

Hjellming, R. M., 1992, *IAUC*, 5473.

Hjellming, R. A., 1995, in *Proceedings of 1994 Padua Conference on Cataclysmic Variables*, ed. A. Bianchini *et al.* Dordrecht: Kluwer, p. 139.

Hjellming, R. A., 1996a, in *Cataclysmic Variables and Related Objects*, ed. A. Evans, & J. H. Wood. Dordrecht: Kluwer, p. 317.

Hjellming, R. A., 1996b, in *Radio Emission from the Stars and the Sun*, ed. A. R. Taylor, & J. M. Paredes. San Francisco: Astronomical Society of the Pacific, p. 174.

Hjellming, R. M., & Wade, C. M., 1970, *Ap. J.*, **162**, L1.

Hjellming, R. M., Wade, C. M., Vandenberg, N. R., & Newell, R. T., 1979, *AJ*, **84**, 1619.

Hjellming, R. M., van Gorkom, J. H., Taylor, A. R. *et al.*, 1986, *Ap. J.*, **305**, L71.

Hutchings, J. B., Smolinski, J., & Grygar, J., 1972, *Publ. Dom. Astrophys. Obs.*, **14**, 17.

Hutchings, J. B., & Fisher, W. A., 1973, *PASP*, **85**, 122.

Ivison, R. J., Hughes, D. H., Lloyd, H. M., Bang, M. K., & Bode, M. F., 1993, *MNRAS*, **263**, L43.

Kohoutek, L., 1981, *MNRAS*, **196**, 87P.

Kwok, S., 1982, *The Nature of Symbiotic Stars*, ed. M. Friedjung, & R. Viotti. Dordrecht: Reidel, p. 17.

Kwok, S., 1983, *MNRAS*, **202**, 1149.

Lang, K. R., 1980, *Astrophysical Formulae*, 2nd edn. Berlin, Heidelberg, New York: Springer.

Lloyd, H. M., Bode, M. F., O'Brien, T. J., & Kahn, F. D., 1993, *MNRAS*, **265**, 457.

Lloyd, H. M., & Bode, M. F., 1996, in *Radio Emission from the Stars and the Sun*, ed. A. R. Taylor, & J. M. Paredes. San Francisco: Astronomical Society of the Pacific, p. 203.

Lloyd, H. M., O'Brien, T. J., & Bode, M. F. 1996, in *Radio Emission from the Stars and the Sun*, ed. A. R. Taylor, & J. M. Paredes. San Francisco: Astronomical Society of the Pacific, p. 200.

O'Brien, T. J., Bode, M. F., Porcas, R. W. *et al.*, 2006, *Nature*, **442**, 279.

Padin, S., Davis, R. J., & Bode, M. F., 1985, *Nature*, **315**, 306.

Pavelin, P. E., Davis, R. J., Morrison, L. V., Bode, M. F., & Ivison, R. J., 1993, *Nature*, **363**, 424.

Porcas, R. W., Davis, R. J., & Graham, D. A., 1987, in *RS Ophiuchi (1985) and the Recurrent Nova Phenomenon*, ed. M. F. Bode. Utrecht: VNU Science Press, p. 203.

Reynolds, S. P., & Chevalier, R. A., 1984, *Ap. J.*, **281**, L33.

Seaquist, E. R., 1989, in *Classical Novae*, 1st edn, ed. M. F. Bode, & A. Evans. New York and Chichester: Wiley, p. 143.

Seaquist, E. R., & Palimaka, J., 1977, *Ap. J.*, **217**, 781.

Seaquist, E. R., Duric, N., Israel, F. P. *et al.* 1980, *AJ*, **85**, 283.

Seaquist, E. R., Taylor, A. R., & Button, S., 1984, *Ap. J.*, **284**, 202.

Seaquist, E. R., Bode, M. F., Frail, D. A. *et al.* 1989, *Ap. J.*, **344**, 805.

Seaquist, E. R., Krogulec, M., & Taylor, A. R., 1993, *Ap. J.*, **260**, 274.

Snijders, M., Batt, P. T., Roche, P. F. *et al.*, 1987, *MNRAS*, **228**, 329.

Sramek, R. A., & Weiler, K. W., 2003, in *Lecture Notes in Physics: Supernovae and Gamma-Ray Bursters*, **598**, ed. K. W. Weiler. Berlin: Springer, p. 145.

Sramek, R. A., Panagia, N., & Weiler, K. W., 1984, *Ap. J.*, **285**, L59.

Strittmatter, P. A., Woolf, N. J., Thompson, R. I. *et al.* 1977, *Ap. J.*, **216**, 23.

Taylor, A. R., Seaquist, E. R., Hollis, J. M., & Pottasch, S. R., 1987, *A&A*, **183**, 38.

Taylor, A. R., Hjellming, R. M., Seaquist, E. R., & Gehrz, R. D., 1988, *Nature*, **335**, 235.

Taylor, A. R., Davis, R. J., Porcas, R. W., & Bode, M. F., 1989, *MNRAS*, **237**, 81.

Turner, K. C., 1985, in *Radio Stars*, ed. R. M. Hjellming, & D. Gibson. Dordrecht: Reidel, p. 283.

Wade, C. M., & Hjellming, R. M., 1971, *Ap. J.*, **163**, L65.

Warner, B., 1989, in *Classical Novae*, 1st edn, ed. M. F. Bode, & A. Evans. New York and Chichester: Wiley, p. 1.

Weiler, K. W., Sramek, R. A., van der Hulst, J. M., & Panagia, N., 1982, in *Supernovae: A Survey of Current Research*, ed. M. J. Rees, & R. J. Stoneham. Dordrecht: Reidel, p. 281.

8

Infrared studies of classical novae

Robert D. Gehrz

8.1 Introduction

Infrared observations have contributed substantially to our understanding of how classical novae participate in the chemical evolution of the Galaxy. We describe how infrared observations, combined with optical measurements, can provide quantitative measurements of the primary physical parameters that characterize the outburst, the abundances of elements that are present in the ejecta, and the properties of the grains that condense in the nova wind. We summarize recent evidence that novae are capable of producing large over-abundances of some metals and that they are potential sources of 'stardust' similar to the small grains that populate comet comae.

8.2 Nova explosions in the context of Galactic chemical evolution

Galactic classical novae take part in a cycle of Galactic chemical evolution in which grains and gas in the ejecta of evolved stars enrich the metal abundance of the Galactic 'ecosystem' (see Figure 8.1). Metals produced in stars during post-main-sequence (PMS) nucleosynthesis are ejected into the interstellar medium (ISM) by stellar winds and explosive events. Some of these metals remain in the gas phase, but others can condense to form grains. In the ISM, the gas and grains may become a component of the giant molecular clouds that give birth to new generations of young stars and planetary systems during the star formation process. Grains believed to be the remnants of the star formation phase are found in our own Solar System as inclusions in meteorites and in interplanetary dust particles (IDPs; Brownlee, 1987) as well as in the debris disks that surround main-sequence stars. Some of the grains in the meteorites appear to have formed in circumstellar outflows of evolved stars (see Bernatowicz and Zinner (1997) and references therein). Novae and supernovae participate in this cycle by expelling copious quantities of metal-rich gas during violent thermonuclear explosions. In these cases, the composition of the ejecta reflects both the PMS evolution of the precursor and nucleosynthesis during the explosion. Although supernovae process far more ISM material than novae, we will argue in this review that the nova explosion may be more efficient in producing some of the chemical anomalies that characterized the primitive Solar System.

Classical Novae, 2nd edition, ed. Michael Bode and Aneurin Evans. Published by Cambridge University Press.

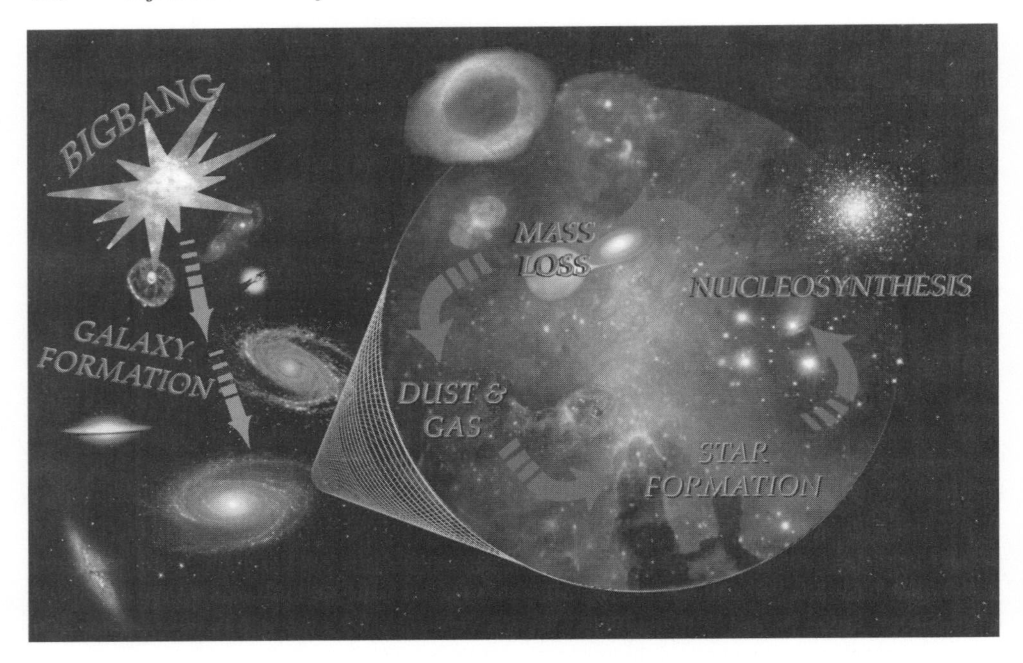

Fig. 8.1. The cycle that enriches the metal content of the ISM. The H/He-rich primordial matter produced in the Big Bang condensed into galaxies during the epoch of galaxy formation. Since then, nucleosynthesis in stars has created metals that are returned to the ISM by winds and explosions of evolved stars. These materials form the building block of planetary systems, debris disks, and life itself in subsequent generations of stars. Novae occupy a special niche in the explosive return of matter to the ISM.

8.3 The infrared temporal development of novae

A classical nova outburst results from a TNR on the surface of a white dwarf accreting matter from a late-type companion through the inner Lagrangian point in a close binary system (Truran, 1982; Starrfield, 1988, 1989, 1990, 1995; Shara, 1989; Livio, 1993; see also Chapter 4 for more recent developments). The partially degenerate base of the accreted layer eventually reaches the critical temperature for hydrogen burning leading to a TNR. The outburst luminosity rapidly approaches or exceeds the Eddington limit, $L_{\rm Edd}$, causing the ejection of a high velocity expanding shell enriched in metals by the TNR itself and by material dredged up from the surface of the WD during the explosion. Continued hydrogen burning in the shell powers the system at approximately constant luminosity for a time, after which the system relaxes to its pre-outburst state. Accretion is re-established and will lead to another outburst after a period of time determined by the steady-state accretion rate. The critical time-scales for the development of the TNR (hours), the ejection of the shell (hours, days, weeks), the constant luminosity phase (months, years), and the period between outbursts (hundreds to thousands of years), as well as the mass, composition, and physical characteristics of the ejected shell are all believed to be a function of the mass of the WD in the binary system.

Infrared observations have led to the identification of two fundamentally different types of novae, referred to hereafter as CO and ONe novae (see Gehrz, Truran & Williams, 1993;

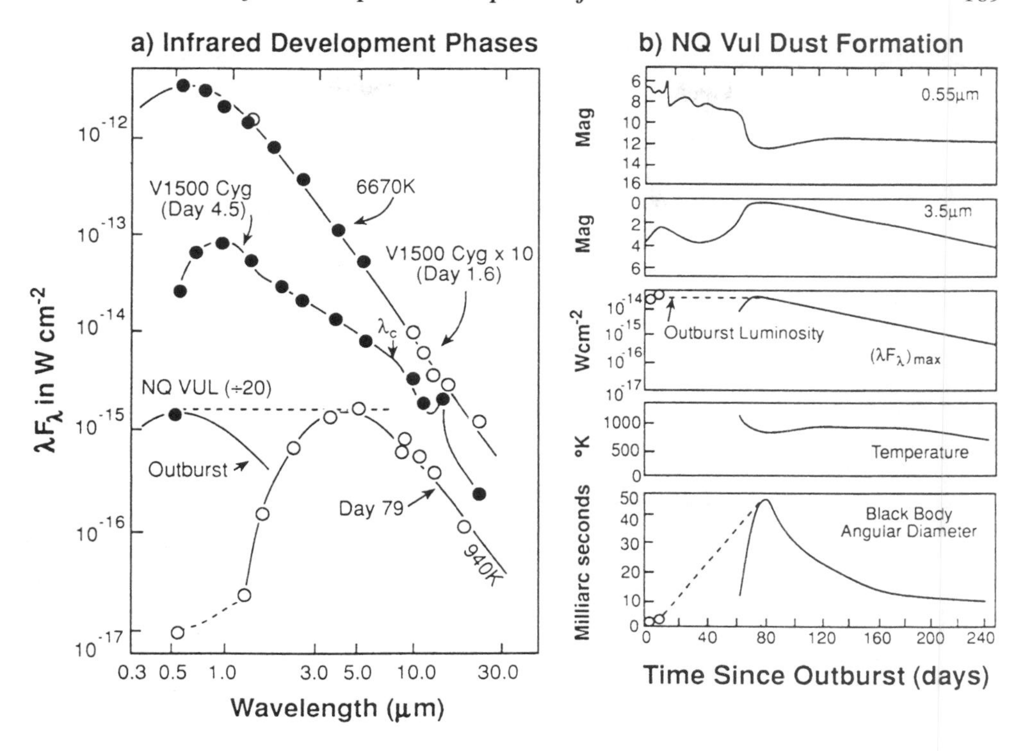

Fig. 8.2. Typical characteristics of several of the development phases of classical novae. a, Infrared broad-band photometric spectral energy distributions of the fireball of V1500 Cyg on day 1.6 (data from Gallagher and Ney (1976) and Ennis *et al.* (1977)), the optically thin free–free emission from V1500 Cyg on day 4.5 (data from Gallagher and Ney (1976)), and the carbon dust emission from NQ Vul on day 79 compared with its outburst fireball (data from Ney and Hatfield (1978)). b, The temporal development of the CO nova NQ Vul illustrating that the carbon dust formed at $\sim$ 1000 K and that the dust shell re-radiated the outburst luminosity (after Ney and Hatfield (1978)). The deep 'transition' in the visible light curve around day 70 coincides with the onset of the dust production as evidenced by the rise in the infrared 3.5 μm light. The quantity λF_λ is the relative flux observed at Earth. From Gehrz (1990).

Gehrz *et al.*, 1998; Gehrz, 1999, 2002). The most extreme examples of these can be readily distinguished in the infrared by their spectral characteristics and photometric light curves (see Gehrz, 1995; Gehrz *et al.*, 1995b). The mass of the WD is the critical factor in determining the nature of the outburst, with CO novae resulting from TNRs on low-mass CO WDs ($M_{\rm WD} \leq 1.2{\rm M}_\odot$) and ONe novae involving high-mass ONe WDs ($M_{\rm WD} \geq 1.2{\rm M}_\odot$). In both cases (Figure 8.2a), the hot gas expelled in the explosion initially appears in the optical/infrared as an expanding photosphere or 'fireball', and free–free and line emission dominate the spectral energy distribution (SED) as soon as the expanding material becomes optically thin. It is following the free–free emission phase that the infrared temporal development of the most extreme CO and ONe novae diverges significantly. The free–free phase in a CO nova is often followed by a dust formation phase (Figure 8.2b), while in ONe novae the free–free phase evolves into an extended coronal emission line phase attended by little or no dust production (Figure 8.3).

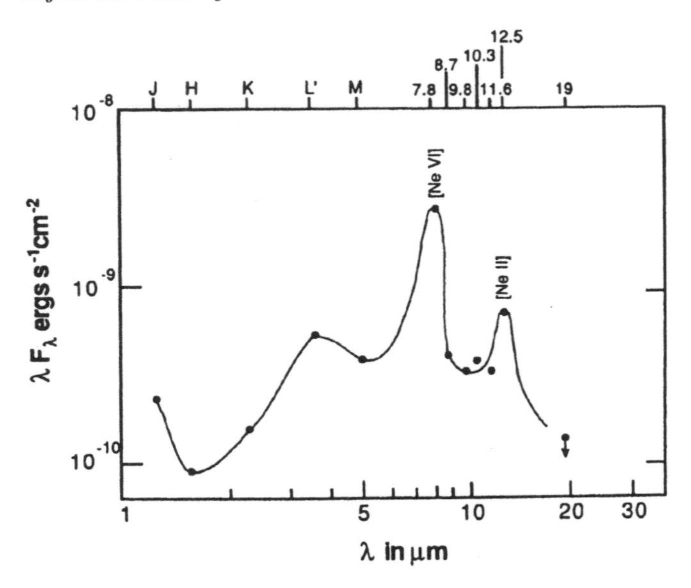

Fig. 8.3. Infrared broad-band photometric SED of QU Vul during its coronal phase. Note that the L' band emission is enhanced by [Al VI] λ3.66 μm and [Si IX] λ3.92 μm, and that forbidden neon emission strongly affects the fluxes in the 7.8 μm and 12.5 μm narrow-band filter set from the NASA Infrared Telescope Facility (IRTF). From Greenhouse *et al.* (1988).

CO novae that produce a significant amount of dust are distinguished by a sudden extinction event in the visible light curve due to obscuration of the central source by the condensation of a dust cloud. The extinction event is accompanied by rising thermal infrared emission from the dust. Dust is usually observed to form on a time-scale t_{cond} that usually varies from 30 to 80 days after outburst, and the condensation temperature of the grains is usually observed to be about 1000–1200 K (see Table 13.3 in Chapter 13). In the most extreme cases, such as NQ Vul (Ney & Hatfield, 1978), LW Ser (Gehrz *et al.*, 1980a) and V705 Cas (Gehrz *et al.*, 1995a), CO novae produce so much dust that a visually optically thick dust 'cocoon' completely obscures the central engine. Since these thick shells re-radiate the entire luminosity of the central engine in the thermal infrared, they act as calorimeters that can be used to assess the temporal development of the luminosity of the post-outburst remnant (Mason *et al.*, 1998).

In extreme ONe novae such as QU Vul (Gehrz, Grasdalen & Hackwell, 1985; Greenhouse *et al.*, 1988, 1990) and V1974 Cyg (Gehrz *et al.*, 1994; Hayward *et al.* 1992, 1996; Woodward *et al.*, 1995), near- and mid-infrared forbidden line radiation from highly ionized 'coronal' atomic states can persist for many years. A prominent early diagnostic of these novae is the presence of strong 12.8 μm [Ne II] emission, which has led to their popular characterization as 'neon' novae. This emission line was still persistent in the ejecta of QU Vul in May 2004, more than 19 years after the eruption (Gehrz *et al.*, 2007a).

Gehrz *et al.* (1995b) presented an analysis of the infrared light curves of recent Galactic classical novae showing that the 2.3 μm (K-band) and 3.6 μm (L-band) photometric light curves can be used to unambiguously identify extreme CO and ONe novae (see Figure 8.4). The K and L light curves of CO novae rise to a sharp peak within several months after outburst because of thermal emission from dust during the condensation phase, and they then decline

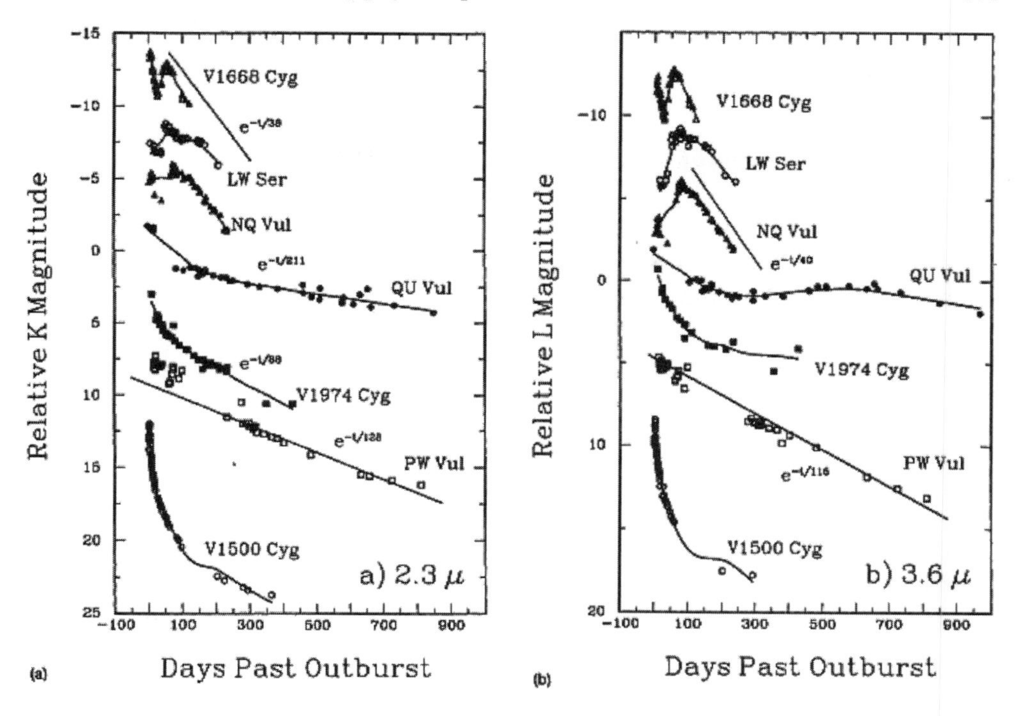

Fig. 8.4. K-band (a) and L-band (b) light curves of seven classical novae. The ONe novae QU Vul (data from Gehrz *et al.* (1995b)) and V1974 Cyg (Woodward *et al.* (1995)) are distinguished by elevated L-band emission during the coronal phase. The CO novae with visually optically thick dust shells (NQ Vul: data from Ney and Hatfield (1978); LW Ser: Gehrz *et al.* (1980a); V1668 Cyg: Gehrz *et al.* (1980b)) peak early during the dust formation phase and decline exponentially thereafter. V1500 Cyg (data from Gallagher and Ney (1976); Ennis *et al.* (1977)) and PW Vul (data from Gehrz *et al.* (1988a)) displayed neither strong coronal nor dust emission phases. From Gehrz *et al.* (1995b).

rapidly after grain growth ceases as the shell density decreases owing to expansion. The K and L light curves of ONe novae, on the other hand, decline rapidly following the outburst as the fireball becomes optically thin, and then show a leveling or even a rise as the 1.96 μm [Si VI], 2.04 μm [Al IX], 2.32 μm [Ca VIII], 2.47 μm [Si VII], 2.88 μm [Al V], 3.02 μm [Mg VIII], 3.66 μm [Al VI] and 3.92 μm [Si IX] forbidden lines strengthen during the coronal emission phase (cf. Woodward *et al.*, 1997). Imaging in the K- and L-band of nearby galaxies like M33 using large aperture ground-based telescopes such as GEMINI and space facilities such as NASA's Spitzer Space Telescope (Werner *et al.*, 2004, Gehrz *et al.*, 2007b; formerly Space Infrared Telescope Facility (SIRTF)) might be expected to provide a very accurate determination of the occurrence rates for various types of novae on a Galactic scale. This work is currently in progress under the author's Spitzer Guaranteed Time Observation (GTO) Program.

8.4 Determination of physical parameters

Infrared photometric and spectroscopic observations of novae in outburst are particularly useful for characterizing the nova event. Derivations of formulae for evaluating specific physical parameters from an analysis of the optical/infrared SEDs, and summaries of results on recent bright novae have been discussed in detail by a number of authors (Clayton & Hoyle, 1976; Clayton & Wickramasinghe, 1976; Clayton, 1979; Bode & Evans, 1989; Gehrz, 1988, 1990, 1993, 1995; Gehrz, 1998, 1999, 2002; Gehrz, Truran & Williams, 1993; Gehrz *et al.*, 1998; Mason *et al.*, 1998). We summarize briefly their arguments and conclusions. Physical parameters derived for a selected sample of the best-studied recent bright novae are given in Table 8.1.

In Table 8.1, the extinction A_V is derived from optical data; V_{ej} is derived from supplementary optical data in many cases. The distance D is derived from optical light-decline where infrared determinations of $d\theta_{BB}/dt$ and/or $d\theta_{dust}/dt$ are unavailable. The mass of ejected gas, M_{gas}, is calculated from the free–free self-absorption cut-off wavelength λ_c when possible, and from κ_T in all other cases. Values of M_{gas} and $d\theta_{gas}/dt$ for QU Vul were obtained from radio data; we assume that the expansion refers to the same gas that formed the fireball ($d\theta_{gas}/dt = d\theta_{BB}/dt$).

Optical/infrared spectrophotometry yields the apparent SED λf_λ of a nova that is t past outburst, where $t = 0$ is the time at which the TNR causes the ejection of the shell. Observations that track the expansion of the optically thick fireball can be used to determine the UT date for $t = 0$. The fireball will emit a black body SED with a temperature T_{BB}, a peak apparent flux density of $(\lambda f_\lambda)_{max}$, and a bolometric apparent flux of

$$F = \int_0^\infty f_\lambda \, d\lambda = \sigma T_{BB}^4 = 1.3586 \, (\lambda f_\lambda)_{max}$$

where $\sigma = 5.6696 \times 10^{-5}\,\text{erg}\,\text{s}^{-1}\,\text{cm}^{-2}\,\text{K}^{-4}$ is the Stefan–Boltzmann constant (see Gehrz & Ney, 1992). Since the black body angular diameter of the shell θ_{BB} (in milliarcsec) can be calculated from the Stefan–Boltzmann law to be:

$$\theta_{BB} = 2.02 \times 10^{14} \left(\frac{(\lambda f_\lambda)_{max}}{T_{BB}^4} \right)^{1/2}, \tag{8.1}$$

(with λf_λ in $\text{W}\,\text{cm}^{-2}$), a plot of $\theta_{BB}(t)$ versus t can be extrapolated back to $\theta_{BB}(0) = 0$ to determine the exact time of the ejection for cases where the angular expansion rate $d\theta_{BB}/dt$ is constant (see Ennis *et al.*, 1977). The same technique can be used to determine the angular size and expansion rates $d\theta_{dust}/dt$ of dust shells that are optically thick in the near- and thermal-infrared, and that emit like a black body at a temperature T_{BB} (Ney & Hatfield, 1978; Gehrz *et al.*, 1980a). In cases where it seems probable that the ejecta that formed dust are moving at the same velocity as the ejecta that earlier formed the photosphere, $d\theta_{BB}/dt = d\theta_{dust}/dt$. When the expansion velocity V_{ej} can be determined from optical/infrared spectroscopy of absorption lines during the photospheric expansion phase and emission lines during the free–free phase, the distance D (in kpc) to the nova can be determined from angular expansion rate $d\theta/dt$ (in milliarcsec day^{-1}; V_{ej} is in $\text{km}\,\text{s}^{-1}$) through the relationships

$$\begin{aligned} D \text{ (in kpc)} &= 1.15 \times 10^{-3} \frac{V_{ej}}{d\theta/dt} \\ &= 1.15 \times 10^{-3} \frac{V_{ej} t}{\theta(t)} \\ &= 5.69 \times 10^{-18} \frac{T_{BB}^2 V_{ej} t}{\sqrt{(\lambda f_\lambda)_{max}}}, \end{aligned} \tag{8.2}$$

Table 8.1. *Physical parameters of selected recent bright novae from optical, infrared and radio measurements*

Parameter	NQ Vul	LW Ser	V1668 Cyg	QU Vul	QV Vul	V838 Her	V1974 Cyg	V705 Cas
Year	1976	1978	1978	1984#2	1987	1991	1992	1993
References[a]	1	2	3	4–7	8	9	10–12	13–14
m_V max (mag)	+6.0	+8.5	+6.2	+5.7	+7.0	+5.3	+4.4	+5.3
A_V (mag)	1.8	~1.0	~0.7	~1	1.0	~1.0	—	1.6
t_2 (days)	42	34	15	27	50	2	17	64
t_3 (days)	64	55	30	40	—	5	35	90
V_{ej}(km s^{-1})	750	1250	1300	1–5000	420,1320	3500	2250	840
$d\theta_{BB}/dt$ (milliarcsec d^{-1})	—	—	—	0.151	0.056	—	—	—
D (kpc)	1.45	5.0	4.6	~3	4.72	~2.8	1.9	2.4
L_o ($L_\odot$)	7×10^4	4×10^4	10^5	10^5	8.6×10^4	1.5×10^5	9×10^4	5.5×10^4
t for κ_T (days)	8	6	6	—	20	5	≤ 10	—
$\lambda_c L$(μm)	—	—	7.5	—	—	—	9	—
t for λ_c (days)	—	—	9.5	—	—	—	10	—
M_{gas} ($M_\odot$)	1×10^{-4}	2×10^{-5}	10^{-4}	3×10^{-4}	3×10^{-5}	3.5×10^{-4}	4×10^{-4}	1×10^{-5}
t_c (days)	48	22	33	9–43	57	15	18	38
t_i (days)[b]	71	51	37	9–47	38–119	12	22	69
t_d (days)[c]	113	51	78	20–101	71–223	35	43	89
t_d, IR max (days)	80	75	57	240	83	18	$\leq$10	$\leq$131
$d\theta_{dust}/d_t$(milliarcsec d^{-1})	0.30	0.15	—	—	0.160	—	—	$\leq$0.06
T_c(K)	1100	1000	1150	—	—	1266	—	1150
T_c, IR max (K)	900	720	1000	—	780	1266	—	670
L_{IR} ($L_\odot$)	7×10^4	2.8×10^4	8×10^3	300	8.6×10^4	7.5×10^3	—	2.8×10^4
L_{IR}/L_0	1.00	0.70	0.08	3×10^{-3}	1.00	0.05	—	0.50
grain type(s)[d]	C	C	C	S	C, SC, S, H	C	—	C, S, H
a_{max} (μm)	0.71	0.50	0.36	—	0.63	0.18	—	0.7
M_d ($M_\odot$)	3.5×10^{-7}	1.6×10^{-6}	2.1×10^{-8}	$\sim 10^{-8}$	1.0×10^{-6}	4.8×10^{-9}	—	8.2×10^{-7}
M_{gas}/M_d	286	13	4.8×10^3	3×10^4	30	7.3×10^4	—	12
ρ_{crit} (g cm^{-3})	6.5×10^{-16}	3.0×10^{-16}	7.6×10^{-17}	1×10^{-15}	1.4×10^{-16}	7.2×10^{-16}	1.6×10^{-1}	9.0×10^{-17}

[a] References: 1. Ney & Hatfield (1978); 2. Gehrz *et al.*(1980a); 3. Gehrz *et al.* (1980b); 4. Gehrz, Grasdalen, & Hackwell (1985); 5. Gehrz *et al.* (1986); 6. Greenhouse *et al.* (1988, 1990); 7. Taylor *et al.* (1988); 8. Gehrz *et al.* (1992); 9. Woodward *et al.* (1992); 10. Hayward *et al.* (1992, 1996); 11. Gehrz *et al.* (1994); 12. Woodward *et al.* (1997); 13. Gehrz *et al.* (1995a); 14. Mason *et al.* (1998).

[b] From t_i (days) $\approx 2.2 \times 10^6 V_{ej}^{-1} L_0^{-1/3}$ (Gallagher, 1977)

[c] From $t_d \approx 320 L_0^{1/2} V_{ej}^{-1}$ (Gallagher, 1977)

[d] Grain types are carbon (C), silicon carbide (SC), silicates (S), and hydrocarbons (H)

where we have used Equation (8.1). The reader should note that this technique for measuring D must be applied with caution, for if the SED is gray, the inferred angular diameter is an upper limit to the true angular diameter, and the resultant distance is therefore a lower limit.

An alternative way of determining distances is to use a combination of infrared/radio imaging and spectroscopy (see Gehrz, 2002). The very high spatial resolution available with the VLA 'A' configuration, MERLIN, the HST infrared camera NICMOS, and infrared imagers and spectrometers operating on the new generation of large ground-based telescopes can resolve details as small as 0.05 to 0.2″. Thus, a typical nova shell expanding at $V_0 = 1000\,\mathrm{km\,s^{-1}}$ at a distance of $D = 2\,\mathrm{kpc}$ will have an angular expansion rate of $\mathrm{d}\theta/\mathrm{d}t \approx 0.1$ arcsec yr^{-1}, and can be well resolved by these facilities after $t = 1$–5 years. Again, the technique uses the concept embodied in Equations (8.2), except that $\mathrm{d}\theta/\mathrm{d}t$ is determined by direct measurement of the angular size of the ejected shell from the images. Expansion parallaxes derived from the radio images are somewhat uncertain because they require the assignment of an expansion velocity from either visual or infrared spectroscopy and some assumptions about the velocity gradients in the ejecta. On the other hand, expansion parallaxes derived from infrared images hold more promise of giving accurate values of D because infrared images can be made in the light of the very hydrogen lines whose profiles at spectral resolutions of $R = \lambda/\Delta\lambda \approx 1000$–2000 (150–300 km s^{-1}) can be used to determine V_{ej} at various positions within the ejecta using infrared long-slit spectroscopy (Starrfield, Truran & Gehrz, 1997). Radio expansion parallaxes from VLA and MERLIN images have been determined for QU Vul (Taylor *et al.*, 1988) and V1974 Cyg (Hjellming, 1995). Infrared expansion parallaxes from HST NICMOS images are available for QU Vul, QV Vul, and V1974 Cyg (Krautter *et al.*, 2002; see also Chapter 12).

A complicating factor in comparing distances obtained from radio and infrared imaging is that radio imaging has sensitivity that shows both the inner, slow-moving, high-density ejecta and outer, fast-moving, low-density ejecta in the 'Hubble flow' of the nova wind. On the other hand, infrared images refer primarily to the inner, slow ejecta. Because of the sensitivity of the radio data to faint flux levels and the high spatial resolution of the VLA and MERLIN, radio imaging is proving especially useful for studying the geometry of the ejecta of novae after one to five years (Bode, 2002). The geometrical non-uniformities that are clues to the physical processes that shaped the remnant are spatially well defined by this time (Taylor *et al.*, 1988; Hjellming, 1995; Heywood *et al.*, 2002).

Given a determination of the distance, it is then straightforward to calculate the optical/infrared luminosity L of the nova as a function of time using:

$$\begin{aligned}\frac{L(t)}{\mathrm{L}_\odot} &= 4.11\times 10^{17}\left(\frac{D}{\mathrm{kpc}}\right)^2\left[\frac{(\lambda f_\lambda(t))_{\max}}{\mathrm{W\,cm^{-2}}}\right]\\ &= 1.33\times 10^{-17}\,T_{\mathrm{BB}}^4\left(V_{\mathrm{ej}}t\right)^2.\end{aligned}\qquad (8.3)$$

Equation (8.3) evaluated at the visual light maximum during the photospheric expansion phase yields the outburst luminosity L_0. In all cases where there have been sufficient infrared data to provide an accurate determination of the outburst luminosity, L_0 has equaled or exceeded the Eddington luminosity $L_{\mathrm{Edd}} = 4\pi cGM\kappa_{\mathrm{T}}^{-1}$ of the WD believed to be responsible for supporting the TNR; L_{Edd} is the luminosity at which a star of mass M and mean Thomson scattering opacity κ_{T} becomes radiatively unstable and is thus susceptible to catastrophic mass-loss. The Thomson scattering opacity for a pure hydrogen atmosphere is $\kappa_T = 0.40\,\mathrm{cm^2\,g^{-1}}$, and $0.20\,\mathrm{cm^2\,g^{-1}}$ for an atmosphere rich in helium and metals. An accreted layer on a WD from material from a companion star with approximately solar composition will have a value of $\kappa_T \approx 0.33\,\mathrm{cm^2\,g^{-1}}$ so that the value of L_{Edd} in terms of solar masses is calibrated by the relationship

$$\frac{L_{\mathrm{Edd}}}{\mathrm{L_\odot}} = 3.8 \times 10^4 \left(\frac{M_{\mathrm{WD}}}{\mathrm{M_\odot}}\right).$$

For classical nova systems, $L_{\mathrm{Edd}} \approx 2.3$–$5.13 \times 10^4\,\mathrm{L_\odot}$, depending on the mass of the WD (Gehrz *et al.*, 1998). Observations of extreme CO novae with optically thick shells have demonstrated that the central engine in many novae maintains a constant luminosity near $L_0 \approx L_{\mathrm{Edd}}$ for many months during the hydrogen-burning shell phase that follows the outburst.

Infrared observations provide several independent methods for determining the ejected gas mass M_{gas}, a crucial parameter for constraining models of the TNR and establishing abundances in the ejecta. In some cases, the expanding photosphere can be observed often enough to establish the time t_T when the SED begins to depart significantly from a black body emission spectrum. The gas temperature at t_T is still high enough that Thomson scattering dominates the shell opacity and M_{gas} can be recovered from the relationship:

$$\frac{M_{\mathrm{gas}}}{\mathrm{M_\odot}} \approx \frac{\pi R^2}{\kappa_T} \approx 3.3 \times 10^{-13} \left(V_{\mathrm{ej}} t_T\right)^2, \tag{8.4}$$

where V_{ej} is in $\mathrm{km\,s^{-1}}$ and t_T is in days. Later, when the shell has cooled sufficiently for radiative transport to be dominated by the Kramers (free–free) opacity κ_{ff}, the hydrogen density n_H can be determined from the wavelength λ_c where the shell becomes optically thick because of free–free self absorption. Gehrz, Hackwell and Jones (1974) showed that λ_c, as defined by the intersection $(f_\lambda)_{\mathrm{ff}} = (f_\lambda)_{\mathrm{BB}}$ of the extrapolations of the Rayleigh–Jeans tails of the optically thin (free–free) and optically thick (free–free self absorbed) regions of the infrared SED, is the point at which the optical depth τ_{ff} of free–free emission spectrum is unity ($\tau_{\mathrm{ff}} = 1$). For a thin, constant-density 10^4 K gas shell of radius R and thickness $\ell = fR$, the information that $(f_\lambda)_{\mathrm{ff}} = (f_\lambda)_{\mathrm{BB}}$ leads to the relationships:

$$n_H^2 \ell \lambda_c^2 = n_H^2 f R \lambda_c^2 \approx 1.3 \times 10^{36}, \tag{8.5}$$

$$\text{and } M_{\mathrm{gas}} \approx 4\pi R^2 \ell n_H m_H \tag{8.6}$$

$$\approx 4\pi R^3 f n_e m_H \approx 8.2 \times 10^{-14} \frac{f^{1/2} \left(V_{\mathrm{ej}} t\right)^{5/2}}{\lambda_c}, \tag{8.7}$$

where $m_H = 1.67 \times 10^{-24}$ g is the mass of the hydrogen atom and $f = \ell/R$ is the ratio of the thickness to the radius of the shell. These equations can be expected to yield a reasonably accurate value for M_{gas} fairly early in the expansion of most nova shells when the clumps that appear later in the extended ejecta of old novae are beginning to evolve from instabilities in a fairly homogeneous medium. Note that Equations (8.6) and (8.7) apply to the case when $f \ll 1$. They reduce to the case of a constant-density sphere for $f = 1/3$. Since the free–free method measures only the mass of the ionized material, cases where the ionization is incomplete yield a lower limit to M_{gas}. It should be noted that the value of M_{gas} derived using the Thompson scattering method and the free–free self absorption technique are usually consistent for a given nova, and that they often substantially exceed the values predicted by TNR theory (see Gehrz *et al.*, 1998). Some ONe nova outbursts may eject shells 10 times as massive as predicted by current TNR theory, and the masses of some CO nova shells are much larger than predicted as well (see Table 8.1). The large ejected mass for the neon nova QU Vul has been confirmed by radio continuum observations (Taylor *et al.*, 1988; see also Chapter 7).

It is instructive to compare the ejected mass given by infrared observations with that given by using radio light curves to model the outflow (see Gehrz, 2002). However, when the radio intensity peaks, some hundred to several thousand days after the outburst depending upon the frequency being observed (see Chapter 7 for a detailed discussion), the interpretation of the data is complicated because velocity gradients have begun to dominate the shell thickness and density distribution. Hjellming (1995) and Hjellming *et al.* (1979) have shown that the expansion is best modeled by a 'Hubble flow' at this point, where the velocity increases outward linearly from the base to the outer edge of the flow, and where the shell density assumes an r^{-3} dependence. The technique is similar to the simplified theory that leads to Equations (8.5) and (8.7), except that one must integrate outward through the shell over the distributions for $V_0(r)$ and $n_\mathrm{H}(r)$ from the inner to the outer radii. These radii are well resolved in VLA and MERLIN images after several years for novae closer than 1–2 kpc.

All three of the methods of mass determination (infrared Thompson scattering, infrared free–free, and radio free–free) yield comparable estimates for the ejected mass for a given nova. Because radio observations detect lower-density, faster-moving, outer ejecta than do infrared measurements, the radio masses are somewhat higher, but by less than a factor of two (see Chapter 7). All methods show that the ionized ejecta of typical CO novae are in the range $1\text{–}10 \times 10^{-5}\,\mathrm{M}_\odot$ and that those of typical ONe novae are in the range $1\text{–}4 \times 10^{-4}\,\mathrm{M}_\odot$. These values are uncomfortably high compared with values predicted by current theories of the TNR (see Chapters 4 and 6). In fact, the problem of reconciling theoretical and observed ejected masses is exacerbated because a large amount of neutral gas may remain undetected by the observations described here. Saizar and Ferland (1994) and Ferland (1998) have suggested the possibility that these large ejected masses make novae potential competitors with supernovae as processors of ISM gas and contributors to ISM abundances. Ferland (1998) has stressed in his analysis of the ionization structure of nova ejecta that both infrared and radio determinations of M_gas, as well as those based on ultraviolet/optical line observations, may substantially underestimate the value of M_gas. In some cases, it is therefore possible that classical novae contribute far more mass to the ISM than heretofore believed.

Yet another conundrum seems to be posed by the infrared shell mass determinations. Theory predicts that the ejected shells of the extreme CO novae will tend to be of high mass and to have expansion velocities at the low end of the range observed in novae, while TNRs on massive ONe WDs should produce lower-mass, high velocity shells (see Gehrz *et al.*, 1998). In fact, the reverse seems to be true with respect to the ejected mass. Extreme neon novae like QU Vul and V1974 Cyg appear to have ejected shells as massive as $M_\mathrm{gas} = 3 \ldots 5 \times 10^{-4}\,\mathrm{M}_\odot$, whereas the masses ejected by some of the most extreme CO novae that produce copious amounts of dust (e.g., LW Ser, QV Vul, and V705 Cas) appear to be about 10 times lower ($M_\mathrm{gas} \approx 1\text{–}3 \times 10^{-5}\,\mathrm{M}_\odot$).

8.4.1 *Coronal line emission in ONe novae*

During the free–free emision phase, the ejecta in extreme neon novae are strongly cooled by near- and mid-infrared forbidden lines from highly ionized isotopes of metals such as CNO, Ne, Mg, Al, Si, and S (Figures 8.5, 8.6, and 8.7). These lines have been termed 'coronal' because of the high excitation conditions under which they arise. In these cases, the relative abundance by number n_X/n_Y of ionization species Y and X can be determined from infrared emission line ratios using the Milne relations (Osterbrock, 1974, 1989), as applied

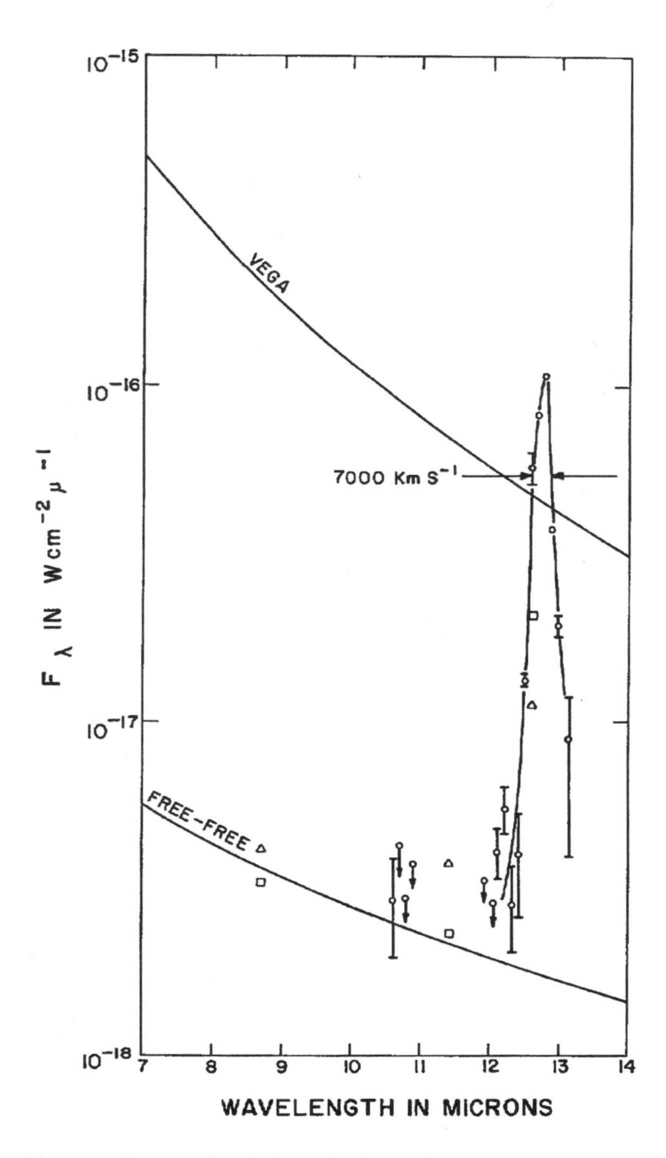

Fig. 8.5. The [Ne II] λ12.8 μm forbidden line in the spectrum of QU Vul on day 140 was the strongest such line relative to the continuum ever observed in an astrophysical source. From Gehrz, Grasdalen and Hackwell (1985).

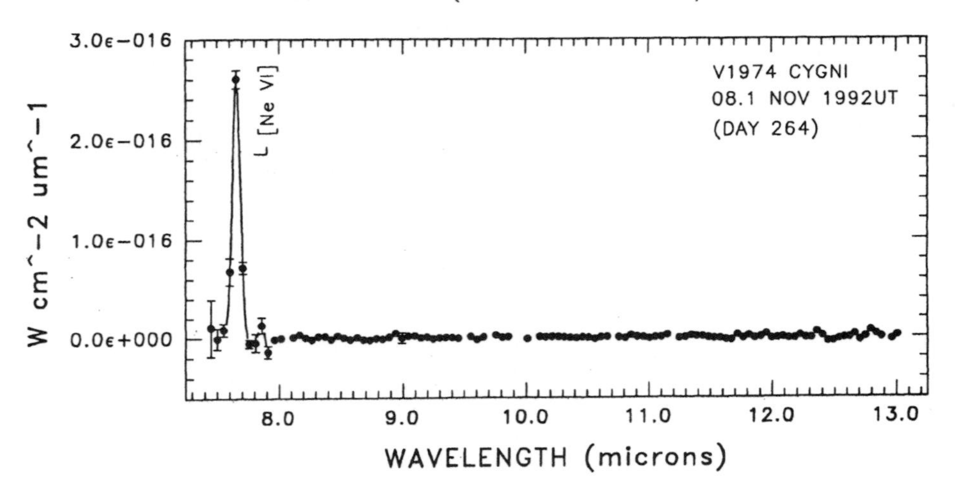

Fig. 8.6. The [Ne VI] $\lambda 7.6$ μm forbidden line in the spectrum of V1974 Cyg on day 264. Note that [Ne II] $\lambda 12.8$ μm is barely detectable at this stage. From Gehrz *et al.* (1994).

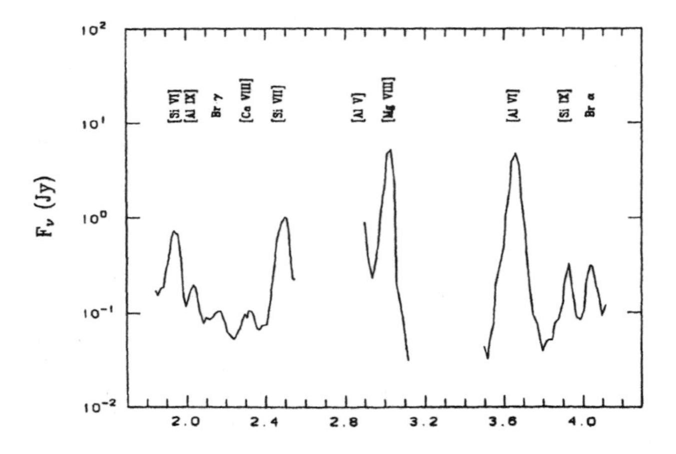

Fig. 8.7. The near-infrared coronal lines in QU Vul. From Greenhouse *et al.* (1988).

by Greenhouse *et al.* (1990), Gehrz *et al.* (1994) and Woodward *et al.* (1995); the radiative-collisional equilibrium code *CLOUDY* (Ferland *et al.*, 1998b; Ferland, 2004) as applied by Hayward *et al.* (1996) and other investigators (see references in Gehrz *et al.* (1998)); or the theory of collisionally excited forbidden line radiation (Osterbrock, 1989) as applied by Gehrz, Grasdalen and Hackwell (1985), Hayward *et al.* (1992) and Gehrz *et al.* (2007a).

When the shell cooling is dominated by free–free emission and the [Ne II] 12.8 μm forbidden emission line, the collisional excitation theory is especially straightforward because the [Ne II] ion is essentially a two-level atom. In these cases, typified by QU Vul (Gehrz, Grasdalen & Hackwell, 1985) and V1974 Cyg (Hayward *et al.*, 1992), the gas number density n_{H} comes from free–free continuum measurements,

$$n_{\mathrm{H}} = 4.67 \times 10^{21} \frac{\exp(7195\lambda T)\, T^{1/4}\, D\, \lambda^{1/2}\, [\lambda f_\lambda]^{1/2}_{\text{free-free}}}{\left(V_{\text{ej}} t\right)^{3/2}}, \tag{8.8}$$

can be compared with the gas number density of [Ne II] ions implied by the intensity of the 12.8 μm [Ne II] emission line $I_{12.8}$ to deduce the abundance of Ne II with respect to hydrogen $n_{\mathrm{NeII}}/n_{\mathrm{H}}$ through the relationship:

$$n_{\mathrm{NeII}}/n_{\mathrm{H}} = 3.48 \times 10^{-11} \frac{I_{12.8}\, n_{\mathrm{H}}}{\lambda f_\lambda \,|_{\lambda=12.8\,\mu\mathrm{m}}}. \tag{8.9}$$

In all cases described above, it is paramount to remember that the total masses of various species in the ejecta derived from a single 'monochromatic' wavelength regime may be severely underestimated. Saizar and Ferland (1994), and Ferland (1998), have pointed out that the very high excitation conditions that occur during the temporal development of novae make it exceedingly difficult to observe optical/infrared line emission from many ionization states of the gas so that abundances derived for a given element by observations of coronal lines must be considered as lower limits. Given the data summarized in Table 8.2, it is clear that classical novae may be important sources for processing of ISM material and for the production of some of the $6 \leq Z \leq 14$ isotopic abundance anomalies. Two, in particular, are the ^{22}Ne (Neon-E) and ^{26}Mg anomalies that are believed to have characterized the primitive solar nebula based upon analysis of meteorites (Gehrz *et al.*, 1998; see also Chapter 6).

Considerable velocity structure is observed in infrared coronal lines at spectral resolutions of $R = \lambda/\Delta\lambda \geq 1000$, and one can, in principle, use this information to construct a detailed model of the density, temperature, and velocity structure of the ejecta. Saizar and Ferland (1994) and Hayward *et al.* (1996) have shown that the structure observed in the [Ne II] and [Ne VI] lines of V1974 Cyg are consistent with an expanding shell composed of cooler ($T \approx 10\,000$ K) dense clumps embedded within a very low density, high temperature ($T \geq 500\,000$ K) gas component (Figure 8.8).

8.4.2 *Grain formation in CO novae*

The first evidence that dust grains could condense in nova ejecta resulted from infrared observations of FH Ser (Geisel, Kleinmann & Low, 1970; Hyland & Neugebauer, 1970) showing an increase in the thermal infrared that coincided with the optical light decline. Clayton and Hoyle (1976), Clayton and Wickramasinghe (1976) and Clayton (1979) used the FH Ser data to lay the foundations of an elementary theoretical characterization of dust formation in nova ejecta. Ney and Hatfield (1978), Gehrz *et al.* (1980a,b), Gehrz (1990), Gehrz *et al.* (1995a), and Mason *et al.* (1996, 1998) have refined this theoretical picture substantially using extensive data sets that detail the infrared temporal development of a number of dust-forming CO novae including NQ Vul, LW Ser, V1668 Cyg, QV Vul, and V705 Cas.

In all cases where the dust-formation episode has been well documented, the primary condensation event is characterized by an abrupt increase in the thermal infrared when the grains reach a critical condensation temperature of $T_{\text{cond}} \approx 1000$–$1200$ K (see Table 13.3 in

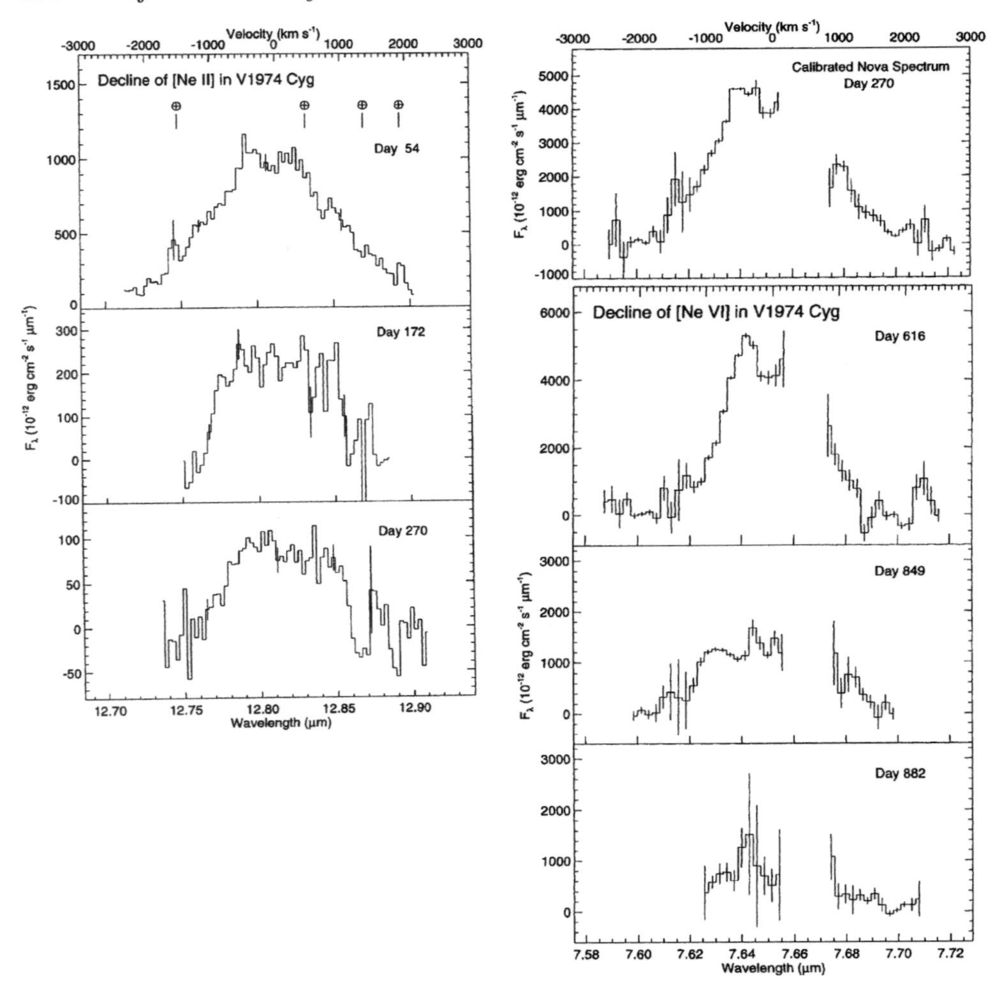

Fig. 8.8. High resolution (250 $\mathrm{km\,s^{-1}}$) spectra of the [Ne II] and [Ne IV] lines showing that there is velocity structure. From Hayward *et al.* (1996).

Chapter 13). Presumably, this event signals the time t_{cond} when the ejecta flowing outward at constant velocity V_{ej} have reached the base of the condensation zone R_{cond} where:

$$R_{\mathrm{cond}} = \left(\frac{L}{16\pi\sigma T_{\mathrm{cond}}^4}\right)^{1/2} = 1.18\times10^{12}\left(\frac{L}{\mathrm{L}_\odot}\right)^{1/2} \quad \mathrm{cm}$$

$$\text{and} \quad t_{\mathrm{cond}} = \frac{R_{\mathrm{cond}}}{V_{\mathrm{ej}}} = 137\left(\frac{L}{\mathrm{L}_\odot}\right)^{1/2}\left(\frac{V_{\mathrm{ej}}}{\mathrm{km\,s^{-1}}}\right)^{-1}.$$

The visual extinction event (transition) produced by the grains coincides with a rise in the infrared caused by thermal re-radiation of the absorbed energy. From the very abrupt onset of the transition phase, one can infer that the grains grow from nucleation to very large sizes within a matter of a few weeks. Shore and Gehrz (2004) have proposed that the photo-ionization of grain surfaces by the hard SED of the central engine is capable of inducing rapid grain growth in nova ejecta. Although infrared observations through the end of the 1970s indicated that the infrared SEDs of dust-forming novae were essentially black and gray bodies (Figure 8.9), a number of dusty novae observed since 1980 show optically thin SEDs that depart significantly from black body behavior. For novae in which the dust shell is optically thin in the infrared as evidenced by the significant departure of the SED continuum from a black body spectrum at long wavelengths, and/or by the presence of emission

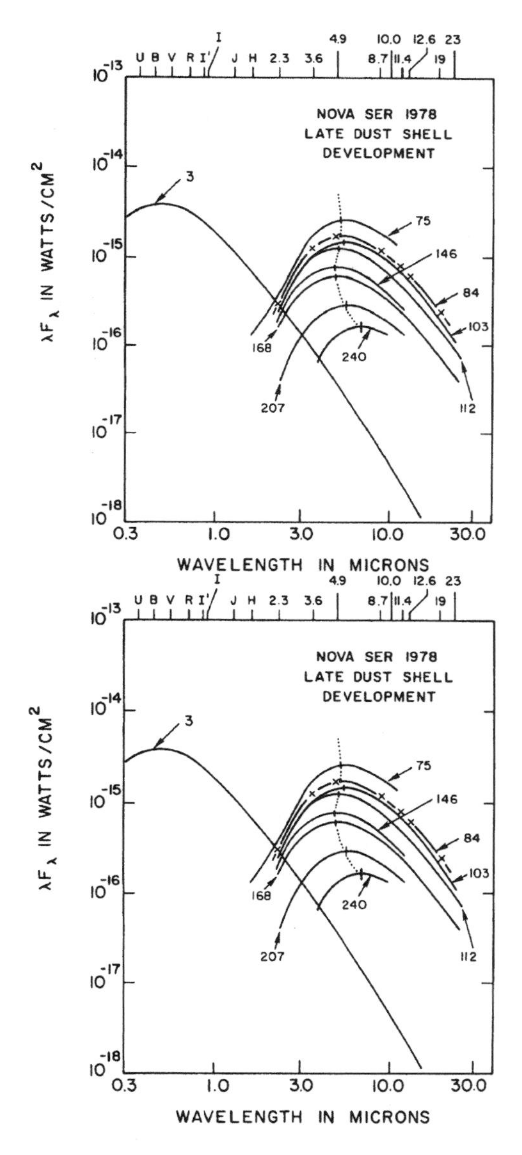

Fig. 8.9. Infrared SEDs during the dust formation phase of LW Ser. The smooth continuum is characteristic of emission from small carbon grains. From Gehrz *et al.* (1980a).

features from silicates, SiC, or hydrocarbons, the physical grain temperature T_{gr} and grain radius a must be recovered by fitting the infrared SED using the function

$$\lambda f_\lambda = \pi \lambda B_{\lambda,T_{gr}}(1 - e^{-\tau_{\lambda,T_{gr}}}) = \frac{2\pi hc^2}{\lambda^4} \frac{(1 - e^{-\tau_{\lambda,T_{gr}}})}{(e^{hc/\lambda k T_{gr}} - 1)}, \tag{8.10}$$

where $B_{\lambda,T_{gr}}$ is the Planck function for temperature T_{gr}, $\tau_{\lambda,T_{gr}} = (\kappa_{\lambda,T_{gr}})\rho\ell$ is the optical depth of a dust shell of thickness ℓ and opacity $\kappa_{\lambda,T_{gr}}$. Opacities for various astrophysical grains as a function of mineral composition and grain radius a are given by Gilman (1974a,b), Draine (1985), and Hanner (1988). Two typical examples of the use of this fitting technique to derive grain size and temperature information are the recent analyses of the infrared SEDs of nova V1425 Aql (Mason *et al.*, 1996) and V705 Cas (Mason *et al.*, 1998).

Almost all of the extreme CO novae that formed visually optically thick dust shells had infrared SEDs that can be fitted by a Planckian energy distribution from 2 to 20 μm at infrared maximum, suggesting that the individual grains and/or dust clumps in the ejecta have a significant optical depth in the infrared. For these cases, the shell black body color temperature T_{BB} and the physical temperature of the grains T_{gr} are essentially equal at infrared maximum. The subsequent temporal evolution of the SEDs of these novae suggests that the grains are carbon with radii $a \leq 1\,\mu$m and a Planck mean emission efficiency of $\langle Q_e \rangle \approx 0.01 a T_{gr}^2$ (where a here is in cm). Assuming that the maximum radius a to which the dust grains grow is signaled by the infrared maximum at t_d (days), it follows that

$$a \approx 2 \times 10^{22} \frac{L_{IR}}{\left(V_{ej} t_d\right)^2 T_{gr}^6}, \tag{8.11}$$

where L_{IR} is the shell luminosity at infrared maximum. In this case, the total mass of carbon dust M_{dust} in the shell is given by

$$\begin{aligned} M_{dust} &= 1.17 \times 10^6 \frac{\rho_d L_{IR}}{T_{BB}^6}\ M_\odot \\ &= 4.81 \times 10^{23} \frac{\rho_d D^2 (\lambda f_\lambda)_{max}}{T_{BB}^6} \\ &= 1.56 \times 10^{-11} \frac{\rho_d \left(V_{ej} t\right)^2}{T_{BB}^6}, \end{aligned}$$

where ρ_d is the density of an individual dust grain. Equation (8.11) might be expected to hold as long as the visual/near-infrared optical depth of the shell is low enough that the grain cross-sections do not cover substantially ($\tau \leq 1$–2, or 37–86% absorption) the light of the central engine.

Based upon a more sophisticated radiative transfer model, *DUSTY* (Ivezic & Elitzur, 1997), Evans *et al.* (2005) have presented an argument that Equation (8.11) may not apply to the case of very dusty novae at infrared maximum in novae where the shell reaches an appreciable visual/near-infrared optical depth, and have concluded that the grains that grow in the ejecta may be much smaller than the sizes predicted by Equation (8.11). There appear to be two counter-arguments to this line of reasoning. First, the expansion velocity of the dust shell indicated by the Evans *et al.* model exceeds the observed Doppler expansion velocity by a factor of 3. If this velocity is inserted into Equation (8.11) for parameters at the onset of transition, the grain size discrepancies between the two models are resolved. Second, the analysis presented by Evans *et al.* refers to a time well past maximum shell optical depth when other processes such as sputtering and sublimation may have already conspired to decrease the average grain size that might have been attained at infrared maximum, as argued by Gehrz *et al.* (1980a,b). Mitchell and co-workers have proposed several plausible mechanisms for grain destruction in the nova wind (Mitchell, Evans & Bode, 1983; Mitchell & Evans, 1984; Mitchell, Evans & Albinson, 1986).

The abundance of the dust grain constituents by mass is given by the ratio $M_{\rm dust}/M_{\rm gas}$. Gehrz and Ney (1987) established that novae that condensed optically thick dust shells prior to 1987 met the criterion that the mass density of the ejecta at the condensation radius $R_{\rm cond}$ exceed the value $\rho_{\rm crit}$, where

$$\rho_{\rm crit} \geq 1.2 \times 10^{-4} \frac{M_{\rm gas}}{L^{3/2}} \approx 3 \times 10^{-16} \ \ {\rm g\ cm^{-3}} \ . \tag{8.12}$$

Clearly, from the more recent data included in Table 8.1, it would appear that several novae that failed to form extensive dust shells meet this criterion, and that there must be some additional factor that inhibits dust production in extreme neon novae like QU Vul and V1974 Cyg. Although Gallagher (1977) has argued that the time $t_{\rm d}$ for the grains to grow to an appreciable size is $t_{\rm d} \approx 320 L^{1/2} V_{\rm ej}^{-1}$ and that dust production will be muted or suppressed if the shell ionization time $t_{\rm i} \approx 2.2 \times 10^6 \, V_{\rm ej}^{-1} L^{-1/3}$ is less than $t_{\rm d}$, in fact $t_{\rm i} < t_{\rm d}$ for almost every nova that forms a significant amount of dust. As suggested by the work of Shore and Gehrz (2004), ionization may actually promote rather than retard grain growth.

In the case of dusty CO novae, the visual extinction properties of optically thick dust shells and analysis of molecular vibrational features from dust emission in optically thin shells can be used to deduce the abundances of the dust constituents (Table 8.2) as well as the gas-to-dust ratio (Table 8.1).

8.5 Abundances in nova ejecta

In order for novae to contribute to ISM chemical evolution on an equal footing with other stellar populations, Gehrz *et al.* (1998) have estimated that they must exhibit abundance elevations of $\sim$250 with respect to the Sun and $\sim$10 with respect to Type II supernovae. It is evident from Table 8.2 that abundance elevations of this magnitude are probably characteristic of certain metals in some CO and ONe novae. In particular, it should be noted that the argument presented by Saizar and Ferland (1994) and Ferland (1998) that emission from many ionization states might go undetected because their line transition energies lie outside the optical/infrared region raises the distinct possibility that the abundances given in Table 8.2 may be conservative lower limits. This statement also holds true for the abundances derived from dust emission from CO novae, since these determinations do not take into account condensibles that may remain in the gas phase. We conclude that classical novae may be significant contributors to Galactic chemical evolution, if not on a global scale, at least on local scales where their Galactic orbital trajectories take nova-producing binary star systems through or near regions of star formation. In particular, CO novae may contribute significantly to the ISM abundances of certain CNO isotopes, and ONe novae may be implicated in producing the ^{26}Mg and Neon-E (^{22}Ne; Black, 1972) anomalies found in solar system meteorites (see Gehrz *et al.*, 1998).

Table 8.2. *Chemical abundances in the ejecta of classical novae from infrared observations; note that the abundances here are presented in a different way from that in Table 6.1*

Nova	X[a]	Y	$\frac{(n_X/n_Y)_{nova}}{(n_X/n_Y)_\odot}$	References
LW Ser	Carbon dust	H	$\geq$15	(1)
QU Vul	Ne	H	$\geq$1.2	(2)
	Ne	H	$\geq$168	(3)
–	Ne	Si	$\geq$6	(4)
–	Mg	Si	~5	(5)
–	Al	Si	~70	(5)
QV Vul	carbon dust	H	$\geq$7	(6)
V1974 Cyg	C	H	~12	(7)
–	N	H	~50	(7)
–	O	H	~25	(7)
–	Ne	H	$\geq$12	(8)
–	Ne	H	~50	(7)
–	Ne	O	~4	(9)
–	Ne	Si	~35	(8)
–	Mg	H	~5	(7)
–	Mg	Si	$\geq$3	(10)
–	Al	H	~5	(10)
–	Al	Si	~5	(10)
–	Si	H	~6	(7)
–	S	H	~5	(7)
–	Ar	H	~5	(7)
–	Fe	H	~4	(7)
V705 Cas	Silicate dust	H	~15	(11)
–	Silicate dust	H	~20	(12)
–	O	H	$\geq$25	(13)
–	Ca	H	~20	(14)
V1425 Aql	C	He	~9	(15)
–	N	He	~100	(15)
–	O	He	~9	(15)
V723 Cas	N	H	~75	(16)
–	O	H	~17	(16)
–	Ne	H	~27	(16)
CP Cru	S	Si	~2.1	(17)
–	Ca	Si	~1.6	(17)
–	Al	Si	~1.5	(17)

[a] The abundances cited for 'dust' are calculated using the amount of gas condensed into grains (see Equation (8.7)).

References: (1) Gehrz *et al.* (1980a); (2) Gehrz, Grasdalen & Hackwell (1985); (3) Gehrz *et al.* (2007a); (4) Greenhouse *et al.* (1990); (5) Greenhouse *et al.* (1988); (6) Gehrz *et al.* (1992); (7) Hayward *et al.* (1996); (8) Gehrz *et al.* (1994); (9) Salama *et al.* (1996); (10) Woodward *et al.* (1995); (11) Gehrz *et al.* (1995a); (12) Mason *et al.* (1998); (13) Salama *et al.* (1999a); (14) Salama *et al.* (1999b); (15) Lyke *et al.* (2001); (16) Evans *et al.* (2003); (17) Lyke *et al.* (2003).

8.6 Nova grain properties

Infrared observations of classical novae are the primary source of data on grain nucleation and growth in circumstellar outflows. Of particular interest are the mineral content of the grains and their size distribution as compared with those of grains found in the outflows of other evolved stellar systems, in the ISM, and in regions of star formation (including the primitive Solar System). Novae have been observed to produce astrophysical dust of virtually every known chemical and mineral composition. In several cases, the signatures of three or four different grain components have been observed during the temporal development of the ejecta of a single outburst. Most of the CO novae that produce optically thick dust shells produce dust that shows a black-body-like thermal continuum. The dust in these novae is believed to be primarily composed of amorphous carbon grains, which are expected to have a smooth and gray emissivity as a function of wavelength in the near- and mid-infrared. In at least three cases, carbon, silicates, and hydrocarbons are required to explain all of the features seen in the spectrum of a single nova. These are V842 Cen (see Gehrz, 1990), QV Vul (see Figures 8.10 and 8.11; Gehrz *et al.*, 1992) and V705 Cas (Mason *et al.*, 1998). Mason *et al.* (1998) and Evans *et al.* (1997) showed in independent spectrophotometric analyses of V705 Cas that hydrocarbon emission can significantly affect the shape of the 10 μm silicate emission feature in some novae (see Figure 8.12).

Furthermore, Mason *et al.* (1998) showed that, to explain the entire thermal SED of V705 Cas in detail, one must consider emission from a carbon continuum component, a number of hydrocarbon vibrational lines with Lorentzian profiles, a silicate grain component, and a free–free emission component from the hot gas. In all cases where silicate grains have formed, the

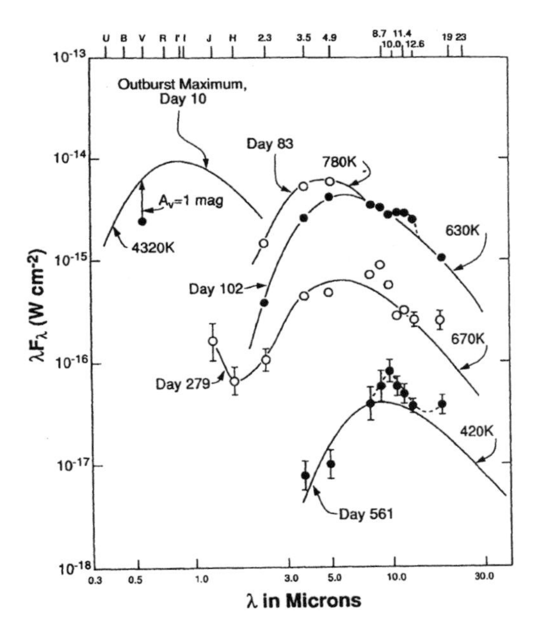

Fig. 8.10. Development of the infrared SED of QV Vul showing evidence for the presence of several types of astrophysical grains in the ejecta. All SEDs have the near-infrared continuum expected for small carbon grains. Silicon carbide appears to have been present on day 102 and silicate on day 561. From Gehrz *et al.* (1992).

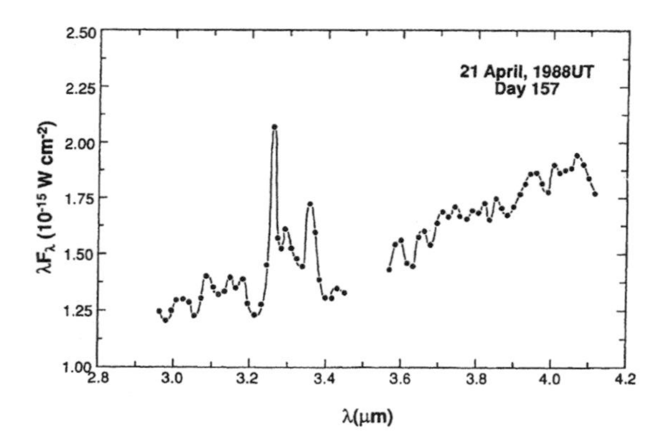

Fig. 8.11. Near-IR SED of QU Vul showing strong 3.28 μm and 3.3–3.4 μm emission from hydrocarbon grains. From Gehrz *et al.* (1992).

silicates appear to form after the carbon forms and fairly late in the shell development process. This may imply either that the ejecta have zones with different C to O ratios or that non-equilibrium conditions govern the condensation process. It is possible that the carbon grains form in the initial fast-moving ejecta and that the subsequent production of silicate grains occurs after the constant luminosity phase when $R_{\rm cond}$ moves inward through the ejecta. These lower, slower-moving zones will have a composition determined by the extended burning phase following the outburst.

Analysis of the visual and infrared light curves of novae suggests that the grains in the dust shells initially grow rapidly to radii as large as 0.2 to 0.7 μm when the extinction event reaches its maximum depth (Table 8.2). Ultraviolet observations of V705 Cas with the International Ultraviolet Explorer (IUE) satellite by Shore *et al.* (1994) provided independent evidence that nova grains grow rapidly to large sizes. Their analysis of the flat ultraviolet extinction curve showed that the grains had grown to radii larger than 0.2 μm in only about 70 days. However, there is evidence that nova grains may be somewhat smaller than the maximum radius to which they initially grow by the time they reach the ISM. Late in the temporal development of the dust phase, the grains in some novae become superheated and less efficient at extinguishing the light from the central engine. Both of these properties are consistent with a decrease in size, which is presumed to occur because of sputtering or evaporation of the more volatile components as the radiation of the central engine hardens (Mitchell, Evans & Bode, 1983; Gehrz, 1988, 1990, 1993, 1995, 1999, 2002).

The physical measurement of the superheat $S = T_{\rm gr}/T_{\rm BB}$ of a grain, where $T_{\rm gr}$ is the grain color temperature observed from infrared photometry and $T_{\rm BB}$ is the temperature that a black sphere would have at the same distance from the central engine, can be used to deduce the radius of the grain given its mineral characteristics (see Gehrz & Ney, 1992). In a constant velocity outflow, black grains should decrease in temperature as $t^{-1/2}$, where t is the time since outburst. In the cases of NQ Vul, LW Ser, and QV Vul, for example, 240 days after the outburst the grains had developed an average superheat of $S = 1.41 \pm 0.03$ with respect to the temperature that would have been predicted based on the condensation time $t_{\rm c}$. Using the relationships given in Table 8.1 and grain opacities calculated for carbon from Mie theory (see Gilman, 1974a,b; Draine, 1985), one can verify that this superheat is consistent with

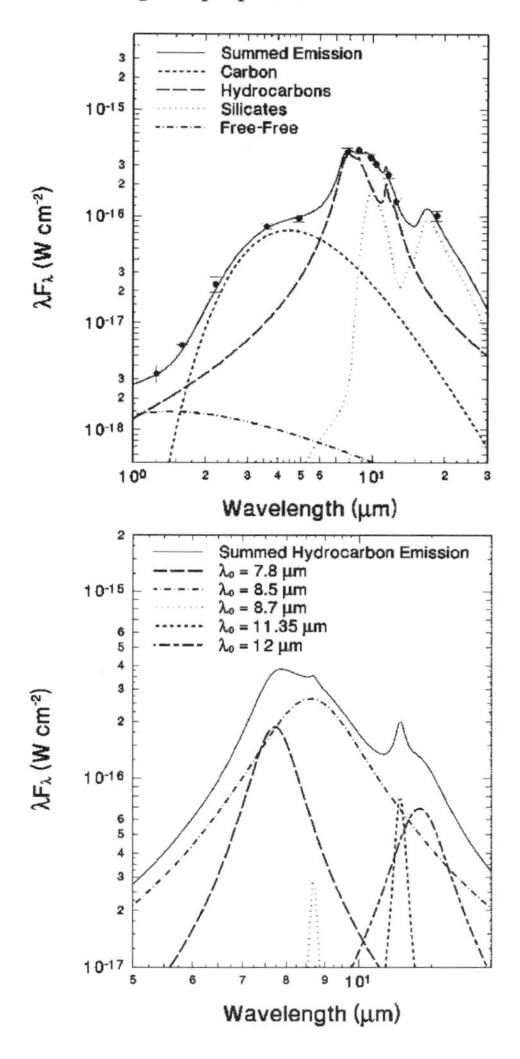

Fig. 8.12. Infrared SED of V705 Cas. Amorphous carbon (short dashed curve), hydrocarbon (long dashed curve), silicate (filled circles), and free–free (dot-dashed curve) emissions combine to fit observations. (b) The hydrocarbon feature is comprised of major emission components at 7.7 and 8.7 μm and broad emission components centered at 8.5 and 12.0 μm. From Mason *et al.* (1998).

grain radii in the range 0.1 to 0.3 μm, quite a bit smaller than the value of $a_{\max}$. It should be noted that although these grains are still several times larger than the small grains believed to be responsible for the general interstellar extinction, they are comparable in size to the small grains in the sub-structures of the IDPs discovered by Brownlee (1987). It is interesting to note that the radii and compositions of the grains that grow in novae (Table 8.1) are similar to those of the grain populations that dominated the thermal emission from comets (Gehrz & Ney, 1992; Williams *et al.*, 1997). Presuming that the grain size parameters given by studies of novae are generally indicative of the size of grains that condense in stellar outflows ('stardust'), the similarities between nova and comet grains can be taken as evidence that stardust was present in the primitive Solar System. The proof that any individual grain from a comet was actually produced in a nova can, of course, only be obtained by chemical analysis of IDPs (Alexander, 1997) and comet grains from sample return missions.

8.7 Spitzer Space Telescope observations of classical novae

The Infrared Spectrometer (IRS; Houck *et al.*, 2004) on the NASA Spitzer Space Telescope (Werner *et al.*, 2004; Gehrz *et al.*, 2007b) with its high sensitivity and extended wavelength capabilities, is proving to be a powerful tool for investigating line emission in classical novae both during outburst and after the accretion phase has been re-established. A number of metallic coronal forbidden emission lines not visible from the ground have been detected, and the high-resolution modes ($R = 600$) have sufficiently high velocity resolution to allow elementary modeling of the gross dynamics of the ejecta. There are currently three Spitzer programs observing target-of-opportunity novae in outburst (Woodward, 2006) and relatively old novae (Evans, 2006; Gehrz *et al.*, 2007a).

Lynch *et al.* (2006) have made the most comprehensive set of ground-based and space-based observations of the early outburst phase yet produced for any classical nova (see Figure 8.13). Their data set on V1187 Sco document nicely the transition from the hydrogen recombination phase to the pure coronal phase in the neon nova. Gehrz *et al.* (2007a) and Schwarz *et al.* (2007) have discovered that there is strong coronal forbidden line emission from the ejecta of the moderately old novae QU Vul and V383 Vel (see Figures 8.14 and 8.15), in the latter case 1613 days after outburst, while the much older novae V1974 Cyg and V705 Cas still show strong coronal forbidden Ne and O emission 4319 and 3874 days

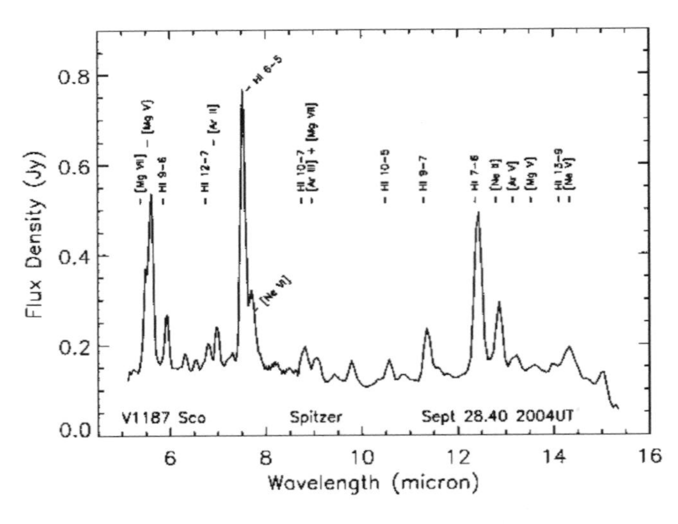

Fig. 8.13. NASA Spitzer IRS short-low IR spectrum of V1187 Sco 53 days (light gray solid line) and 227 days (black solid line) after outburst showing that hydrogen emission was fading as coronal forbidden lines increased in strength with time. Figure reproduced by permission of L. A. Helton *et al.* (in preparation).

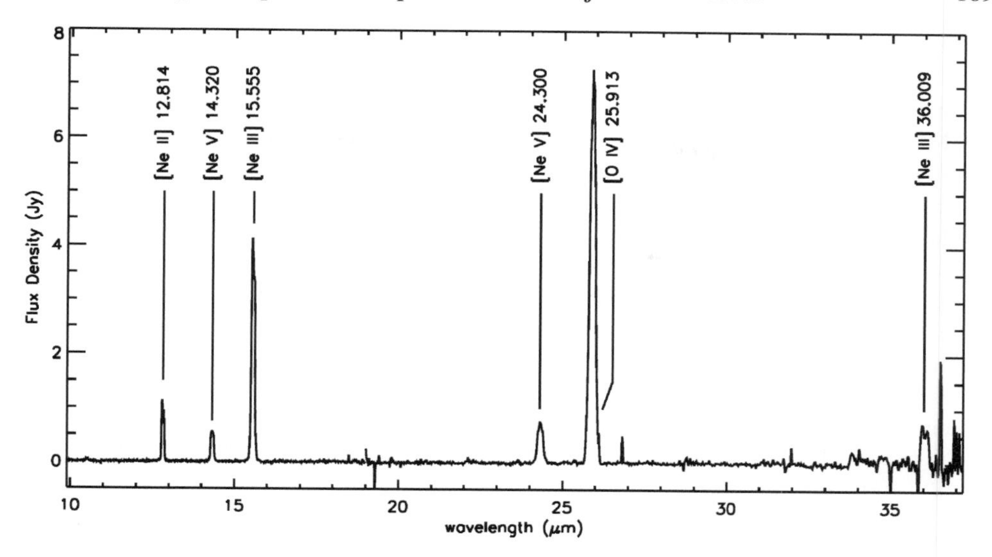

Fig. 8.14. Left and right panels are NASA Spitzer IRS short-high and long-high IR spectrum of V382 Vel 1613 days after outburst. Strong forbidden line emission was still persistent at this late date. The separation of the castellated peaks is consistent with an expansion velocity of 600 km s^{-1}. Figure reproduced by permission of L. A. Helton *et al.* (in preparation).

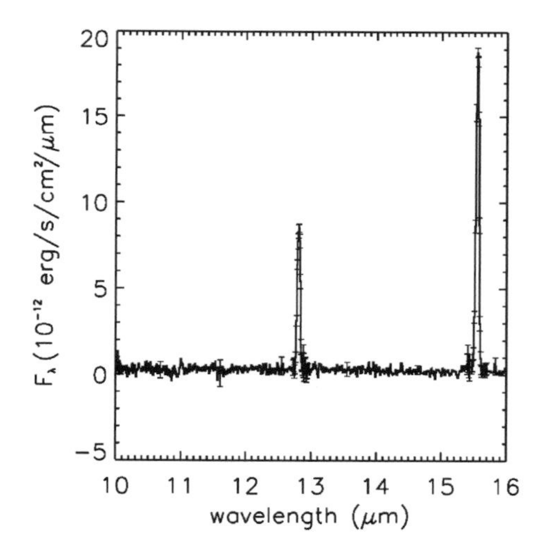

Fig. 8.15. NASA Spitzer IRS short-high IR spectrum of QU Vul 7077 days after outburst, showing that strong forbidden lines ([Ne II] λ12.8 μm and [Ne III] λ15.5 μm) were still persistent. From Gehrz *et al.* (2007a).

after outburst. In the case of QU Vul, coronal emission is seen an astonishing 7077 days after outburst. For this object, initial calculations based on detailed level balancing suggest that very high abundances (>168 times solar) are required to explain the strengths of some of these lines (Gehrz *et al.*, 2007a).

8.8 Summary and conclusions

Infrared measurements of classical novae can be used to quantify parameters describing the outburst and to determine metal abundances in their ejecta. The overabundances in CNO, Ne, Mg, Al, and Si in some novae are high enough to suggest that novae are potential contributors to some chemical anomalies on local and Galactic scales. K- and L-band light curves can be used to distinguish the temporal development of extreme CO and ONe, presenting the possibility that IR photometry can be used for determining their occurrence rates and global influence in nearby galaxies. Novae can produce every known type of astrophysical grain, in some cases showing the signatures of three or four different grain components during the temporal development of a single outburst. Neon novae may be capable of producing dust grains that carry the Ne-E and ^{26}Mg anomalies. Nova grains grow as large as 0.2 to 2 μm in radius. The similarities between nova and comet grains suggest that stardust may have been present in the primitive Solar System.

Acknowledgements

Major portions of this manuscript were excerpted with permission from Elsevier Science from *Physics Reports* **311**, 'Infrared studies of classical novae and their contributions to the ISM' by Robert D. Gehrz, pages 405–418, Copyright 1999. C. E. Woodward read the manuscript with a critical eye and offered substantive suggestions. The author acknowledges support from the NASA, the NASA SIRTF Project Office, the NSF, the University of Minnesota Graduate School, and the University of Minnesota Institute of International Studies and Programs.

References

Alexander, C. M. O'D., 1997, in *Astrophysical Implications of the Laboratory Study of Presolar Materials*, ed. T. J. Bernatowicz, & E. Zinner. New York: American Institute of Physics, p. 567.

Bernatowicz, T. J., & Zinner, E., 1997, eds., *Astrophysical Implications of the Laboratory Study of Presolar Materials*. New York: American Institute of Physics.

Black, D. C., 1972, *Geochim. Cosmochim. Acta*, **36**, 377.

Bode, M. F., 2002, in *Classical Nova Explosions*, ed. M. Hernanz, & J. José. New York: American Institute of Physics, p. 497.

Bode, M. F., & Evans, A., 1989, in *Classical Novae*, 1st edn, ed. M. F. Bode, & A. Evans. New York and Chichester: Wiley, p. 163.

Brownlee, D. E., 1987, *Phil. Trans. R. Soc.* Series A, **323**, 305.

Clayton, D. D., 1979, *Ap. &SS*, **65**, 179.

Clayton, D. D., & Hoyle, F., 1976, *Ap. J.*, **203**, 490.

Clayton, D. D., & Wickramasinghe, N. C., 1976, *Ap. &SS*, **42**, 463.

Draine, B. T., 1985, *Ap. JS*, **87**, 587.

Ennis, D., Bedewith, S., Gatley, I. *et al.*, 1977, *Ap. J.*, **214**, 478.

Evans, A., 2006, Spitzer Classical Nova GO Program PID 20262, http://ssc.spitzer.caltech.edu/geninfo/go/abs-go2/20262.txt.

Evans, A., Geballe, T. R., Rawlings, J. M. C., Eyres, S. P. S., & Davies, J. K. 1997, *MNRAS*, **292**, 192.

Evans, A., Gehrz, R. D., Geballe, T. R. *et al.*, 2003, *AJ*, **126**, 1981.

Evans, A., Tyne, V. H., Smith, O. *et al.* 2005, *MNRAS*, **360**, 1483.

Ferland, G. J., 1998, in *Wild Stars in the Old West: Proceedings of the 13th North American Workshop on Cataclysmic Variables and Related Objects*, ed. S. Howell, E. Kuulkers, & C. E. Woodward. Astronomical Society of the Pacific, **137**, p. 165.

Ferland, G. J., Korista, K. T., Verner, D. A. *et al.*, 1998, *PASP*, **110**, 761.

Ferland, G. J., 2004, http://www.nublado.org/.

Gallagher, J. S., 1977, *AJ*, **82**, 209.

Gallagher, J. S., & Ney, E. P., 1976, *Ap. J.*, **204**, L35.

Gehrz, R. D., 1988, *ARA&A*, **26**, 377.

Gehrz, R. D., 1990, in *Physics of Classical Novae*, ed. A. Cassatella, & R. Viotti. Berlin: Springer, p. 138.

Gehrz, R. D., 1993, *Annals of the Israel Physical Society*, **10**, 100.

Gehrz, R. D. 1995, in *Proceedings of the Padua (Abano-Terme) Conference on Cataclysmic Variables*, ed. A. Bianchini, M. Della Valle, & M. Orio. Dordrecht: Kluwer, p. 29.

Gehrz, R. D., 1998, in *Wild Stars in the Old West: Proceedings of the 13th North American Workshop on Cataclysmic Variables and Related Objects*, ed. S. Howell, E. Kuulkers, & C.E. Woodward. Astronomical Society of the Pacific, **137**, p. 146.

Gehrz, R. D., 1999, in *Processes in Astrophysical Fluids*, ed. O. Regev, & D. Prialnik, *Phys. Rep.*, **311**, 405.

Gehrz, R. D., 2002, in *Classical Nova Explosions*, ed. M. Hernanz, & J. José. New York: American Institute of Physics, p. 198.

Gehrz, R. D., 2006a, NASA Spitzer Space Telescope M33 GTO program, PID 5, http://ssc.spitzer.caltech.edu/geninfo/gto/abs/pid5.

Gehrz, R. D., 2006b, NASA Spitzer Space Telescope Nova GTO program, PID 122, http://ssc.spitzer.caltech.edu/geninfo/gto/abs/pid122.

Gehrz, R. D., & Ney, E. P., 1987, *PNAS*, **84**, 6961.
Gehrz, R. D., & Ney, E. P., 1992, *Icarus*, **100**, 162.
Gehrz, R. D., Hackwell, J. A., & Jones, T. W., 1974, *Ap. J.*, **191**, 657.
Gehrz, R. D., Grasdalen, G. L., & Hackwell, J. A., 1985, *Ap. J.*, **298**, L47; erratum 1986, *Ap. J.*, **306**, L49.
Gehrz, R. D., Grasdalen, G. L., Hackwell, J. A., & Ney, E. P., 1980a, *Ap. J.*, **237**, 855.
Gehrz, R. D., Hackwell, J. A., Grasdalen, G. L. *et al.*, K., 1980b, *Ap. J.*, **239**, 570.
Gehrz, R. D., Grasdalen, G. L. Greenhouse, M. A. *et al.*, 1986, *Ap. J.*, **308**, L63.
Gehrz, R. D., Harrison, T. E., Ney, E. P. *et al.*, 1988a, *Ap. J.*, **329**, 894.
Gehrz, R. D., Jones, T. J., Woodward, C. E. *et al.*, 1992, *Ap. J.*, **400**, 671.
Gehrz, R. D., Truran J. W., & Williams, R. E., 1993, in *Protostars & Planets III*, ed. E. H. Levy, & J. I. Lunine. Tucson: University of Arizona Press, p. 75.
Gehrz, R. D., Woodward, C. E., Greenhouse, M. A. *et al.*, 1994, *Ap. J.*, **421**, 762.
Gehrz, R. D., Greenhouse, M. A., Hayward, T. L. *et al.*, 1995a, *Ap. J.*, **448**, L119.
Gehrz, R. D., Jones, T. J., Matthews, K. *et al.*, 1995b, *AJ*, **110**, 325.
Gehrz, R. D., Truran, J. W., Williams, R. E., & Starrfield, S. G. 1998, *PASP*, **110**, 3.
Gehrz, R. D., Woodward, C. E., Helton, L. A. *et al.*, 2007a, *Ap. J.*, in press.
Gehrz, R. D., Roellig, T. L., Werner, M. W. *et al.*, 2007b, *Rev. Sci. Instrum.*, **78**, 011302.
Geisel, S. L., Kleinmann, D. E., & Low, F. J., 1970, *Ap. J.*, **161**, L101.
Gilman, R. C., 1974a, *Ap. JS*, **28**, 397.
Gilman, R. C., 1974b, *AJ*, **188**, 87.
Greenhouse, M. A., Grasdalen, G. L., Hayward, T. L., Gehrz, R. D., & Jones, T. J., 1988, *AJ*, **95**, 172.
Greenhouse, M. A., Grasdalen, G. L., Woodward, C. E. *et al.* 1990, *Ap. J.*, **352**, 307.
Hanner, M. S., 1988, *NASA Conf. Pub.* **3004**, 22.
Hayward, T. L., Gehrz, R. D., Miles, J. W., & Houck, J. R., 1992, *Ap. J.*, **401**, L101.
Hayward, T. L, Saizar, P., Gehrz, R. D. *et al.* 1996, *Ap. J.*, **469**, 854.
Heywood, I., O'Brien, T. J., Eyres, S. P. S., Bode, M. F., & Davis, R. J., 2002, in *Classical Nova Explosions*, ed. M. Hernanz, & J. José. New York: American Institute of Physics, p. 242.
Hjellming, R. M., 1995, in *Cataclysmic Variables*, ed. A. Bianchini *et al.* Dordrecht: Kluwer, p. 139.
Hjellming, R. M., Wade, C. M., Vandenberg, N. R., & Newell, R. T., 1979, *AJ*, **84**, 1619.
Houck, J. R., Roellig, T. L., van Cleve, J. *et al*, 2004, *Ap. JS*, **154**, 18.
Hyland, A. R., & Neugebauer, G., 1970, *Ap. J.*, **160**, L177.
Ivezic, Z., & Elitzur, M., 1997, *MNRAS*, **287**, 799.
Krautter, J., Woodward, C. E., Schuster, M. T. *et al.*, 2002, *AJ*, **124**, 2888.
Livio, M., 1993, in *22nd SAAS-Fee Advanced Course, Interacting Binaries*, ed. H. Nussbaumer & A. Orr. Berlin: Springer, p. 135.
Lyke, J. E., Gehrz, R. D., Woodward, C. E. *et al.*, 2001, *AJ*, **122**, 3305.
Lyke, J. E., Koenig, X. P., Barlow, M. J. *et al.*, 2003, *AJ*, **126**, 1005.
Lynch, D. K., Woodward, C. E., Geballe, T. R., *et al.*, 2006, *Ap. J.*, **638**, 987.
Mason, C. G., Gehrz, R. D., Woodward, C. E. *et al.*, 1996, *Ap. J.*, **470**, 577.
Mason, C. G., Gehrz, R. D., Woodward, C. E. *et al.*, 1998, *Ap. J.*, **494**, 783.
Mitchell, R. M., & Evans, A., 1984, *MNRAS*, **209**, 945.
Mitchell, R. M., Evans, A. & Bode, M. F., 1983, *MNRAS*, **205**, 1141.
Mitchell, R. M., Evans, A., & Albinson, J. S., 1986, *MNRAS*, **221**, 663.
Ney, E. P., & Hatfield, B. F., 1978, *Ap. J.*, **219**, L111.
Osterbrock, D. E., 1974, *Astrophysics of Gaseous Nebulae*. San Fransisco: Freeman.
Osterbrock, D. E., 1989, *Astrophysics of Gaseous Nebulae and Active Galactic Nuclei*. Mill Valley: University Science Books.
Saizar, P., & Ferland, G. J., 1994, *Ap. J.*, **425**, 755.
Salama, A., Evans, A., Eyres, S. P. S. *et al.*, 1996, *A&A*, **315**, L209.
Salama, A., Eyres, S. P. S., Evans, A., Geballe, T. R., & Rawlings, J. M. C., 1999a, *MNRAS*, **304**, L20.
Salama, A., Gehrz, R., Woodward, C. E., & Barlow, M., 1999b, in *The Universe as Seen by ISO*. ESA Publications SP-427, **1**, p. 233.
Shara, M. M., 1989, *PASP*, **101**, 5.
Shore, S. N., & Gehrz, R. D., 2004, *A&A*, **417**, 695.
Shore, S. N., Starrfield, S., Gonzalez-Riestra, R., Hauschildt, P. H., & Sonneborn, G., 1994, *Nature*, **369**, 539.
Starrfield, S. G., 1988, in *Multi-wavelength Studies in Astrophysics*, ed. F. Cordova. Cambridge: Cambridge University Press, p. 159.

Starrfield, S. G., 1989, in *Classical Novae*, ed. M. F. Bode, & A. Evans. New York and Chichester: Wiley, p. 123.
Starrfield, S. G., 1990, in *Physics of Classical Novae*, ed. A. Cassatella, & R. Viotti. Berlin: Springer, p. 127.
Starrfield, S., 1995, in *Physical Processes in Astrophysics*, ed. I. Roxburgh, & J. L. Masnou. Heidelberg: Springer, p. 99.
Starrfield, S. G., Truran, J. W., & Gehrz, R. D., 1997, in *Astrophysical Implications of the Laboratory Study of Presolar Materials*, ed. T. J. Bernatowicz, & E. Zinner. New York: American Institute of Physics, p, 203.
Taylor, A. R., Hjellming, R. M., Seaquist, E. R., & Gehrz, R. D., 1988, *Nature*, **335**, 705.
Truran, J. W., 1982, in *Essays in Nuclear Astrophysics*, ed. C. A. Barns, D. D. Clayton, & D. N. Schramm. Cambridge: Cambridge University Press, p. 467.
Werner, M. W., Roellig, T. L., Low, F. J. *et al.*, 2004, *Ap. JS*, **154**, 1.
Williams, D. M., Mason, C. G., Gehrz, R. D. *et al.*, 1997, *Ap. J.*, **489**, 91.
Woodward, C. E., 2006, Spitzer Target of Opportunity Nova GO Program PID 2333, http://ssc.spitzer.caltech.edu/geninfo/go/abs-go1/2333.txt.
Woodward, C. E., Gehrz, R. D., Jones, T. J., & Lawrence, G. F., 1992, *Ap. J.*, **384**, L41.
Woodward, C. E., Greenhouse, M. A., Gehrz, R. D. *et al.*, 1995, *Ap. J.*, **438**, 921.
Woodward, C. E., Gehrz, R. D., Jones, T. J., Lawrence, G. F., & Skrutskie, M. F., 1997, *Ap. J.*, **477**, 817.

9

Optical and ultraviolet evolution

Steven N. Shore

9.1 Introduction

Astrophysical spectroscopy, and with it our understanding of the cause and progression of the nova event, has progressed apace in the past two decades. When the ink was drying on the first edition, the International Ultraviolet Explorer (IUE) satellite was still in its heyday and many new phenomena related to the outburst were still to be discovered. Some glimpses had been provided by the Copernicus satellite (for V1500 Cyg), but the Hubble Space Telescope, and with it the Goddard High Resolution Spectrograph (GHRS) and the Space Telescope Imaging Spectrograph (STIS), were waiting to be launched and CCD technology was just being developed. Historically, much of the early work on ultraviolet spectra was undertaken with the aim of determining abundances through analyses of the nebular spectra. This is understandable since, before the early 1980s, the optically thick stage was impossible to model. Theoretical models have guided a shift in methodology, as did the development of spectrophotometric capabilities. One point should, however, be emphasized: no nova – classical or recurrent – was observed panchromatically before the Copernicus and IUE satellites were launched in the 1970s and the classical analyses were based entirely on data longward of the atmospheric cut-off. As we will discuss, we now know the role played by the ultraviolet in the details of spectrum formation at all wavelengths, a view that has changed dramatically since the first edition of *Classical Novae*, so it is this connection that will be stressed throughout this chapter.

It is well known, and has been since the time of H. D. Curtis at the start of the twentieth century, that novae in outburst show phenomenological systematics that make them useful cosmological distance indicators. The calibration between the rate of decline and their peak absolute magnitude in selected bandpasses (specifically Johnson B and V), coupled with their high brightness at maximum and ease of identification in large surveys, makes them attractive for distance determinations. They also sample the line of sight interstellar – and intergalactic – medium along different directions than the brighter stars in a galaxy and are not compromised by their environments. With increasing red shift, novae will be observable throughout their spectrum with the new generation of space telescopes, since their maximum bolometric magnitudes, $M_{\rm bol} \approx -10$, mean they can be observed to about a few hundred Mpc to 1 Gpc without confusion. But for nearby novae, especially within the Local Group and Virgo cluster,

Classical Novae, 2nd edition, ed. Michael Bode and Aneurin Evans. Published by Cambridge University Press.

the ultraviolet is below the atmospheric cut-off and, as we will emphasize throughout this chapter, the optical spectral and photometric development of a nova outburst is governed by the vacuum ultraviolet and may therefore show systematic biases due to metallicity and stellar evolution. The reader should keep in mind, however, that this relatively limited historical ultraviolet archive now provides the only data on the details of the spectrum formation. So we will discuss how to go backwards, from the optical and infrared to the shorter-wavelength behavior with particular emphasis on the uncertainties.

9.2 The ultraviolet–optical connection and the production of the visible spectrum

Since novae originate as thermonuclear runaways that lead to explosions, it is appropriate to contrast their behavior with the radiative and dynamical processes of a terrestrial counterpart: thermonuclear explosions. We will see that in the spectral development of both classical and some recurrent novae there are some striking useful similarities.

In a bomb, the energy release is *very* fast, less than a few tens or hundreds of microseconds, because of the unconfined initiation of the fusion reaction and small size of the igniting zone. This releases a hard radiative pulse that propagates into the surrounding air. A time-dependent ionized region forms that resembles a Strömgren sphere whose radius initially increases on a characteristic e-folding time-scale $t_S \sim (n_e \alpha)^{-1}$, where n_e is the electron density and α is the recombination coefficient. This stage ends when the environment exhausts the supply of ionizing photons and the now overpressured hot sphere expands supersonically, a stage called the *fireball*. We will use this terminology when describing initial evolution of the ultraviolet spectrum when the ejecta are completely opaque. On expansion, the combined effects of recombination and decreasing gas temperature and density decouple the shock from the radiating surface, called *breakout*. The optically thick surface, essentially the *pseudo-photosphere*, first stalls and then recedes relative to the compressional front. The blast ultimately turns transparent, revealing the more slowly moving debris – the actual ejecta – within the fireball. During this early stage, within the first few milliseconds in a bomb, structures form at the fireball surface, because of acceleration-induced instabilities on the front (Rayleigh–Taylor, Richtmyer–Meshkov) regardless of the means of detonation (ground/airburst); detonation-produced instabilities (e.g. Darrieus–Landau instability) could also occur during the initiation of the runaway, since the expanding front also ignites fuel, but it is not clear whether the growth times are fast enough to produce observable structure. Novae display almost every temporal step in sequence. The ejecta are optically thick in the earliest stages of the expansion in both the continuum and lines. The principal difference between a bomb (and Type Ia supernovae) and a nova is the survival of the central site for the explosion. This is destroyed in a nuclear explosion, and the evolution of the ejecta depends only on the total energy injected at the instant of the explosion powered by the expanding shock (and internal radioactivity);[1] novae are not energized from within the expanding medium, and neither radioactive products nor a propagating shock-driven precursor complicate the picture. In contrast, nova shells are actually passive photon converters whose spectra reflect the reprocessing of the incident ultraviolet light emitted from the remnant hot white dwarf. Because of this, the spectra depend on the geometry, filling factor of density fluctuations,

[1] The nova explosion is actually triggered by the radionuclides produced during the runaway but these provide negligible energy during the free expansion stage of the ejecta.

and the velocity gradient, but their origin is comparatively simple to understand – if not to interpret. Also, in the later stages (once the radiation pressure is low enough and this stage can be quantified) the dynamics are particularly simple. This is why the analogy with a time-dependent H II region is so interesting. The white dwarf remains active as a nuclear engine for some time and continues to irradiate the ejecta even as they freely expand: the white dwarf continues some envelope nuclear burning. The duration of this phase depends sensitively on M_{WD}. Its surface temperature increases as the star readjusts while maintaining stable burning. This induces a radiative temperature gradient that dominates the spectrum formation.

9.2.1 Radiative processes and spectrum formation

The kinetic temperature of the ejecta falls at first since the expansion is fast compared with the radiative diffusion time-scale. This adiabatic behavior is the origin of the infrared fireball stage (see Chapter 8). When the ejecta are sufficiently dense, above about 10^{10} cm^{-3}, approximate collisional equilibrium holds and the radiation, ionization, and kinetic temperatures are about the same but this does not last long. When T_{rad} falls below $\sim$ (1–2) $\times$ 10^4 K, recombination accelerates and *increases* the opacity, which is due to line absorption by relatively low ionization species (such as Cr II, Fe II and other heavy metals) and continuum absorption by CNO. This is the *iron curtain* stage (see also Chapter 5). Strong ultraviolet absorption bands, around 1500–1800 Å and 2300–2600 Å, redistribute the flux toward less opaque (longer) wavelengths and optical Fe II and [Fe II] emission appears (absorption from the upper levels of optically thick ultraviolet transitions also appear in absorption). This stage persists as long as the ultraviolet transitions remain thick, and the outer ejecta can become cold enough to show molecular lines (e.g. DQ Her showed CN absorption within the first few days after outburst; see Chapter 13). With continuing expansion, however, the optical depth decreases, $\tau_{UV} \sim t^{-2}$ or faster, and the irradiating spectrum becomes progressively harder as the pseudo-photosphere moves inward toward higher T_{rad}. This exacerbates the ionization of the shell, further reducing its opacity, and the iron curtain finally disappears (see Figure 9.1). Ultimately, since the ejecta have fixed mass in the absence of a wind, the pseudo-photosphere 'drops out the bottom': the entire shell becomes transparent and completely ionizes. This is the *nebular* stage and is marked by the appearance of nebular and coronal forbidden lines from very highly ionized species such as Ne VII and Fe VII to Fe XIV. It is also perhaps the best evidence that the spectrum is produced by a shell – not a wind contiguous with the stellar surface – after the fireball stage.

The characteristic development thereafter is remarkably generic (see Chapter 2). The emission decreases in strength as the density declines but the line profiles remain essentially invariant except for large density fluctuations and local recombination (see below). Through the combined effects of large density gradients (because of the velocity gradient) and the evolution of the central source (see the discussion below on the turn-off stage), the ionization temperatures obtained from simple line ratio nebular analyses (using mean quantities averaged over the profiles) may indicate much harder radiation than that which is illuminating the ejecta at the moment of observation. Nova ejecta resemble H II regions, but they are more like fossil regions that persist for a long time after they are no longer actively powered by a central engine.

Observationally it is clear that the ejecta velocity field is set by processes that occur during the explosion and not significantly altered during the expansion. The only exceptions seem to be the superluminous sources (but this remains conjectural), recurrents in symbiotic-like

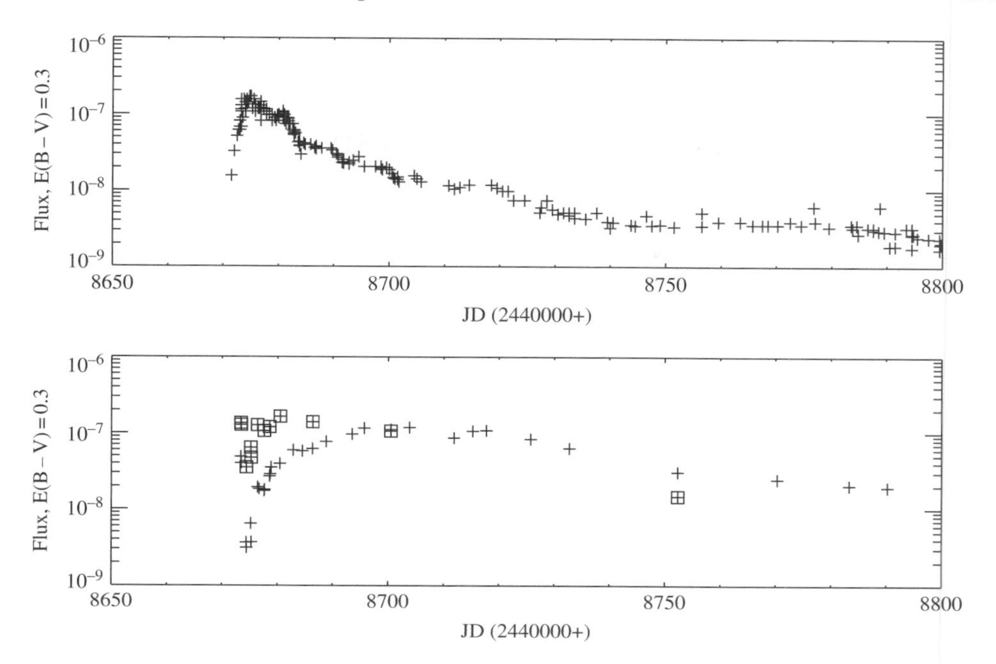

Fig. 9.1. Photometry for V1974 Cyg comparing the integrated optical ($U + B + V$; top) with the ultraviolet (bottom: 1200–2000 Å, crosses; 2000–3200 Å, cross-squares).

systems, and one anomalous event during the expansion of LMC 1990 #1 (see below). Along with any density structures there is a (nearly) frozen record of nucleosynthesis from the explosion event (which was the original justification for starting ultraviolet observations, to measure the resonance lines of dominant ions early in the outburst). Models indicate that the explosion rapidly becomes nearly ballistic (homologous) with a nearly constant velocity gradient, called a 'Hubble flow' in analogy with cosmology. This is very different from a normal stellar wind-like outflow where the gradient in the most rapidly expanding outermost layers of the wind asymptotically approaches zero. The expansion velocity is $V = V_{\max}(r/R_{\max})$, where r is the radial distance. The ejecta expand self-similarly, so r and t, the time since outburst, are interchangeable. Since M_{ej} is constant, the density then varies as $\rho(r) \sim r^{-3} \sim t^{-3}$. Dynamically, then, novae are comparatively simple but spectrum modeling is still extremely computationally intensive (see Chapter 5). Although the fireball spectrum seems to require unusually steep velocity gradients, indications are that this initial stage is quickly replaced by the linear law and by the appearance of the *transition* spectrum; this latter dominates the line formation.

9.2.2 *Nova spectroscopic taxonomy*

As emphasized repeatedly in this volume (see, for example, Chapters 2, 8), the distinction between different taxonomic subgroups is based more on their *abundances* than their photometric behavior or spectral development. Particular spectroscopic markers are used early in the emission line stage, first in the infrared and later at optical and ultraviolet wavelengths, to 'call the type', so we will begin with a very brief description of the two principal types.

The *CO novae* present the most diverse light-curve behavior and spectral development (contrasting, for example, VSNET and AAVSO light curves of HR Del and V723 Cas (large brief optical flares), DQ Her and V705 Cas (opaque dust) and OS And, LMC 1988#1, and LMC 1992 (boringly uniform)). The abundances are enhanced in CNO but not generally heavier elements and the He/H ratio is near solar. As found for other carbon-rich objects (i.e. AGB stars), dust formation – carbonaceous and/or silicate – is a notable but not exclusive property of this class (e.g. Gehrz, 1988; Shore *et al.*, 1994; Mason *et al.*, 1996; Gehrz *et al.*, 1998; Shore & Gehrz, 2004, and references therein; see also Chapters 8 and 13). The CO class may also show evidence for something resembling a wind phase, based on ultraviolet and optical spectra (e.g. Hauschildt *et al.*, 1994), but in general, as we will discuss, the ejecta can be adequately treated by finite shells.

The *ONe novae* are principally distinguished by their infrared spectra marked by the early appearance of [Ne II] 12.8 μm. The abundances for these novae vary but usually show some evidence of carbon depletion relative to solar, with associated enhancements of oxygen and neon (and in one case, V838 Her, over-abundances of sulfur and significantly subsolar oxygen). Given the expected rarity of massive progenitors, a surprising number of recent novae have been of this subclass, including two of the brightest novae of the last century (V1974 Cyg and V382 Vel). The class is remarkably uniform in its characteristics, even novae observed in the Large Magellanic Cloud (LMC 1990#1, LMC 2000) have shown nearly identical spectral evolution to their Galactic counterparts. A somewhat puzzling, but general, phenomenology is a phase of strong P Cygni profiles on all ultraviolet lines, first detected for V693 CrA, following the disappearance of the iron absorption spectrum that persists for days or up to one week before the transition to the nebular spectrum.

9.3 Stages of the spectral development and the light curve

To date, enough outbursts have been studied panchromatically that the interpretation is probably robust and it is on observations of several novae for which early high-resolution ($\geq 10^4$) data were secured – in particular OS And, V1974 Cyg, and V382 Vel – that much of our discussion will be based. Partial data sets exist in the combined ultraviolet–visible range for about 20 additional novae. In a few cases, although bright, the systems were not observed early enough in the outburst to study the rise to maximum visible light at all wavelengths.

9.3.1 Optical spectrum

The optical spectrum passes through more-or-less well-marked stages whose timing is usually indicated relative to the visual maximum of the light curve. Definitions and extensive reviews are provided by Payne-Gaposchkin (1957) and McLaughlin (1960) that still frame the descriptions of the optical development; see the discussion in Chapter 2 for further details. There is little need to update this phenomenology based on more recent novae, although it does not directly correspond to the line formation processes, and we will summarize it here for historical reasons as well as completeness. The *principal spectrum*, which is a stage dominated by absorption lines, shows relatively low ionization species such as Mg I and Mg II, O I, C I, and singly ionized and neutral iron peak species (especially Fe II and Ti II). Neutral helium absorption lines are also seen at this stage in some novae and also P Cygni profiles for the optical Na I lines (especially in the dust-formers such as DQ Her and V705 Cas; see Figure 9.2). The Balmer lines, particularly Hα, also show P Cygni profiles at this stage. The strongest emission lines are O I 5577 Å, 6300 Å, and 6364 Å. There are a series

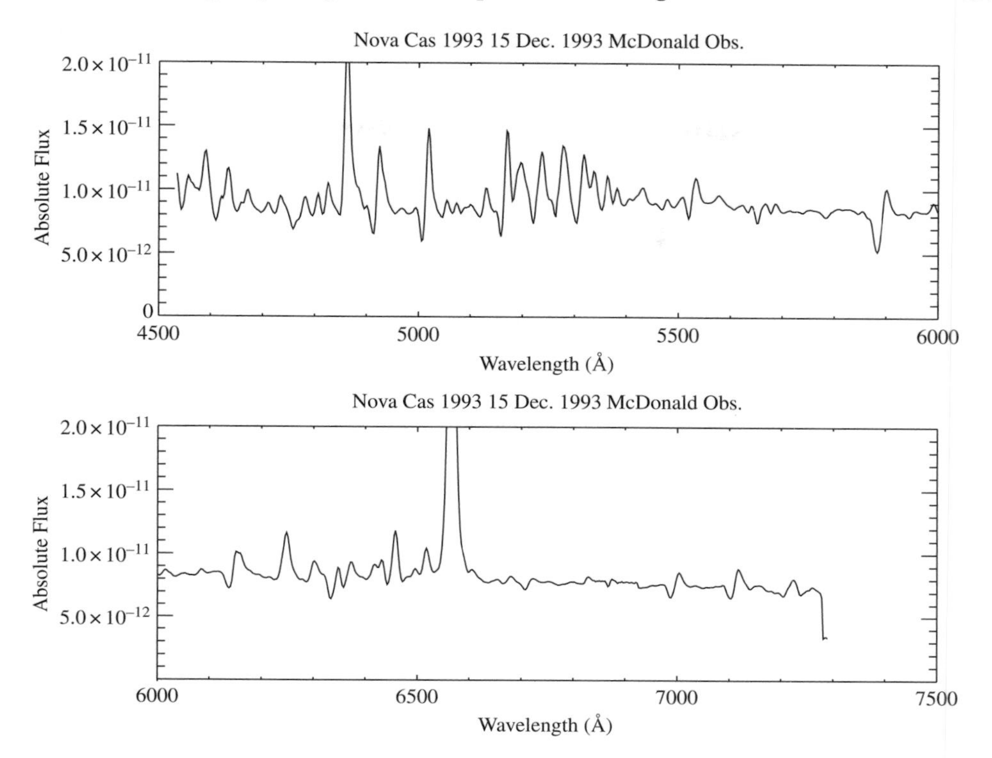

Fig. 9.2. Early optical spectrum of the CO dust-forming nova V705 Cas taken within two weeks of discovery (McDonald Observatory spectrum).

of 'flashes' at relatively well-marked points in the light curve. About $\Delta V \approx 2.6$ magnitudes below maximum, O I suddenly brightens, followed at about 3.3 magnitudes by N II and, subsequently, at about 3.6 magnitudes, the 'helium flash' occurs with a rapid increase in the He I optical emission. Forbidden lines of Fe II appear in emission at the end, along with the emergence of He II 4686 Å emission; at this stage, these [Fe II] lines are mainly fluorescent (see the discussion of the ultraviolet below), later they are caused by recombination and collisional excitation in the extended, low density ejecta.

The *diffuse enhanced spectrum* appears in some novae, mainly CO type, and strengthens to a maximum at about 2 magnitudes below maximum. These lines are, as the name implies, secondary and broader than the *principal* absorption system, displaced to negative radial velocities relative to the latter, and reaching high radial velocities (in classical studies about twice that of the principal system). The *Orion spectrum*, preserving the antique distinction among spectral types from the pre-Harvard classification days, displays N II/N III, O II/O III, and He I – as in O stars – that are displaced to very high velocities, about -2000 km s^{-1} (see Figures 9.2, 9.3). The *nebular stage* is the best-established – and most consistently defined – transition in the optical spectrum. Its entry is marked by the emergence of forbidden lines, beginning with those from relatively low ionization stages such as the auroral [O I] and [N II], followed by the appearance of [O III] and [N III], and culminating with the presence of coronal type emission lines such as [Fe X] 6347 Å and [Fe XIV] 5303 Å.

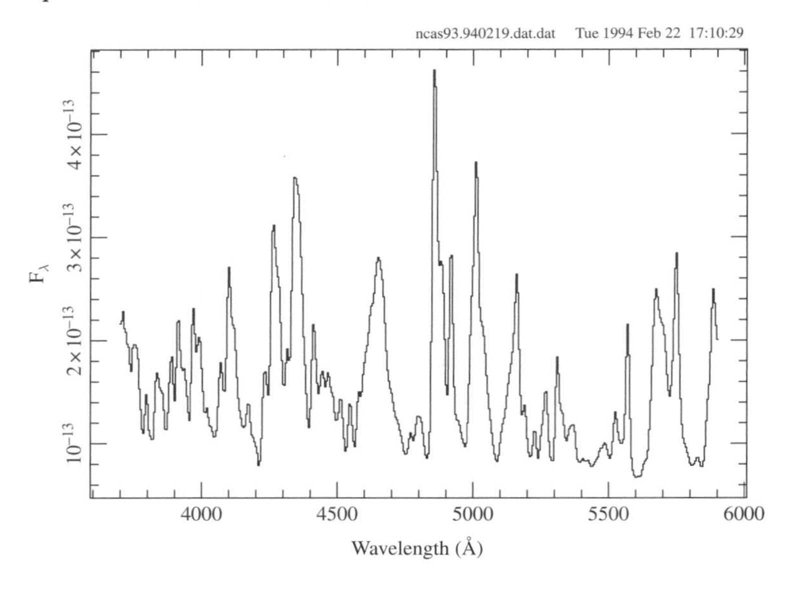

Fig. 9.3. A sample medium-resolution optical spectrum of V705 Cas immediately before the dust-forming event showing the complex emission features. See text for details.

9.3.2 Ultraviolet spectrum

In the ultraviolet ($\lambda \leq 3300$ Å), these stages correspond, more or less, to the iron curtain, the transition and/or pre-nebular, and nebular stages, respectively. The ultraviolet is really driving everything in the early stage, once the line opacity kicks in. V1974 Cyg showed most clearly what is happening since it was the first nova to be completely covered during the rise to optical maximum. The fireball stage resembles Type II supernovae; the rise at longer wavelengths is simply redistributed flux and the time-scale is fast because of the sensitivity of the ionization to the local conditions in the cooling ejecta. The two characteristic optical time-scales, t_2 and t_3 (see Chapter 2), are also linked to the ultraviolet. The region longward of 2000 Å becomes progressively less opaque, followed by the shorter wavelengths, and t_3 is approximately the time when the 1200–2000 Å integrated flux becomes about equal to that between 2000–3300 Å. The optical, infrared, and radio spectral domains turn transparent progressively earlier in the outburst. The radio and infrared regions are the best tracers of the true fireball stage since the principal opacity source at these long wavelengths is due to thermal bremsstrahlung which depends only on the change in the column density, but the same essential features follow for all higher frequencies. The spectral turn-over frequency (where $\tau_\nu = 1$) decreases as the expansion progresses. Recall that, for a stationary isothermal stellar wind, the frequency dependence of the opacity means the solid angle is a function of frequency which leads to the observed spectral curvature (see, for example, Chapter 7). The flux at each wavelength declines as t^{-3} and, because of the time delay, there is a characteristic decrease in the peak flux with decreasing frequency. For the optical and ultraviolet, this is not so simple since the combined effects of photo-ionization and line transitions make these wavelengths much more sensitive to both the filling factor of the dense gas and the ionizing flux of the central source (the higher energy, X-ray and γ-ray, emissions are discussed in Chapters 10 and 11).

As shown in Figures 9.4–9.7, virtually all novae appear to evolve in the same way: it is almost always possible to find two novae that display the same spectrum at a similar stage in outburst by scaling the expansion velocities and t_2 or t_3 (see Figures 9.8, 9.9).

One recent development that was not possible using only uncalibrated, or only relatively calibrated, photographic spectra is the separation of emission and absorption spectra during the complex optical stages of the iron curtain . Two optical phenomenological spectral classes have been distinguished (Williams *et al.*, 1991; Williams, Phillips & Hamuy, 1994): the Fe II-type, which show strong Balmer and iron emission in the optical (the Fe II and [Fe II] lines are the more transparent transitions from highly excited states pumped from the ultraviolet), and the He-type (or He/N), which almost immediately display He I/He II and lines from more excited or ionized species. These sub-groups, which also have (as always in our business) transitional forms, have been used to frame much of the discussion in the recent literature so it is important to note the connections with the ultraviolet and ejecta opacity: if the shell fragments early, or if the ejecta are completely photo-ionized to near nebular conditions very quickly, the iron curtain stage may not occur or lines may appear simultaneously from regions with very different densities in the ejecta. In particular, it has been noted historically that many slower novae develop [Fe II] emission at a comparatively late stage in the outburst. Recall that these lines arise from coupled transitions in the ultraviolet and far-ultraviolet and therefore a full NLTE treatment is required to reproduce the spectral features. In general, however, we can say that *virtually all* novae pass through some [Fe II]-like stage.

During the most opaque stage in the ultraviolet, the optical spectrum is already formed by more extended gas lying outside the pseudo-photosphere. Thus, the Balmer emission lines will already be strong and may show P Cygni profiles. These are rarely saturated and may even show structure with a generally strong Balmer decrement for the absorption depth. The highest velocities seen, in the very earliest appearances of the H I emission, will often reach nearly the terminal velocity of the ultraviolet resonance profiles. But in general, the optical line formation is more heavily weighted toward the higher density parts of the ejecta, because they are formed by recombination, so are systematically narrower than the ultraviolet lines at the same stage.

The development during this transition to the nebular spectrum subsequently divides between the CO and ONe sub-classes, although the reasons are far from clear. Essentially all ONe novae enter a phase of P Cygni profiles associated with the principal resonance lines (Si IV 1400 Å, C IV 1550 Å, Al III 1860 Å and Mg II 2800 Å) just before the appearance of He II 1640 Å emission and following the disappearance of the iron curtain, which persists for days or up to one week before the transition to the nebular spectrum (see Figure 9.10). The observed terminal velocity is nearly always larger than that obtained from optical absorption troughs (up to twice) and always exceeds the velocity width of the subsequent emission profiles at *all* wavelengths. The ultraviolet resonance lines are often saturated and nearly black in the absorption troughs, implying large covering factors as well as large column densities; these narrow with time and $V_{\rm max}$ decreases (seen best for V1974 Cyg). The best examples are V693 CrA, LMC 1990#1, V1974 Cyg and V382 Vel. Two fast recurrent novae, LMC 1990#2 and U Sco in its 1979 eruption, also showed this stage. The P Cygni absorption edges show the highest velocities, ranging from -5000 to $-12\,000$ km s^{-1}. One interpretation is that, because the opaque surface reaches temperatures that are above about 3×10^4 K, the ejecta completely ionize leaving the P Cygni components on the resonance lines while removing the overlying, cooler iron curtain absorption. This would be consistent with the higher $L_{\rm bol}$ of

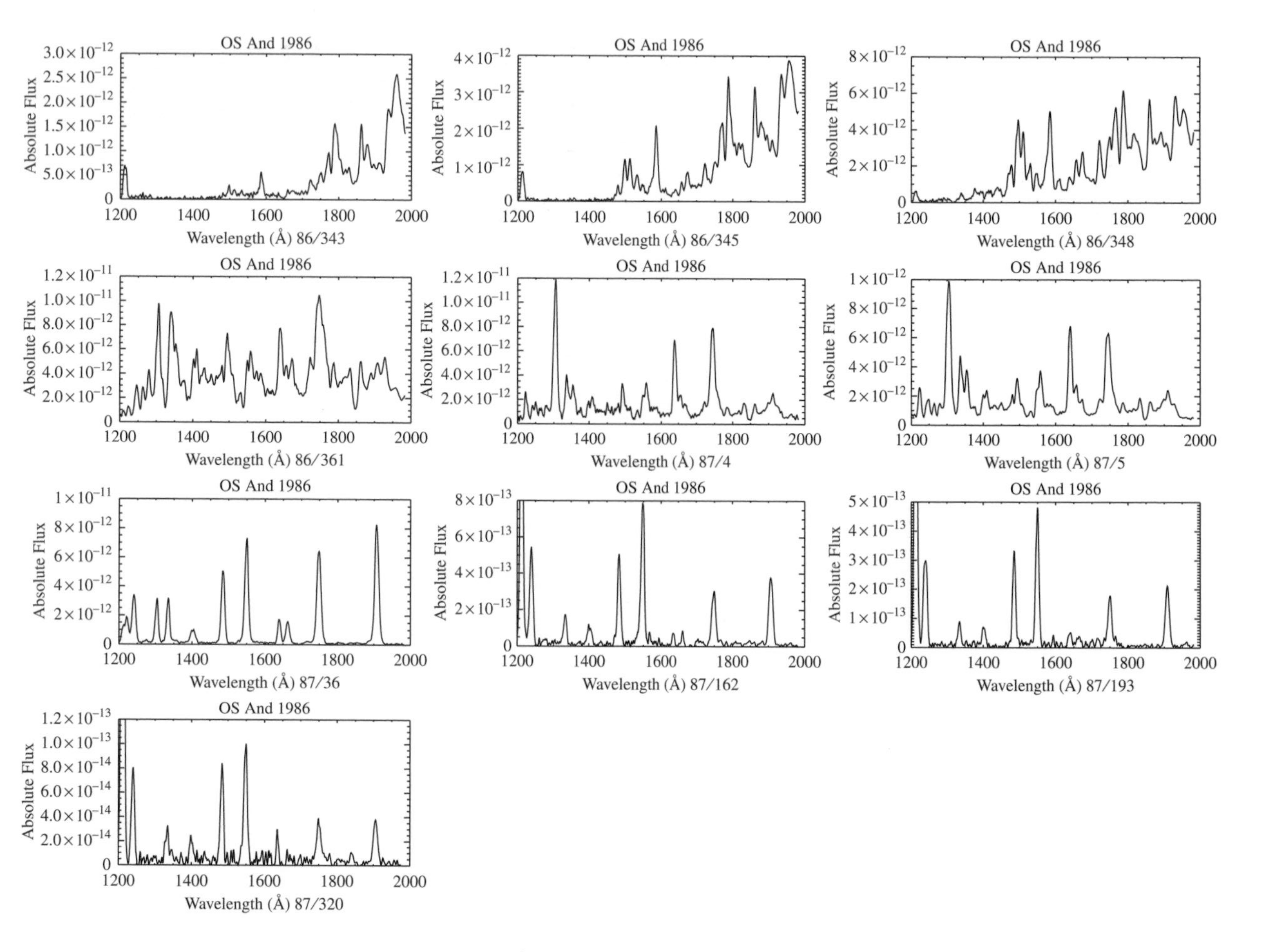

Fig. 9.4. Development of the iron curtain, 1200–2000 Å, for the CO nova OS And during early outburst.

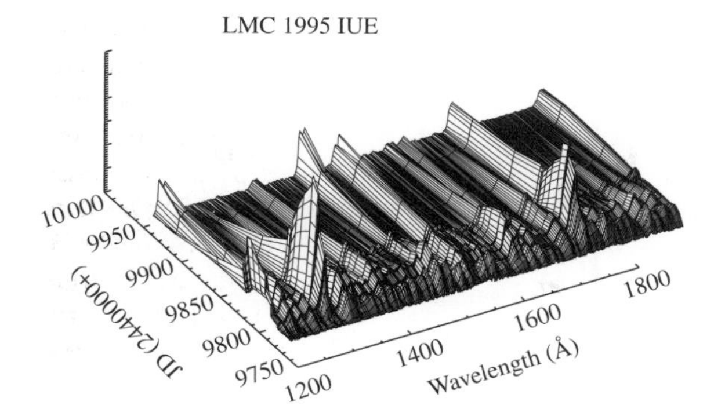

Fig. 9.5. Spectral development for the CO nova LMC 1995 (1200–2000 Å) with time showing the transition from the iron curtain to pre-nebular stage. Relative flux, linear scale, is shown on the vertical axis.

the ONe class. The interval of appearance is nearly independent of the speed of the outburst, and only V838 Her did not display it.

9.3.3 *Bolometric evolution, the origin of the light curve, and the Maximum Magnitude–Rate of Decline relation*

The iron curtain stage occurs at constant bolometric luminosity L_{bol}. Its duration is still an open question but is probably at least t_2. While longer than currently predicted by models, this constant bolometric luminosity interval is the most important for determining the nova's physical properties, and as we have seen, the constant bolometric luminosity phase greatly simplifies the picture of how individual spectral regimes develop. The infrared, having the lowest opacity, is the first to experience a rise from both the increase in ultraviolet opacity and the color temperature of the optically thick surface. It is here too that emission will inevitably first appear. Remember, the ejecta are quite distended and emission appears as soon as there is sufficient excitation of the appropriate transition and a low enough optical depth for photon escape; in NLTE comparatively high ionization can be seen in the infrared before it is observed in other wavelength regimes.

An atmosphere is dynamically unstable when the photon flux produces a radiation pressure gradient that overbalances gravity. Thus a spherical critical atmosphere cannot remain hydrostatic if the star's luminosity exceeds

$$L_{\mathrm{Edd}} = 3 \times 10^4 \left(\frac{M}{\mathrm{M}_\odot}\right) \left(\frac{0.4\,\mathrm{cm}^2\,\mathrm{g}^{-1}}{\kappa}\right) \mathrm{L}_\odot, \tag{9.1}$$

the Eddington limit. Here κ is the Rosseland mean opacity per gram, and the maximum value of L_{Edd} comes from electron scattering for which κ is smallest; any line scattering only decreases this critical luminosity. Why would we expect this to be a limit for novae? Note the caveats here. Stability is out of the question during the explosive phase. It is only required for the later phase of stable nuclear burning, when the bolometric luminosity remains roughly constant. Were the white dwarf to be generating its luminosity centrally, then L_{Edd} would indeed present a barrier to stability of the whole star. But the atmosphere is a different matter. The nuclear energy generation is shallow and only requires a small departure from isotropy to

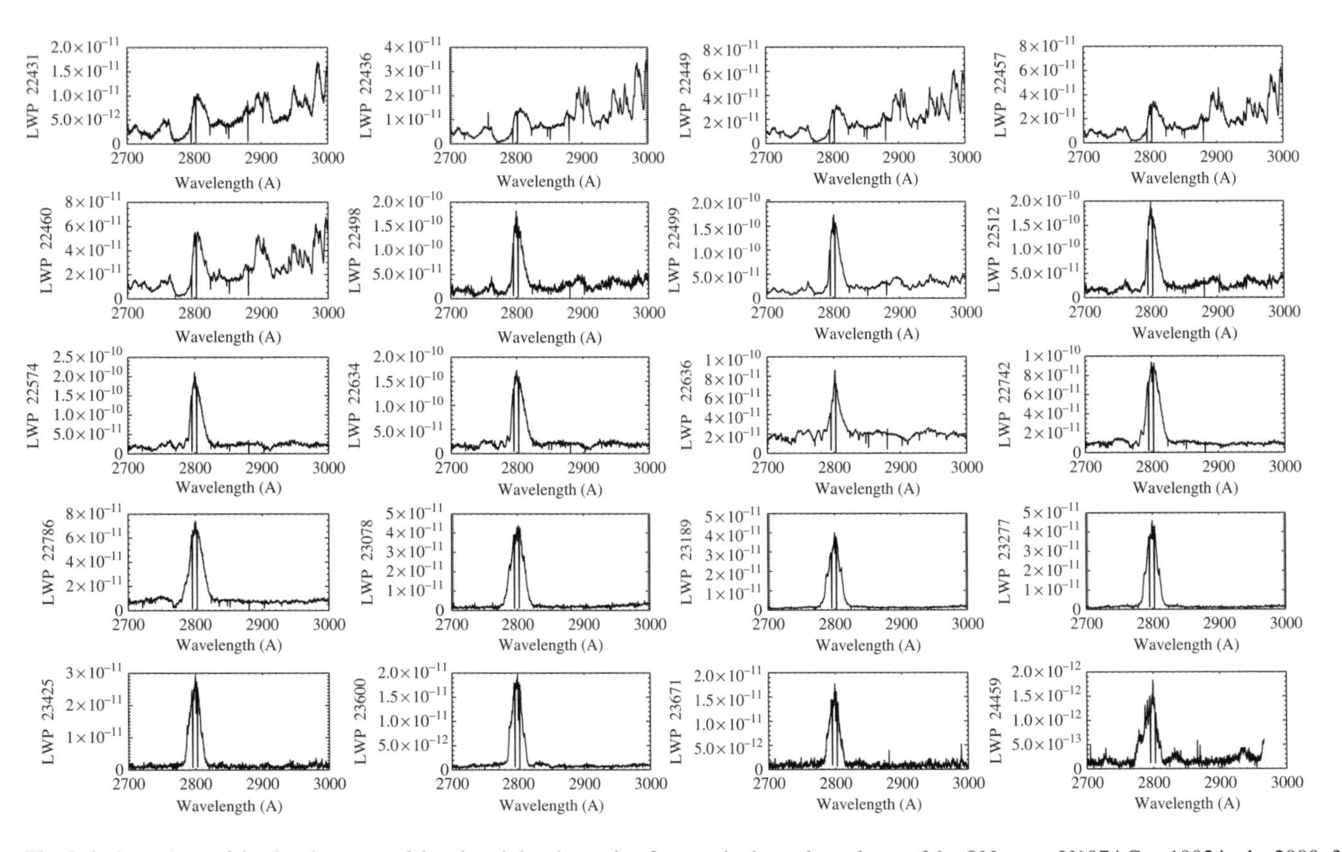

Fig. 9.6. Snapshots of the development of the ultraviolet absorption features in the early outburst of the ONe nova V1974 Cyg 1992 in the 2000–3300 Å region.

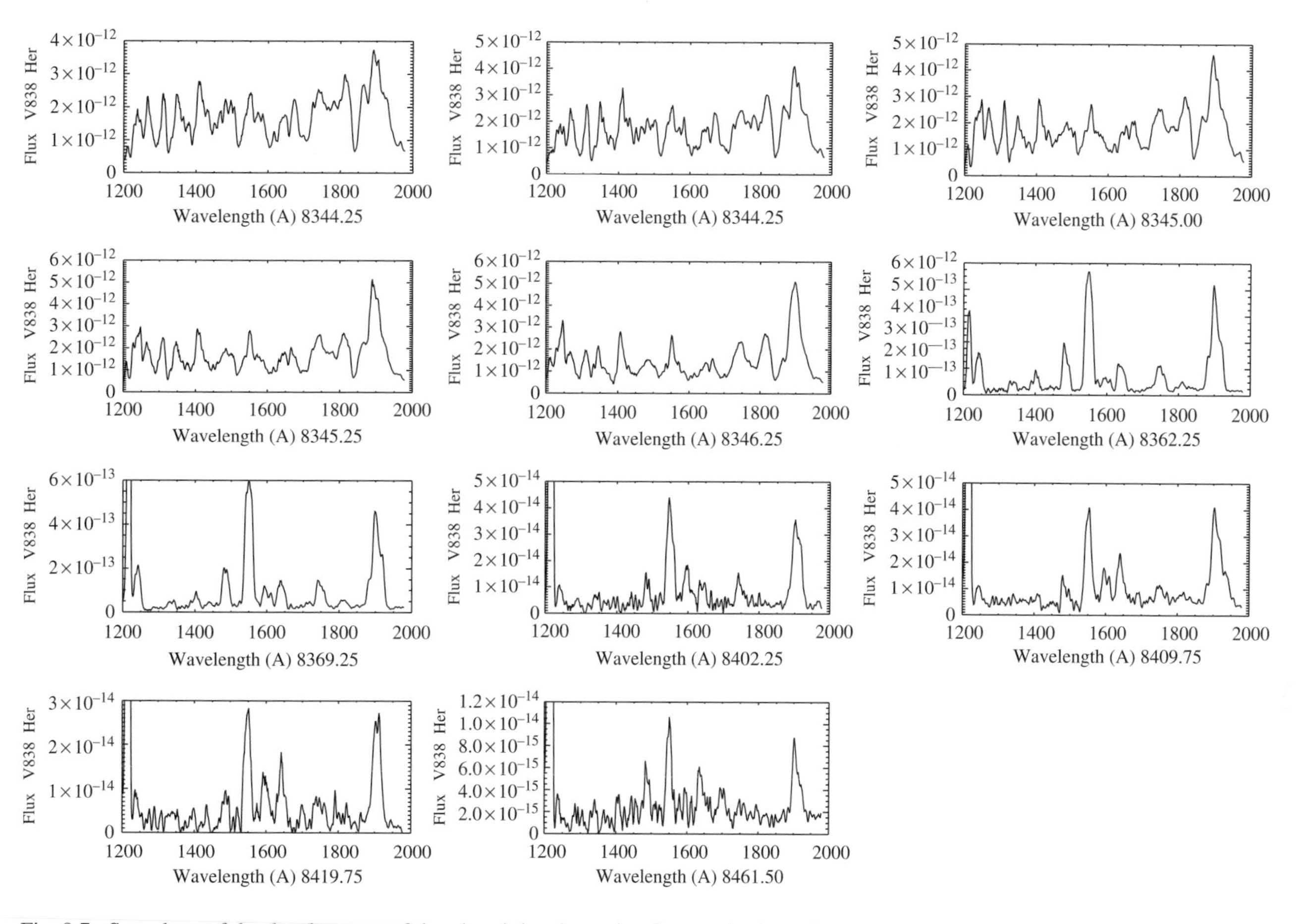

Fig. 9.7. Snapshots of the development of the ultraviolet absorption features in the early outburst of the very fast ONe nova V838 Her 1991 in the 1200–2000Å region.

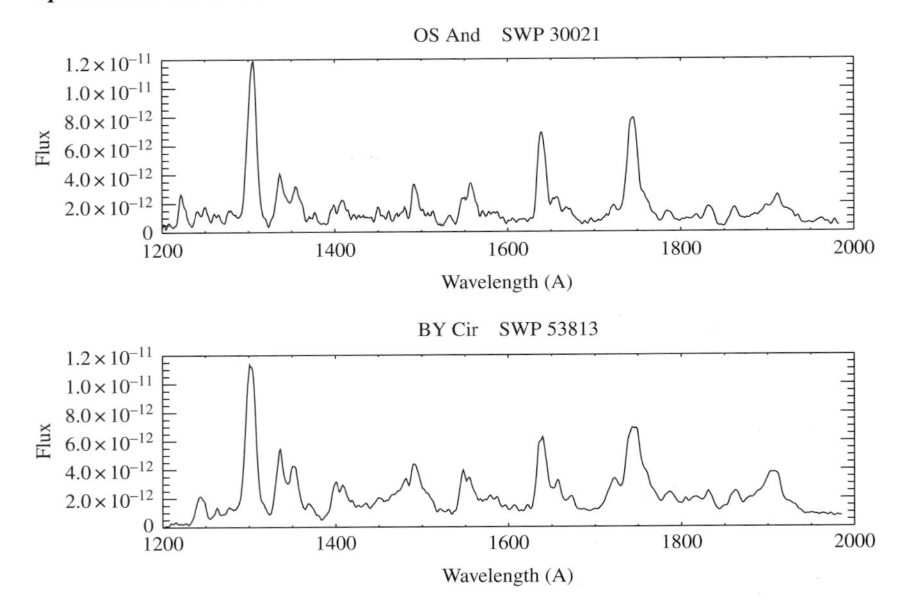

Fig. 9.8. Comparison of two CO novae, OS And and BY Cir, at the same stage in the iron curtain phase.

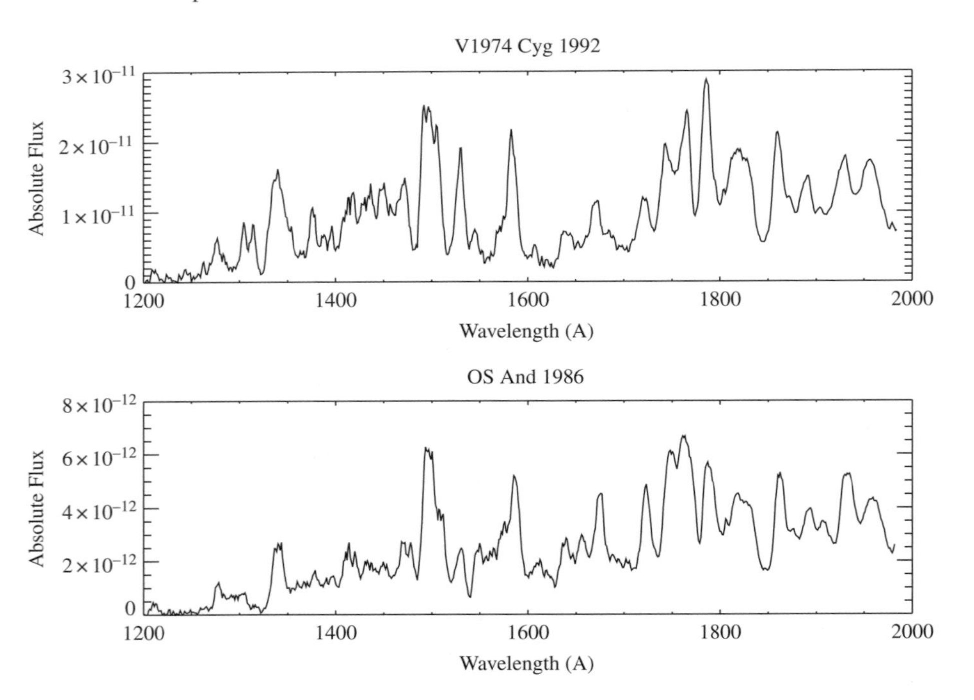

Fig. 9.9. Comparison between OS And and V1974 Cyg, an ONe nova, at the same stage in the iron curtain development in early outburst.

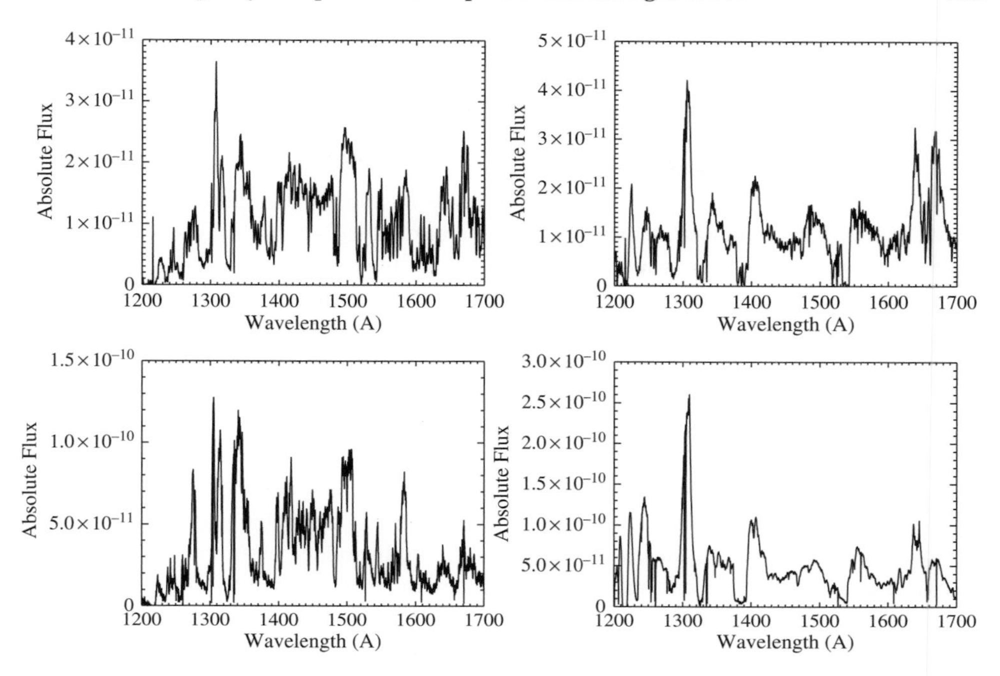

Fig. 9.10. Comparison of the ultraviolet spectra for two Galactic ONe novae at similar stages during the early outburst: V1974 Cyg (top) and V392 Vel (bottom); iron curtain (left), P Cygni (right).

reduce the effective opacity and remain stable (e.g. Shaviv, 2002). A few anomalous classical novae, termed *superluminous*, that constitute not so much a class as a group of extreme objects spanning the primary types (e.g. Della Valle *et al.*, 2002; Schwarz *et al.*, 2001) have been far more luminous than any reasonable mass permits. The one certain case is LMC 1991 for which the distance is certain and where the fireball and iron curtain stages were observed bolometrically (Schwarz *et al.*, 2001). In this one case, a firm *lower* limit of $1.5 \times 10^5\ L_\odot$ was obtained. Samples of extragalactic novae (e.g. Shara & Zurek, 2002; Shafter, 2002; see also Chapter 14) contain a small number of exceedingly bright sources, but since these surveys are optical the actual peak luminosity depends on shell geometry and filling factors and may be higher than inferred. The full examination of ultraviolet and visual data for the LMC novae remains to be completed, however, and there are no comparable observations for any other extragalactic novae.

If $L_{\rm max} \leq L_{\rm Edd}$ always, we find some interesting observational consequences, particularly for the *cosmological* utility of classical novae. Unlike supernovae, but resembling ionization bounded nebulae, classical nova ejecta are essentially photon converters. How that matter responds determines the flux observed at wavelengths longer than the peak of the central star's spectrum. This constant bolometric luminosity phase accounts in a general way for the success of the t_2 and t_3 predictors for the maximum brightness, via the *maximum magnitude versus rate of decline* (MMRD) relation (Della Valle & Livio, 1995; Downes & Duerbeck, 2000; see also Chapters 2 and 14) – but with an important caveat: this method works with visible light (or infrared, or any other limited bandpass) only if the ejecta actually are calorimeters, so at some point during the decline the bolometric correction must vanish in any bandpass. This places very severe constraints on the opacity, as well as the geometry, of the ejecta in the earliest stages of outburst. If the central source remains constant, the radiative flux is *exactly* (in the usual astrophysical sense) redistributed – the *bolometric* luminosity of the ejecta should remain constant – provided the ejecta are equally optically thick in all directions (see Figures 9.11 and 9.12). The same remark holds for the P Cygni absorption troughs. A deficit of lower-frequency emission relative to the ultraviolet is an indication of density and geometric variations in the expanding material. The problem is, naturally, that this cannot be determined uniquely and requires computationally elaborate and intensive models. Note, for example, that novae with the shortest t_2 and t_3 intervals may also have lower mass ejecta, or may be more fragmented, or more rapidly expanding, or any combination of these. With comprehensive spectrophotometry the assumption can be checked, but one or another spectral domain is often lacking (and in the long term there will continue to be lacunae depending on satellite availability).

The constant bolometric luminosity stage can also be used to obtain the extinction since A_λ is strongly wavelength-dependent. Combining differentially corrected wavelength regimes, constrained by $L_{\rm bol}$ = constant, yields $E(B-V)$ and the added assumption $L_{\rm bol} \leq L_{\rm Edd}$ gives a distance, except for rare superluminous cases, independent of alternative calibrations.

At this moment in our discussion, one point should be emphasized: novae show two completely different behaviors depending on their circumstellar environment. The epithet *classical* excludes the recurrent novae living in symbiotic-like binary systems. For those, the extended wind of the companion dominates the emission line evolution at late stages, in effect creating a fossil H II region that slowly recombines following the initial ultraviolet/X-ray pulse from the ejecta and is powered by both the shock effects of the ejecta and the continuing emission from the white dwarf. Their composite spectra arise from essentially independent regions, much as seen in active galactic nuclei, and the emission lines from the ejecta may

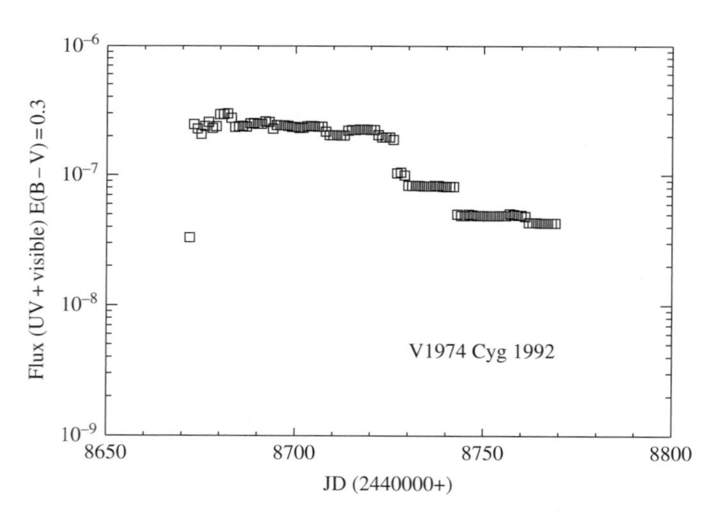

Fig. 9.11. Bolometric flux for V1974 Cyg during the first few months of outburst binned to 1-day intervals.

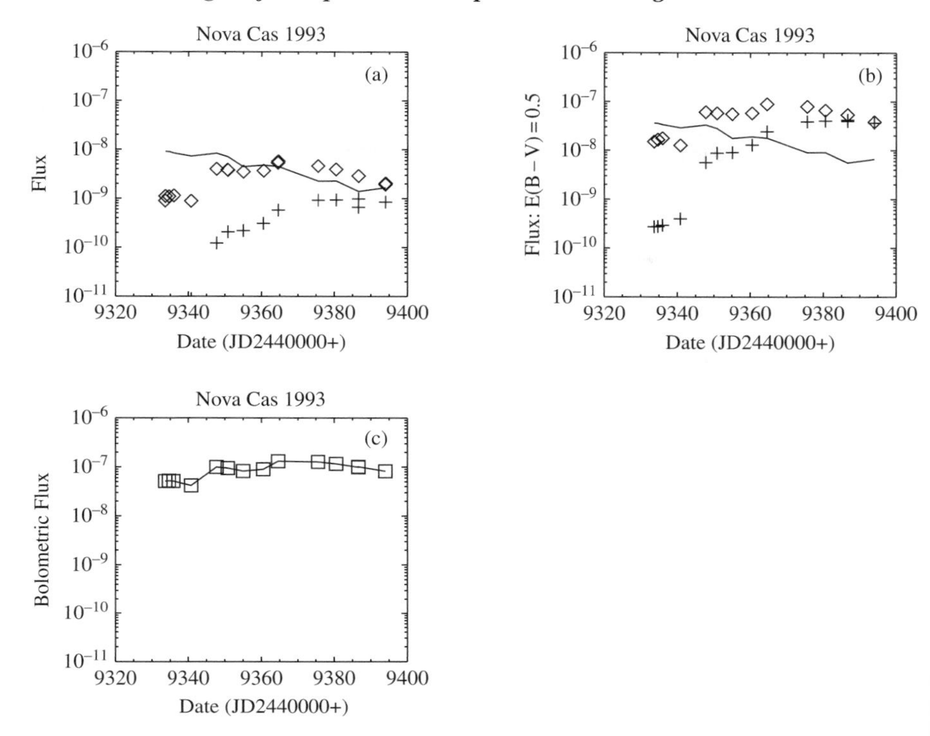

Fig. 9.12. Bolometric development of the CO nova V705 Cas before the onset a, b and c of dust formation. The panels a, b and c show the uncorrected photometry (crosses, 1200–2000 Å; boxes, 2000–3200 Å; line, optical), corrected for $E(B - V) = 0.5$, and the bolometric flux, respectively.

be completely masked by the larger volume of the extended red giant atmosphere. Few instances have been observed panchromatically, but the 1985 and 2006 outbursts of RS Oph have been studied in sufficient detail at high resolution (see Figure 9.13). Among the known symbiotic-like recurrent systems, single epoch observations early in outburst suffice to check the picture. Recurrents found in compact systems develop more or less identically with classical novae, except for their low ejected masses which cause the various stages to occur more quickly. Classical novae behave essentially uniformly, independent of their subtype, so the conclusions below are generic for all.

9.3.4 The fireball stage

The ejecta initially resemble a very hot stellar atmosphere, albeit one that may have a large density gradient depending on the velocity profile, as discussed in Chapter 5. The initial instant of the explosion has never been observed. This stage, when both the lines *and continuum* are optically thick, does not last very long but has been serendipitously observed in a few novae (e.g. OS And, V723 Cas, V1974 Cyg, and LMC 1991 – see Figure 9.14). As we have said, like the first stages of a nuclear explosion, this is called the fireball spectrum, a term also used in the infrared for the moment when the bremsstrahlung continuum displays a hot black body (although this is actually a later stage than seen at shorter wavelengths).

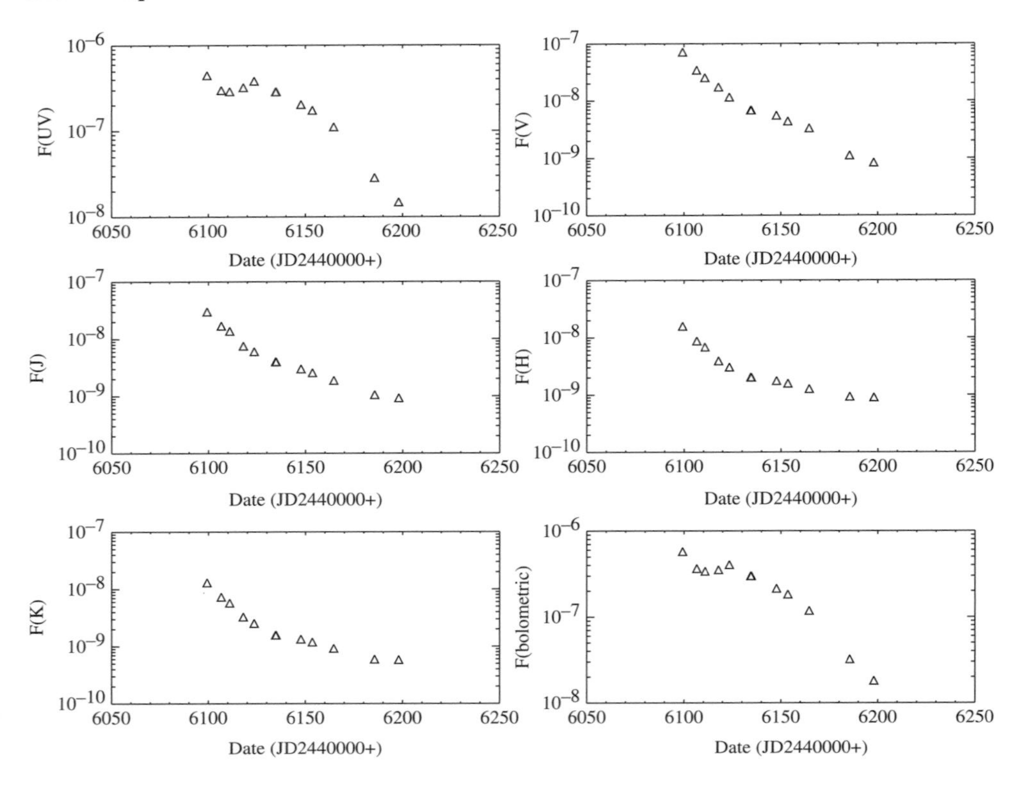

Fig. 9.13. Photometric development from X-ray through infrared of the symbiotic-like recurrent nova RS Oph during the 1985 outburst.

During this stage the color temperature is $>2 \times 10^4$ K and the spectrum extends to high energy, above about 6 eV (continuum emission is detected shortward of Lyα). The principal line opacities come from third spectrum metal lines (i.e. doubly ionized) and similar species and the CNO group, so if this stage can be observed it provides a good first check on the abundances. On continuing expansion, the temperature falls below a critical threshold at which the ions recombine. The resulting rapid increase in *line* opacity subsequently dominates the spectral development at all wavelengths. As noted above, this stage has been called the iron curtain because the principal contributors are heavy elements of the iron group, especially Fe I/Fe II, Ti I/Ti II, and Cr I/Cr II, but all species with relatively low second ionization potentials
contribute.

9.4 Structure of the unresolved ejecta

The structure of the ejecta can be studied long before they are spatially resolved using line profiles (see, for example, Figures 9.15–9.27). It has been known since the first photographic studies of novae at the beginning of the last century that nearly all nova emission lines show components ranging over several hundred to several thousand, kilometers per second. These are distinct from the absorption systems, especially the diffuse enhanced

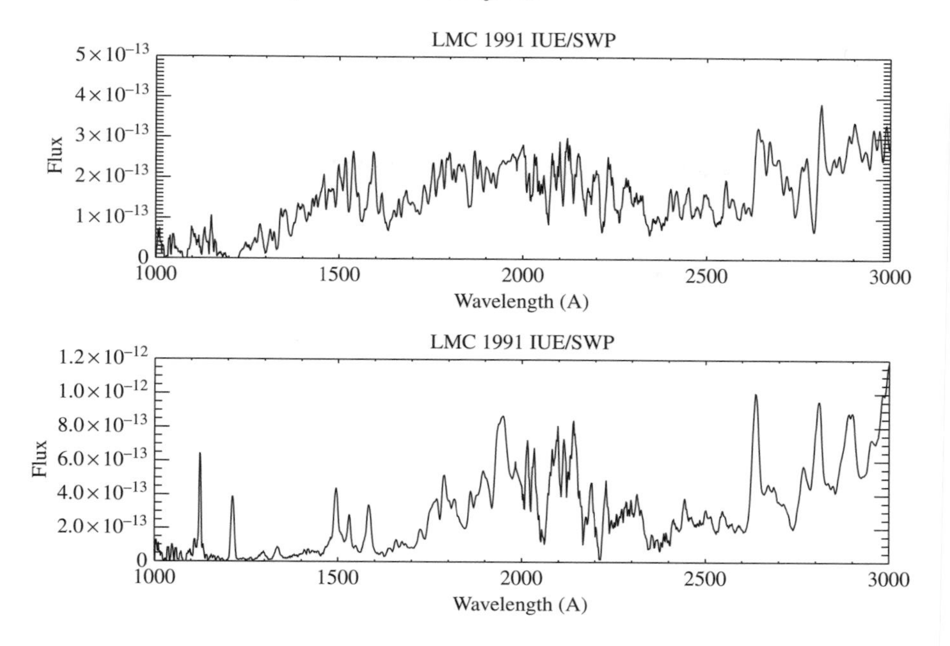

Fig. 9.14. Two stages of the development of the superluminous nova LMC 1991: the top spectrum shows the initial fireball stage, the bottom is about a week later, showing the fully developed 'iron curtain'. In neither spectrum are individual lines observed; the structure is entirely due to the combined effects of the intrinsically banded distribution of the atomic transitions and the smearing resulting from the depth dependence of the expansion velocity. Several gaps, particularly the one at around 1700 Å, are especially sensitive to the maximum ejection velocity.

series, and remain stationary as the profile evolves. A feature of the optical series is that the absorption components generally appear at progressively higher velocities and in some novae (HR Del, for instance) evolve toward lower velocity with time. A universal feature of classical novae at all stages, and on essentially all transitions, is the appearance of knots and broad, non-Gaussian wings as soon as the emission appears. These *may* depend on the stage of the outburst and the specific transition (excitation energy and ion) since the emission comes from a specific density and temperature domain in the shell. The earliest optical and infrared profiles are nearly always broader than those emerging after t_3, and in some cases at a single epoch the profile can depend on the depth of formation of the transition (Della Valle *et al.*, 2002).

The structure generally becomes progressively more discernible as the emission line weakens but, unlike the absorption systems, the emission knots remain stationary throughout the decline stage. The emission peaks are rarely symmetrically distributed about line center, although ring-like distributions have been observed and modeled in ways consistent with the subsequent observations of resolved ejecta (Gill & O'Brien, 1999; see also Chapter 12). The structures detected in different transitions, and in different wavelength regimes, also correspond (see, for example, Hayward *et al.* 1996). Again, the peaks are rarely symmetrically distributed about line center, although ring-like distributions have been observed (Shore *et al.*, 1993, 2003; Gill & O'Brien, 1999). The optical line profiles are often less symmetric than their ultraviolet counterparts, especially the He I and Balmer series. What

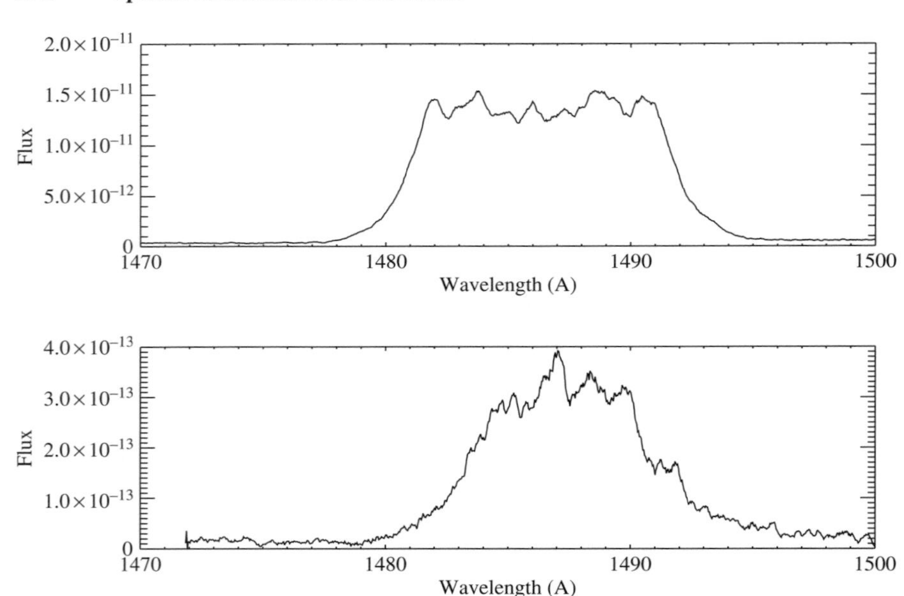

Fig. 9.15. Comparison of V1974 Cyg (top) and LMC 1992 (bottom) of the N IV] λ1486 Å line at the same stage in early outburst. Note the appearance of the line profile, indicating at least two relatively distinct components: the featureless high-velocity material responsible for the line wings, and the slower-moving knots of emission.

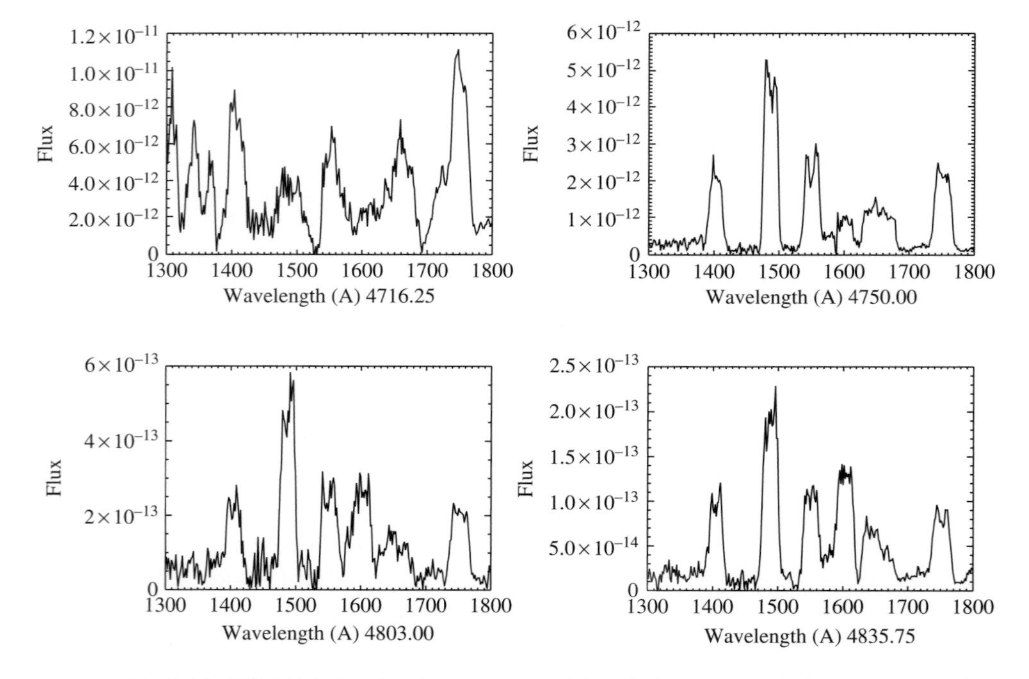

Fig. 9.16. V693 CrA showing the emergence of the nebular spectrum and structure observable with low-resolution ($R \sim 1000$) IUE spectra. The JD are indicated immediately after the abscissa label (2440000+E).

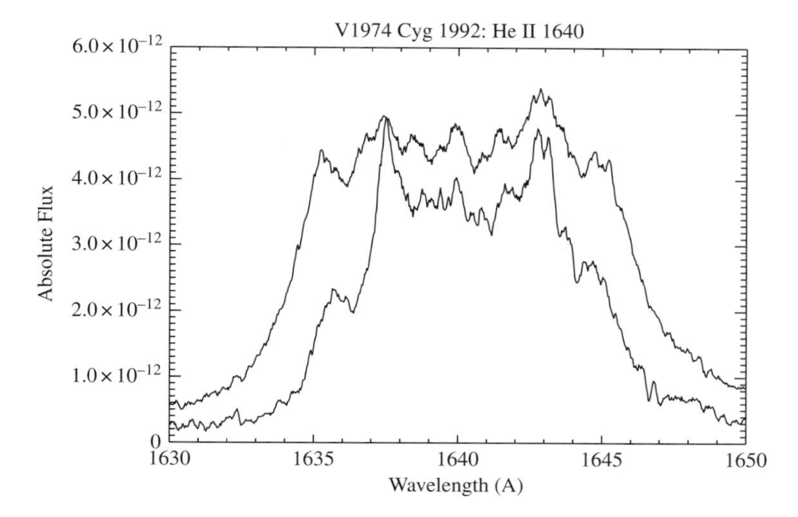

Fig. 9.17. He II 1640 Å line profile at two epochs during the nebular stage of V1974 Cyg, showing the invariance of the structure. The broader profile, obtained about one month earlier, was closer to the transition interval following the P Cygni stage of this ONe nova.

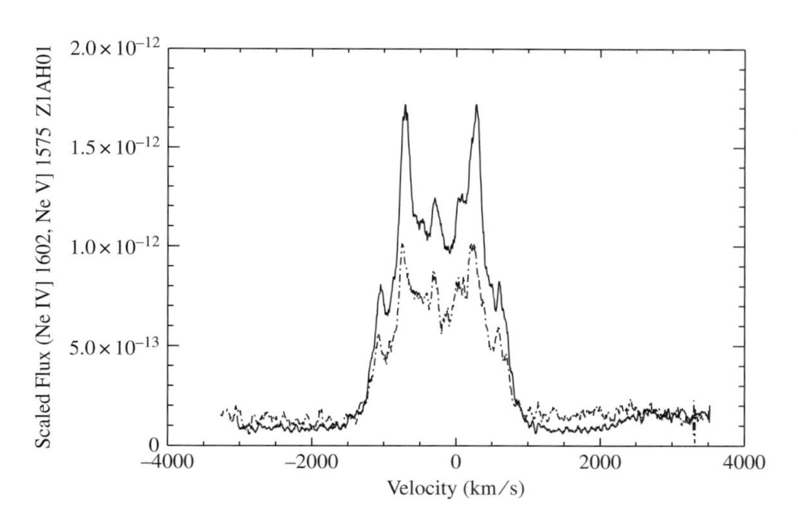

Fig. 9.18. Ne IV] 1602 Å (solid) and Ne V] 1575 Å (dashed line; multiplied by factor of 2.0) for V1974 Cyg from 1993 April 1, during the nebular stage, showing the identical structures with different ions. Compare with Figure 9.17 for the He/N ratio.

the asymmetries mean is unclear since they cannot be simply caused by changing optical depths along the line of sight this late in the outburst (Shore *et al.*, 2003). Often, when the emission first appears, the profiles display almost featureless cores and broad wings, extending to the same terminal velocities as the P Cygni troughs on ultraviolet resonance lines. These fade with the profile and the residual emission shows the knots (e.g. V1974 Cyg; Shore *et al.*, 1993).

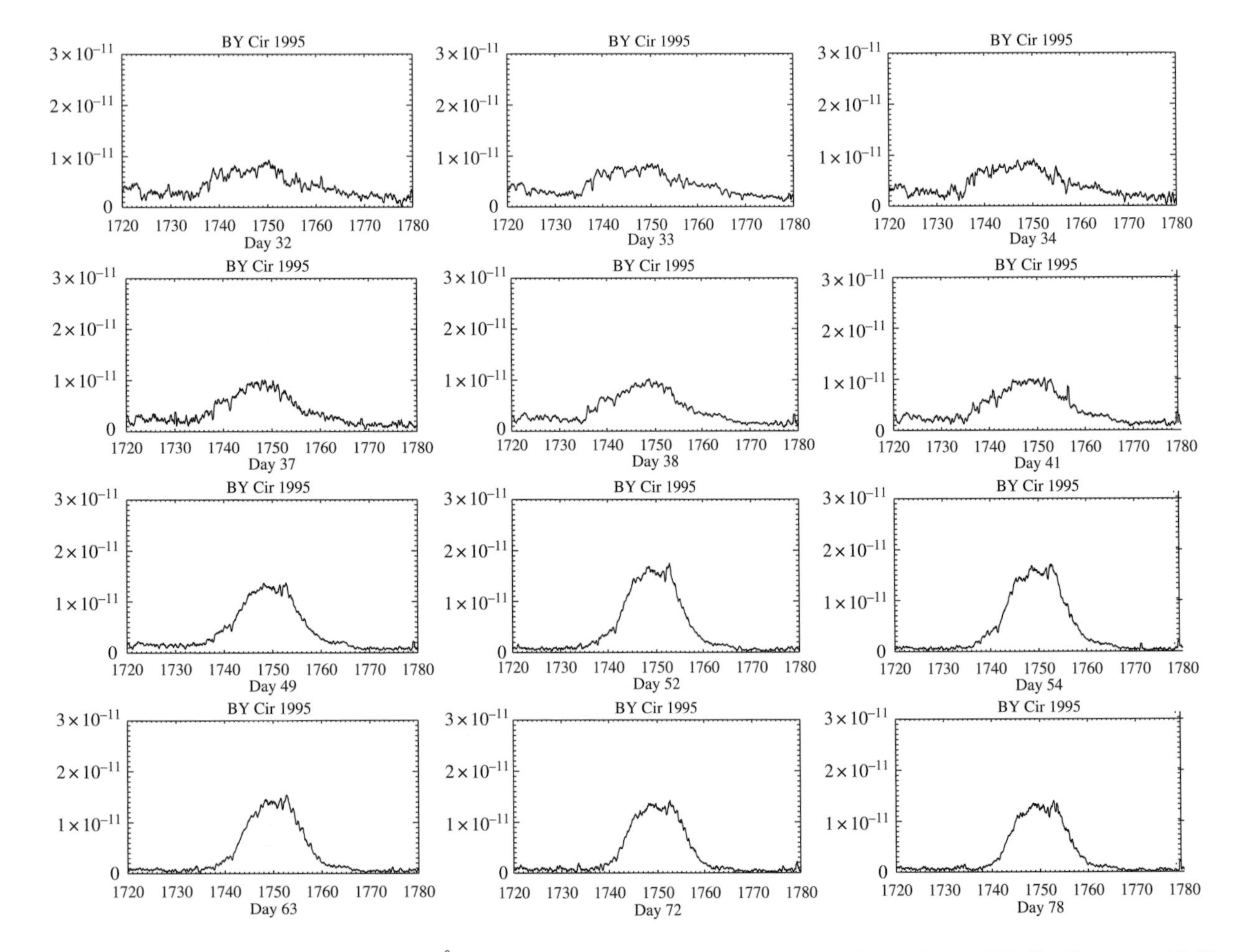

Fig. 9.19. Development of the N III] 1750 Å emission feature during the early outburst of the CO nova BY Cir. Compare with Figure 9.15 for the profiles of V1974 Cyg and LMC 1992.

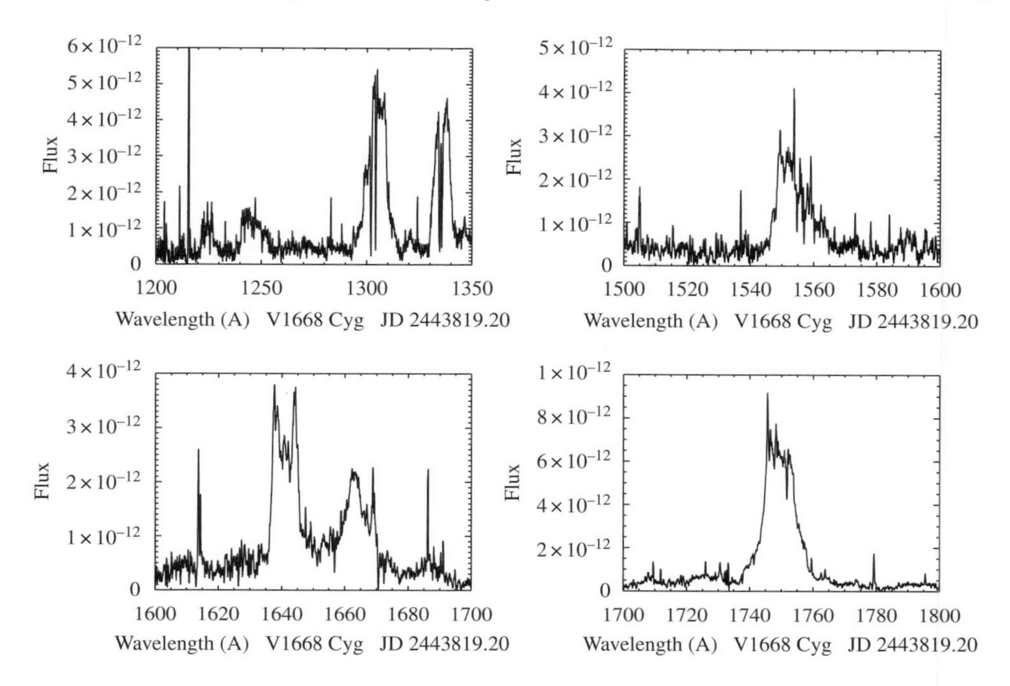

Fig. 9.20. Development of the high-resolution IUE line structures during the nebular phase for the CO nova V1668 Cyg 1978. Notice the He II 1640 Å line showing the 'canonical' narrow emission spikes indicative of a ring-like spectrum.

A fundamental consideration for modeling and interpreting the light curve is therefore clear from the profiles: the ejecta are neither spherical nor simply axisymmetric. A superb collection of high-resolution optical line profiles during the nebular phase is provided by Andrea, Drechsel and Starrfield (1994) that systematically display large, possibly low filling fraction, density fluctuations. This is confirmed by observations of the resolved ejecta (see Chapter 12 and, for example, Wade, Harlow and Ciardullo (2000)). Imaging of η Car and other extremely luminous stars suggests such structure is generic for stars evolving at or possibly even above $L_{\rm Edd}$ at outburst. Line profile analyses of several novae for which high-resolution ultraviolet spectra were obtained in the earliest stages show the emergence of structure as the fastest-moving material becomes transparent. The picture that emerges is of a more nearly spherical (or elliptical) fast component of low density, within which is imbedded a more fragmented, axisymmetric (or simply less spherical) slower-moving, higher-density component (e.g. Shore *et al.* 1993, 2003). In all outbursts, the lines become progressively narrower with time as the emission region moves inward within the ejecta toward slower-moving regions of higher density (e.g. Cassatella *et al.* 2004). It is also clear, from the persistence of O I and other low ionization species (whose ionization potentials are near or just slightly above H I) in optical and ultraviolet spectra throughout the constant bolometric luminosity phase that far-ultraviolet-shielded dense regions coexist with tenuous hot gas long after the onset of the nebular stage (Williams, 1992). As discussed in Chapters 7 and 12, radio images also show temporal evolution of the symmetry of the resolved ejecta. To date, this structure has not been adequately included in abundance studies of the early spectra.

9.4.1 P Cygni profiles and multiple absorption and emission line systems

When the iron curtain is completely opaque, all line radiative transfer is strongly local since the mean free path for photons is short relative to the size of the shell. If, however, we wait a while, as the ejecta expand, a transition must occur and this should be at around the same time as the lifting of the curtain. The lines from the outer ejecta may still have sufficient optical depth that the pumping of their lower levels, and the scattering of photons through lines, produce a system of high-velocity lines corresponding to the structures seen in the star emission spectra. These are not necessarily only in absorption – there should also be emission depending on the orientation of the structures relative to the observer. In the co-moving frame, photons that would otherwise have been trapped within the line cores can now excite the lower levels of transitions whose velocity shift takes their line centers out of the core region. This means many systems of lines will be visible, depending on the overall structure of the ejecta, and the line systems, if they behave similarly to the emission profiles, will remain stationary and simply fade. In addition, excitation of lines progressively closer to the core will be possible and the absorption system will shift to lower velocity *in some cases* over time. In the vast majority of instances, about a dozen novae, this behavior was observed.

A ubiquitous feature of the spectral development at all wavelengths is the appearance of multiple structures on the line profiles (e.g. Figures 9.21–9.25). Multiple absorption components have been reported for many novae, especially in the pre-CCD era (Payne-Gaposchkin, 1957). These have shown a general development toward progressively more negative radial velocities through the initially optically thick phase; examples include DQ Her (McLaughlin, 1954) and HR Del 1967 (e.g. Hutchings, 1972; Rafanelli & Rosino, 1978). Among recent novae, FH Ser (Rosino, Ciatti & Della Valle, 1986), GQ Mus (Krautter *et al.*, 1984), V443 Sct

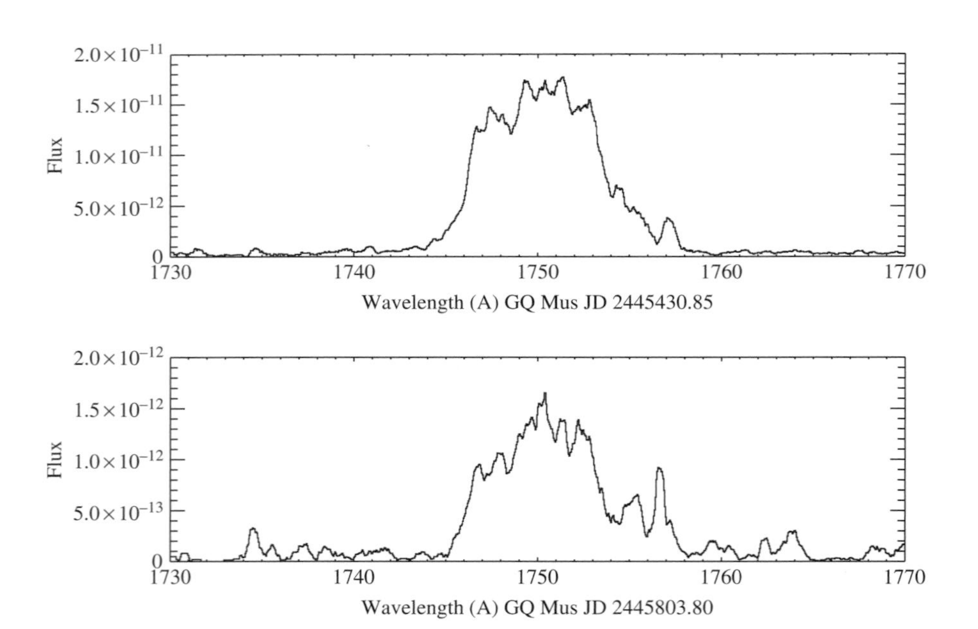

Fig. 9.21. Development of the N III] 1750 Å emission feature during the early outburst of the CO nova GQ Mus showing fine structure at high resolution.

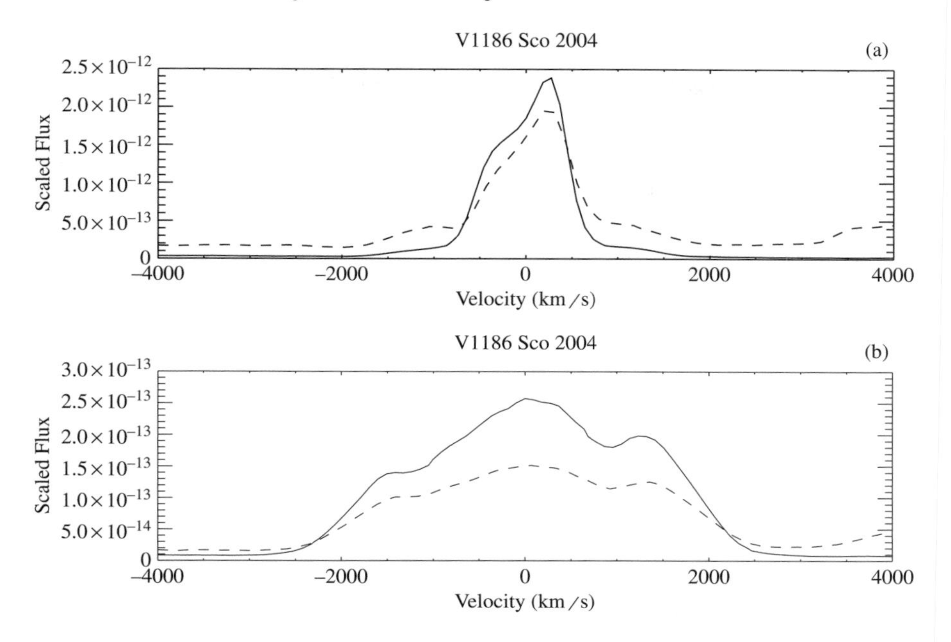

Fig. 9.22. Balmer line profiles early in outburst for two recent novae. a, V1186 Sco (probable CO nova); b, V1187 Sco (probable ONe); solid, Hα, dashed, Hβ. These MMT spectra, obtained around 81 days and 49 days post-discovery respectively, courtesy of C. E. Woodward and G. Ruch, show emission knots and broad wings.

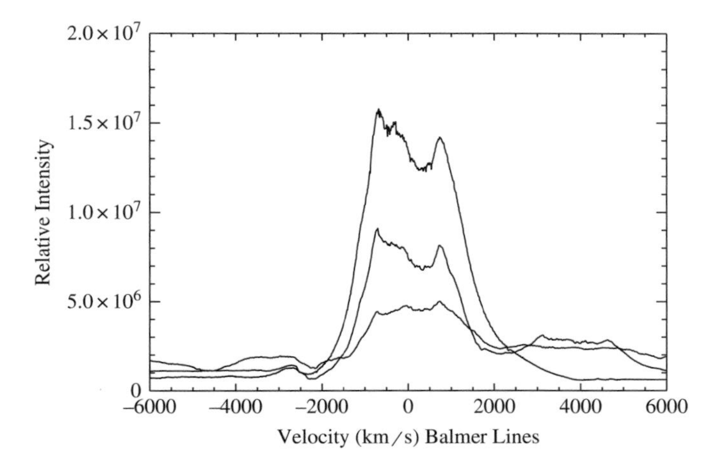

Fig. 9.23. Optical Balmer line profiles early in the outburst of V382 Vel. European Southern Observatory spectra, courtesy of M. Della Valle, obtained approximately 29 days post-outburst and showing a P Cygni profile on Hα and the Balmer decrement (the lines in order of decreasing strength are Hα, Hβ, and Hγ; note also the change in asymmetry with increasing depth of line formation).

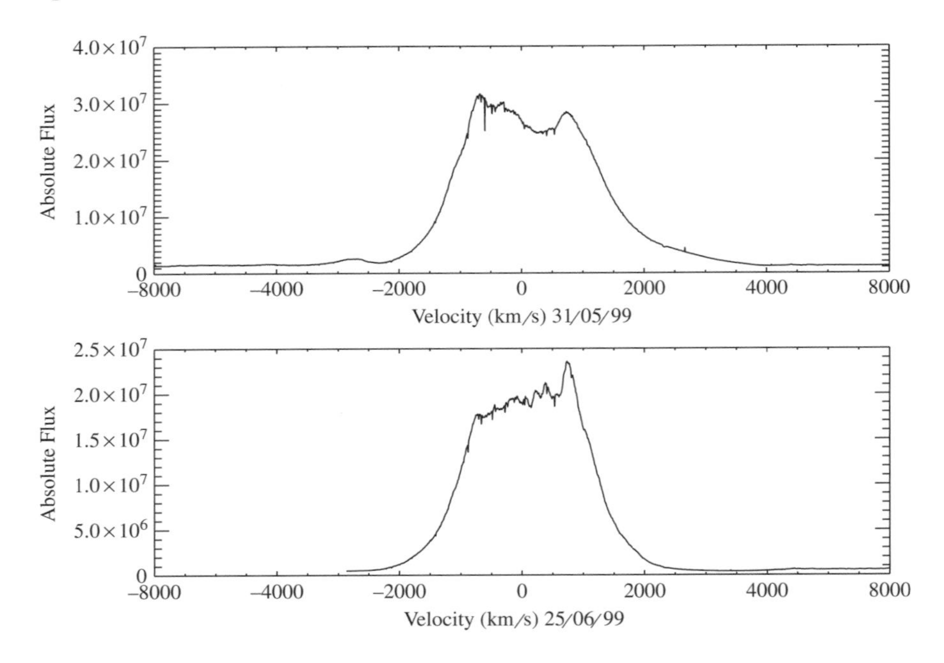

Fig. 9.24. Comparison of two Hα profiles for V382 Vel (European Southern Observatory spectra); the first was obtained within 10 days of outburst, the second about one week later. Note the striking change in the profile asymmetry, also noted in historical profiles (see Della Valle *et al.*, 2002; Shore *et al.*, 2003).

(Rosino *et al.*, 1991), V1974 Cyg (Rafanelli, Rosino & Radovich, 1995), V888 Cen (Yan Tse *et al.*, 2001) have all shown absorption systems during the diffuse enhanced and Orion stages of the optical development with velocities up to 2000 km s^{-1}. In a few cases, GQ Mus in particular, once the emission structures appear (see below) there are some correspondences between absorption and emission narrow components at negative radial velocity. Multiple P Cygni absorption components are also reported for the Balmer lines but never to such high velocities. For V705 Cas, the Na I P Cyg components displayed broader components than the Fe II spectrum of other CO novae at velocities that were also far lower, about −500 km s^{-1} (Shore & Starrfield, 1993) but with velocity widths of order a few hundred km s^{-1}. In the ultraviolet, there is some weak evidence for narrow absorption systems (Cassatella *et al.*, 2004) but the complexity of the absorption bands precludes definite identification. There are also examples of sudden changes in the line profiles during the P Cygni stage, should it occur. For instance, during the ultraviolet development of LMC 1990#1 (see Figure 9.26), the regular decrease of the terminal velocity and intensity of the P Cygni troughs on the ultraviolet resonance lines was interrupted for only one day by their sudden reappearance as fully saturated absorption. The event is still unexplained (see Sonneborn, Shore & Starrfield, 1990; Vanlandingham *et al.*, 1999).

Wind-like profiles, in the form of P Cygni lines, are almost always seen in the earliest stages of the outburst in the optical. The ultraviolet is another matter to which we will return in a moment. The strongest optical P Cygni lines on the Balmer series show V_{max} decreasing

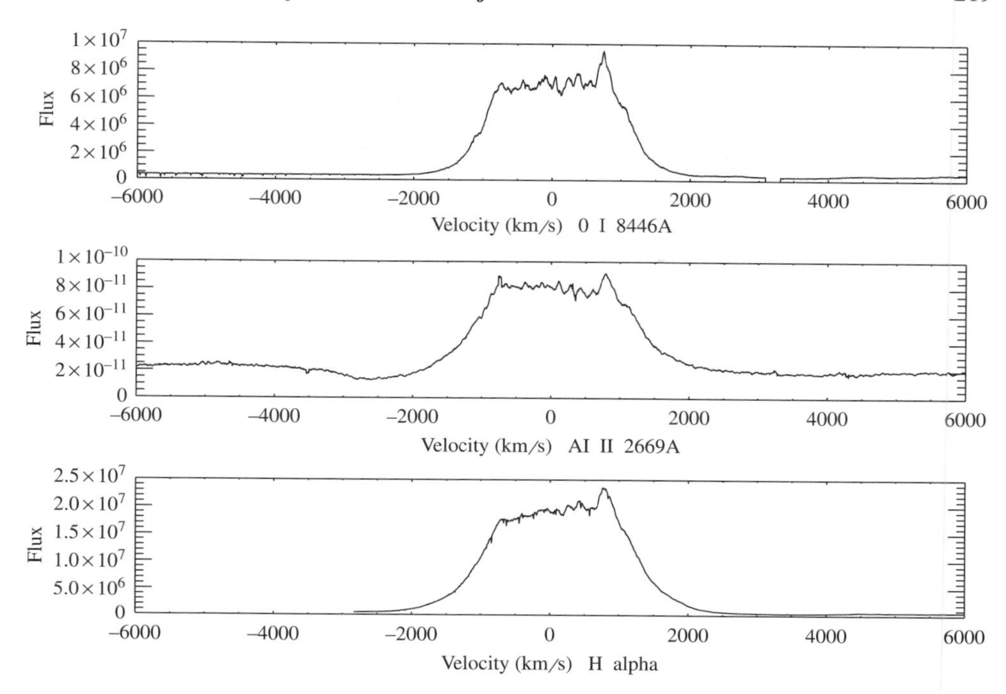

Fig. 9.25. V382 Vel: comparison of Hα, Al II 2669 Å and O I 8446 Å lines during the beginning of the nebular phase showing the similarity of emission profile fine structures and differences in the line asymmetries depending on ionization. This is an unexplained effect seen in other classical novae (see text).

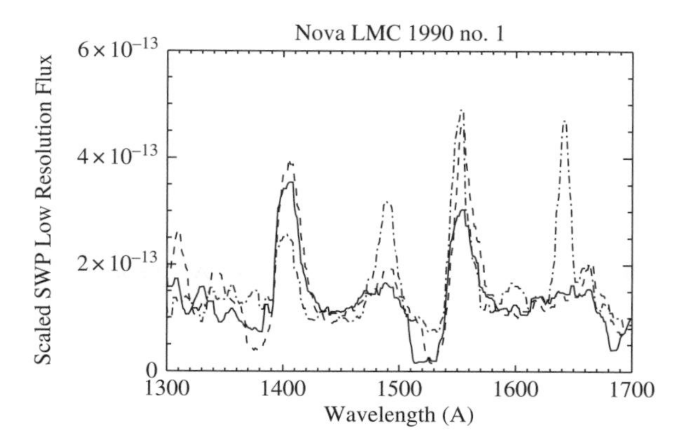

Fig. 9.26. Time development of the P Cygni absorption trough for LMC 1990#1 showing the progressive narrowing and weakening of the absorption (individual profiles are plotted using different lines; the sequence is weaker absorption at later time). As discussed in the text, this progression is consistent with a constant velocity gradient, finite thickness shell and not with a standard wind-like profile. From the IUE Short-Wavelength Prime Camera (SWP).

with increasing upper level, similar to Be stars. It is important, though, to note the difference in the velocity law in the ejecta compared with a continuous outflow: novae will not show the same saturation effects at large r. For a wind, the optical depth at the terminal velocity is limited by the intrinsic line width, $\Delta\nu$, which is usually interpreted as 'turbulent' flows, while for a nova the constant gradient produces a power law decrease in τ. This behavior appears to confirm the basic prediction of models – the ejection event is sudden, impulsive, and nearly ballistic with an imposed approximately constant velocity gradient and the resulting density law required to explain the ionization behavior during the X-ray turn-off stage.

Because of the details of line formation, maximum velocities differ depending on the spectral region. For recurrents, the ultraviolet and optical velocities agree and are quite high, of order 10^4 km s^{-1}. The highest velocities are determined from the ultraviolet resonance lines. For ONe novae, the peak ultraviolet velocity is about 7000 km s^{-1} (V838 Her) and generally between 4000 km s^{-1} (V1974 Cyg) and 5500 km s^{-1} (V382 Vel, LMC 1990#1, LMC 2000). CO novae show lower velocities, generally around 2000–3000 km s^{-1} (OS And, LMC 1988#1). These are typically about a factor of 1.5 to 2 higher than seen from Hα or other optical lines at the same stage (e.g. Figure 9.27). The line profiles, as we have mentioned, narrow with time as the outermost regions contribute progressively less to the emission measure. Unlike a stationary outflow, the relative covering factor of either the ejecta's pseudo-photosphere or the actual white dwarf surface compared with the increasingly distended optically thin volume causes the P Cygni absorption trough to weaken rapidly as well as decrease in width. In general, the emission wings for the P Cygni stage extend nearly to the maximum absorption velocity but these too narrow with time.

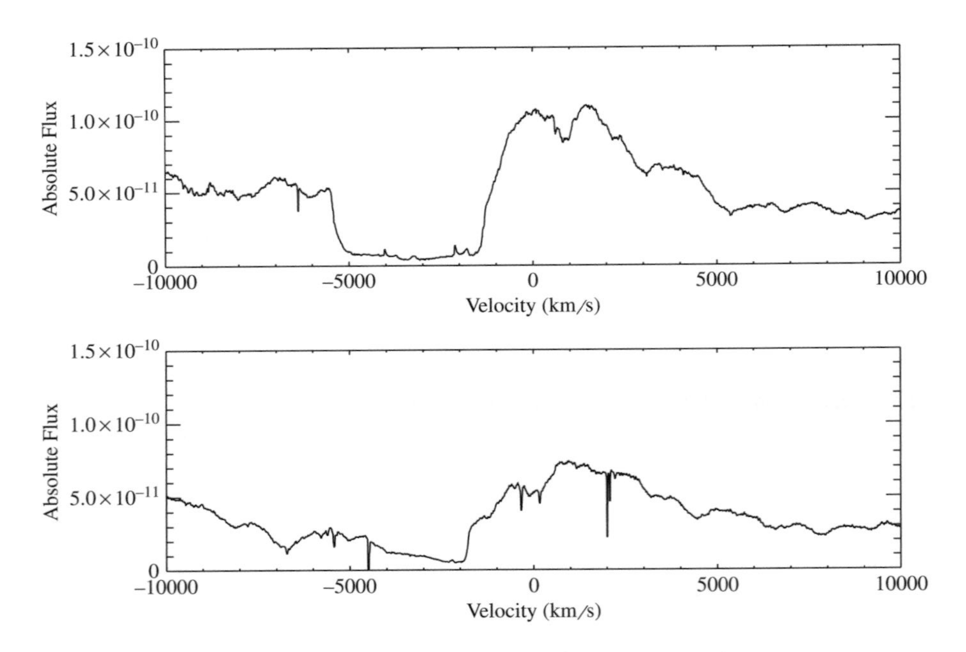

Fig. 9.27. P Cygni profiles of the Si IV 1400 Å and C IV 1550 Å resonance doublets for V382 Vel about 1.5 months after outburst. The terminal velocities are higher than those seen for the Hα transition at a slightly earlier stage.

The absorption line systems may indicate structure but these filaments, shells, and/or knots must be along the line of sight so, depending on the geometry of the outburst, it is not certain they will be seen. However, once emission lines appear, the knots are transparent and all of the gas becomes visible. It is important to note this difference: not every absorption feature finds a counterpart among the emission line structures even when only optical and infrared data are available. For V1974 Cyg and V705 Cas, for which high-resolution infrared and ultraviolet profiles were obtained, the infrared line profiles show the same density knots observed much later on in the ultraviolet emission lines, so evidently something imposes the structure invariantly very early in the outburst. The optical absorption and emission systems maintain constant radial velocity once they appear (cf. McLaughlin, 1954). Many novae display large flare-like transitions during their early decline (generally CO type), often before t_3, whose origin is unknown. But these may be individual ejection events while the nova is at or even exceeding the Eddington luminosity. Whether these are also the originating events for the shells and knots seen later in the line profiles remains an open question.

9.4.2 Spectropolarimetry

This method has been little used in the analysis of structure during the earliest stages of outburst but it is potentially a very valuable tool for structure (Whitney & Clayton, 1989; Bjorkman *et al.*, 1994; Evans *et al.*, 2002; S. Potter & B.-G. Andersson, private communication). Differential polarization of emission lines has been observed in the ultraviolet relative to the continuum for BY Cir (Johnson *et al.*, 1997) although *UBV* observations (Evans *et al.*, 2002;) obtained within the first week after outburst are consistent with an initially spherical ejection. Non-sphericity can be inferred from the position angle measurements if some adequate estimate of the interstellar contribution is available, and changes in the structure of the ejecta might be observable as a function of velocity in the profiles at high resolution. Large (VLT-class) telescopes are needed for this to provide spectra with high signal-to-noise and high resolution (tens of km s^{-1}). This technique, applied to specific transitions, would be especially valuable in light of the changes observed in the Balmer line asymmetries. For instance, in V842 Cen (Sekguchi *et al.*, 1989) and V382 Vel (Della Valle *et al.*, 2002; Shore *et al.*, 2003), the hydrogen lines (especially Hα) change while the [O III] lines retain constant asymmetry for months after outburst (as do the ultraviolet recombination and intercombination lines). Polarization measurements could be very valuable in sorting out these details early in the outburst.

9.4.3 Evidence for abundances and mixing in the ejecta

The homogeneity of the ejecta is hard to assess uniquely from spectra since it requires completely spatially resolved shells or very high signal-to-noise data. At least for V1974 Cyg (Shore *et al.*, 1997), there is strong evidence for a dispersion in the Ne/C ratio between the knots based on small-aperture GHRS spectra of the brightest knot in the resolved ring, but tracing backwards it is difficult to see evidence for this in the pre-resolved data. It should be noted that the spatially resolved spectra structure as observed within the He II 1640 Å line indicates even finer fragmentation, perhaps a fractal-like distribution of densities. Similar results have been obtained for the knots in V382 Vel but without the high spatial resolution. Individual knots have, in general, not been analyzed and the usual approach to photo-ionization modeling (using integrated fluxes) precludes such analyses. This is necessary since these regions have a wide range of recombination times before the turn-off (e.g. Shore *et al.*, 1996, 1997). Instead, one could use the $\tau(v)$ approach now employed for interstellar and Lyα forest absorption lines (e.g. Savage & Sembach, 1996) by taking line ratios for individual profiles to obtain abundance ratios as a function of velocity since the internal velocity dispersion of the material is small relative to the gradient between filamentary structures in the ejecta.

9.5 Mass determinations for the ejecta

This is perhaps the greatest area of uncertainty when confronting observations with theory (see also Chapter 6). The problem is that one never actually measures the mass of the ejecta, only the column density in any particular transition or ion, and must infer the properties of the ejecta from models. Early mass estimates were based on incomplete analyses, for instance ionization correction factors, that attempted to extend the observed ionization states to the total elemental abundances. In general, these produced masses lower than about $10^{-5}\,M_\odot$ (e.g. Stickland *et al.*, 1981; Snijders *et al.*, 1987; Andrea *et al.*, 1991). These can be considered lower bounds because they also had exceptionally low filling factors (not directly modeled from the line profiles) and also required very high temperatures and only nebular state diagnostics. The introduction of model atmosphere methods and comprehensive photoionization modeling has altered this picture. Using the optically thick stage, the masses now come back about an order of magnitude higher, about $10^{-4}\,M_\odot$. But this is actually quite reasonable as a simple consideration of the opacity demonstrates. With the exception of the recurrents, in the optically thick stage almost all classical novae look alike which can be explained by the same mechanism that produces the spectrophotometric light curve. Flux redistribution universally requires that the material reaches column densities in excess of $10^{24}\,\mathrm{cm}^{-2}$, which means the masses (assuming large filling factors) are generally at least 10^{-5} to $10^{-4}\,M_\odot$. Only a few novae have been observed early enough in the ultraviolet to catch the fireball spectrum, as the iron curtain drops – V723 Cas, LMC 1991, V1974 Cyg – and the density gradient required for the models seems to be steeper than r^{-3}, resulting from a linear velocity law. With this column density one achieves an approximate scaling,

$$M_{\mathrm{ej}} \sim 6 \times 10^{-7}\,\phi\,N_{\mathrm{H},24}\,v_3^2\,t_3^2\;M_\odot, \tag{9.2}$$

where ϕ is the filling factor (usually 0.01 to 0.1), v_3 is in $10^3\ \mathrm{km\,s^{-1}}$, t_3 is in days, and the subscript 24 indicates units of $10^{24}\,\mathrm{cm}^{-2}$. This estimate agrees with the results from infrared measurements (see Chapter 8) and also almost all mass determinations from more precise atmosphere and photo-ionization analyses. Thus we have a problem, even with comparatively low filling factors: although the abundance patterns of all types of novae are remarkably well produced by models, none yields such large ejecta masses. Oddly, one class – the recurrent novae-conform remarkably well to theoretical predictions: their ejected masses range from 10^{-7} to $10^{-6}\,M_\odot$ *regardless of the sub-class*. For one system, RS Oph, the white dwarf mass is about $1.2\,M_\odot$ based on the bolometric luminosity, while another, T CrB, has a large dynamical mass around M_{Ch}, the Chandrasekhar mass. This may be the clue: the recurrent novae are presumed to be repeaters on such relatively short time-scales because of the mass of the white dwarf so it is possible that the currently available models for the highest-mass systems, along with the enhanced He/H ratio in the accreted material, render less important the (still unknown) processes that complicate the prediction for lower-mass accretors.

9.6 The X-ray turn-off seen from the ultraviolet

The rate of decline of recombination lines is a proxy measure of when the X-ray emission has ceased for the white dwarf based on the assumption that when the luminosity and T_{rad} of the white dwarf fall below the critical ionization level, the ejecta recombine with the rate asymptotically slowing down due to their free expansion (cf. Shore *et al.*, 1996; Vanlandingham *et al.*, 2001; see also Hernanz & Salas, 2002). The ionization level is fixed, reflecting the radiative conditions at the time of turn-off of the central source, and the ionization structure asymptotically freezes with time as the recombination time-scale increases beyond the expansion time-scale, which for a Hubble-type flow remains constant. Before direct X-ray detections of the super-soft phase (see Chapter 10), the ultraviolet strongly pointed to the onset of this ionizing continuum and, ultimately, signaled its exhaustion (for V1974 Cyg it was possible to show that individual knots differ in density by examining the disappearance of specific profile features with time. See also Figure 9.28). Again, without continuing simultaneous ultraviolet and X-ray observations, optical and infrared surrogates will have to suffice.

The densities are initially high (>10^7 cm^{-3}) and the recombination times are short enough (relative to the expansion-driven decrease) to establish a simple local equilibrium (at least for the ionization). However, eventually the ultraviolet dominates since the recombination time per ion depends on n_e and, in this phase, there is a systematic increase in the ionization of the ejecta. As the density continues to fall, the ionization should eventually 'freeze' and the ejecta

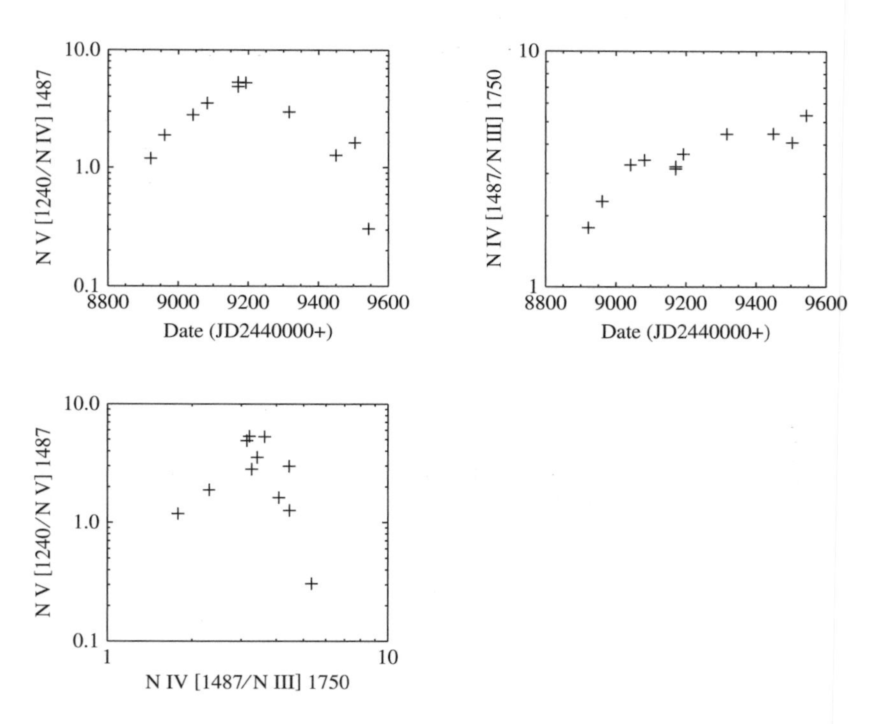

Fig. 9.28. Time development of the ultraviolet nitrogen lines from the nebular phase of V1974 Cyg during and leading to the X-ray turn-off stage. Notice the initial increase in ionization as the ejecta expand.

should reflect the incident radiation from an earlier epoch when the highest ionization is set by the ultraviolet. Thereafter the spectrum *should* remain unaltered and simply fade as t^{-3} because of decreasing emission measure. Yet at the same time, the central source is changing. The temperature increases as the star shrinks and this drives a further increase in the ionization.

The He II 1640 Å line, for instance, follows a simple evolution during the outburst (Figure 9.29). After its first appearance, which marks the transition stage of the ultraviolet spectrum and is accompanied by inter-combination line emission (N IV], Si III], C III]), its strength decreases in a simple way during the phase of turn-off. Since helium appears to be uniformly distributed throughout the ejecta, and since the 1640 Å line is produced only by recombination and is visible long after the hydrogen has faded because of the high ionization, it is a good indicator of any change in the photoionization conditions in the ejecta. In many novae (e.g. Vanlandingham *et al.*, 2001) there is a break in the rate of decline of the He II emissivity after which it is much slower. Assuming that the photoionizing source had turned off at time t_0, the emission measure varies as:

$$\ln\left[j_{1640}(t)/j_b\right] = \frac{1}{2}\,\alpha\, n_{e,b} t_b \left[\left(\frac{t_b}{t}\right)^2 - 1\right], \tag{9.3}$$

where j_b is the emission measure at time t_b when the break occurs, α is the He II recombination coefficient, and $n_e \sim t^{-3}$ is the electron density from the linear velocity law. In fact, the recombination line development provides an argument against any continuous wind source for the ejecta. The observed recombination requires a specific density stratification, hence acceleration law, which would otherwise require a fine tuning of the rate of decrease in wind intensity with time to mimic the observed features.

This behavior also agrees with the ejecta structure required to explain the initial rise in the soft X-ray emission. If the central source remains at constant bolometric luminosity and the ejecta simply extinguish the source with a variable optical depth $\tau \sim N_H(t) \sim t^{-2}$ then the

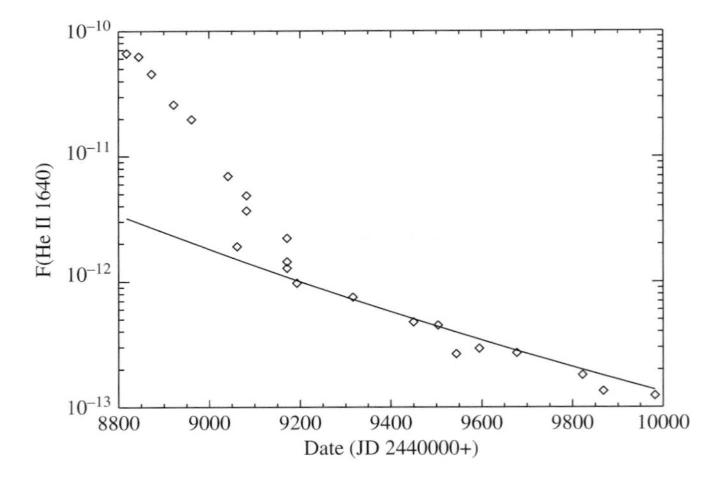

Fig. 9.29. He II 1640 Å decay during the nebular phase of V1974 Cyg. The transition marks the onset of the X-ray turn-off of the central source and the solid line indicates the solution from Equation (9.3) (see text).

result is generic. Note that the initial column density, $N_{H,0}$, must be $> 10^{22}$ cm^{-2} to extinguish the source (Hauschildt *et al.*, 1994; Krautter *et al.*, 1996), which agrees with estimates from the iron curtain. For the one nova that was completely covered at all wavelengths, V1974 Cyg, the transition to the He II turn-off coincided with the X-ray turn-off and the final decrease in the luminosity of the central star was very rapid. The X-ray evolution also probes the ejecta homogeneity by looking at *when* the X-rays are detectable relative to the predictions of the ultraviolet and optical spectrum. González-Riestra, Orio and Gallagher (1998) have examined the change in the bolometric luminosity based on the ultraviolet and X-ray decline and their time-scales agree with those derived from the recombination analysis: with the exceptions of QU Vul and V705 Cas, the typical time-scale is $\leq$4 years. Augmented by non-detections of X-ray emission, all 10 novae they examined have turn-off times of $\leq$7 years. The LMC novae show the same behavior. For the ONe nova V351 Pup, Saizar *et al.* (1996) obtained $n_e \approx 10^8$ cm^{-3} as late as 400 days after outburst, arguing that the column density was still rather high and probably sufficient to block the underlying source. The nebular spectrum of V1974 Cyg appears to require at least two distinct density regions (Vanlandingham *et al.*, 2005). Similarly, QU Vul, a slower ONe nova, also maintained high densities well into the expansion (Schwarz, 2002). In contrast, the very rapid expansion of the ejecta for V838 Her would have permitted the earlier visibility of the X-rays from the central star, which appears to have turned off almost immediately.

9.7 Recurrent novae as special cases

Recurrent novae are defined essentially historically: they have experienced repeated *explosive* outbursts (here we must distinguish between mass ejection and photometric brightening) within the past few centuries of comparatively systematic observations. In other words, what would be called a 'recurrent' depends on the archives and it is not impossible that some could have been missed during the past millennium because of unfavorable placement relative to the Sun (e.g. Robinson, Clayton & Schaefer, 2006), bad weather, lack of funding for observers, or other extrinsic selection effects. Anupama (2002, and references therein) has reviewed the basic characteristics of recurrents, so we will make only some remarks about their development. There are two distinct types of recurrent novae, with orbital periods less than one day or greater than 200 days, and oddly with no 'transitional form'. The recurrence times span a wide range independent of the type of system, from about a decade to about a century. This is the basic problem with identification of these systems; the comprehensive data base for novae from all-sky coverage only postdates the surveys of the mid nineteenth century, so there may still be recurrent systems with longer intervals between outburst yet to be discovered. But we will examine the spectral behavior of the known systems since they clearly divide phenomenologically.

The short-period systems resemble classical novae; IM Nor (1920, 2000), CI Aql (1917, 2000), LMC 1990#2 (1968, 1990), V394 CrA (1949, 1987), and U Sco (most recently 1979, 1999) all have compact companions. What is not clear is whether these systems actually contain an *unevolved* dwarf star since the He/H ratio is usually enhanced by a factor of at least two above solar (e.g. Shore *et al.*, 1991; Anupama & Sethi, 1994; Anupama, 2002). Their optical spectra resemble those of classical novae although they pass through the successive stages very rapidly, and at least the late nebular stage is optically not observed (e.g. Barlow *et al.*, 1981; Williams *et al.*, 1981; Sekguchi *et al.*, 1989; Iijima, 2002). In the ultraviolet, and usually also in the optical, when first detected they show P Cygni profiles on the Balmer lines and ultraviolet resonance lines (C IV, Si IV) and strong He I and often He II emission. They *must* pass through an iron curtain stage to produce the rise to maximum light, but their ejecta turn transparent very quickly with extremely short t_2 and t_3. This is strong, independent evidence for low ejected masses. It is also certainly due to their very high expansion velocities: V394 CrA, U Sco, and LMC 1990#2 displayed ultraviolet resonance line P Cyg troughs with terminal velocities between 8000 and 12 000 km s^{-1}, higher than *any* observed for classical systems. It should be noted that the absorption is not always saturated (even allowing for effects of low spectral resolution) so it is not clear that the ejecta are completely covering.

An interesting, apparently systematic, behavior is seen only in the ultraviolet spectra for these systems. In the earliest stages of outburst after the iron curtain lifts and including the P Cygni stage, the feature at 1400 Å is the Si IV doublet. Later, however, once the pre-nebular stage is passed, this feature is actually the O IV] resonance multiplet. Its disappearance in the three compact recurrent systems, U Sco, V394 CrA, and LMC 1990#2, immediately after the appearance of the He II 1640 Å emission suggests that these systems may have low oxygen abundances. This apparently systematic property of the compact recurrents contrasts sharply with the behavior of ONe novae. In V1974 Cyg, this feature – clearly O IV] – was visible more than two years after outburst and it also persisted for V382 Vel and LMC 1990#1. For the fastest ONe nova, V838 Her, the line was visible for the first three weeks of the outburst but disappeared when the He II reached maximum strength; before this, when Si IV was visible, O III] 1667 Å was also stronger than He II. It should be noted that, based on nebular phase photoionization analyses, this nova was both O deficient, relative to solar, and also S enhanced.

In contrast, the somewhat less studied long-period systems resemble symbiotic novae but with explosive outbursts (cf. Anupama & Mikołajewska, 1999). T CrB, RS Oph, V745 Sco, and V3890 Sgr all have red giant companions (see Figure 9.30) and periods, when known, longer than 200 days. As of this writing, the latest discovery of a recurrent in the LMC, YY Dor (called a nova by Hoffmeister in 1949 and discovered in outburst on 2004 October 20.2 by Liller at $V = 11$), may signal a new member of the long-period group, but little is now known about the optical outburst spectrum or the pre-nova except its possible Mira-like variability during the 1970s (the red star associated with this position has $B = 18.8$, $(B - R) \approx 1$, inconsistent with an accretion disk). The white dwarf appears to be massive and they also show enhanced He/H ratio in their ejecta. The principal difference here is the effect of the environment on their spectral and photometric development of the outburst, one of the reasons for discussing nuclear explosions early in this chapter. The initial hard radiative pulse (i.e. γ, X-ray, and ultraviolet) of a thermonuclear weapon triggers an ionization-induced vapor cloud that dissipates quickly, just as the initial radiative pulse of the nova ionizes the red giant wind. In a bomb, the expansion shock also generates an ionizing precursor and so do the nova ejecta as they move through the comparatively dense wind.

Most of our knowledge of the physics comes from one system, RS Oph, that was observed panchromatically throughout its 1985 and 2006 outbursts, but the behavior of the related systems is very similar (of these, only T CrB has not been observed beyond the visible in outburst; see González-Riestra (1992); Shore, Starrfield and Sonneborn (1996), and references therein). Unlike classical novae, the initial spectrum is difficult to interpret: the stellar wind has about the same absorption features as the iron curtain, as shown by the analysis of T CrB in quiescence (Shore *et al.*, 1993), and this blends with and partially masks the broad lines from the ejecta. Again, there must be an opaque phase for the ejecta since the rise to

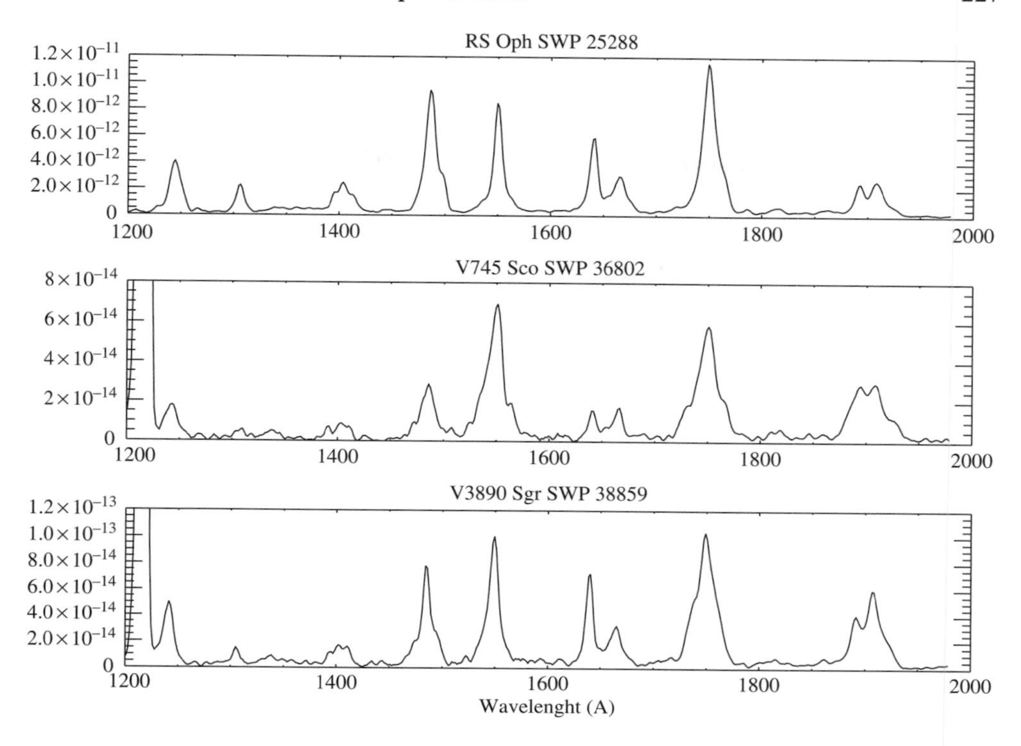

Fig. 9.30. Low-resolution IUE spectra of three symbiotic-like recurrent novae, RS Oph, V745 Sco, and V3890 Sgr during the same stage in outburst. The narrow emission features from the red giant, discernible in the line profiles, are flanked in each case by the high-velocity emission from the ejecta.

maximum light is similar to that seen in the more compact systems, but this has not been detected. The observed emission lines consist of two distinct components that develop on different time-scales. A broad featureless profile with a full width at half maximum (FWHM) of about 2000–3000 km s^{-1} appears first; this must be from the ejected gas. Narrow components having velocity widths of tens of km s^{-1} appear later and persist after the broad line has faded (Figure 9.31). X-ray emission was detected by EXOSAT from RS Oph in 1985 as the ejecta moved through the circumstellar environment whose steep density gradient is produced by the red giant stellar wind. When it ceased, a stage that can be thought of as *breakout* (analogous to a nuclear explosion), the broad component had all but disappeared. For this nova, which never showed P Cygni components at the ejecta velocities, the evidence is that the matter was axisymmetrically erupted. Spatially resolved radio (centimeter) interferometric observations showed that the material resembled two counter-moving lobes (it is best to not call these 'jets' since they were not continually powered; see also Chapter 7). X-ray and radio observations early in the 2006 outburst have tended to confirm the basic conclusions from observations at these wavelengths in 1985 (Bode *et al.*, 2006; Sokoloski *et al.*, 2006; O'Brien *et al.*, 2006). The persistent low-density emission region produced lines that faded over time and may acount for the slowing of the optical photometric decline in this and other systems.

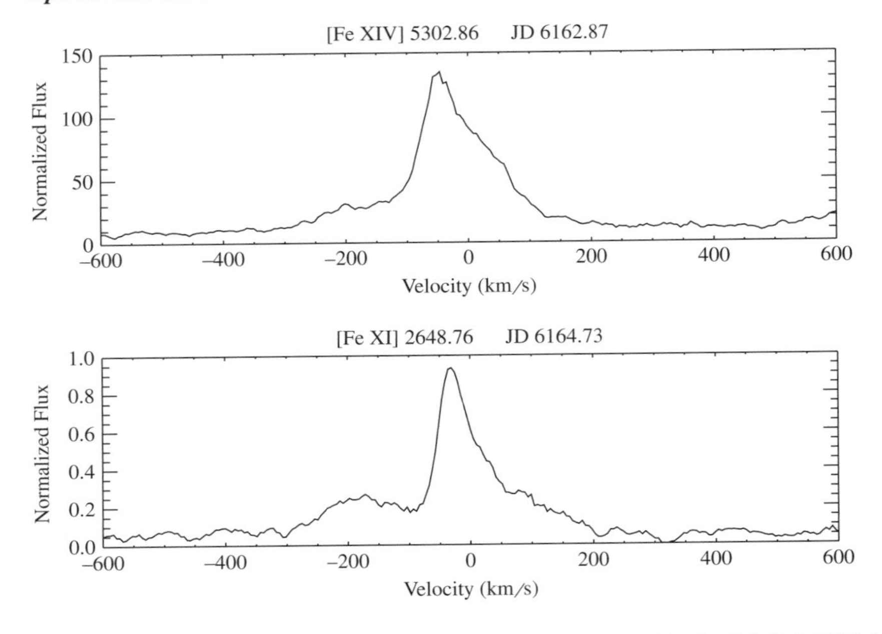

Fig. 9.31. Comparison of optical and ultraviolet coronal line profiles for RS Oph 1985 during the later nebular stage when the narrow line emission from the stellar wind H II region dominates the emission.

The ejecta mass problem of classical outbursts does not seem to hold for recurrent systems. For RS Oph in 1985, the ejected mass was about 10^{-7} $M_{\odot}$; for U Sco (most recently 1979 and 1999) and LMC 1990#2 it was about the same. These originated in very different binary systems, the first having a long ($\sim$ year) period and the others being in short-period systems. In general, He/H is at least solar and usually higher, up to a factor of 2 above solar, which may be a clue that the explosion is triggered from material with a different composition, or mixed differently, from that in classical systems.

9.8 Summary

As we have emphasized in this chapter, atomic radiative processes in the ultraviolet dominate the line formation and spectrophotometric evolution at longer wavelengths because of flux redistribution throughout the evolution of novae in outburst. The demise in particular of high-spectral-resolution ultraviolet capability means that, at the time of writing, the archival satellite observations provide the only such data with which to understand spectral phenomena at longer wavelengths. These have, however, been exploited in a timely manner, although much remains to be mined from the optical–ultraviolet comparisons aided by continuing improvements in model atmospheres and photoionization and shock codes. Substantial uncertainties remain in critical properties, especially in the masses of the ejecta and the possible presence of a wind phase during the opaque stages of the outburst. Many novae, especially in the LMC and M31 for which distance is not an important uncertainty, appear to exceed the Eddington luminosity for long periods, some by more than a factor of 3, and the phase of constant bolometric luminosity persists for longer than most models now predict. Although no model to date can address the wealth of structure and lack of spherical symmetry evinced in the line profiles and, later, from the resolved remnants, it is just a matter of time until three-dimensional hydrodynamic computations become feasible in light of the rapid growth in the past decade in computational performance of parallelized codes. Whether a wind accompanies the ejection is still a matter of debate although there is some indication for it in the ultraviolet line profiles. If any further justification is necessary for the coninued study of novae in outburst, it should be kept in mind that anything we learn is applicable to a wide range of objects including the Luminous Blue Variables, proto-planetary nebulae, AGB stars and supernovae. This chapter may now be closed, but only for this edition.

Acknowledgements

I wish to thank Jason Aufdenberg, Angelo Cassatella, Massimo Della Valle, Bob Gehrz, Charo Gonzalez-Riestra, Joachim Krautter, Peter Hauschildt, Jordi Josè, Greg Schwarz, George Sonneborn, Sumner Starrfield, Karen Vanlandingham, Mark Wagner, and Chick Woodward for many valuable discussions during the past decade, and for providing data during the course of this writing. I especially thank the editors of this volume, Mike Bode and Nye Evans, for their invitation to write this chapter and their valuable suggestions, and Jason Aufdenberg and Jordi Josè for comments on the entire manuscript. This work has been supported by NASA through the IUE, HST and FUSE guest observer programs. I also wish to acknowledge the substantial contribution made by the then project scientist for IUE, Yoji Kondo, and the directors of the STScI, Bob Williams and Steve Beckwith, for allocating directors' discretionary time for some of the observations discussed here.

References

Andreae, J., Drechsel, H., Snijders, M. A. J., & Cassatella, A., 1991, *A&A*, **244**, 111.

Andrea, J., Drechsel, H., & Starrfield, S., 1994, *A&A*, **291**, 869.

Anupama, G. C., 2002, in *Classical Nova Explosions*, ed. M. Hernanz, & J. José, New York: American Institute of Physics, p. 32.

Anupama, G. C., & Mikołajewska, J., 1999, *A&A*, **344**, 177.

Anupama, G. C., & Sethi, S., 1994, *MNRAS*, **269**, 105.

Barlow, M. J., Brodie, J. P., Brunt, C. C. *et al.*, 1981, *MNRAS*, **195**, 61.

Bjorkman, K. S., Johansen, K. A., Nordsieck, K. H., Gallagher, J. S., & Barger, A. J., 1994, *Ap. J.*, **425**, 247.

Bode, M. F., O'Brien, T. J., Osborne, J. P. *et al.*, 2006, *Ap. J.*, **652**, 629.

Cassatella, A., Lamers, H. J. G. L. M., Rossi, C., Altamore, A., & González-Riestra, R., 2004, *A&A*, **420**, 571.

Della Valle, M., 2001, *A&A*, **252**, L9.

Della Valle, M., & Livio, M., 1995, *Ap. J.*, **452**, 704.

Della Valle, M., Pasquini, L., Daou, D., & Williams, R. E., 2002, *A&A*, **390**, 155.

Downes, R. A., & Duerbeck, H. W., 2000, *AJ*, **120**, 2007.

Evans, A., Yudin, R. V., Naylor, T., Ringwald, F. A., & Koch Miramond, L., 2002, *A&A*, **384**, 504.

Gehrz, R. D., 1988, *ARA&A*, **26**, 377.

Gehrz, R. D., Truran, J. W., Williams, R. E., & Starrfield, S., 1998, *PASP*, **110**, 3.

Gill, C. D., & O'Brien, T. J., 1999, *MNRAS*, **307**, 677.

González-Riestra, R., 1992, *A&A*, **265**, 71.

González-Riestra, R., Orio, M., & Gallagher, J., 1998, *A&A*, **129**, 23.

Hauschildt, P. H., Starrfield, S., Austin, S. *et al.*, 1994, *Ap. J.*, **422**, 831.

Hayward, T. L., Saizar, P., Gehrz, R. D. *et al.*, 1996, *Ap. J.*, **469**, 854.

Hernanz, M., & Sala, G., 2002, *Science*, **298**, 393.

Hutchings, J. B., 1972, *MNRAS*, **158**, 177.

Iijima, T., 2002, *A&A*, **387**, 1013.

Johnson, J. J., Anderson, C. M., Bjorkman, K. S. *et al.*, 1997, *AJ*, **113**, 2200.

Krautter, J., Beuermann, K., Leitherer, C. *et al.*, 1984, *A&A*, **137**, 307.

Krautter, J., Ögelman, H., Starrfield, S., Wichmann, R., & Pfeffermann, E., 1996, *Ap. J.*, **456**, 788.

Liller, W., 2004, *IAUC* 8422.

Mason, C. G., Gehrz, R. D., Woodward, C. E. *et al.*, 1996, *Ap. J.*, **470**, 577.

McLaughlin, D. B., 1954, *Ap. J.*, **119**, 124.

McLaughlin, D. B., 1960, in *Stars and Stellar Systems*, Vol. VI, *Stellar Atmospheres*, ed. J. L. Greenstein. Chicago: University of Chicago Press, p. 585.

O'Brien, T. J., Bode, M. F., Porcas, R. W. *et al.*, 2006, *Nature*, **442**, 279.

Payne-Gaposchkin, C., 1957, *The Galactic Novae*. Amsterdam: North Holland Publishing Co.

Rafanelli, P., & Rosino, L., 1978, *A&AS*, **31**, 337.

Rafanelli, P., Rosino, L., & Radovich, M., 1995, *A&A*, **294**, 488.

Robinson, P. B., Clayton, G. C., & Schaefer, B. E., 2006, *PASP*, **118**, 385.

Rosino, L., Ciatti, F., & Della Valle, M., 1986, *A&A*, **158**, 34.

Rosino, L., Benetti, S., Iijima, T., Rafanelli, P., & Della Valle, M., 1991, *AJ*, **101**, 1807.

Saizar, P., Pachoulakis, I., Shore, S. N. *et al.*, 1996, *MNRAS*, **279**, 280.

Savage, B. D., & Sembach, K. R., 1996, *ARA&A*, **34**, 279.

Schwarz, G. J. 2002, *Ap. J.*, **577**, 940.
Schwarz, G. J., Shore, S. N., Starrfield, S. *et al.*, 2001, *MNRAS*, **320**, 103.
Sekiguchi, K., Feast, M. W., Fairall, A. P., & Winkler, H., 1989, *MNRAS*, **241**, 311.
Shafter, A. W., 2002, in *Classical Nova Explosions*, ed. M. Hernanz, & J. José. New York: American Institute of Physics, p. 462.
Shara, M. M., & Zurek, D., 2002, in *The Physics of Cataclysmic Variables and Related Objects*, ed. B. T. Gänsicke, K. Beuermann, & K. Reinsch. Astronomical Society of the Pacific Conference Series, **261**. San Francisco, p. 661.
Shaviv, N. J., 2002, in *Classical Nova Explosions*, ed. M. Hernanz, & J. José. New York: American Institute of Physics, p. 259.
Shore, S. N., & Aufdenberg, J. P., 1993, *Ap. J.*, **416**, 355.
Shore, S. N., & Gehrz, R. D., 2004, *A&A*, **417**, 695.
Shore, S. N., & Starrfield, S., 1993, *Sky & Tel.*, **87**, 42.
Shore, S. N., Sonneborn, G., Starrfield, S. G., *et al.*, 1991, *Ap. J.*, **370**, 193.
Shore, S. N., Sonneborn, G., Starrfield, S., González-Riestra, R., & Ake, T. B., 1993, *AJ*, **106**, 2408.
Shore, S. N., Starrfield, S., González-Riestra, R., Hauschildt, P. H., & Sonneborn, G., 1994, *Nature*, **369**, 539.
Shore, S. N., Starrfield, S., & Sonneborn, G., 1996, *Ap. J.*, **463**, 21.
Shore, S. N., Starrfield, S., Ake, T. B., & Hauschildt, P. H., 1997, *Ap. J.*, **490**, 393.
Shore, S. N., Schwarz, G., Bond, H. E. *et al.*, 2003, *AJ*, **125**, 1507.
Snijders, M. A. J., Batt, T. J., Roche, P. F. *et al.*, 1987, *MNRAS*, **228**, 329.
Sokoloski, J. L., Luna, G. J. M., Mukai, K., & Kenyon, S. J., 2006, *Nature*, **442**, 276.
Sonneborn, G., Shore, S. N., & Starrfield, S. G., 1990, in *Evolution in Astrophysics : IUE Astronomy in the Era of New Space Missions*, ed. E. J. Rolfe. ESA Publications, ESA SP-310, p. 439.
Stickland, D. J., Penn, C. J., Seaton, M. J., Snijders, M. A. J., & Storey, P. J., 1981, *MNRAS*, **197**, 107.
Vanlandingham, K. M., Starrfield, S., Shore, S. N., & Sonneborn, G., 1999, *MNRAS*, **308**, 577.
Vanlandingham, K. M., Schwarz, G. J., Shore, S. N., & Starrfield, S., 2001, *AJ*, **121**, 1126.
Vanlandingham, K. M., Schwarz, G. J., Shore, S. N., Starrfield, S., & Wagner, R. M., 2005, *Ap. J.*, **624**, 914.
Wade, R. A., Harlow, J. J. B., & Ciardullo, R., 2000, *PASP*, **112**, 614.
Whitney, B. A., & Clayton, G. C., 1989, *AJ*, **89**, 297.
Williams, R. E., Sparks, W. M., Gallagher, J. S. *et al.*, 1981, *Ap. J.*, **251**, 221.
Williams, R. E., Hamuy, M., Phillips, M. M. *et al.*, 1991, *Ap. J.*, **376**, 721.
Williams, R. E., 1992, *AJ*, **392**, 725.
Williams, R. E., Phillips, M. M., & Hamuy, M., 1994, *Ap. JS*, **90**, 297.
Yan Tse, J., Hearnshaw, J. B., Rosenzweig, P. *et al.*, 2001, *MNRAS*, **324**, 553.

10

X-ray emission from classical novae in outburst

Joachim Krautter

10.1 Introduction

X-ray observations of novae in outburst have turned out to be a very powerful tool for the study of novae. The X-ray regime is best suited to study the hot phases in a nova outburst. Unfortunately, the present status of our knowledge is still rudimentary, quite different from that of other spectral ranges such as the ultraviolet, visible, or infrared. Few novae have been observed so far in X-rays and essentially all of them lack observations over a complete outburst cycle. In the visible spectral range novae have been systematically observed over more than a century, and in the infrared and ultraviolet many systematic observations which cover the whole outburst cycle have been obtained. So the picture which has emerged from X-ray observations is far less systematic than that from the abovementioned spectral regions (cf. for example Chapters 9 and 8). However, although by no means plentiful, the X-ray observations carried out so far have provided many fundamental and sometimes totally unexpected new results.

GQ Mus was the first classical nova from which X-ray emission was detected during the outburst (Ögelman, Beuermann & Krautter, 1984). In 1984 April GQ Mus was detected at about 460 days after outburst at a 4.5σ level with the low-energy telescope (0.04–2 keV) and the Channel Multiplier Array (CMA) detector aboard EXOSAT. Since the CMA had *no* energy resolution, no spectral information could be obtained. The count rate was compatible with either a shocked shell of circumstellar gas emitting 10^7 K thermal bremsstrahlung radiation at 10^{35} erg s^{-1} luminosity, or with a white dwarf remnant of 1 $M_\odot$ emitting 3.5×10^5 K blackbody radiation at 10^{37} erg s^{-1} luminosity. Subsequent observations of GQ Mus, PW Vul, and QU Vul gave similar results (Ögelman, Krautter & Beuermann, 1987). The data indicate an initial increase of the X-ray flux, followed by a period of more or less constant X-ray flux. EXOSAT also detected intense soft X-ray emission from the recurrent nova RS Oph approximately two months after its optical outburst in 1985 (Mason *et al.*, 1986). The X-ray radiation was interpreted as emission from the collision of the nova shell with circumstellar material previously emitted by the red giant component (see also Chapter 9 for a discussion

Classical Novae, 2nd edition, ed. Michael Bode and Aneurin Evans. Published by Cambridge University Press.

of multi-frequency observations of the 2006 outburst). An extensive discussion of the early observations can be found in Ögelman (1990).

A big step forward came with the launch of ROSAT (Trümper, 1983). With its Position Sensitive Proportional Counter (PSPC), which had an energy range from 0.1–2.4 keV, X-ray data of much higher sensitivity could be obtained than from any comparable X-ray telescope/detector prior to this. The energy resolution of the PSPC, even if very poor, gave some spectral information. With ROSAT many basic X-ray properties of novae could be derived. In addition, the ROSAT All-Sky Survey (RASS), the first all-sky survey in the soft energy range, allowed the acquisition of data over the entire sky.

The next step towards higher data quality came by the beginning of the twenty-first century with Chandra and XMM-Newton. Both these facilities have grating spectrometers which allow for X-ray observations with spectral resolution of up to a thousand, along with a higher sensitivity than ROSAT, and an energy range extending, at the hard end, far beyond ROSAT's 2.4 keV. With ASCA, RXTE, Beppo-Sax and later Swift, a few, but nonetheless valuable, X-ray observations were obtained.

Reviews on X-ray observations of novae in outburst have been given by Ögelman (1990); Ögelman and Orio (1995); Orio (1999); Krautter (2002); and Orio (2004).

10.2 Sources of X-rays

There are in principle four different mechanisms which might lead to X-ray emission from novae in outburst:

(1) Thermal emission from the hot white dwarf remnant. There are two distinct phases in the nova outburst when we expect X-ray radiation. The first opportunity is in the very early phases of the outburst. The thermonuclear runaway (TNR) which causes the nova outburst starts at the bottom of the accreted hydrogen-rich envelope on top of the white dwarf (see Chapter 4). Once the energy created by the thermonuclear reactions has reached the surface of the white dwarf its effective temperature rapidly increases. With a luminosity of the order of $L_{\rm Edd}$ and the radius of a white dwarf, temperatures at peak will exceed 10^6 K, depending on the mass of the white dwarf (Starrfield *et al.*, 1996). The nova becomes a very strong X-ray emitter with a soft spectral energy distribution (SED) of a hot stellar atmosphere. After ejection of the nova shell, which is expanding and cooling adiabatically, the temperature of the expanding pseudo-photosphere rapidly drops, and the X-ray flux decreases, since the expanding envelope becomes opaque to X-rays within a few hours. Because the duration of this 'fireball' phase is very short, it is very unlikely that a nova will be caught during this phase with X-ray observations and indeed, to date no X-ray observations of a nova during the fireball phase have been made. Furthermore, since X-ray satellites prior to Swift were not real-time observatories, it has been very difficult to explore the detection of X-rays from a nova in this phase.

A second phase of X-ray emission from the white dwarf occurs during later phases of the outburst. According to the TNR nova models (and also confirmed in the meantime by ultraviolet and infrared observations) only part of the ejected envelope reaches velocities higher than escape velocity. The remaining material soon returns into quasi-static equilibrium forming an envelope around the white dwarf with, initially, the dimensions of a giant star (e.g. Starrfield, 1989). This material provides sufficent fuel that during the so-called 'phase of constant bolometric luminosity' hydrogen burning ensues in a thin shell on top of the white dwarf until the hydrogen supply has been used up. As the evolution proceeds, the shell mass, and hence the radius, decreases because of nuclear reactions, radiation-driven winds and dynamical friction. With decreasing radius at constant luminosity of $\sim L_{\rm Edd}$, the surface temperature increases to several 10^5 K. The nova again becomes a strong X-ray emitter with a soft SED of a hot stellar atmosphere. The duration of the phase of constant bolometric luminosity is an inverse function of the white dwarf mass (Starrfield *et al.*, 1991; Krautter *et al.*, 1996). Determination of the length of the hydrogen-burning phase can, hence, provide an estimate of the white dwarf mass. After depletion of the envelope, hydrogen burning turns off, and the photosphere of the white dwarf is expected to cool at constant radius. As a consequence of this turn-off the soft X-ray luminosity drops.

(2) X-ray emission originating in the circumstellar medium surrounding the nova system where the expanding nova shell and/or a nova wind interact with each other, or with pre-existing circumstellar material. In such a regime one expects strong shocks giving rise to a thermal bremsstrahlung spectrum with temperatures of up to several keV (Brecher, Ingham & Morrison, 1977). Balman, Krautter and Ögelman (1998) propose three different ways to produce hard X-ray emission from the nova ejecta:

  (a) shocks within the nova wind, which should resemble X-ray emission from the winds of hot OB stars;
  (b) shocks at a circumstellar interaction region where nova ejecta collide with an old shell or a red giant wind; and
  (c) shocks from the collision of a fast wind with pre-existing slow wind material from the white dwarf remnant.

  The luminosity of any such X-rays from the nova surroundings will depend on the temperature and density of the emitting region. In addition to the continuous bremsstrahlung radiation, emission lines produced in the hot plasma are also expected.

(3) Re-establishment of accretion via an accretion disk or magnetic accretion. One expects the typical X-ray emission of a cataclysmic variable in quiescence with a thermal bremsstrahlung spectrum.

(4) Hard X-rays resulting from Compton downgradation of γ-rays (i.e. scattering leading to a reduction in their energy) produced by the radioactive decay of ^{22}Na (Livio *et al.*, 1992; see also Chapter 11). No evidence for X-rays originating by this mechanism has yet been found.

10.3 X-ray light curves

So far the most completely covered X-ray light curve of a classical nova is that of V1974 Cygni, where X-ray observations have been obtained on 18 occasions over a period of $\sim$2 yr, i.e. with reasonably spaced observing epochs from early rise to decline (Krautter *et al.*, 1996). V1974 Cyg, a moderately fast ($t_3 \sim 35$ days) ONe nova, was discovered in outburst on 1992 February 20 (Collins, 1992). With a maximum brightness $V_{\rm max} \sim 4.4$ mag, it was the brightest nova since V1500 Cyg. Because of its brightness, V1974 Cyg was observed over all wavelength ranges, from radio to γ-rays (see Chapters 7, 8, 9, 11, and 12). All X-ray observations were carried out using the PSPC on ROSAT. Observations started on day 63 after outburst, as soon as V1974 Cyg had entered the ROSAT observing window (Krautter *et al.*, 1993). Three separate phases can be distinguished in the light curve presented in Figure 10.1.

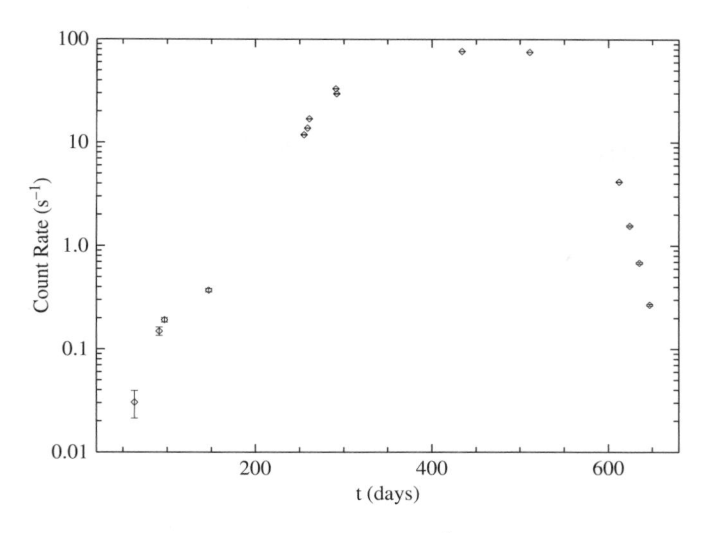

Fig. 10.1. X-ray light curve (log counts s^{-1} vs. days after outburst) of V1974 Cyg in the ROSAT PSPC 0.1–2.4 keV energy band. From Krautter *et al.* (1996).

The rise phase from days 63 to 147 is characterized by an increase of the count rate, from 0.03 to 0.37 counts s^{-1}. Around three months later, on day 255, the beginning of the second phase (the 'plateau phase', as it was referred to by Krautter *et al.* (1996)), the count rate had strongly increased to 11.8 counts s^{-1}. During this plateau phase the count rate increases to a peak of ~75 counts s^{-1}, which occurred between days 434 and 511. The decline of the X-rays down to 0.2 counts s^{-1} starts between days 511 and 612.

The three phases are reflected in the spectral behaviour of V1974 Cyg. Figure 10.2 shows the SEDs at one epoch during each of the three phases (Balman, Krautter and Ögelman, 1998). During the rise phase V1974 Cyg had a very hard spectrum with essentially no photons below 0.7 keV. The first indication of a rising soft component was found on day 97. At the beginning of the plateau phase the spectral behaviour had changed dramatically, since the SED of V1974 Cyg this time exhibits all the spectral characteristics of a super-soft source (SSS) (e.g. Hasinger, 1994; Kahabka & van den Heuvel, 1997). With a maximum count rate of 76.5 counts s^{-1}, V1974 Cyg was by far the strongest SSS ever observed by ROSAT . The hard component was still present but could barely be seen on days 434 and 511, when the soft component had reached its maximum. During the decline phase the count rate decreased strongly, but the general character of the observed SED remained essentially the same as during the plateau phase, even if it did become slightly harder. Both the soft and much weaker hard components were still present.

The absence of soft X-ray flux during the rise phase is due to absorption by hydrogen in the expanding nova shell. Between days 97 and 147 the envelope starts to become transparent for the soft radiation and subsequently the soft flux increases. The start of the decline phase should correspond to the turn-off of the hydrogen burning. The energy radiated during this phase comes from gravitational contraction only. Using the Kelvin–Helmholtz time-scale and assuming a total cooling time of 6 months, Krautter *et al.* (1996) estimated a mass of $\sim 10^{-5}$ $M_{\odot}$ for the hydrogen-exhausted, remnant envelope on the white dwarf.

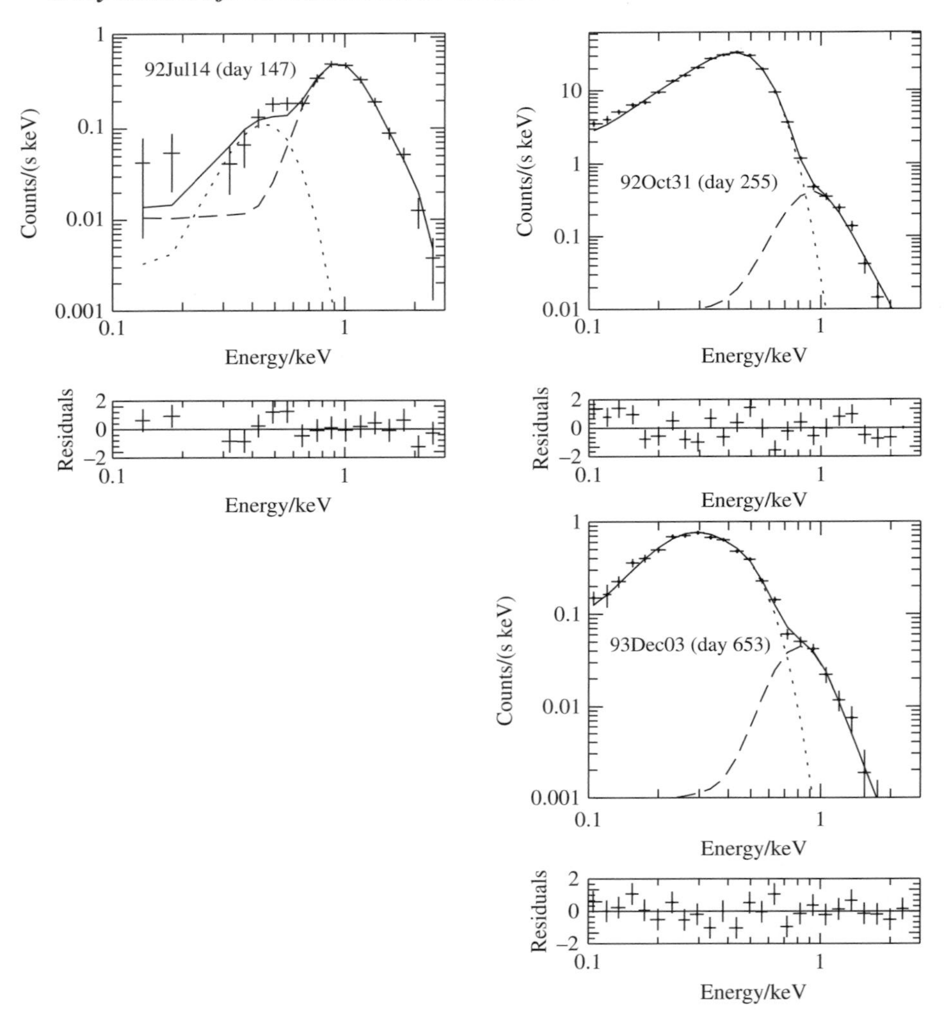

Fig. 10.2. Spectral energy distributions of V1974 Cyg in the ROSAT PSPC band at three different epochs. The actual PSPC data are indicated with crosses. The data are fitted with ONe-enhanced white dwarf atmosphere models (soft component) and Raymond–Smith plasma emission models (hard components). The lower figures show the residuals between the data and the models in standard deviations. From Balman, Krautter and Ögelman (1998).

10.4 The soft component

The most prominent spectral feature is the soft component, which is a clear signature of the ongoing hydrogen burning on top of the white dwarf. Its origin and behaviour can be best interpreted in terms of the TNR model of the nova outburst (see, for example, Starrfield (1989) and Chapter 4). Fits to the SEDs of the super-soft phase allow temperature, luminosity and the absorbing column density N_{H} to be derived. However, one general problem which affects all spectral fits is the absorption of the soft X-rays by the interstellar and, during the early phases, by the circumstellar hydrogen. Hence, only the high-energy tail of the radiation emitted in the soft X-ray regime is actually observable. Since most of the emitted photons are absorbed, small deviations of the SED used for the fit from the real energy distribution will give rise to large errors in the resulting fit parameters.

For the spectral fits it is absolutely crucial to apply realistic hot stellar atmospheres. As Krautter *et al.* (1996) demonstrated, black-body energy distributions are unsuitable to fit the data because the model fits give very unreliable results. With column density N_{H}, temperature T_{eff} and luminosity L as free parameters, super-Eddington values of several thousand L_{Edd} for a 1 $M_{\odot}$ white dwarf were obtained. On the other hand, if one assumes $L = L_{\mathrm{Edd}}$, the model fits do not converge. This result can be generalized for any other SSS (Kahabka & van den Heuvel, 1997).

In the following sections, classical novae with observed super-soft phases will be discussed in detail. We also note that similar behaviour is apparent from observations of the 2006 outburst of the recurrent nova RS Oph (Bode *et al.*, 2006; Osborne *et al.*, 2006a) where short-period oscillations in the soft X-ray flux have also been detected (Osborne *et al.*, 2006b).

10.4.1 V1974 Cyg

An analysis of the V1974 Cyg data with real white dwarf atmospheres was carried out by Balman, Krautter and Ögelman (1998) using model atmosphere continua provided by J. MacDonald (private communication). Best results arose from the models with enhanced O-Ne-Mg abundances (cf. the fits to the soft components in Figure 10.2). Figure 10.3

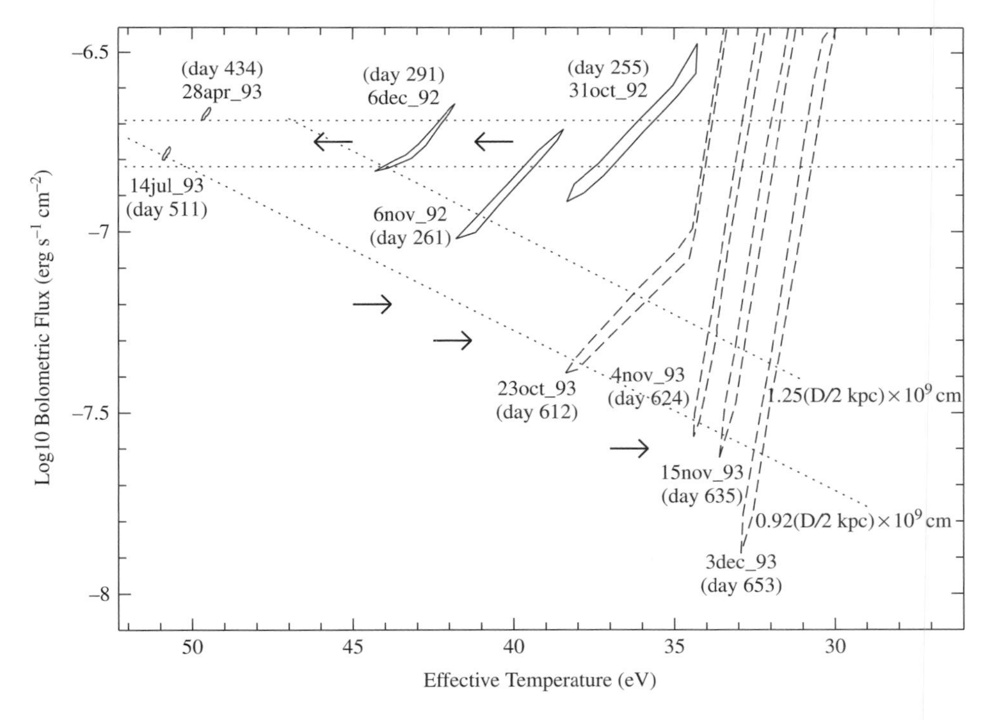

Fig. 10.3. Effective temperature vs. bolometric flux of V1974 Cyg. The H–R diagram shows the spectral evolution of the post-outburst WD. The figure includes nine 3σ confidence contours from fits with ONe-enhanced atmosphere emission models. The solid line contours belong to the constant bolometric luminosity phase, the dashed contours show the cooling phase after turn-off of H burning. The leftward arrows indicate the constant bolometric flux evolution, the rightward arrows indicate a constant radius path for the WD stellar remnant. From Balman, Krautter and Ögelman (1998).

shows the effective temperature vs. the bolometric flux for the evolution of V1974 Cyg. As predicted by the TNR model (Starrfield, 1989) the $T_{\rm eff}$ of the hot white dwarf increases at constant bolometric luminosity up to $\sim 5 \times 10^5$ K while the white dwarf radius decreased by more than a factor of 2 from 2.2 to 0.9 $(D[\text{kpc}]/2) \times 10^9$ cm. In the last four observations $T_{\rm eff}$ decreased to a range of $4.5-3.8 \times 10^5$ K, indicating that the stellar remnant started cooling. This, of course, implies that the hydrogen burning must have turned off. However, since the luminosities were not well defined it was not clear whether the photosphere was cooling at constant radius.

During its bright phase the point source image of V1974 Cyg was surrounded by an extensive halo caused by scattering from interstellar grains. Mathis *et al.* (1995) and Draine and Tan (2003) used this halo to study the nature of the interstellar dust grains.

10.4.2 GQ Mus

As mentioned above, this was the first nova in outburst to be discovered in X-rays (by EXOSAT) and is of particular importance because of the unusual length of its hydrogen-burning phase. GQ Mus was observed on five different occasions with the ROSAT PSPC. At the first observation during the RASS it was detected on day 3118, i.e. ~8.5 years after outburst, with a count rate of 0.143 counts s^{-1} and a low signal-to-noise ratio (Voges *et al.*, 1999). On subsequent observations carried out in the pointed mode the count rate dropped from 0.127 counts s^{-1} on day 3322 (Ögelman *et al.*, 1993) over 0.007 counts s^{-1} on day 3871 to <0.003 (3σ) counts s^{-1} (Shanley *et al.*, 1995).

From these data it is obvious that GQ Mus turned off more than 8.5 years after maximum. However, since both Ögelman *et al.* and Shanley *et al.* used black-body fits for their analysis (and, as noted above for the case of V1974 Cyg, this does not give reliable results), it was not clear when exactly the nova turned off. Balman and Krautter (2001) re-analysed the GQ Mus data, again using MacDonald's white dwarf models. Best results were obtained with CO white dwarfs. Because of the low number of counts in the RASS observation on day 3118 (the exposure time was just 150 s), only an upper limit $kT < 54$ eV ($< 6.2 \times 10^5$ K) could be estimated. The results obtained by Balman and Krautter, however, show some evidence that hydrogen burning was still going on as of day 3118. For day 3322, an effective temperature in the range 38.3–43.3 keV (4.4–5.1 $\times$ 10^5 K) was derived. On day 3641 $T_{\rm eff}$ and the unabsorbed flux had further decreased. Whereas Ögelman *et al.* found from their analysis that GQ Mus was still undergoing hydrogen burning on day 3322, Balman and Krautter found that the hydrogen burning had most likely already ceased by day 3322, and had turned off at some date between day 3118 and day 3322. With a duration between 8.5 and 9.1 years, GQ Mus exhibits an extraordinarily long phase of hydrogen burning, the longest known so far. On the basis of the duration of the hydrogen burning Balman and Krautter estimated a lower limit for the mass of the white dwarf as $M_{\rm WD} \geq 0.8\,{\rm M}_\odot$.

It should be noted that GQ Mus also had some unusual properties in the optical spectral range (Krautter & Williams, 1989). For many years its emission line spectrum exhibited very high ionization, with the coronal lines [Fe XI] λ7892 Å as strong, and [Fe X] λ6374 Å about 2.5 times as strong, respectively, as Hα; this is unprecedented and has not, to the knowledge of the author, been observed in any other astronomical object. The high ionization lines were due to photo-ionization of a hot (several 10^5 K) source. In a spectrum taken on 1993 March 10, after the turn-off of the hydrogen burning, the coronal lines had disappeared and the ionization was much lower. This shows that the ionization stage of the optical spectrum could, in some cases, be used as a qualitative indicator for the turn-off of the hydrogen burning.

10.4.3 LMC 1995

A hydrogen-burning phase of slightly shorter duration (6–8 yr) than was the case for GQ Mus was found for nova LMC 1995. This nova, which was discovered in early March 1995(Liller, 1995), was observed with ROSAT on several occasions after the outburst from 1995 to 1998 February (Orio & Greiner, 1999). Unfortunately, only for the observations obtained in 1998 February was the PSPC available: all other observations had to be carried out with the HRI which had no energy resolution. In the PSPC observations the nova appeared as a bright super-soft source whose SED could be satisfactorily fitted with a LTE CO atmosphere model for a 1.2 $M_\odot$ white dwarf with an effective temperature of about 3.3×10^5 K. The increase of the X-ray flux from 1996 to 1998 by a factor of 2 suggests that the photosphere was still shrinking and that the remnant was becoming hotter. Nova LMC 1995 was observed again in December 2000, nearly 6 years after the outburst, with XMM-Newton and the European Photon Imaging Camera (EPIC; Orio *et al.*, 2003). The nova was still a bright super-soft source; from NLTE atmospheric models a temperature in the range 4–4.5 $\times$ 10^5 K and a bolometric luminosity $L_{\rm bol} \sim 2.3 \times 10^{37}$ erg s^{-1}, close to Eddington luminosity, could be derived. Best fits were obtained with models that did not have enhanced C abundances. However, in an observation carried out with XMM-Newton 8 years after outburst the super-soft source had faded (Orio, 2004).

10.4.4 V382 Vel

For V382 Vel, a very fast ONe nova with a decay time $t_3 \sim 10$ days, a bright soft component was for the first time detected by Beppo-Sax observations on day 184 by Orio *et al.* (2002), who found that fits to the soft component with NLTE models by Hartmann and Heise (1997) and Hartmann *et al.* (1999) could be significantly improved if at least one emission feature was superimposed. Chandra ACIS-I observations on day 222 showed the soft component still to be very strong but only 46 days later, on day 268, it had totally disappeared when the nova was observed for the first time in a high-resolution mode with Chandra-LETG+HRC (Starrfield *et al.*, 2000; Burwitz *et al.*, 2002). Hydrogen burning on top of the white dwarf must have turned off between days 222 and 268, probably shortly after day 222, since even the longest possible cooling time of about 6 weeks is extremely short. The total duration of the hydrogen burning was, at only 7.5–8 months, very short. This indicates that the mass of the white dwarf in V382 Vel is very high, consistent with its nature as an ONe nova and the short decay time (10 days) of the visual light curve.

10.4.5 V1494 Aql

This was a fast CO nova with a t_3 of 13 days. Three Chandra ACIS-I observations were carried out on days 134, 187, and 247 after outburst. The spectra obtained on the first two occasions showed emission lines only, but on day 247 the X-ray spectrum had evolved into that characteristic of a super-soft source with a peak around 0.5 keV (Starrfield *et al.*, 2001). Two Chandra-LETG+HRC observations with a resolution of 600 were carried out on days 300 and 303 with exposure times of 8 and 17 ks, respectively. V1494 Aql was the first nova of which high-resolution spectra were obtained during the super-soft phase and new features, never previously observed, showed up. At first glance emission features seem to be superimposed on the strong soft continuum component. However, none of these features could be identified with any known emission lines, so it cannot be excluded that the spectrum is in reality an absorption spectrum where the apparent emission features are only those parts with less local absorption. Spectra with such a character are observed in the ultraviolet during the early fireball phase (e.g. Hauschildt, 1992; Hauschildt *et al.*, 1997; see also Chapter 5 for a discussion of this). On the other hand, the spectra of V1494 Aql were taken 10 months after outburst, when the situation was very different from the fireball phase where one observes the initial expansion of the opaque nova shell. This is certainly not the case 10 months into the outburst. The application of suitable NLTE models to the observed SED would help to clarify this situation; however, this has not yet been done. During the super-soft phase very unusual short-term variations of the X-ray flux were found which will be discussed below (Drake *et al.*, 2003). The soft component had disappeared by day 726. Since no other X-ray observations had been obtained in the meantime, the duration of the hydrogen-burning phase, between 10 months and <2 years, is rather poorly defined.

10.4.6 V4743 Sgr

The first Chandra-ACIS-S observation of this fast nova ($t_3 < 15$ days) on day 150 did not exhibit any soft component; only weak hard X-rays were present. High-resolution Chandra-LETG+HRC observation on day 180 yielded a spectrum with a strong super-soft continuum and absorption features which could be identified with H- and He-like lines of highly ionized ions of C V, C VI, N VI, N VII, as well as O VII (Ness *et al.*, 2003). The spectrum is presented in Figure 10.4. The presence of these ions implies a temperature between 1 and 2×10^6 K. The absorption lines were blue-shifted by $\sim$2400 km s^{-1} because of the expanding envelope. Interestingly, this velocity is twice the expansion velocity of 1200 km s^{-1} found by Kato, Fujii and Ayani (2002) for the material ejected early in the outburst. Model calculations

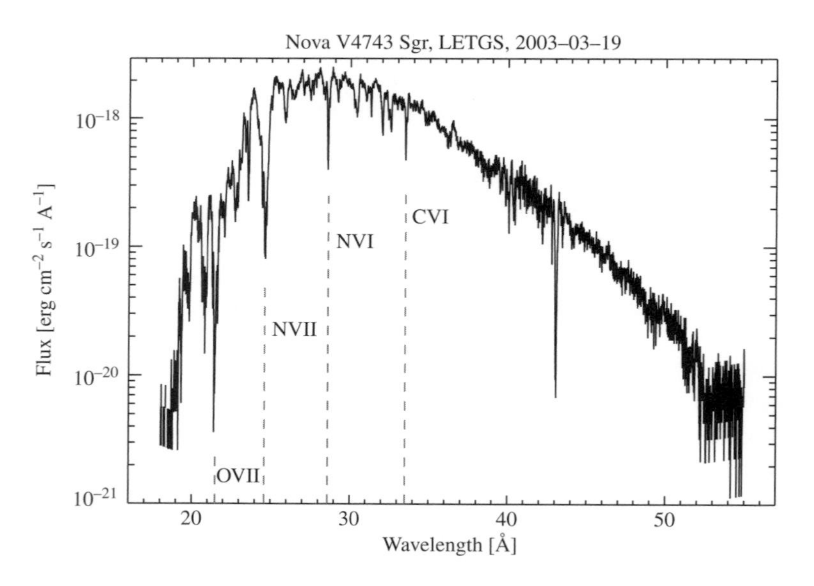

Fig. 10.4. Chandra-LETGS spectrum of V4743 Sgr. The strongest absorption lines are indicated. Weak emission lines are probably also present. From Ness *et al.* (2003).

of the day 180 spectrum were carried out by Petz *et al.* (2005) with *PHOENIX*, a 1-D spherical, expanding, line blanketed, full NLTE model atmosphere (Hauschildt *et al.*, 1992; Hauschildt & Baron, 1999; see also Chapter 5). A best fit was provided with an effective temperature of $T_{\rm eff} = 5.8 \times 10^5$ K, an expansion velocity of 2500 km s^{-1} and a bolometric luminosity of 2×10^{38} erg s^{-1}. On subsequent Chandra-LETG observations on days 301 and 369, the general character of the spectra did not change; there was, however, a decrease of the continuum level on day 369 by about 50% (J. Ness, S. Starrfield & J. Krautter, in preparation). Also, some of the absorption lines had somewhat weakened. Between days 180 and 301 the SED got slightly harder, probably owing to a reduction of the free–free emission.

10.4.7 X-ray survey of super-soft novae

A systematic search for super-soft nova sources was carried out by Orio *et al.* (2001a), who analysed 350 pointed and serendipitous observations of classical and recurrent novae in outburst and quiescence, contained in the ROSAT archive. Only three super-soft X-ray sources were detected among the 132 novae in the Galaxy and in the Magellanic Clouds, specifically those already mentioned: V1974 Cyg, GQ Mus and Nova LMC 1995. Thirty Galactic and nine LMC novae, observed up to 10 years after explosion, were included in the ROSAT sample. The question now is whether this ‘missing soft X-ray emission’ is real, i.e., due to a short hydrogen-burning time, or due to a selection effect. One reason for a selection effect could be that the interstellar hydrogen column density is so high for most novae that the soft X-ray emission is absorbed. However, as Nickel (1995) showed, 21 Galactic novae, which were observed in the RASS within 10 years after the outburst and for which $E(B-V)$ data were available, have an average interstellar extinction $E(B-V) \sim 0.5$ mag. With this value, only a few novae should have hydrogen column densities high enough to absorb the whole of the soft X-ray emission. Even more striking is that, with the exception of Nova LMC 1995, no soft X-rays were detected from any other nova in the LMC where the mean interstellar extinction $E(B-V)$ is well below 0.2. Circumstellar extinction should play no role, since most novae were observed in the RASS months or even years after maximum, when the expanding envelopes should have become optically thin.

On the basis of these data a selection effect can be excluded. One can conclude that the majority of the novae turn off hydrogen burning after a relatively short time. No precise number can be given, but it seems safe to assume that most novae turn off after less than 2 years. This implies that a very efficient mass-loss mechanism must be at work for the nova to get rid of the envelope within such a short time. Time-scales for nuclear burning are much longer, of the order of years to decades (e.g. Truran, 1982), and dynamical friction only works if the envelope is still big enough that the secondary star in the nova system is moving within the envelope. The most promising mass-loss mechamism for ejecting the envelope within a short time is for the mass-loss to be driven by radiation pressure. Starrfield *et al.* (1991) calculated the turn-off time-scale as a function of the white dwarf mass assuming a radiation-pressure-driven mass-loss as described by Castor, Abbott & Klein (1975). They find that the turn-off time-scale strongly decreases with increasing white dwarf mass. For instance, a $1.25\,M_\odot$ white dwarf has a turn-off time of about 2 years, whereas a nova with a $1.10\,M_\odot$ white dwarf should turn off after about 100 years. The short turn-off time found from the X-ray observations is clear evidence that most novae have white dwarfs with relatively high masses above $1.2M_\odot$. Such a result has been predicted by the TNR model, since according to this model the critical mass to start the runaway decreases strongly with increasing white dwarf mass (cf. Starrfield, 1989; and Chapter 4). Statistically, novae with high-mass white dwarfs should be seen much more frequently than those with white dwarfs with lower masses.

10.5 The hard component

The first detection of a hard component came from observations of V838 Her which were carried out 5 days after dicovery with the PSPC aboard ROSAT (Lloyd *et al.*, 1992). Only hard X-rays, with no counts below 0.7 keV, were found, at a count rate of $0.16\,\mathrm{s}^{-1}$. The spectrum was best fitted with a thermal bremsstrahlung model with $kT \sim 10\,\mathrm{keV}$ and column density $N_\mathrm{H} = 3.2 \times 10^{21}\,\mathrm{cm}^{-2}$. Observations carried out 1 year (O'Brien, Lloyd and Bode, 1994) and 19 months past outburst (Szkody & Hoard, 1994) showed that the count rate had dropped well below $10^{-2}\,\mathrm{s}^{-1}$. O'Brien, Lloyd and Bode interpreted these results in terms of shock heating due to interaction of different components within the ejecta.

The so far most detailed analysis of the hard component and its temporal evolution was carried out by Balman, Krautter and Ögelman (1998) for V1974 Cyg. As already mentioned above, a hard X-ray tail was detected in the spectrum of V1974 Cyg for all ROSAT observations (Krautter *et al.*, 1996). Examples for three different epochs can be seen in Figure 10.2. Balman, Krautter and Ögelman did a detailed analysis of the hard component using Raymond–Smith thermal plasma emission models (Raymond & Smith, 1977). The development of the plasma temperature, the unabsorbed flux (corrected for hydrogen absorption), and the hydrogen column density are presented in Figure 10.5.

The plasma temperature decreased from ~10 keV and 5 keV on days 63 and 91, respectively, to a more or less constant level around 1 keV. However, the temperatures for the first two epochs were not very well defined, since the peak temperatures were outside the ROSAT energy range (0.1–2.4 keV). The middle panel shows the light curve of the hard component. After an initial rise from day 63 to day 147 the flux declined from days 255–653. The maximum was extrapolated to be about 150 days after outburst. The emission measures (EM) had a similar time evolution, with a peak EM of

$$(27.0\text{–}40.0)\left(\frac{D}{2\,\mathrm{kpc}}\right)^2 \times 10^{55}\,\mathrm{cm}^{-5},$$

corresponding to an electron density $n_\mathrm{e} \sim (0.6\text{–}2.0) \times 10^5\,\mathrm{cm}^{-3}$. The peak unabsorbed flux was $\sim 2.0 \times 10^{-11}\,\mathrm{erg\,s}^{-1}\,\mathrm{cm}^{-2}$, which corresponds to a hard X-ray luminosity $L_\mathrm{hx} \sim (0.8\text{–}2.0) \times 10^{34}\,\mathrm{erg\,s}^{-1}$ for a source distance in the range 2–3 kpc. The total shocked mass necessary to produce the detected spectral evolution in X-rays is $M_\mathrm{S} > 1 \times 10^{-6}\,\mathrm{M}_\odot$. The X-ray temperatures indicate large shock speeds of 1500–3000 $\mathrm{km\,s}^{-1}$, and the cooling time-scale has been estimated to be ~3 yr.

While the time evolution of the hard X-ray flux and the decreasing plasma temperatures strongly indicate emission from shock-heated gas in the nova outflow, the data did not permit discrimination between the three possibilities mentioned in Section 10.2. Balman, Krautter and Ögelman favour a model in which a fast wind from the nova collides with either a circumstellar pre-existing gaseous shell, material ejected at the initial expansion phase of the nova, a wind expelled during the common envelope phase because of dynamical friction or a pre-existing slower massive outflow from the nova.

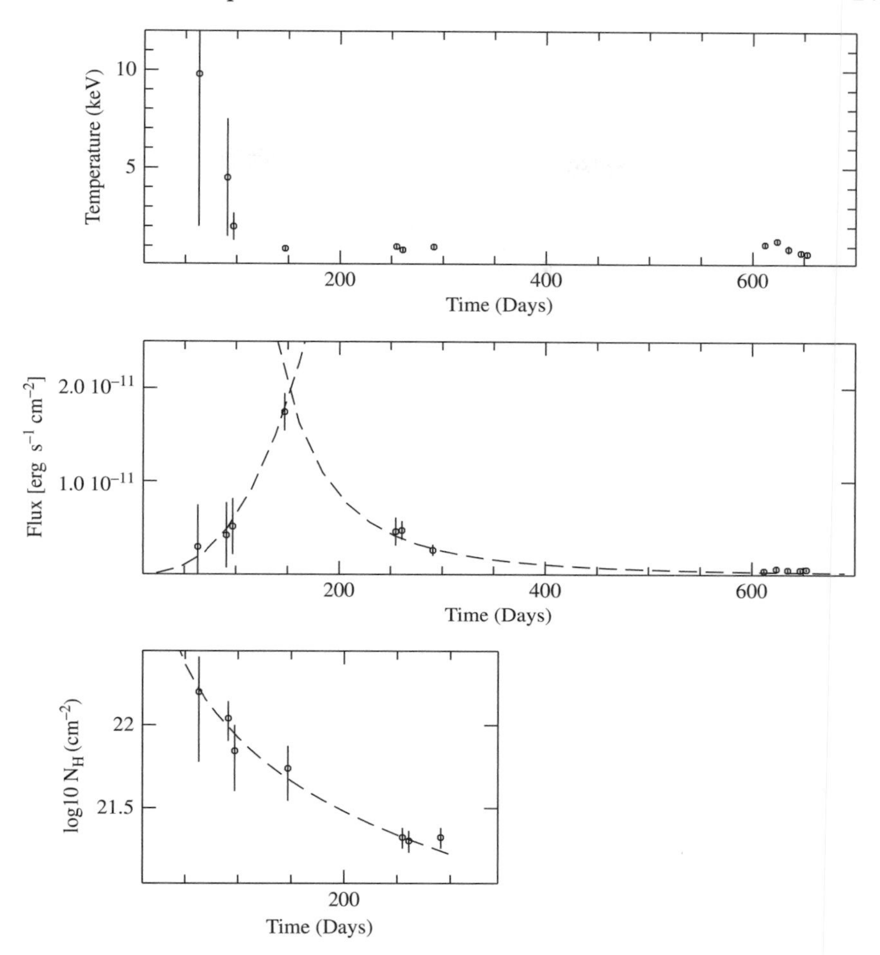

Fig. 10.5. Development of temperature vs. time of the hard component of V1974 Cyg (upper panel), light curve of the hard unabsorbed X-ray flux, corrected for neutral hydrogen absorption (middle panel), and the hydrogen column density N_H (lower panel). The results are derived using a Raymond–Smith plasma emission model. Error bars correspond to a 1σ confidence level; dashed curves are fits to the data. From Balman, Krautter and Ögelman (1998).

Of particular interest is the evolution of the hydrogen column density N_H which dropped by about a factor of 10 from an initial high value of more than 10^{22} cm^{-2} to a rather constant level around $\sim 2 \times 10^{21}$ cm^{-2}. This temporal evolution is clear evidence for decreasing circumstellar absorption in the expanding shell. According to Krautter *et al.* (1996), N_H should decrease in an expanding shell as

$$N_H = \frac{3M_H}{4\pi m_H V_{ej}} t^{-2}.$$

The initial high circumstellar hydrogen density is, of course, a natural explanation for the absence of any soft X-ray emission before day 97.

V382 Vel was observed extensively during early phases with RXTE and ASCA (Mukai & Ishida, 2001) and Beppo-Sax (Orio *et al.*, 2001b). While the RXTE observations on day 5.6 did not give a statistically significant result, Mukai and Ishida found with ASCA observations on day 20 a highly absorbed ($N_H \sim 10^{23}\,cm^{-2}$) hard spectrum with $kT \sim 10$ keV. In subsequent RXTE observations the spectrum changed to be consistent with $kT \sim 2.5$ keV and $N_H \sim 2 \times 10^{23}\,cm^{-2}$ about two months after the outburst. With an assumed distance of 2 kpc, V382 Vel maintained an X-ray luminosity of $\sim 7.5 \times 10^{34}$ erg s^{-1} for at least 20, and perhaps as long as 40 days. Mukai and Ishida attributed the hard X-ray emission to internal shocks within the nova ejecta, caused by a non-smooth outflow with varying velocities. They concluded that the evolving spectral characteristics could not be explained by a collision of a post-maximum wind with pre-existing material. Similar results were found by Orio *et al.* (2001b) from Beppo-Sax observations on day 15 while in Beppo-Sax observations on day 184 the hard flux had decreased by about a factor of 40 as compared with day 15.

10.6 Re-establishment of accretion

V2487 Oph, which had its outburst on 1998 June 15, is so far the only convincing case of the re-establishment of accretion in a nova system which had a recent outburst. From observations carried out with XMM-Newton on days 986 (2.7 years after discovery) and 1187, Hernanz and Sala (2002a) found emission over the whole spectral range 0.3–8.0 keV, with the hard part quite similar to that of a typical cataclysmic variable. In addition, there is a positional correlation with a previous source discovered by the RASS in 1990, 8 years before the nova eruption. Three other novae 2–4 years after outburst, V4633 Sgr, V1142 Sco, and LZ Mus, were detected by Hernanz and Sala (2002b) with XMM-Newton. However, owing to the lack of information below 0.3 keV, together with low signal-to-noise ratios, it was not possible to decide whether the novae were still undergoing hydrogen burning or had re-established accretion.

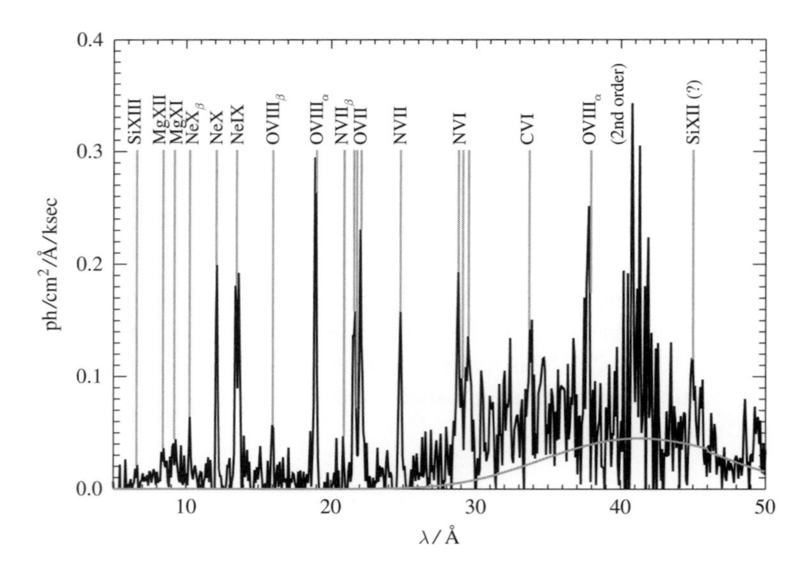

Fig. 10.6. Chandra-LETG flux spectrum of V382 Vel (background subtracted). A continuum model is overplotted representing an arbitrarily scaled thermal spectrum with T = 270 000 K. From Ness *et al.* (in preparation).

10.7 Emission line spectra

With Chandra and XMM-Newton it became possible, for the first time, to obtain spectra with high resolution (up to 1000). Figure 10.6 shows a Chandra-LETG spectrum of V382 Vel which was obtained on day 268, immediately after the hydrogen burning had turned off (Ness *et al.*, 2005). Only emission lines can be seen, with very little continuum emission. Emission lines from Si, Mg, Ne, O, N, and C, but no Fe lines, could be detected. The lack of Fe lines, which are expected in this temperature range in which the other lines are formed, can only be explained by an actual under-abundance of Fe with respect to the other elements. Iron probably has solar abundance, whereas the other elements are usually strongly over-abundant in an ONe nova. From density-sensitive lines Ness *et al.* could determine electron densities around $2 \times 10^9\,\mathrm{cm}^{-3}$. They also found that the emission lines were broadened. The structure of the line profiles differs for different elements, showing a rather compound broadened profile for oxygen, a clear double profile for neon, and highly structured profiles for the nitrogen lines.

10.8 Short-term variability

Classical novae are highly variable sources, both on longer time-scales of weeks and months, and on short time-scales of minutes to hours and days. Several unusual features were found in the X-ray light curve of V1494 Aql. After the expanding shell had become transparent to the soft X-ray radiation from the white dwarf, V1494 Aql was observed on days 300 and 303, when hydrogen burning was still going on, in two segments of 8 and 17 ks, respectively, using the LETG+HRC-S on Chandra (Drake *et al.*, 2003). The two X-ray light curves which were computed by Drake *et al.* from the bright zeroth order of the LETF+HRC-S spectrum are shown in Figure 10.7. Three different types of variability are found in these light curves:

(1) A stochastic variability on time-scales of minutes. This kind of irregular short-term variability seems to be present in all nova X-ray observations where the super-soft component is present. For instance, Orio *et al.* (2002) found, from Beppo-Sax of V382 Vel, irregular flickering and a decrease of the flux in the 0.1–0.7 keV range by a factor of two in less than 1.5 hours. The flux remained faint for some 15 minutes; no significant spectral changes were found. So far no convincing explanation for this variability has been found. Absorption due to an ejected clump of matter can be excluded, since in this case the hardness ratio of the spectrum should have changed.

(2) A short time-scale X-ray burst. For about 1000 s the X-ray count rate showed a complex rise and fall. From Figure 10.8, which shows an enlargement of the region around the time of the X-ray burst, it is clear that the burst is not a single isolated event, but exhibits two main flares with possibly a precursor and a trailer. At its peak the count rate of the burst was about a factor of six higher than the mean level before and after the burst. The spectrum obtained during the burst is slightly harder than during the rest of the observations. Drake *et al.* offer no explanation for this burst. No comparable burst has so far been observed in any other X-ray light curve of a nova.

(3) Periodic variations. A timing analysis of the combined 25 ks observation shows a peak at ~2499 s with an amplitude of 15% of the mean count rate level. The 2499 s period was found in an independent analysis of both the zeroth-order and the dispersed spectrum but is not present in similar analyses of grating data for HZ 43 and Sirius B, showing that the period is not related to the spacecraft motion or dithering. Drake *et al.* conclude that the observed periods are not instrumental.

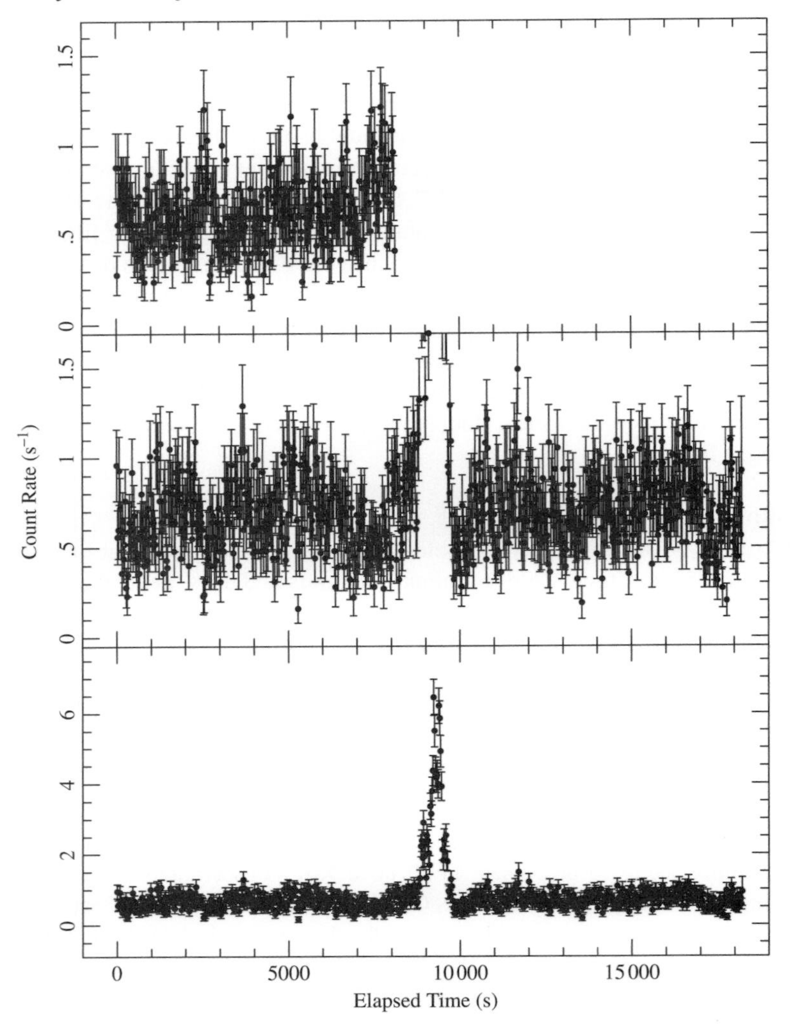

Fig. 10.7. Chandra-LETG+HRC-S X-ray light curve of V1494 Aql. Top: light curve obtained on day 300. Middle: light curve obtained on day 303. Bottom: same as middle panel but data scaled by the full range of the count rate to highlight the burst. From Drake *et al.* (2003).

Additional periods were found in both the zeroth order and the dispersed spectrum. This suggests that one is not observing the rotation of the white dwarf. Drake *et al.* interpret this result in terms of non-radial g^+-mode pulsations in the hot, rekindled white dwarf, driven by the κ/γ effects in the partial ionization zones of C and O near the surface of the white dwarf. The hot, luminous white dwarf in a nova evolving from explosion to quiescence has a structure which resembles that of the central stars of planetary nebulae. The power spectrum and the X-ray light curve of V1494 Aql are very similar to the hot central star of several planetary nebulae (e.g. Bond *et al.*, 1996; Ciardullo & Bond, 1996).

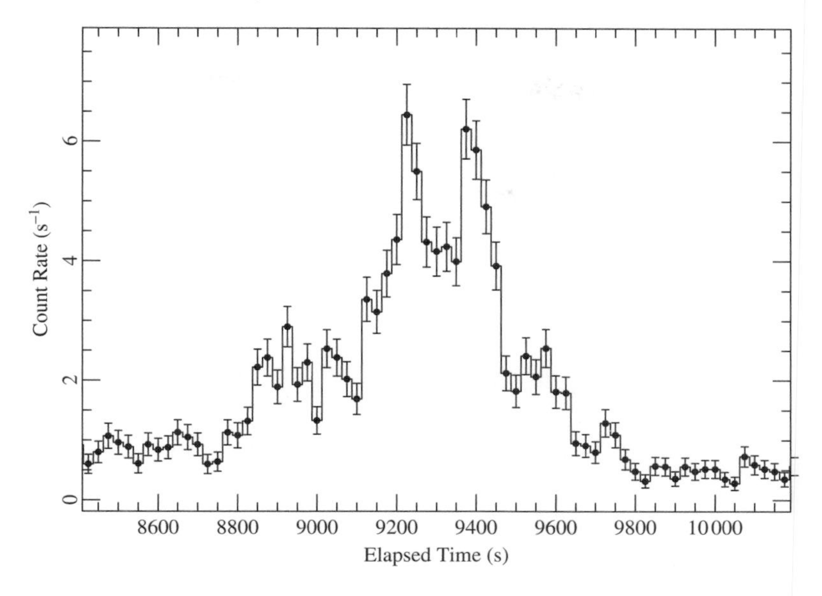

Fig. 10.8. Enlargement of the region around the time of the X-ray burst of V1494 Aql on day 303, sampled in 25 s bins. From Drake *et al.* (2003).

Even stronger variations were exhibited by V4743 Sgr. Figure 10.9 shows the light curve obtained with a 25 ks Chandra LETGS observation on 2003 March 19 (day 180 after outburst), when the nova had entered the super-soft state (Ness *et al.*, 2003). The top panel shows the complete wavelength range in zeroth and first order. Immediately obvious is the strong periodic variability, with a range from $\sim$30 to 60 counts s^{-1}. As the middle and lower panels show, the hardness ratio changes during the oscillations, the nova being slightly harder or, equivalently, the white dwarf slightly hotter, when it is brighter. A timing analysis reveals a period of 1324 s with two harmonic overtones at 668 and 448 s. XMM-Newton observations taken about 2 months after the Chandra observations yielded a rich power spectrum with two main periods of 1308 and 1374 s (Orio, 2004). The interpretation of the oscillations is difficult, since their large amplitude is not easily reconciled with pulsations, as in the case of V1494 Aql. Rotation of the white dwarf might be a plausible alternative in this case. However, it cannot be excluded that both rotation and pulsation cause the complex light curve. A second very obvious feature of the light curve is the decline of the X-ray flux, which starts slowly at $\sim$13 ks into the outburst, and drops to essentially zero within 6 ks, where it stays for the rest of the observations. As the middle panel of Figure 10.9 shows, the hard flux declines first, while the soft component declines later, but more rapidly. The interpretation of this phenomenon is not straightforward. If it were an eclipse, the long duration of the decline in count rate is difficult to explain with the typical orbital period (a few hours) of a cataclysmic binary. In that case V4743 Sgr should be a CV with a longer period of $\sim$2 days, like GK Per, which is an intermediate polar with hard X-ray variations (Watson, King & Osborne, 1985). A possible solution could be a third stellar component in the system. The knowledge of the binary orbital period could help to clarify this question. During this 'dark' phase the spectrum was dominated by strong emission lines of highly ionized carbon, nitrogen and oxygen. It is clear that the region

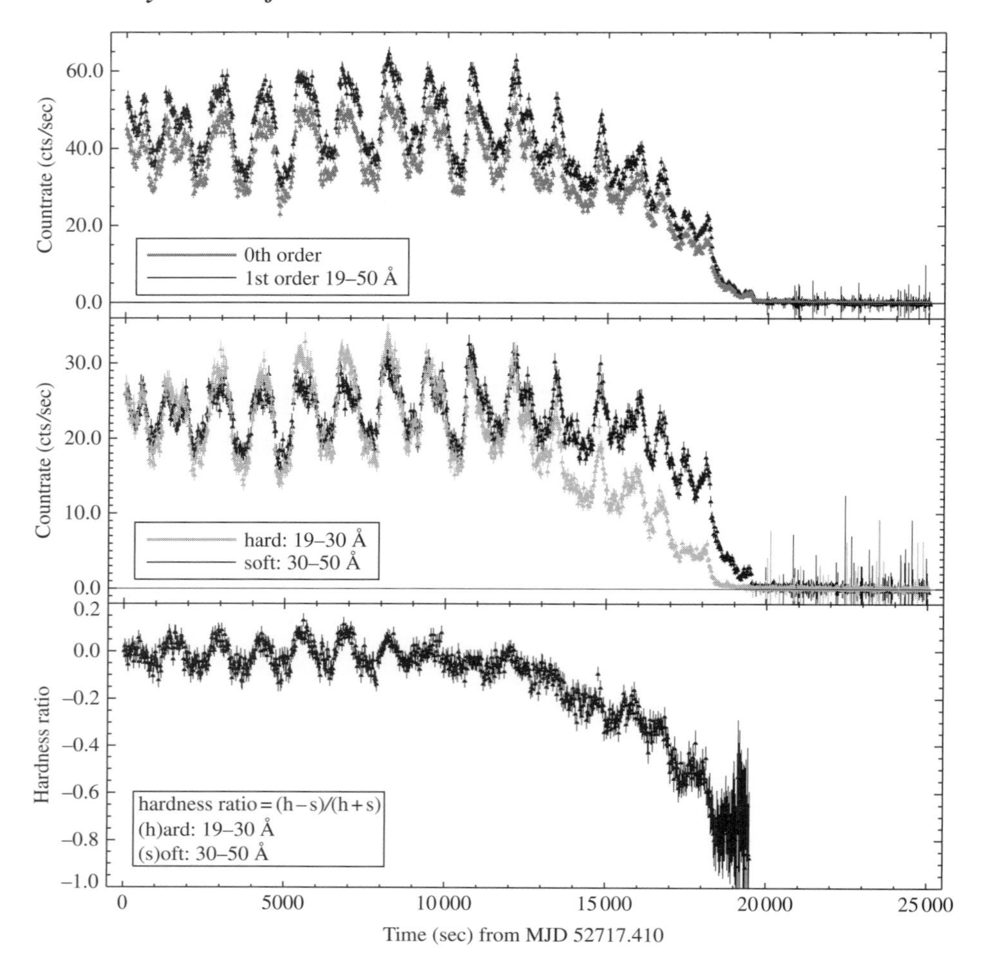

Fig. 10.9. Light curve of a 25 ks exposure of V4743 Sgr extracted in the designated wavelength intervals. Top: complete wavelength ranges in zeroth and first order. Middle: light curve broken into 'hard' and 'soft' components of the spectrum. Bottom: time evolution of the hardness ratio. From Ness *et al.* (2003).

where the emission lines originate must be much larger than the source of the continuum emission.

At subsequent observations of V4743 Sgr with XMM-Newton in 2003 April, and Chandra in 2003 July, the count rate was at about the same level as during the first part of the March observation. It had dropped again in 2003 September, and even more in 2004 February (Ness *et al.*, in preparation). The oscillations were always present, however, with lower amplitude. It is not clear whether X-ray oscillations are a general phenomenon. For instance, Krautter *et al.* (1996) performed fast Fourier transformations on their extensive X-ray data of V1974 Cyg to search for short-term variations, with a negative result.

10.9 Concluding remarks

Even if few X-ray observations of classical novae have been carried out so far compared with other wavelength regions, they have not only given us important new insights into the outburst physics of these exciting objects, but have also raised many new questions because of very unexpected results which challenge our understanding of the nova outburst. So far, X-ray observations of each new nova have revealed new facets, and no nova has, as yet, shown the same behaviour as preceding ones.

X-ray observations offer a wide variety of opportunities of studying the outbursts of classical novae. They allow us to study the hot white dwarf and shocks in the expanding envelope, and to derive physical parameters such as $T_{\rm eff}$ and L, the duration of the hydrogen burning on top of the white dwarf, and chemical abundances; while grating spectroscopy will allow us to carry out plasma diagnostics for the line emitting regions. Through the scattered light of bright super-soft sources the interstellar medium can also be studied.

Certainly problems exist. During the early phases only very limited information can be obtained, owing to the high opacity in the expanding envelope. In addition, the interpretation of the results is often not straightforward. However, X-ray observations offer the unique possibility of studying directly the regions where the energy of the nova outburst is produced – all other wavelength regimes can yield only indirect evidence from these regions.

References

Balman, S., & Krautter, J., 2001, *MNRAS*, **326**, 1441.
Balman, S., Krautter, J., & Ögelman, H., 1998, *Ap. J*, **499**, 395.
Bode, M. F., O'Brien, T. J., Osborne, J. P. *et al.* 2006, *Ap. J*, **652**, 629.
Bond, H. E., Kawaler, S. D., & Ciardullo, R., 1996, *AJ*, **112**, 2699.
Brecher, K., Ingham, W. H., & Morrison, P., 1977, *Ap. J*, **213**, 492.
Burwitz, V., Starrfield, S., Krautter, J., & Ness, J.-U., 2002, in *Classical Nova Explosions*, ed. M. Hernanz, & J. José. New York: American Institute of Physics, p. 377.
Castor, J. I., Abbott, D. C., & Klein, R. I., 1975, *Ap. J*, **195**, 157.
Ciardullo, R., & Bond, H. E., 1996, *AJ*, **111**, 2332.
Collins, P., 1992, *IAUC* 5454.
Draine, B. T., & Tan, J. C., 2003, *Ap. J*, **594**, 347.
Drake, J., Wagner, R. M., Starrfield, S. *et al.*, 2003, *Ap. J*, **548**, 448.
Hartmann, H. W., & Heise, J., 1997, *A&A*, **322**, 591.
Hartmann, H. W., Heise, J., Kahabka, P., Motch, C., & Parmar, A. N., 1999, *A&A*, **346**, 125.
Hasinger, G., 1994, *Rev. Mod. Astrophys.*, **7**, 129.
Hauschildt, P., 1992, *J. Quant. Spectrosc. Radiative Transfer*, **47**, 433.
Hauschildt, P. H., & Baron, E, 1999, *J. Comput. Appl. Math.*, **109**, 41.
Hauschildt, P. H., Wehrse, R., Starrfield, S., & Shaviv, G., 1992, *Ap. J*, **393**, 307.
Hauschildt, P. H., Shore, S. N., Schwarz, G. *et al.*, 1997, *Ap. J*, **490**, 803.
Hernanz, M., & Sala, G., 2002a, *Science*, **298**, 393.
Hernanz, M., & Sala, G., 2002b, in *Classical Nova Explosions*, ed. M. Hernanz, & J. José. New York: American Institute of Physics, p. 381.
Kahabka, P., & van den Heuvel, E. P. J., 1997, *ARA&A*, **35**, 69.
Kato, T., Fujii, M., & Ayani, K., 2002, *IAUC* 7795.
Krautter, J., & Williams, R. E., 1989, *Ap. J*, **341**, 968.
Krautter, J., 2002, in *Classical Nova Explosions*, ed. M. Hernanz, & J. José. New York: American Institute of Physics, p. 345.
Krautter, J., Ögelman, H., Starrfield, S., Trümper, J., & Wichmann, R., 1993, *Annals. Israel. Phys. Soc.*, **10**, 28.
Krautter, J., Ögelman, H., Starrfield, S., Wichmann, R., & Pfeffermann, E., 1996, *Ap. J*, **456**, 788.
Liller, W., 1995, *IAUC* 6143.
Livio, M., Mastichiadis, A., Ögelman, H., & Truran, J. W., 1992, *Ap. J*, **394**, 217.
Lloyd, H. M., O'Brien, T. J., Bode, M. F. *et al.*, 1992, *Nature*, **356**, 222.
Mason, K. O., Córdova, F. A., Bode, M. F., & Barr, P., 1986, in *RS Ophiuchi (1985) and the Recurrent Nova Phenomenon*, ed. M. F. Bode. Utrecht: VNU Science Press, p. 167.
Mathis, J. S., Cohen, D., Finley, J., & Krautter, J., 1995, *Ap. J*, **449**, 320.
Mukai, K., & Ishida, M., 2001, *Ap. J*, **551**, 1024.
Mukai, K., & Swank, J., 1999, *IAUC 7206.*
Ness, J.-U., Starrfield, S., Burwitz, V. *et al.*, 2003, *Ap. J*, **594**, L127.
Ness, J.-U., Starrfield, S., Jordan, C., Krautter, J., & Schmitt, J. H. M. M., 2005, *MNRAS*, **364**, 1015.
Nickel, U., 1995, Examensarbeit, Ruprecht-Karls-Universität Heidelberg.
O'Brien, T. J., Lloyd, H. M., & Bode, M. F., 1994, *MNRAS*, **271**, 155.
Ögelman, H., 1990, in *Physics of Classical Novae*, ed. A. Cassatella, & R. Viotti. Berlin: Springer, p. 148.

Ögelman, H., & Orio, M., 1995, in *Cataclysmic Variables*, ed. A. Bianchini, M. della Valle, & M. Orio, *Astrophys. Space Sci. Lib.*, **205**, p. 11.
Ögelman, H., Beuermann, K., & Krautter, J., 1984, *Ap. J*, **287**, L31.
Ögelman, H., Krautter, J., & Beuermann, K., 1987, *A&A*, **177**, 110.
Ögelman, H. B., Orio, M., Krautter, J., & Starrfield, S., 1993, *Nature*, **361**, 331.
Orio, M., 1999, *Phys. Rep.*, **311**, 419.
Orio, M., 2004, *Rev. Mex. AA* (Serie de Conferencias), **20**, 182.
Orio, M., & Greiner, J., 1999, *A&A*, **344**, L13.
Orio, M., Covington, J., & Ögelman, H., 2001a, *A&A*, **373**, 542.
Orio, M., Parmar, A., Benjamin, R. *et al.*, 2001b, *MNRAS*, **326**, L13.
Orio, M., Parmar, A. N., Greiner, J. *et al.*, 2002, *MNRAS*, **333**, L11.
Orio, M., Hartmann, W., Still, M. & Greiner, J., 2003, *Ap. J*, **594**, 435.
Osborne, J., Page, K., Beardmore, A. *et al.*, 2006a, *A. Tel.*, 838.
Osborne, J., Page, K., Beardmore, A. *et al.*, 2006b, *A. Tel.*, 770.
Petz, A., Hauschildt, P. H., Ness, J.-U., & Starrfield, S., 2005, *A&A*, **431**, 321.
Raymond, J. C., & Smith, B. W., 1977, *Ap. JS*, **35**, 419.
Shanley, L., Ögelman, H. Gallagher, J., Orio, M., & Krautter, J., 1995, *Ap. J*, **438**, L95.
Starrfield, S., 1989, in *Classical Novae*, 1st edn, ed. M. Bode, & A. Evans. New York and Chichester: Wiley, p. 39.
Starrfield, S., Truran, J. W., Sparks, W. M., & Krautter, J., 1991, in *Extreme Ultraviolet Astronomy*, ed. R. Malina, & S. Bowyer. New York: Pergamon, p. 168.
Starrfield, S., Truran, J. W., Wiescher, M. C., & Sparks, W. M., 1996, in *Cosmic Abundances*. San Francisco: Astronomical Society of the Pacific, p. 242.
Starrfield, S. *et al.*, 2000, *BAAS*, **32**, 1253.
Starrfield, S., Drake, J., Wagner, R. M. *et al.*, 2001, *BAAS*, **33**, 804.
Szkody, P., & Hoard, D. W., 1994, *Ap. J*, **429**, 857.
Trümper, J., 1983, *Adv. Space Res.*, **2**, 241.
Truran, J., 1982, in *Essays in Nuclear Astrophysics*, ed. C. A. Barnes, D. D. Clayton, & D. N. Schramm. Cambridge: Cambridge University Press, p. 467.
Voges, W., Aschenbach, B., Boller, T. *et al.*, 1999, *A&A*, **349**, 389.
Watson, M. G., King, A. R., & Osborne, J., 1985, *MNRAS*, **212**, 917.

11

Gamma-rays from classical novae

Margarita Hernanz

11.1 Introduction

The most energetic photons from the whole electromagnetic spectrum are the γ-rays, with energies larger than about 10 keV, i.e. wavelengths shorter than 1.2 Å and frequencies larger than 2.4×10^{18} Hz. These photons trace the most energetic phenomena of the Universe and, in particular, stellar explosions such as classical novae or supernovae.

The potential of novae as γ-ray emitters was first pointed out by Clayton and Hoyle (1974). In that paper the authors stated that observable γ-rays from novae would come from electron–positron annihilation, with positrons from ^{13}N, ^{14}O, ^{15}O, and ^{22}Na decays, as well as a result of the decay of ^{14}O and ^{22}Na to excited states of ^{14}N and ^{22}Ne nuclei, which de-excite by emitting photons at 2.312 and 1.274 MeV respectively. The need for fast mixing within the nova envelope, in order to have a favorable interplay between the transparency of the expanding envelope and the short lifetimes of the radioactive nuclei (^{22}Na excluded), was emphasized. The main ideas presented in that seminal work have remained unchanged; but some aspects have changed in the past 20 years, mainly related to new detailed nucleosynthesis studies of novae, as will be explained in this review.

Seven years later, in a new important paper, Clayton (1981) noted that another γ-ray line could be expected from novae when ^{7}Be transforms (through an electron capture) to an excited state of ^{7}Li, which de-excites by emitting a photon of 478 keV. The original idea came from Audouze and Reeves (1981) and both works were inspired by the contemporaneous papers mentioning the possibility of ^{7}Li synthesis in novae (Arnould & Norgaard, 1975; Starrfield *et al.*, 1978a). In fact, ^{7}Li production in novae was, and continues to be, a crucial topic, since Galactic ^{7}Li is not well accounted for by other sources, whether stellar (asymptotic giant branch (AGB) stars), interstellar (spallation reactions by cosmic rays) or cosmological (Big Bang).

In these pioneering papers, the predictions of detectability were quite optimistic, both for the contemporaneous short mission High Energy Astrophysics Observatory 3 (HEAO 3) and for the future Compton Gamma-Ray Observatory (CGRO), already planned by NASA at that epoch. Unfortunately, later computations, more sophisticated numerically and with updated input physics (especially that related to nuclear physics aspects), have provided lower

Classical Novae, 2nd edition, ed. Michael Bode and Aneurin Evans. Published by Cambridge University Press.

yields and consequently less optimistic expectations, as will be explained later. That is the reason for the undetectability, to date, of γ-rays from classical novae. In fact, an unfortunate combination of low emission fluxes and low sensitivity of γ-ray instruments makes detection of novae a really hard task, even with the instruments on board the International Gamma-ray Laboratory (INTEGRAL), launched on 2002 October 17.

The γ-rays emitted by novae have a radioactive origin, i.e. they ultimately depend on the synthesis of radioactive nuclei during the explosion (see reviews by Leising, 1991, 1993, 1997; Hernanz, 2002). The thermonuclear runaway scenario of classical novae (see Chapter 4) guarantees that such is the case, since the explosive hydrogen burning through the CNO (carbon–nitrogen–oxygen) cycle out of equilibrium produces some β^+-unstable nuclei, such as ^{13}N, ^{14,15}O, and ^{17,18}F. In addition, other radioactive nuclei that are synthesized in novae through proton–proton chains, ^{7}Be, or through the NeNa or MgAl 'cycles', ^{22}Na and ^{26}Al, decay (or, for ^{7}Be, suffer an electron capture) to excited states of daughter nuclei (i.e. ^{7}Li, ^{22}Ne, and ^{26}Mg) which de-excite to their ground state emitting γ-ray photons. The second condition to be fulfilled is that the lifetimes of these nuclei are not shorter than the time-scale of envelope transparency, in order to allow the γ-rays to escape; this practically excludes ^{14}O, ^{15}O, and ^{17}F (lifetimes 102, 176, and 93 s respectively). From these conditions it can be easily understood that γ-rays give a direct insight not only into the nucleosynthesis during the nova explosion, but also into the global properties of the expanding envelope, mainly dependent on density, temperature, chemical composition, and velocity profiles, which determine its transparency to γ-rays. Gamma-rays can directly trace isotopes, whereas observations at other wavelengths give only elemental abundances, except some exceptional measurements of CO molecular bands in the infrared, where ^{12}CO and ^{13}CO can be distinguished, thus giving the ^{13}C/^{12}C ratio (Ferland *et al.*, 1979; Rudy *et al.*, 2003; Harrison, Osborne & Howell, 2004; Banerjee, Varricatt & Ashok, 2004; see also Chapter 13).

In summary, novae emit γ-rays because some of the radioactive nuclei they synthesize during the hydrogen thermonuclear runaway either experience electron captures or are β^+-unstable (thus emitting positrons), decaying in some cases to nuclei in excited states which de-excite to their ground states by emitting photons at particular energies in the γ-ray range. The positrons themselves annihilate with electrons and therefore also produce γ-ray emission (Leising & Clayton, 1987). Comptonization and photoelectric absorption take place in nova envelopes, but other typical mechanisms of γ-ray production and absorption (such as the inverse Compton effect, or collisions with energetic particles, to mention only a few), responsible for the emission in other astrophysical scenarios, do not play any role.

In addition to the emission from individual novae, it is worth noting that the integrated emission of several novae in the Galaxy should give rise to a kind of diffuse emission, whenever the lifetime of the corresponding isotope is longer than the typical time interval between two successive nova outbursts; this is the case for ^{7}Be, ^{22}Na, and ^{26}Al. The cumulative emission of ^{22}Na is particularly relevant for novae, because this isotope is mostly produced by them and has a lifetime long enough (3.75 yr) to allow the accumulation of a non-negligible amount of ^{22}Na, with respect to the quantity emitted by a single nova. Therefore, the integrated emission of ^{22}Na from all recent Galactic novae should map the distribution of novae in the Galaxy (at least that of novae producing ^{22}Na). For ^{7}Be there is not much accumulation, whereas for ^{26}Al it is large, but novae are not the main contributors to ^{26}Al.

This review is organized as follows: first the origin of the γ-rays is analyzed, by presenting the most relevant trends of nova nucleosynthesis, with special emphasis on the nuclei relevant

Table 11.1. *Radioactivities in nova ejecta*

Isotope	Lifetime	Type of emission	Main disintegration process	Nova type
^{13}N	862 s	511 keV line & continuum	β^+-decay	CO & ONe
^{18}F	158 min	511 keV line & continuum	β^+-decay	CO & ONe
^{7}Be	77 days	478 keV line	e^--capture	CO
^{22}Na	3.75 years	1275 & 511 keV lines	β^+-decay	ONe
^{26}Al	10^6 years	1809 & 511 keV lines	β^+-decay	ONe

for the γ-ray emission, as well as the main aspects of the propagation of γ-rays in nova envelopes. Then the γ-ray output of novae is displayed through some examples of theoretical calculations, in the form of spectral evolution with time and of light curves of the most relevant γ-ray features. Some words about the cumulative emission of novae in the Galaxy follow, with a particular analysis of the integrated emission from ^{22}Na and ^{26}Al. A detailed summary of the observations of all the lines expected from novae ensues, followed by the prospects for detectability with the INTEGRAL satellite and with the future envisioned missions.

11.2 Origin of gamma-rays from novae

We can classify γ-ray emission from novae according to its spectral (line and continuum) or its temporal (prompt and long-lasting) characteristics. A summary of the most relevant radioactive species produced by novae is given in Table 11.1, with their lifetimes and the associated type of emission. The β^+-unstable nuclei with relatively short lifetimes (^{13}N and ^{18}F) produce positrons which then annihilate with electrons giving photons of 511 keV (line) and in a range between some tens of keV and 511 keV (continuum; cf. Section 11.3.1 below). Positrons from the longer-lived ^{22}Na also contribute to annihilation radiation. On the other hand, ^{7}Be and ^{22}Na are responsible for emission of photons of 478 keV and 1275 keV, respectively (cf. Section 11.3.2 below). Concerning the temporal behavior of the emission, annihilation radiation is emitted promptly, since it comes from short-lived nuclei, whereas line emission from ^{7}Be and ^{22}Na is long-lasting, as dictated by their lifetimes (from months to years). Finally, ^{26}Al cannot be detected in individual objects, either in novae or in other stellar sources (except eventually for a very close object, like the Wolf–Rayet star γ^2 Velorum), because its lifetime is much longer than the typical time elapsed between two successive events. Therefore, only the integrated emission of many novae (and other production sites) can be observed.

In this section we will first present the most important aspects of nucleosynthesis of the relevant radioactive nuclei mentioned above. Then we will describe γ-ray production and propagation.

11.2.1 Nucleosynthesis of radioactive nuclei

An overview of the radioactive isotope content of nova ejecta a short time after peak temperature at the base of the envelope is presented in Table 11.2, for a handful of nova models

Table 11.2. *Properties of nova ejecta: ejected masses of the most relevant radioactive isotopes and total ejected mass. Values for ^{13}N and ^{18}F are given at 1 h after T_{peak}*

Nova	M_{WD} ($M_\odot$)	M_{ej} ($M_\odot$)	^{13}N ($M_\odot$)	^{18}F ($M_\odot$)[a]	^{7}Be ($M_\odot$)	^{22}Na ($M_\odot$)	^{26}Al ($M_\odot$)
CO	0.8	6.2×10^{-5}	1.5×10^{-7}	1.8×10^{-9}	7.4×10^{-11}	7.3×10^{-11}	1.7×10^{-10}
CO	1.15	1.3×10^{-5}	2.3×10^{-8}	2.6×10^{-9}	8.9×10^{-11}	1.1×10^{-11}	6.0×10^{-10}
ONe	1.15	2.6×10^{-5}	2.9×10^{-8}	5.9×10^{-9}	1.6×10^{-11}	6.4×10^{-9}	2.1×10^{-8}
ONe	1.25	1.8×10^{-5}	3.8×10^{-8}	4.5×10^{-9}	1.4×10^{-11}	5.8×10^{-9}	1.1×10^{-8}

[a] According to current nominal rates for the ^{18}F + p and ^{17}O + p reactions, these yields should be reduced by a factor of ~50.

computed with a 1-D Lagrangian implicit hydrodynamical code that follows the whole nova evolution from the onset of mass accretion up to mass ejection (José & Hernanz, 1998). Some degree of mixing between the accreted envelope (with solar composition) and the underlying white dwarf core, carbon–oxygen (CO) or oxygen–neon (ONe), has been adopted (50% in the cases shown in the table), to reproduce the observed abundances in many novae; the accretion rate is 2×10^{-10} $M_\odot$ yr^{-1} (José & Hernanz, 1998; Hernanz *et al.*, 1999; Hernanz & José, 2004). Although not completely satisfactory, this is by now the usual way to mimic the explosion (see Politano *et al.*, 1995; Starrfield *et al.*, 1998), to be compared with the calculations by Prialnik and Kovertz (1997), which include diffusion.

It is important to distinguish between novae occurring on CO and on ONe white dwarfs. Stars originally less massive than ~10–12 $M_\odot$ end their lives as white dwarfs, made of helium, of carbon and oxygen, or of oxygen and neon. The exact mass interval leading to the different types of white dwarfs is not completely well determined, since it depends on details of stellar evolution and, specially, on the single or binary nature of the progenitor (see Chapter 3). The most common case are CO white dwarfs; helium white dwarfs can only exist in binary systems, since single evolution takes longer than the Hubble time to produce a white dwarf made of pure helium (i.e. with a very low mass preventing helium ignition). On the other end of the range of stellar masses ending up as white dwarfs, i.e. massive progenitors with masses around 10 $M_\odot$, non-degenerate carbon ignition leads to the formation of a degenerate core made mainly of oxygen and neon, with traces of magnesium and sodium. These cores were thought to be made of oxygen, neon and magnesium in similar amounts (the so-called ONe white dwarfs) some years ago, when parametrized calculations of hydrostatic carbon burning were adopted (Arnett & Truran, 1969) before self-consistent models of AGB following the thermally pulsing phase were available (Domínguez, Tornambé & Isern, 1993; Ritossa, García-Berro & Iben, 1996). In Table 11.2 it is shown that the short-lived radioactive nuclei ^{13}N and ^{18}F are produced in both types of novae (CO and ONe), whereas ^{7}Be is mainly synthesized in CO novae, and ^{22}Na and ^{26}Al in ONe novae.

There is an important issue concerning the chemical composition in the outer envelope of ONe cores, since stellar evolutionary models predict that the core is not naked, but overlaid with a *buffer* rich in carbon and oxygen and a transition zone intermediate between the core and the buffer (García-Berro, Ritossa & Iben, 1997). The thickness of the CO buffer depends on the efficiency of mass-loss during the late stages of the AGB phase and on the number of previous nova eruptions which could have eroded it. So the composition of the zone where mixing presumably takes place is not necessarily that of the bare ONe core (José *et al.*, 2003).

^{13}N and ^{18}F

The main sources of positrons in novae are ^{13}N and ^{18}F, and to a lesser extent ^{22}Na. The synthesis of ^{13}N and ^{18}F occurs during the operation of the CNO cycle of hydrogen burning (see also Chapter 4). Actually, the dominant reaction at the beginning of the thermonuclear runaway leading to the explosion is ^{12}C(p,γ)^{13}N, so that ^{13}N synthesis is guaranteed. The exact amount transported to the outer envelope, and contributing to γ-ray emission once transparency allows for the escape of photons, depends on the detailed evolution, especially on convection. In some models of CO and ONe novae (José & Hernanz, 1998; Hernanz & José, 2004) shown in Table 11.2, ^{13}N is produced in similar amounts in CO and in ONe novae, but this result is somewhat model-dependent since different convection efficiencies could lead to a different result. Therefore, detection of positrons from ^{13}N (through the associated annihilation emission: see Section 11.3.1 below) would provide an important diagnostic about the dynamics of nova explosions.

The synthesis of ^{18}F in novae proceeds through the hot CNO cycle. Since the initial abundance of ^{16}O is large both in CO and ONe novae, ^{16}O is the main source for ^{18}F formation, either through the chain of reactions ^{16}O(p,γ)^{17}F(p,γ)^{18}Ne(β^+)^{18}F or through ^{16}O(p,γ)^{17}F(β^+)^{17}O(p,γ)^{18}F (there is a competition between (p,γ) reactions and β^+-decays). However, ^{18}F yields are severely constrained by its destruction mode, whatever the production channel is. During the thermonuclear runaway, ^{18}F destruction by β-decays can be neglected when compared to its destruction by proton captures. Because of the low α-emission threshold, ^{18}F(p,α)^{15}O is faster than ^{18}F(p,γ)^{19}Ne, and hence it is the main destruction channel of ^{18}F (Hernanz *et al.*, 1999). Other nuclear reactions affecting ^{18}F synthesis are proton captures on ^{17}O, i.e. ^{17}O(p,γ)^{18}F and ^{17}O(p,α)^{14}N.

A detailed analysis of the influence of some key reaction rates on ^{18}F production in novae (Coc *et al.*, 2000) showed that the rate of the nuclear reaction ^{18}F(p,α)^{15}O was uncertain by factors between 100 and 1000 at nova temperatures, because of the very limited spectroscopic data available for ^{19}Ne in the domain of interest. In addition, it was shown that ^{17}O + p reactions have an impact, though smaller, on ^{18}F synthesis as well. Fortunately, updated nuclear physics measurements have largely reduced the uncertainties in both these rates. De Séréville *et al.* (2003) have obtained a nominal rate similar to that in Coc *et al.* (2000) and have reduced the uncertainty by a factor of around 5, for the ^{18}F(p,α)^{15}O reaction. Parallel reductions in the uncertainties of the ^{17}O+p reactions have been reported (Fox *et al.*, 2004). However, in a subsequent paper by Chafa *et al.* (2005), a previously unknown resonance in the ^{17}O(p, α)^{14}N reaction has been discovered, which reduces the final yield of ^{17}O and ^{18}F. In summary, ^{18}F yields computed with the available nuclear reaction rates are about a factor of 50 smaller than those adopted before (see footnote in Table 11.2), but they are less uncertain, by a factor of about 20 instead of 300 (Hernanz & José, in preparation).

It is important to stress that the main uncertainty affecting ^{18}F synthesis in novae, and in particular the amount of ^{18}F contributing to γ-ray emission, is the nuclear one, in contrast with ^{13}N. The reason is that ^{18}F has a longer lifetime than ^{13}N and, therefore, the whole amount of ^{18}F in the nova envelope can contribute to γ-ray emission because the envelope becomes transparent before it decays.

^{7}Be and ^{7}Li

The production of ^{7}Li during nova explosions proceeds through the so-called *beryllium transport mechanism* (Cameron, 1955); this process relies on the previous formation of ^{7}Be, which transforms into ^{7}Li through an electron capture ($\tau = 77$ days; see Table 11.1) and releases a γ-ray photon of 478 keV (see Figure 11.1). The ^{7}Be has to be transported to cooler zones than those where it is formed, with a time-scale shorter than its electron capture characteristic time, in order to preserve its fragile daughter ^{7}Li from destruction. This mechanism requires a dynamic situation like that encountered in novae.

The first studies of ^{7}Li production in novae based on parametrized one-zone models (Arnould & Norgaard, 1975) were followed by hydrodynamical computations by Starrfield *et al.* (1978a), which still did not follow the accretion phase (i.e. they had an initial envelope already in place). They concluded that the final amount of ^{7}Li synthesized depended on the initial abundance of ^{3}He and on the treatment of convection. Other works based either on one- or two-zone models (Boffin *et al.*, 1993) or on semi-analytical models reached similar conclusions about the critical role played by convective mixing. A complete hydrodynamical

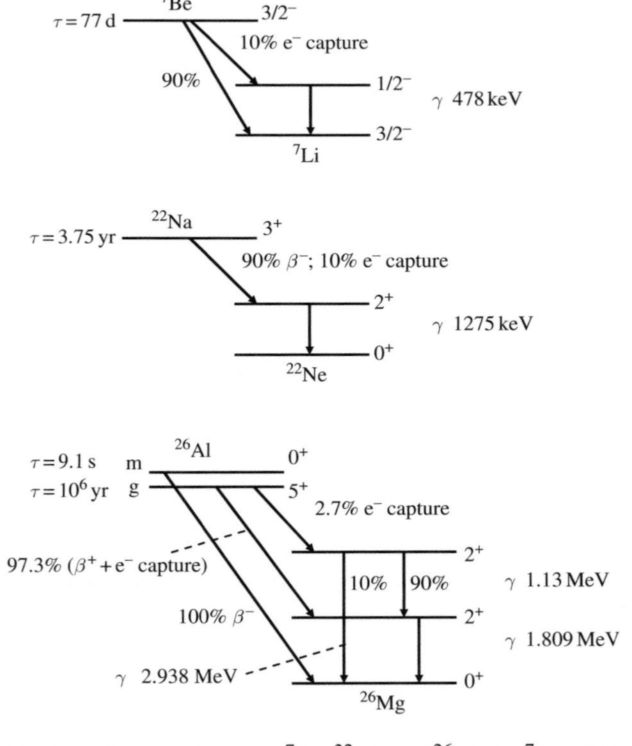

Fig. 11.1. Decay schemes of ^{7}Be, ^{22}Na, and ^{26}Al. For ^{7}Be, 10% of the decays occur through an electron capture and a 478 keV photon is emitted. For ^{22}Na, there is a β^{+}-decay in 90% and an electron capture in 10% of the cases, both leading to the emission of a 1275 keV photon. For ^{26}Al in the isomeric state, no γ-ray photon is emitted, since it directly decays to the ground state of ^{26}Mg, whereas for ^{26}Al in the ground state, in 97% of the decays (either β^{+}: + 82%; or electron capture: 15%) there is emission of a 1809 keV photon.

study, following the accretion phase as well, was performed by Hernanz *et al.* (1996). It showed that ^{7}Li formation is favored in CO novae with respect to ONe novae. Formation of ^{7}Be proceeds through ^{3}He(α, γ)^{7}Be, since (p, γ) reactions cannot bridge the $A = 5$ gap. Destruction occurs via ^{7}Be(p, γ)^{8}B; however, at high temperatures ($T \approx 10^8$ K) there is a quasi-equilibrium between the proton capture destructive channel, i.e. ^{7}Be(p, γ)^{8}B, and its inverse reaction, i.e. photo-disintegration of ^{8}B, ^{8}B(γ, p)^{7}Be. Since ^{7}Be is more efficiently destroyed than produced below $T \approx 10^8$ K, and since it comes only from ^{3}He, it can be formed only if enough ^{3}He survives the initial burning phase. Survival of ^{3}He is favored if the evolution proceeds rapidly, which is the case in CO novae. The reason for the faster evolution of CO novae is the following: the release of nuclear energy is mainly dominated by proton captures on ^{12}C (^{12}C(p,γ)^{13}N(β^+)^{13}C(p,γ)^{14}N; cold CNO cycle); as CO novae have more carbon (^{12}C) in the envelope, if mixing between core and accreted material is accepted as a necessary condition for the explosion (see Chapter 4), they evolve faster (Hernanz *et al.*, 1996; José & Hernanz, 1998). For instance, a CO nova of 1.15 $M_\odot$ requires 0.72×10^6 s for a temperature rise from 3×10^7 K to 10^8 K, whereas an ONe nova of the same mass would require 13×10^6 s.

It is relevant at this point to comment that novae hosted by ONe white dwarfs, where mixing occurs in the CO buffer or in the transition zone mentioned above (see Section 11.2.1), would behave more like CO novae regarding ^{7}Li synthesis, because of their larger content of carbon (José *et al.*, 2003).

Overproduction factors with respect to the solar values around 1000 are obtained in CO nova models, meaning that novae can be important contributors to the Galactic ^{7}Li (Hernanz *et al.* 1996; see also Chapter 6); detailed models of Galactic chemical evolution show that novae are required to reproduce the steep rise of the observed lithium abundance between the formation of the Solar System and the present time (Romano *et al.*, 1999; Alibés, Labay & Canal, 2002).

A good confirmation of ^{7}Be synthesis in novae would come from the detection of ^{7}Li in the optical, through the Li I doublet at 6708 Å. However, this line is not easy to detect, and for many decades only upper limits to the ratio Li/Na were available (Friedjung, 1979). It is evident that ^{7}Be should not be completely ionized, in order to undergo the electron capture leading to ^{7}Li formation and to γ-ray emission, but this condition is fulfilled in the envelope zones where most of ^{7}Be decays to ^{7}Li. Della Valle *et al.* (2002) have reported the detection of a short-lived (less than 1 week) feature at 6705–6715 Å in V382 Vel, 10 days after maximum; a tentative identification of this feature could be the Li I doublet at 6708 Å, but the authors stated that the interpretation of the observed feature was not unambiguous. In a later paper, Shore *et al.* (2003) suggested that the feature could instead correspond to a doublet from neutral nitrogen (see also Chapter 6).

^{22}*Na*

Since the pioneering work by Clayton and Hoyle (1974), it has been mentioned in the literature that novae are potential emitters of the γ-ray line at 1275 keV resulting from ^{22}Na decay (see Figure 11.1). In that work, Clayton and Hoyle assumed that the ^{22}Na mass fraction in nova envelopes was around 10^{-3}, coming from the conversion of all ^{20}Ne to ^{22}Na. It has been seen in the past 15 years that this conversion is not so efficient, but interestingly the currently accepted ^{22}Na yields in the most prolific novae are not far from those historic predictions.

In fact, the amount of ^{22}Na synthesized during nova explosions took a long time to be clearly determined. The first hydrodynamical models of nova outbursts did not include complete nuclear reaction networks, reaching the Ne–Na and Al–Mg regions; since the most common white dwarfs are made of CO, the CNO cycle was the basic nuclear network needed for the simulations. In fact, as discovered by Starrfield *et al.* (1972), it is in the CNO cycle itself where the origin of the explosion energy resides, since its operation out of equilibrium together with the role played by convection are the necessary conditions for a nova outburst (see Chapter 4 for details). In addition, it was already noted in the 1980s that some uncertain nuclear reaction rates played a crucial role in the final amount of ^{22}Na (and ^{26}Al, see below). So it was not until the 1980s that extensive nucleosynthesis in nova outbursts was computed, first with parametrized models, i.e. through simplified one-zone models with representative temperature–density temporal profiles taken from evolutionary nova models (Starrfield, Truran & Sparks 1978b). Estimates of ^{22}Na production ranged from 2×10^{-7} to 10×10^{-7} in mass fraction (Hillebrandt & Thielemann, 1982), much lower than the abovementioned values. In this work it was already pointed out that some nuclear reactions which were crucial for ^{22}Na synthesis had very uncertain rates (e.g. ^{21}Na(p,γ)^{22}Mg and ^{22}Na(p,γ)^{23}Mg). A later study (Wiescher *et al.*, 1986) incorporating an updated nuclear reaction network and adopting the same evolutionary profiles yielded smaller ^{22}Na mass fractions, in the range 2×10^{-8} to 5×10^{-7}. In summary, small yields of ^{22}Na (and ^{26}Al) were obtained in CO nova models and the main conclusion from these one-zone model studies was that novae were not important Galactic producers of ^{22}Na (and ^{26}Al) but might explain some of the abundance anomalies observed in meteorites, i.e. excesses of ^{22}Ne (Eberhardt, 1974, 1979; Lewis *et al.*, 1979; Clayton, 1975) and ^{26}Mg (Lee, Papanastassiou & Wasserburg, 1977). These anomalies could be attributed to the decay of ^{22}Na and ^{26}Al in nova ejecta respectively;[1] however, some predicted isotopic ratios did not fit the observations, e.g. (^{20}Ne+^{21}Ne)/^{22}Na was much greater than 1, in contradiction with observed ratios in the corresponding meteoritic grains.

In the 1990s, new one-zone models for nova nucleosynthesis were developed, adopting various initial compositions which included the possibility of mixing with massive white dwarf cores, made of oxygen, neon and magnesium (with abundances as expected at the epoch, based on hydrostatic carbon-burning models by Arnett and Truran (1969)). These models (Weiss & Truran, 1990; Nofar, Shaviv & Starrfield, 1991) were aimed at studying the synthesis of ^{22}Na and ^{26}Al, in view of the recent detection of Galactic ^{26}Al and non-detection of ^{22}Na (see Section 11.5.1 below), and adopted ONe white dwarfs because significant enrichments of neon had been found which could not be explained through standard CO novae nucleosynthesis (see Chapter 4). Interestingly, Weiss and Truran (1990) obtained ^{22}Na yields as large as 10^{-4}, which combined with envelope masses of 2×10^{-5} M$_\odot$ gave 4×10^{-9} M$_\odot$ of ^{22}Na. In the same paper, they gave a formula to derive the expected γ-ray flux at 1275 keV, which is the application of

$$f(t) = \frac{M_{\rm ej} N_{\rm A}}{A\tau} \frac{1}{4\pi D^2} \exp(-t/\tau)$$

[1] The ^{22}Ne yield was null in the Hillebrandt and Thielemann (1982) and Wiescher *et al.* (1986) models.

to the particular case of ^{22}Na. Here $M_{\rm ej}$ is the ejected mass, $N_{\rm A}$ is Avogadro's number, A the atomic mass, τ the lifetime and D the distance to the nova; therefore

$$f_{1275\,\rm keV} = 4 \times 10^{-5} \frac{M_{\rm ej}}{10^{-8}\,{\rm M}_\odot} \frac{\exp(-t/\tau)}{(D/1\,{\rm kpc})^2}\ {\rm phot\ cm^{-2}\,s^{-1}}.$$

With this expression and their predicted yields, Weiss and Truran predicted a flux of 6×10^{-5} phot cm^{-2} s^{-1} for a nova at $D = 500$ pc, not far from current predictions (see Section 11.3.2 below). However, a few years after that paper, Starrfield and collaborators performed the first hydrodynamical calculations of ONe novae with full nucleosynthesis included (Politano *et al.*, 1995). They obtained larger ^{22}Na (and ^{26}Al) yields than in Weiss and Truran's model; the question of the ejected mass continued to be a problem, as deeply analyzed in a later work by the same group (Starrfield *et al.*, 1998). Other contemporaneous complete hydrodynamic models of nova outbursts came from the Israeli group, which also has a long tradition in nova modeling (Prialnik, Shara & Shaviv, 1978, 1979), mainly focused on CO novae not able (in principle) to synthesize either ^{22}Na or ^{26}Al (Prialnik & Kovetz, 1997). But interestingly enough, a mechanism to produce moderate-mass white dwarfs with Ne- and Mg-enriched outer shells in cataclysmic variables was suggested (Shara, 1994; Shara & Prialnik, 1994; Prialnik & Shara, 1995): rapid accretion, at a rate $\sim 10^{-6}$ M$_\odot$ yr^{-1}, of hydrogen onto a CO white dwarf before a nova outburst occurs. The advantage of this scenario was that it could simultaneously explain massive ejecta and the presence of neon without the need of very massive white dwarfs. The idea was very appealing, but it required a fine tuning between succesive mild He flashes and mass-loss between them, leading to the accumulation of a massive neon- and magnesium- rich envelope and a final nova eruption (when the mass accretion rate decreased). In any case, it is the only known way to produce massive ejecta enriched in neon, as some observations indicate.

The synthesis of ^{22}Na proceeds through ^{20}Ne(p,γ)^{21}Na followed either by the decay of ^{21}Na to ^{21}Ne, i.e. ^{21}Na(β^+)^{21}Ne(p,γ)^{22}Na, or by a proton capture on ^{21}Na, i.e. ^{21}Na(p,γ)^{22} Mg(β^+)^{22}Na. Some nuclear reactions which have an important impact on the final nova yields of ^{22}Na, such as ^{21}Na(p,γ)^{22}Mg, ^{22}Na(p,γ)^{23}Mg (see Coc *et al.* (1995) and José, Coc and Hernanz (1999) for an extensive review of the nuclear uncertainties in the NeNa–MgAl 'cycles') have deserved much recent experimental work: Bishop *et al.* (2003) and D'Auria *et al.* (2004) have made the first direct measurements of the ^{21}Na(p,γ)^{22}Mg reaction, using the TRIUMF-ISAC radioactive beams facility located in Vancouver, Canada, whereas Jenkins *et al.* (2004) have carried out an indirect measurement of the ^{22}Na(p,γ)^{23}Mg rate in Argonne National Laboratory. From these measurements the ^{22}Na yields from novae have been slightly reduced (i.e. from 3.5×10^{-4} to 2.8×10^{-4} in a particular ONe nova of 1.25 M$_\odot$, the fourth model in Table 11.2) but, more important, the uncertainty has been largely reduced and thus the ensuing predictions of γ-ray emission are now more secure. It is important to recall that the impact of any new nuclear reaction rate on the nova yields of any isotope should be accounted for by means of a complete model instead of one-zone models, because the role played by convection (e.g. transporting nuclei from hot to cool regions where they are preserved from destruction) is crucial.

The most recent hydrodynamic models of ONe (or ONeMg) nova outbursts on masses larger than 1.0 M$_\odot$ (José & Hernanz, 1998; José, Coc & Hernanz, 1999; Politano *et al.*, 1995; Starrfield *et al.*, 1998) provide ^{22}Na yields in the range 10^{-4} to 10^{-3} (José & Hernanz, 1998; José, Coc & Hernanz, 1999) or 6×10^{-4} to 5×10^{-3} (Politano *et al.*, 1995; Starrfield *et al.*, 1998). The discrepancy is not very large and it is mainly related to the different initial compositions adopted: smaller yields correspond to mixing with underlying ONe white dwarfs (almost devoid of magnesium and with neon abundance around half that of oxygen, as stellar evolutionary models predict), whereas the larger ones consider ONeMg white dwarfs, with O:Ne:Mg ratios from hydrostatic carbon-burning calculations, i.e. much larger Ne and Mg proportions than in the ONe case. It is worth mentioning that if mixing occurs with the CO buffer on top of the bare ONe core, no ^{22}Na would be expected (José *et al.*, 2003).

In Table 11.2 are shown the ^{22}Na ejected masses for some of the models in José and Hernanz (1998) and José, Coc and Hernanz (1999), which range around 6×10^{-9} M$_\odot$ and (1–7) $\times 10^{-11}$ M$_\odot$ for ONe and CO novae, respectively. A broader range was obtained by Starrfield *et al.* (1998), for ONe novae, from 8×10^{-10} M$_\odot$ to 3×10^{-8} M$_\odot$ (but see results from more recent calculations in Chapter 4).

^{26}Al

The role played by the radioactive decay of ^{26}Al as a heat source in the early Solar System was first suggested by Urey (1955). Later on, this suggestion was confirmed by the discovery of large excesses of ^{26}Mg (the daughter nucleus of ^{26}Al), up to 10%, in a Ca–Al inclusion of the Allende meteorite. The strict correlation found between ^{26}Mg excess and the ^{27}Al/^{24}Mg isotopic ratio in four co-existing phases with distinct chemical compositions demonstrated that ^{26}Al must have been present in the inclusions when they condensed in the early Solar System (Lee, Papanastassiou & Wasserburg, 1977). An independent discovery of ^{26}Al in the interstellar medium has come from observations of the 1.809 MeV line. Both discoveries have led to active research on the origin of this fascinating isotope (Clayton & Leising, 1987; Prantzos & Diehl, 1996; see further details in Section 11.5.4).

Synthesis of ^{26}Al requires moderate peak temperatures, around 2×10^8 K, and a fast decline from them. Its production is in fact complicated by the presence of a short-lived (half-life 6.3 s) spin isomer (^{26}Al^g and ^{26}Al^m are the usual notations for ^{26}Al in ground and isomeric state). When the temperature is smaller than 4×10^8 K (as in novae), the ground and isomeric states must be treated as two separate isotopes, because they do not reach thermal equilibrium (Ward & Fowler, 1980). The γ-ray line at 1.809 MeV is a consequence of the decay from ^{26}Al^g to the first excited state of ^{26}Mg, which de-excites to its ground state by emitting a 1.809 MeV photon (see Figure 11.1).

The first calculations of ^{26}Al synthesis during explosive hydrogen burning were the same as mentioned above for ^{22}Na, i.e. simplified one-zone models which adopted density and temperature profiles thought to mimic explosive hydrogen burning (Hillebrandt & Thielemann, 1982; Weischer *et al.*, 1986). They concluded that ^{26}Al could be produced in novae, but not in very large amounts; these computations used solar or CNO-enhanced white dwarf envelopes. Later computations, again one-zone models, demonstrated that the envelope's matter should initially be enriched in O, Ne and Mg, dredged up from the white dwarf cores, to get larger amounts of ^{26}Al (Weiss & Truran, 1990; Nofar, Shaviv & Starrfield, 1991).

The major seed nuclei for ^{26}Al synthesis are ^{24,25}Mg (José, Hernanz & Coc, 1997). At the early phases of the thermonuclear runaway (burning shell temperatures around 5×10^7 K), the dominant reaction is ^{25}Mg(p,γ)^{26}Al^{g,m}; the subsequent reaction ^{26}Al^m(β^+)^{26}Mg(p,γ)^{27}Al produces the stable isotope ^{27}Al. At higher temperatures ($\sim 10^8$ K), the nuclear path ^{24}Mg(p,γ)^{25}Al(β^+)^{25}Mg dominates, with again ^{25}Mg(p,γ)^{26}Al^{g,m}. When the temperature reaches 2×10^8 K, (p,γ) reactions proceed very efficiently and reduce the amount of ^{25}Al, leading to the formation of ^{26}Si (^{25}Al(p,γ)^{26}Si) which decays into ^{26}Al^{m}, thus bypassing ^{26}Al^{g} formation. Also ^{26}Al itself (in both states) is destroyed to ^{27}Si which decays into ^{27}Al. It is worth noting that ^{24}Mg is originally in the envelope (coming from the underlying ONe core), but also comes from ^{23}Na(p,γ)^{24}Mg. In summary, the final amount of ^{26}Al^{g} and the ratio ^{26}Al^{g}/^{27}Al mainly depend on the competition between the two nuclear paths ^{24}Mg(p,γ)^{25}Al(β^{+})^{25}Mg(p,γ)^{26}Al^{g,m} and ^{24}Mg(p,γ)^{25}Al(p,γ)^{26}Si. The first channel is the only one producing ^{26}Al^{g}, whereas both channels produce ^{27}Al (through ^{26}Al^{g,m}(p,γ)^{27}Si(β^{+})^{27}Al or ^{26}Si(β^{+})^{26}Al^{m}(β^{+}) ^{26}Mg(p,γ)^{27}Al).

The main nuclear uncertainties relative to ^{26}Al synthesis are ^{25}Al(p,γ)^{26}Si and the confirmation of a resonance of ^{26}Al^{g}(p,γ)^{27}Si (see José, Coc and Hernanz (1999) for more details). The ^{23}Na(p,γ)^{24}Mg has been reinvestigated and its uncertainty, which has a direct impact on the ^{24}Mg and hence the ^{26}Al yields, has been reduced from about a factor of 3 to approximately 25% (Rowland *et al.*, 2004).

The final ^{26}Al yields from novae largely depend on the initial mass of the white dwarf and on the degree of mixing between the accreted envelope and the core. The most recent hydrodynamic models of ONe (or ONeMg) nova outbursts on masses larger than 1.0 $M_\odot$ provide ^{26}Al yields in the range 2×10^{-4} to 2×10^{-3} (José & Hernanz, 1998; José, Coc & Hernanz, 1999) or 5×10^{-4} to 10^{-2} (Politano *et al.*, 1995; Starrfield *et al.*, 1998); as in the ^{22}Na case, the discrepancy is related to the different initial compositions adopted, but also to different nuclear reaction rates. It is worth mentioning that if mixing occurs with the CO buffer on top of the bare ONe core (or in the transition zone), some ^{26}Al would be expected (but no ^{22}Na), since there is a non-negligible amount of the seed nucleus ^{25}Mg both in the CO buffer and in the transition zone (José *et al.*, 2003).

In Table 11.2 are shown the ^{26}Al ejected masses for some of the models in José and Hernanz (1998) and José, Coc and Hernanz (1999), which range around 10^{-8} $M_\odot$ and (2–6) $\times$ 10^{-10} $M_\odot$ for ONe and CO novae, respectively. Again a broader range was obtained by Starrfield *et al.* (1998) for ONeMg novae, from 10^{-10} $M_\odot$ to 4×10^{-8} $M_\odot$ (but see results from more recent calculations in Chapter 4).

11.2.2 Gamma-ray production and propagation

The origin of the γ-rays emitted by novae is the radioactive decay of the unstable isotopes synthesized during the explosions and described above. However, the shape and intensity of the γ-ray output of novae, as well as its temporal evolution, depend not only on the amount of γ-ray photons produced, but also on how they propagate through the expanding envelope and ejecta.

The first step in computing the spectrum is to generate γ-rays according to the decay schemes of the corresponding radioactive isotopes (see Figure 11.1). The amount of photons generated in a particular nova depends on the relative isotopic abundances and rates of disintegration of the above-mentioned nuclei. In addition to these *direct* γ-ray photons, positrons emitted as a consequence of the radioactive decays of ^{22}Na and ^{26}Al, as well as a result of the decays of the shorter-lived ^{18}F and ^{13}N, should be traced.

Once photons are generated, their trip across the expanding ejecta should be simulated by taking into account the various interaction processes affecting their propagation, i.e. Compton scattering, $e^{-}-e^{+}$ pair production and photoelectric absorption. The cross-section for Compton scattering can be computed according to the usual Klein–Nishina expression, whereas photoelectric absorption and pair-production cross-sections can be taken from experimental data bases, such as those from Brookhaven National Laboratory.

The treatment of positron annihilation deserves particular attention. When a positron is emitted, it can either escape without interacting with the expanding nova envelope or annihilate with an ambient electron. It can be safely assumed that in nova envelopes positrons thermalize before annihilating. This approximation is wrong in less than 1% of cases in an electronic plasma, according to Leising and Clayton (1987). In a neutral envelope, the excitation cross-section dominates any other interaction at energies above ~100 eV (Bussard, Ramaty & Drachman, 1979), and thus positrons lose energy until they reach this extremely low value. In order to reproduce this braking effect, positrons should be propagated until they cross an equivalent column mass density of ~0.2 g cm^{-2}, measured in a straight line (Chan & Lingenfelter, 1993). This is the mean range expected for a 0.6 MeV positron braked down to energies ~100 eV through elastic collisions with the surrounding medium, when the effect of magnetic fields on its propagation is neglected. Once thermalized, the positron covers a negligible distance and then annihilates.

For densities and temperatures typical of nova envelopes, positrons form positronium (positron–electron system) in ~90% of annihilations (Leising & Clayton, 1987), while in the remaining 10% of cases they annihilate directly. Positronium is formed in singlet state 25% of the time, leading to the emission of two 511 keV photons, and in triplet state 75% of the time, leading to a three-photon annihilation. The continuum spectrum of photons produced by the triplet state was obtained by Ore and Powell (1949). In summary, once a positron is produced, its trip should be followed until it escapes or covers the average energy-loss distance. In the latter case it produces positronium 90% of the time, leading to triplet to singlet annihilations in 3:1 proportion, while in 10% of the cases it annihilates directly.

11.3 Gamma-ray spectra and light curves of individual novae

The γ-ray signatures of classical novae mainly depend on their yields of radioactive nuclei. As explained above (Section 11.2.1), CO and ONe novae differ in their production of ^{7}Be, ^{22}Na, and ^{26}Al, whereas they synthesize similar amounts of ^{13}N and ^{18}F. Therefore, CO novae should display line emission at 478 keV related to ^{7}Be decay, whereas for ONe novae line emission at 1275 keV related to ^{22}Na decay is expected. In both nova types, there should also be line emission at 511 keV related to $e^{-} - e^{+}$ annihilation, and a continuum produced by Comptonized 511 keV emission and positronium decay (see Table 11.1). Finally, line emission at 1.809 MeV related to ^{26}Al decay is expected from ONe novae, but since the lifetime of ^{26}Al (10^{6} years) is longer than the typical time interval between two succesive novae in the Galaxy, this emission cannot be detected in individual objects (see Section 11.4 for details).

A Monte Carlo code, based on the method described by Pozdniakov, Sobolev and Sunyaev (1983) and Ambawni and Sutherland (1988) has been developed by Gómez-Gomar *et al.* (1998) to compute the γ-ray output of novae. The temporal evolution of the whole γ-ray spectrum of four representative models (see Table 11.2) is shown in Figures 11.2 and 11.3. The most prominent features of the spectra are the annihilation line at 511 keV and the continuum at energies between 20–30 keV and 511 keV (in both nova types), the ^{7}Be line

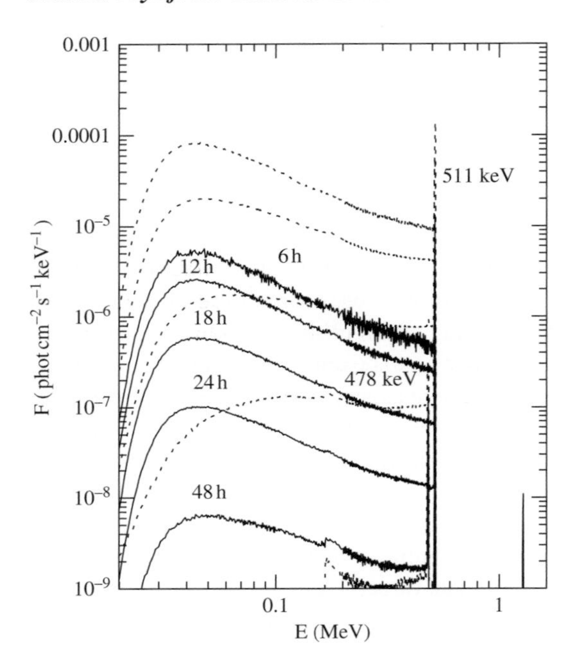

Fig. 11.2. Spectra of CO novae of masses 0.8 (solid) and 1.15 $M_\odot$ (dotted) at different epochs after T_{peak} (labels for dotted lines follow the same sequence as those for solid lines: from top to bottom 6, 12, 18, 14, and 48 h).

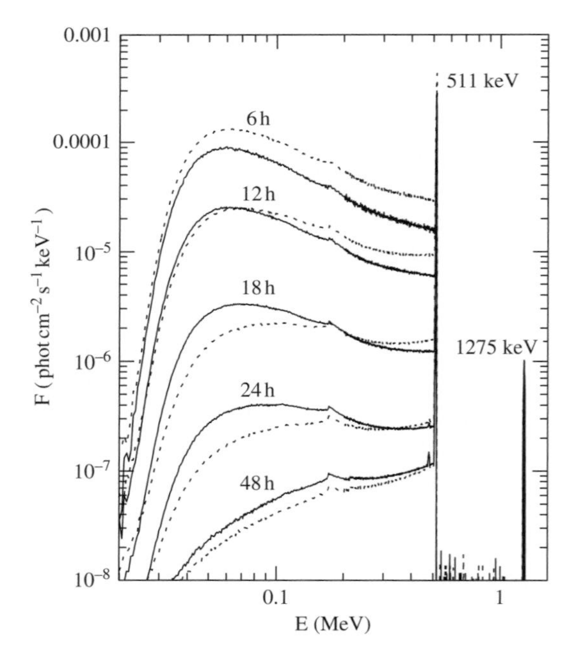

Fig. 11.3. Spectra of ONe novae of masses 1.15 (solid) and 1.25 $M_\odot$ (dotted) at different epochs after T_{peak}.

at 478 keV in CO novae, and the ^{22}Na line at 1275 keV in ONe novae. A few hours after the outburst, when transparency increases, the back-scattering of the 511 keV photons produces a feature at 170 keV. The main difference between spectra of CO and ONe novae is the long-lived lines: 478 keV in CO novae as compared with 1275 keV in ONe novae, which directly reflect the different chemical composition of the expanding envelope (^{7}Be-rich in CO novae and ^{22}Na-rich in ONe novae).

11.3.1 Positron annihilation radiation: 511 keV line and continuum

The early γ-ray emission, or *prompt* emission, of novae is related to the disintegration of the very short-lived radioisotopes ^{13}N and ^{18}F. The radiation is emitted as a line at 511 keV (direct annihilation of positrons and singlet state positronium), plus a continuum. The continuum is related with both the triplet state positronium continuum and the Comptonization of the photons emitted in the line (see Section 11.2). There is a sharp cut-off at energies 20–30 keV (the exact value depending on the envelope composition) because of photoelectric absorption. A better insight into the early γ-ray emission is obtained from inspection of the light curves shown in Figures 11.4, 11.5, and 11.6. The largest flux is emitted in the (20–250) keV range, since the continuum has its maximum at $\sim$60 keV (ONe novae) and at $\sim$45 keV (CO novae), followed by the flux in the (250–511) keV range (excluding the 511 keV line) and the flux in the 511 keV line. The two maxima in the light curves of the 511 keV line correspond to ^{13}N and ^{18}F decays, but the first maximum is difficult to resolve because its duration is really short; in addition, it is very model-dependent: only ^{13}N in the outermost zones of the envelope can be seen in γ-rays because of limited transparency at

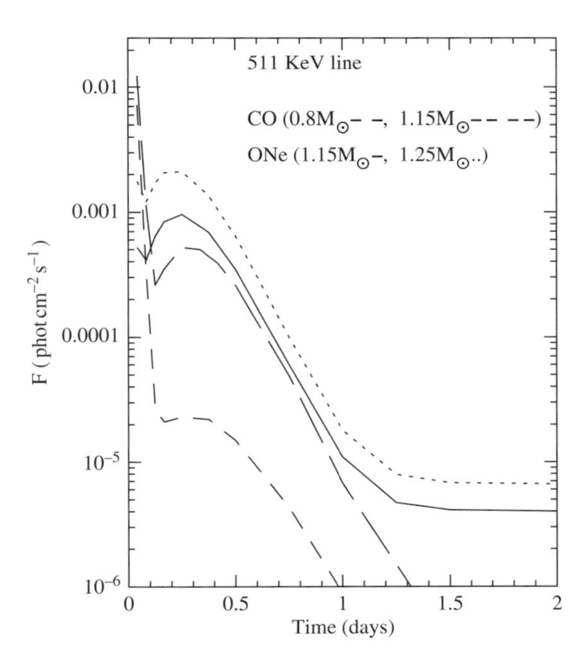

Fig. 11.4. Light curve of the 511 keV line for CO and ONe novae (from bottom to top, CO nova of 0.8 and 1.15 M$_\odot$, and ONe nova of 1.15 and 1.25 M$_\odot$). From Hernanz, Gómez-Gomar & José (2002).

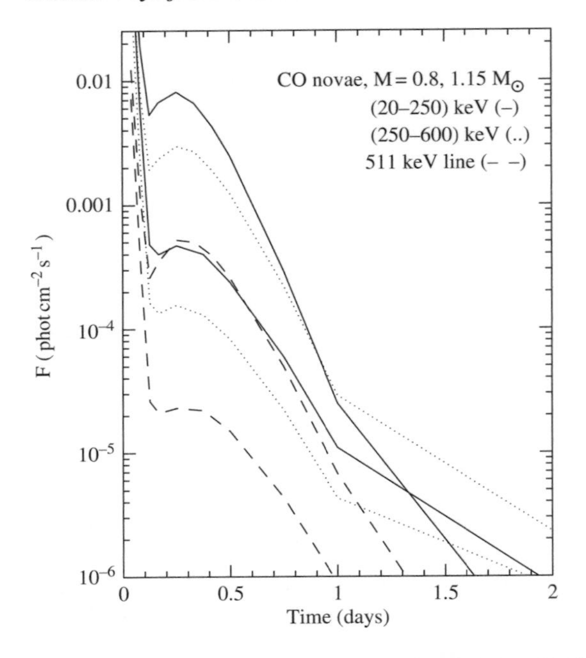

Fig. 11.5. Light curve of two continuum bands below 511 keV for CO novae. The light curve of the 511 keV line is also shown for comparison. The upper curves correspond to the more massive novae.

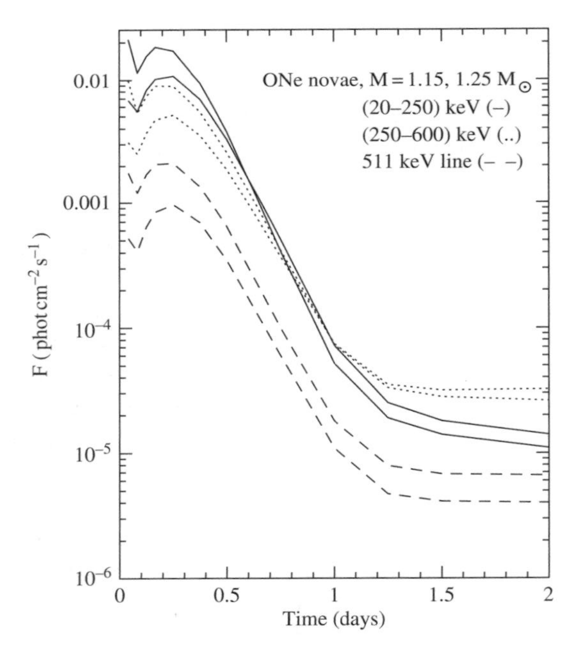

Fig. 11.6. Same as Figure 11.5 but for ONe novae. Again the upper curves correspond to the larger masses, at early times; but at later epochs the most massive nova emits a slightly smaller flux, except for the 511 keV line, because of larger transparency.

very early epochs and, therefore, the intensity of the first maximum depends on the efficiency of convection (see Section 11.2.1). This first maximum thus gives an important insight into the dynamics of the envelope after peak temperature is attained at its base.

The behavior of the continuum as time elapses differs from model to model, directly reflecting the different opacities of the expanding envelopes to γ-rays: more energetic novae become transparent earlier. For the particular cases shown in Figures 11.2 and 11.3, a CO nova of 1.15 $M_\odot$ becomes transparent before an ONe nova of the same mass: at 48 hours its continuum is already out of scale (flux lower than 10^{-9} phot cm^{-2} s^{-1}). The reason is that the CO nova has more specific kinetic energy than the ONe nova, mainly because the total available energy is similar in both novae but the CO nova ejects less mass (see Table 11.2). That is because the CO nova attains the explosive conditions earlier (i.e. with less accreted mass) since there the dominant nuclear reaction, ^{12}C(p,γ), proceeds faster because it has a higher ^{12}C content than the ONe case. However, for different masses of the underlying cores, the effect of core mass is even more important than that of composition: i.e. the 0.8 $M_\odot$ CO nova has a larger ejected mass than the 1.15 $M_\odot$ CO nova, because the smaller the gravity at the base of the envelope, the larger the accreted mass to reach the degeneracy necessary for an explosion. Therefore, the envelope of the less massive 0.8 $M_\odot$ CO nova is more opaque to γ-rays at a given epoch than that of the 1.15 $M_\odot$ CO nova (see Figure 11.2).

For detectability purposes it is important to know that the 511 keV line is somewhat blue-shifted and not symmetric at the earliest phase after peak temperature, but it becomes more symmetric and less blue-shifted as the ejecta become more transparent and both the approaching and the receding material contribute. At around 12 hours after peak temperature the full width at half maximum is around 8 keV, except for the 0.8 $M_\odot$ case where the line is somewhat narrower, $\sim$3 keV, because of the smaller expansion velocities.

The annihilation emission is the most intense γ-ray feature expected from novae, but unfortunately it has a very short duration, because of the short lifetime of the main positron producers (^{13}N and ^{18}F). There are also positrons available from ^{22}Na decay in ONe novae, but these contribute much less (they are responsible for the *plateau* at a low level, between 10^{-6} and 10^{-5} phot cm^{-2} s^{-1} for $d = 1$ kpc; see Figure 11.4). These positrons do not contribute all the time; after roughly one week the envelope is so transparent that ^{22}Na positrons escape freely without annihilating. In summary, annihilation radiation lasts only $\sim$1 day at a high level, and one to two weeks at a lower-level plateau (the latter only in ONe novae).

Another important fact is that annihilation radiation is emited well before the visual maximum of the nova, i.e. before the nova is discovered optically. This early appearance of γ-rays from electron–positron annihilation makes their discovery through pointed observations almost impossible. Only wide field of view instruments, monitoring the sky continuously in the appropriate energy range could detect it (see Section 11.5.3 below).

The flux of the annihilation radiation emitted by a nova depends both on the amount of ^{18}F (and ^{13}N) ejected and on the dynamical conditions in the expanding envelope, which determine its transparency. Some models can be computed for illustrative purposes, although not in a self-consistent way, where ejected masses or velocity profiles are varied, taking as starting models the realistic models shown before (computed with a hydrocode for nucleosynthesis and dynamical properties plus a Monte Carlo code for γ-ray production and transfer; see Section 11.2.2 above). The result is that the effect of the ejected mass is twofold: the larger the ejected mass the larger the amount of ^{18}F but the larger the opacity; therefore the emission is less intense at the beginning because of the larger opacity, but more intense at the end when

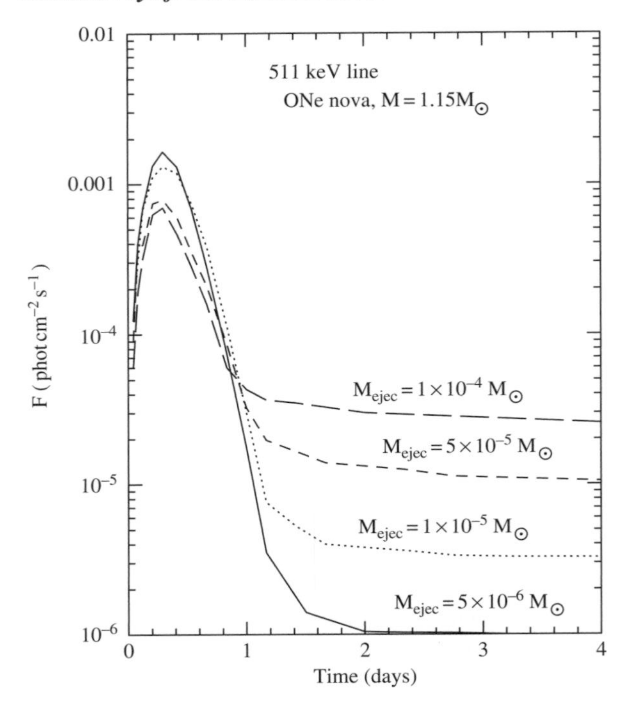

Fig. 11.7. Light curves for the 511 keV line of an ONe nova of $1.15\,M_\odot$, for a range of parametrized ejected masses. From Hernanz, Gómez-Gomar and José (2002).

the envelope is almost completely transparent (see Figure 11.7). On the other hand, the effect of the velocity is the following: the larger the velocity (for an unchanged ejected mass), the larger the transparency and thus the larger the flux. However, for ONe novae at *late* epochs (from 1 day until 1 to 2 weeks) there are still positrons contributing to annihilation radiation, coming from ^{22}Na; then, the larger the velocity, the larger the transparency and the earlier the epoch when positrons begin to escape freely without annihilating. This means that for larger velocities, the low flux plateau has shorter duration (see Figure 11.8). These results show again that the annihilation emission provides a wealth of information about the properties of the expanding envelope, beyond the mere content of radioactive species (Hernanz, Gómez-Gomar & José, 2002).

11.3.2 Line emission from ^{7}Be and ^{22}Na at 478 and 1275 keV

Line emission at 478 keV, related to de-excitation of the ^{7}Li which results from an electron capture on ^{7}Be, is the most distinctive feature in the γ-ray spectra of CO novae (besides the annihilation line and continuum, common to all types of novae), as shown in Figure 11.2. The reason is that these novae are more prolific producers of ^{7}Be than ONe novae (see Section 11.2.1). The light curves of the 478 keV line are shown in Figure 11.9: the flux reaches its maximum at day 13 and 5 in the more and less opaque models, with total masses 0.8 and $1.15\,M_\odot$, respectively. The width of the line is 3 and 8 keV for the 0.8 and $1.15\,M_\odot$ CO novae, respectively. The maximum flux is around 10^{-6} phot cm^{-2} s^{-1}, for $d = 1$ kpc.

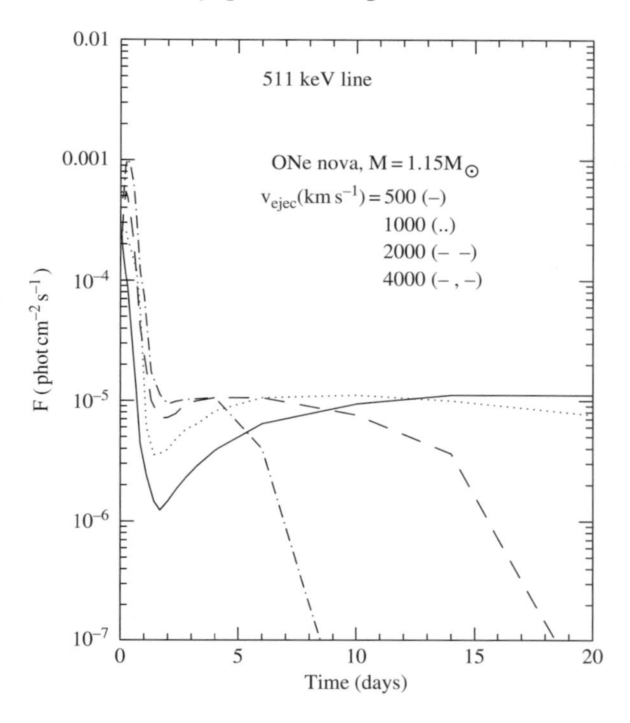

Fig. 11.8. Light curves for the 511 keV line of an ONe nova of 1.15 $M_\odot$, for a range of parametrized average velocities of the ejecta. From Hernanz, Gómez-Gomar and José (2002).

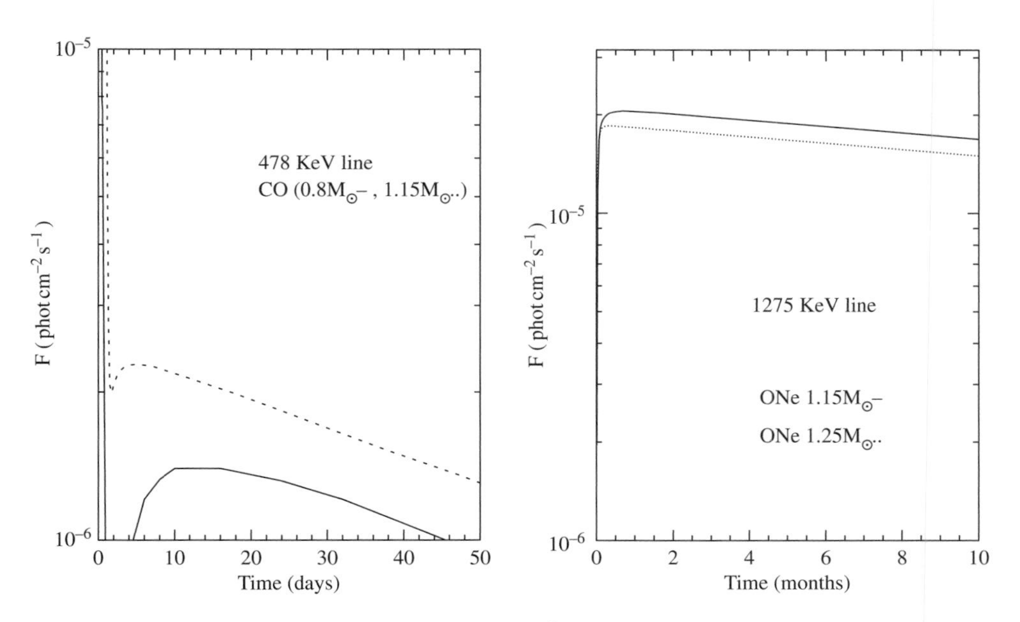

Fig. 11.9. Left: light curve of the 478 keV ^{7}Li line for two CO nova models. Right: light curve of the 1275 keV ^{22}Na line for two ONe nova models.

There is an earlier maximum, which has nothing to do with the envelope's content of ^{7}Be, but with the strong continuum related to the annihilation of ^{13}N and ^{18}F positrons.

The ^{22}Na line at 1275 keV appears only in ONe novae, because CO novae do not synthesize this isotope (see Section 11.2.1). The rise phase of the 1275 keV line light curves (see Figure 11.9) lasts between 10 (1.25 $M_\odot$) and 20 days (1.15 $M_\odot$). Soon after the maximum, the line reaches the stable decline phase dictated by the lifetime of ^{22}Na, 3.75 years; during this phase, the line intensities directly reflect the amount of ^{22}Na ejected mass. The corresponding fluxes at maximum are around 10^{-5} phot cm^{-2} s^{-1} at $d = 1$ kpc, for the typical models shown in Table 11.2. The width of the line is around 20 keV, which is a handicap for its detectability with instruments having high energy resolution (see Section 11.6 below).

11.4 Cumulative Galactic gamma-ray emission of novae

Up to now we have analyzed the γ-ray emission from individual novae, but there is in addition the emission resulting from the cumulative effect of all Galactic novae *active in γ-rays* at a given epoch. This kind of emission should trace the Galactic distribution of classical novae and, if detected, would give very valuable information, not easy to obtain from observations at other wavelengths because of the effects of interstellar extinction. For the same reason, the Galactic nova rate is also very poorly known. Estimates of the Galactic nova rate are mainly based on observations of novae in other Galaxies or on extrapolations of observations in our Galaxy, taking into account the distribution of extinction related to interstellar dust (Della Valle & Livio, 1994; Shafter, 1997, 2002; see Chapter 14 for a full discussion).

It is straightforward to understand that a condition must be fulfilled in order to allow the accumulation of γ-rays from multiple novae: the average time interval between two successive Galactic novae should be shorter than the lifetime of the corresponding radioactive species. Since around 35 novae are expected to explode per year in the Milky Way (Shafter, 1997), ^{7}Be, ^{22}Na, and ^{26}Al fulfil the condition, even when taking into account that only a fraction of the novae are ^{7}Be-rich (CO novae, i.e. around 2/3 of all novae) or ^{22}Na and ^{26}Al-rich (ONe novae, i.e. around 1/3 of all novae; Livio & Truran, 1994; Gil-Pons *et al.*, 2003). For ^{22}Na, there is the additional advantage that only novae are expected to contribute to its Galactic content, whereas for ^{7}Be AGB stars can also contribute and for ^{26}Al, massive stars and AGBs clearly contribute. Therefore, the most interesting case of cumulative emission is the Galactic 1275 keV emission from ^{22}Na, because it traces directly the spatial nova distribution; but as analyzed below, the predicted emission is too low for the performances of the current instruments. An analysis of the nova contribution to ^{26}Al is also very relevant since its emission has been detected in the Galaxy.

11.4.1 Galactic ^{22}Na emission at 1275 keV from novae

The global flux at 1275 keV in the Galaxy depends on the amount of ^{22}Na ejected per explosion and on the distribution and rate of ONe novae in the Galaxy (since only ONe novae produce ^{22}Na). A detailed study of the diffuse Galactic 1275 keV line emission from the cumulative distribution of individual nova contributions was performed by Higdon and Fowler (1987). Through a Monte Carlo simulation, the authors modeled the expected emission as a function of the Galactic longitude, based on some assumed nova spatial distributions and frequencies, accounting for disk plus bulge (spheroid) populations. The main result was that contributions from a few young and close novae dominate, yielding a very irregular distribution versus Galactic longitude. A comparison with the upper limits from HEAO 3 observations (Mahoney *et al.*, 1982) gave $5.6 \times 10^{-7}\,M_\odot$ as the upper limit to the mean ^{22}Na yield per nova, for a disk nova population. It was clear from this work that the results were subject to many uncertainties because there are many unknowns, such as the real nova Galactic distribution, the bulge/disk ratio, the nova rate and the fraction of ONe versus CO novae.

A more recent extended analysis of the cumulative emission at 1275 keV from novae has been performed by Jean *et al.* (2000). Unfortunately, the ejected ^{22}Na masses needed for a detection of this emission with the SPI spectrometer on board the INTEGRAL satellite are far above what current theoretical models predict ($\sim 10^{-7}\,M_\odot$, versus a few times $10^{-9}\,M_\odot$).

11.4.2 Contribution of novae to Galactic ^{26}Al emission at 1.809 MeV

The production of ^{26}Al by classical novae occurs mainly in ONe novae, as explained in Section 11.2.1. It is important to stress that (ONe) novae with low-mass white dwarfs are more prolific producers of ^{26}Al than massive novae. Therefore, the evaluation of the global contribution of novae to the ^{26}Al content in the Galaxy is not straightforward, but a crude estimate can be made. Let us assume that all novae contribute with the same amount of ^{26}Al, $M_{ej}(^{26}\mathrm{Al})$, and that ^{26}Al is active during a time equal to τ (its lifetime). Then the Galactic mass of ^{26}Al coming from novae would be (Weiss & Truran, 1990; José, Hernanz & Coc, 1997)

$$\frac{M_{nova}(^{26}\mathrm{Al})}{M_\odot} = \left(\frac{M_{ej}(^{26}\mathrm{Al})}{M_\odot}\right) \tau\ \mathcal{R}(\mathrm{novae\,yr^{-1}})\ f_{\mathrm{ONe}}$$

$$= 0.12\left(\frac{M_{ej}}{10^{-8}\,M_\odot}\right)\left(\frac{\mathcal{R}}{35\ \mathrm{novae\,yr^{-1}}}\right)\left(\frac{f_{\mathrm{ONe}}}{0.33}\right),$$

where $\mathcal{R}$ is the total Galactic nova rate and f_{ONe} is the proportion of ONe novae. Adopting typical ^{26}Al ejected masses ($2 \times 10^{-8}\,M_\odot$; see Table 11.2), the contribution of novae to Galactic ^{26}Al can be estimated as $\sim 0.2\,M_\odot$, more than a factor of 10 below the observed mass, in agreement with the current idea (deduced from the observed 1.809 MeV distribution; see Section 11.5.4) that Galactic ^{26}Al comes mainly from massive stars (predictions of the expected angular distribution of ^{26}Al for a variety of plausible origins were studied by Leising and Clayton (1985) and are reviewed in Prantzos and Diehl (1996)). A complete analysis of the global contribution of novae to the ^{26}Al in the Galaxy was done by Kolb and Politano (1997), applying Galactic nova population models, adopting the ^{26}Al yields from Politano *et al.* (1995) and taking very large ejected masses (larger than those from typical hydrodynamical calculations). In any case, they concluded that the nova contribution could range between 0.15 and $3\,M_\odot$, but this number largely depended on the unknown degree of mixing in novae, which largely influences their ^{26}Al yield, in addition to other parameters of the population synthesis code, as for instance the mass ratio distribution in zero-age main sequence binaries.

In spite of the shape of the observed distribution of the Galactic ^{26}Al (Diehl *et al.*, 1995; see Section 11.5.4 below), the significant contribution of novae to Galactic ^{26}Al has not been completely ruled out, since they can play a role in some zones where massive stars are absent. A more detailed ^{26}Al map is needed before more precise conclusions can be reached.

11.5 Observations

There have been many unsuccessful attempts to detect γ-rays from novae. The main efforts have been focused on the 1275 keV line from ^{22}Na, but there have also been observations of the 478 keV line from ^{7}Be. Observations have concentrated on individual objects, but searches of the cumulative emission of ^{7}Be and ^{22}Na have also been performed. Finally, the annihilation line has been searched for whenever wide field of view instruments were available, scanning zones of the sky where novae had exploded.

11.5.1 The 1275 keV line from ^{22}Na

The first observations of γ-rays with possible origin in novae were performed by a joint Bell Laboratories–Sandia Laboratories group, with a balloon-borne γ-ray telescope using a large volume lithium-drifted-germanium [Ge(Li)] crystal as primary detector (Leventhal, MacCallum & Watts, 1977). During the second flight of the instrument, in 1976 May 10–11 and with around 20 hours of useful astronomical data acquisition, two novae were targeted, V1500 Cyg and FH Ser, to search for their 1275 keV emission. The 2σ flux upper limit was 1.0×10^{-3} phot cm^{-2} s^{-1} for V1500 Cyg, and 1.1×10^{-3} phot cm^{-2} s^{-1} for FH Ser. These fluxes were smaller than, but close to, the very optimistic predictions of the epoch, not yet based on computed models: ejected masses of ^{22}Na around 2×10^{-6} M$_\odot$, coming from *assumed* ^{22}Na mass fractions of 2×10^{-2}, neither based on incipient models nor on Clayton and Hoyle (1974) predictions, and total ejected masses 10^{-4} M$_\odot$.

Some years later, in 1979 September, NASA's HEAO 3 satellite was launched; it performed a sky survey in the 50 keV–10 MeV range, getting a 3σ upper limit of 4.4×10^{-4} phot cm^{-2} s^{-1} rad^{-1} on diffuse emission at 1275 keV from the Galactic plane in the vicinity of the Galactic center, i.e. Galactic longitudes within $\pm 30°$ of the Galactic center (Mahoney *et al.*, 1982). This limit translated into an upper limit of ^{22}Na production of 6.5×10^{-8} M$_\odot$ for the particular assumptions made (all novae contributing by the same amount of ^{22}Na and with an ejected mass of 10^{-4} M$_\odot$ per nova). Therefore the upper limit to the average mass fraction was 6.5×10^{-4}.

Another attempt to detect the ^{22}Na line from novae came with the Solar Maximum Mission (SMM), a spacecraft with a γ-ray spectrometer (GRS) on board. Data accumulated from 1980 to 1987 were searched for the line at 1275 keV without success, but upper limits on the emission by individual neon novae (V1500 Cyg, V693 CrA, V1370 Aql, QU Vul) and on the whole Galactic center were placed (Leising *et al.*, 1988): $f(10^{-4}\,\text{phot cm}^{-2}\,\text{s}^{-1}) \leq 25$, 3.2, 2.5, 3.0, and 1.2 respectively (see Table 1 in Leising *et al.* (1988)). The corresponding limits on the ^{22}Na ejected mass were (in units of 10^{-6} M$_\odot$) 1.5, 8.6, 1.6, 0.7, and 3.1 respectively, once distances to these objects and to the Galactic center were adopted. The upper limit of 1.2×10^{-4} phot cm^{-2} s^{-1} on a steady 1275 keV flux from the Galactic center direction was more than a of factor 3 below the only previous measurement by HEAO 3, mentioned above (4.4×10^{-4} phot cm^{-2} s^{-1}, by Mahoney *et al.* (1982)).

The Oriented Scintillation Spectrometer (OSSE) on board the CGRO also searched for the 1275 keV line. For individual novae, the 2σ upper limits obtained were (Leising, private communication): 7.5×10^{-5} (V838 Her), 6.8×10^{-5} (V1974 Cyg) and 8.2×10^{-5} phot cm^{-2} s^{-1} (V382 Vel). The 3σ upper limit for the central radian of the Galaxy (Harris, 1996) was 2.25×10^{-4} phot cm^{-2} s^{-1} rad^{-1}, worse than the abovementioned limit obtained with SMM. However, this limit led to constraint on the Galactic distribution of novae, with the then current predictions of ^{22}Na production in novae by Politano *et al.* (1995).

But the most recent and deep observational work on the 1275 keV line from novae has been performed by Iyudin *et al.* (1995), with the COMPTEL instrument on board the CGRO. COMPTEL observed a number of recent novae during the period 1991–1993, five of which were of the neon type (i.e. those expected to emit the 1275 keV line): V838 Her, V4160 Sgr, V444 Sct, V351 Pup, and V1974 Cyg. None of these (or the other 'standard', i.e. CO) novae observed was detected. The average 2σ upper limit for any nova of the ONe type in the Galactic disk was around 3×10^{-5} phot cm^{-2} s^{-1}, which translated into an upper limit of the ejected ^{22}Na mass around 3.7×10^{-8} M$_\odot$, for the adopted distances to novae observed. This limit was constraining for the currently available models at the epoch (Starrfield *et al.*, 1992, 1993; Politano *et al.*, 1995), but not for the present models (José & Hernanz, 1998). There are two main reasons for the discrepancy between models of different groups (José & Hernanz (1998) versus Politano *et al.* (1995) and Starrfield *et al.* (1998)). First of all, older models were based on the explosion on ONeMg white dwarfs, with some mixing between the accreted H-rich matter and the underlying material, whereas later models adopt ONe white dwarfs as underlying cores, because more recent evolutionary calculations of stellar evolution predict lower neon (and magnesium) abundances (see Section 11.2.1), which make ^{22}Na synthesis much less favored. In addition, it should be kept in mind that to compute the ejected mass of a given isotope, both the yield (in mass fraction) and the total ejected mass are needed. It is a well-known problem in the nova community that theoretical models predict ejected masses which are much smaller than some of the observed ones (i.e. 10^{-3} M$_\odot$: see Chapters 4, 6, and 8). Therefore, comparison between model predictions of the ejected mass of any given isotope (i.e. ^{22}Na or ^{26}Al), which is the relevant quantity as far as γ-ray emission is concerned, should be made carefully. Ejecta composition (in mass or number fractions) can be similar but flux predictions be very different if (a) ejected masses (theoretical) differ from one model to another or (b) the observed ejected mass, if available, is adopted instead of the theoretical one. Option (b) is in fact not self-consistent but has been often adopted to compare observational upper limits and models (Starrfield *et al.*, 1992). This was partially the reason why predictions were more optimistic than models really were and why the non-detection of the 1275 keV line by COMPTEL from V1974 Cyg or V838 Her was considered a surprise.

Concerning the cumulative emission of novae, an analysis of COMPTEL data by Iyudin *et al.* (1999) gave a 2σ upper limit for 1275 keV emission in the bulge (defined as a circle of 10° radius around the Galactic center) of 3.0×10^{-5} phot cm^{-2} s^{-1}. The corresponding upper limit depends a lot on how the distribution and the rate of novae are handled: Iyudin *et al.* (1999) obtained 3.6×10^{-9} M$_\odot$ for a bulge nova (but adopting an unrealistically large rate of 40 yr^{-1}). If one accounts for the fact that only ONe novae can emit the 1275 keV line and that spatial distributions follow some laws (see for instance Jean *et al.* (2000) and the review paper by Shafter (2002)), the upper limit is much larger (by a factor $\sim$170, see Jean *et al.* (2001)). Thus, unfortunately we are still far from obtaining restrictions on nova models from γ-ray observations. It is interesting to mention another analysis of COMPTEL data by Iyudin *et al.* (2005), where a possible detection of the 1275 keV line from the Galactic bulge is claimed, but according to the authors, systematic uncertainties affect the flux estimate.

11.5.2 The 478 keV line from ^{7}Be

The first search for the 478 keV line from the Galactic center and from some particular novae was performed with the Gamma-Ray Spectrometer (GRS) on the SMM by Harris, Leising and Share (1991). Three individual novae, V1370 Aql, QU Vul, and V842 Cen, were observed, and the upper limits derived were 2.0×10^{-3}, 8.1×10^{-4} and 1.1×10^{-3} phot cm^{-2} s^{-1} respectively (or 1.2×10^{-3}, 7.5×10^{-4} and 9.6×10^{-4} phot cm^{-2} s^{-1}, as referenced in a later paper by Harris *et al.* (2001)). Adopting distances of 3.5, 3.0, and 1.1 kpc to the three novae, respectively, the upper limits of ^{7}Be ejected masses were 6.3×10^{-7}, 3.1×10^{-7}, and 5.2×10^{-8} M$_\odot$ respectively (Harris, Leising & Share, 1991). These fluxes and masses are well above the current predictions (see Table 11.2 and Figure 11.9).

More recent observations of the 478 keV line from novae come from the Transient Gamma-Ray Spectrometer (TGRS) on board the WIND satellite, which pointed continuously toward the southern ecliptic pole. Five novae were observable with TGRS during the period 1995–1997: BY Cir, V888 Cen, V4361 Sgr, CP Cru, and V1141 Sco. The 3σ upper limits were 6.8×10^{-5}, 6.3×10^{-5}, 1.1×10^{-4}, 8.8×10^{-5}, and 1.6×10^{-4} phot cm^{-2} s^{-1} respectively, implying 3σ upper limits of ^{7}Be ejected masses 3.0×10^{-8}, 6.4×10^{-8}, 2.2×10^{-7}, 3.9×10^{-8}, and 2.7×10^{-7} M$_\odot$ respectively, for adopted distances of 3.2, 4.8, 6.7, 3.2, and 6.1 kpc (Harris *et al.*, 2001). The flux limits from TGRS were a factor of 10 better (smaller) than those from SMM observations, but the upper limits on ^{7}Be ejected masses did not improve by the same factor, mainly because novae observed with TGRS were at larger distances than those observed with SMM. Upper limits on the integrated emission flux at 478 keV from the Galactic center with SMM and TGRS were 1.5×10^{-4} and 7.7×10^{-5} phot cm^{-2} s^{-1} respectively; again the improvement in the limit from SMM to TGRS is not very noticeable, mainly because the *equivalent aperture* of TGRS (occulted region $16° \times 90°$) is smaller than that of SMM ($\sim$130°).

Finally, CGRO/OSSE observations of nearby novae have provided the following 2σ upper limits to the 478 keV emission (M. D. Leising, private communication): 4.4×10^{-5} (V838 Her), 5.6×10^{-5} (V1974 Cyg) and 3.4×10^{-5} phot cm^{-2} s^{-1} (V382 Vel).

11.5.3 The 511 keV annihilation line

As mentioned above (Section 11.3.1), the emission resulting from $e^- - e^+$ annihilation is the most intense γ-ray outcome of classical novae, and it consists of a 511 keV line plus a continuum. However, these γ-rays are emitted well before the visual maximum of the nova, i.e. before the nova is discovered, and have a very short duration. Therefore, they cannot be detected through observations pointing to a particular nova already discovered. Wide field of view instruments monitoring the sky in the appropriate energy range, like the Burst and Transient Source Experiment (BATSE) on board CGRO or TGRS on board WIND, are therefore the best-suited instruments for the search of the 511 keV line and the continuum below it.

TGRS was very convenient to use in searching for the 511 keV line for several reasons (Harris *et al.*, 1999). Its very elliptical orbit avoided the larger γ-ray background levels that low Earth orbit satellites suffer, and its germanium detectors had enough spectral resolution to separate the cosmic 511 keV line from the nova line, provided that the latter is a bit blue-shifted (in fact this happens only at the beginning of the emission phase, when material is not yet completely transparent). TGRS's large field of view contained five new novae during the period 1995–1997 (see Section 11.5.2 above). For each nova, the line at 6 hours and at 12 hours was modeled according to the theoretical models of Gómez-Gomar *et al.* (1998). Then a comparison with background spectra during periods encompassing some days before the discovery date of each nova provided upper limits to the 511 keV line flux in 6 hours;

these were 2.2×10^{-3}, 2.0×10^{-3}, 2.8×10^{-3}, 2.3×10^{-3}, 2.9×10^{-3} phot cm^{-2} s^{-1}, for BY Cir, V888 Cen, V4361 Sgr, CP Cru, and V1141 Sco, respectively (see Section 11.5.2 above for the distances, but it should be borne in mind that these are largely uncertain, as in almost all novae). Harris *et al.* (1999) deduced that their method was sensitive enough to detect novae occurring out to about 2.5 kpc. This number has to be corrected, since it was based on the models current at that time (Gómez-Gomar *et al.*, 1998), which predicted more ^{18}F than now, because there has been a drastic change in a nuclear reaction rate affecting ^{18}F synthesis (see Section 11.2.1 for a detailed discussion). There was a reduction of ^{18}F synthesis in novae by a factor around 10 with respect to the yields adopted in Gómez-Gomar *et al.* (1998) and taken as model templates by Harris *et al.* (1999); so that TGRS would be sensitive enough to detect novae at around 0.8 kpc, of any type (CO and ONe); (but see the comment at the end of this section).

Another instrument that was well suited for the detection of the prompt γ-ray emission from novae was BATSE on board CGRO. Before the launch of CGRO in 1991, Fishman *et al.* (1991) had already made a prediction of the detectability of low-energy γ-rays from novae with the BATSE instrument, based on the models of γ-ray emission from Leising and Clayton (1987). BATSE had the advantage of covering the whole sky all the time, but on the other hand it was not very sensitive and it had poor energy resolution. Data analysis techniques that have been applied for BATSE observations of 511 keV transients of short duration were applied.

A posteriori analyses were made of the background data at the explosion epoch of all the classical novae discovered optically during the whole period of CGRO operation (1991–2000), searching for some signal (Hernanz *et al.*, 2000). Intervals of 12 hours were analyzed at 6 hour spacing, to make sure that the peak of the outburst was not split. Background data were taken 24 hours before the period of interest.

The predicted detectability distance of novae with BATSE was around 2 kpc, but this issue will be better known after the data analysis techniques are optimized for our particular type of study. Since novae distances are known only with great uncertainty, it was considered safe to include all novae that had exploded since CGRO launch which were at distances smaller than 3–4 kpc. The distances adopted were either those from Shafter (1997) or those computed from the data in the IAU circulars and the MMRD relationship (see Chapters 2 and 14). The adopted periods to analyze ranged from 15 to 20 days before maximum visual luminosity, since the outburst in γ-rays with $E \leq 511$ keV occurs some hours after peak temperature, which happens well before the maximum in visual luminosity.

We have evaluated the 3σ upper limits to the fluxes for the three most promising (in principle) novae of the whole sample, i.e. V1974 Cyg, V992 Sco, and V382 Vel. For the derivation of the 511 keV line fluxes, two methods were applied: (a) data in the whole (250–511) keV range with the assumed Comptonization from theoretical models were used to determine the flux in the line, between 510 and 520 keV; (b) the upper limits were derived from the data in the 510–520 keV range only. All upper limits obtained were compatible with theory, except for V992 Sco, which should either have $M < 1.5\,M_{\odot}$, $d > 0.8$ kpc or both. The 3σ sensitivity using the 511 keV data only is similar to that of Harris *et al.* (1999) with WIND/TGRS, but the sensitivity of Harris *et al.* (1999) requires a particular line blue shift, whereas ours is independent of it. The 3σ sensitivity using the (250–511) keV data with assumed Comptonization is a little more than a factor of 2 better than that of Harris *et al.* (1999) (see Table 11.3 and Hernanz *et al.* (2000)).

Table 11.3. *Upper limits (3σ, in photons $cm^{-2}\,s^{-1}$) to the fluxes for the novae V1974 Cyg, V992 Sco, and V382 Vel from BATSE observations, compared to model fluxes*

	$f(3\sigma$ limit)	f(model)
V1974 Cyg (model: 1.25 $M_\odot$ ONe nova at $d = 1.7$ kpc)		
(250–511) keV	5.2×10^{-3}	2.3×10^{-3}
511 keV line[a]	1.0×10^{-3}	4.8×10^{-4}
511 keV line[b]	2.4×10^{-3}	4.8×10^{-4}
V992 Sco (model: 1.15 $M_\odot$ CO nova at $d = 0.8$ kpc)		
(250–511) keV	3.6×10^{-3}	5.3×10^{-3}
511 keV line[a]	7.1×10^{-4}	1.0×10^{-3}
511 keV line[b]	2.3×10^{-3}	1.0×10^{-3}
V382 Vel (model: 1.25 $M_\odot$ ONe nova at $d = 2$ kpc)		
(250–511) keV	5.3×10^{-3}	1.7×10^{-3}
511 keV line[a]	1.0×10^{-3}	3.5×10^{-4}
511 keV line[b]	1.6×10^{-3}	3.5×10^{-4}

[a] Using (250–511) keV data with assumed Comptonization.
[b] Using 511 keV data only.

It is important to mention that again (as with TGRS in Harris *et al.* (1999)) detectability distances in Hernanz *et al.* (2000) were computed with the *old* γ-ray spectra models (Gómez-Gomar *et al.*, 1998) prior to new determinations of ^{18}F+p reaction rates (see Section 11.2.1), which reduced the ^{18}F yields by a factor of $\sim$10. So predicted fluxes should be reduced by a factor of $\sim$10 and detectability distances by a factor of $\sim\sqrt{10}$. Therefore, the inconsistency with V992 Sco disappears and upper limits from observations still do not constrain theoretical models. (In fact, with the newest rates mentioned in Section 11.2.1 above, fall 2005, the reduction factor should be $\sim$50 in flux, rather than $\sim$10, leading to detectability distances shorter by factors $\sim\sqrt{50}$ instead. But the ^{13}N flux peak is not altered, so the expected change will not be so drastic (Hernanz & José, in preparation).

An additional observational search for the annihilation radiation from novae has been performed with the Reuven Ramaty High Energy Solar Spectroscopic Imager (RHESSI) in 2006. A subset of all novae that had their outburst since the launch of RHESSI in 2002 February has been analyzed, including all objects with distances around 2 kpc or closer, and in particular the bright recurrent nova RS Oph, which erupted, for the first time since 1985, in 2006 February. The novae analyzed are: V4742 Sgr, V2573 Oph, V1187 Sco, V476 Sct, and RS Oph. Theoretical predictions of the annihilation radiation spectrum during the first peak, related to ^{13}N decay, show a blue-shifted 511 keV line plus a Comptonized continuum. The challenge is to pick out the blue-shifted nova spectrum from the background 511 keV line. No positive signal for any of the novae has been obtained (Matthews *et al.*, 2006). RHESSI is no longer very sensitive and will not be until its Ge detectors are annealed (end 2006 or beginning 2007). It is also planned to do a search of the lower energy continuum, where the flux is larger (see Section 11.3.1), but the corresponding sensitivity is not yet known.

Finally, it is worth mentioning that the Burst Alert Telescope (BAT) instrument on board the Swift satellite (Gehrels *et al.*, 2006), primarily designed to detect γ-ray bursts, is sensitive up to 190 keV, so it could detect a fraction of the nova continuum. The most promising candidate has been RS Oph, for which detection up to 50 keV (but after the 2006 eruption) has been reported (Bode *et al.*, 2006). This emission most probably has nothing to do with the annihilation emission (which should appear earlier), but rather with the hard X-ray tail of the X-ray bremsstrahlung emission, from the shock-heated ejecta.

11.5.4 The 1.809 MeV line from ^{26}Al

In the 1970s Ramaty and Lingenfelter (1977) suggested that ^{26}Al decay should give rise to a detectable γ-ray line in the interstellar medium, since its narrowness (less than 3 keV, related to the long lifetime of ^{26}Al) made it easily detectable. This theoretical prediction was nicely confirmed by the γ-ray spectrometer on board the HEAO 3 satellite, which detected for the first time emission of ^{26}Al from the Galactic center, as a narrow line at 1809 $\pm$ 0.41 keV (Mahoney *et al.*, 1984). The flux in the vicinity of the Galactic center was $(4.8 \pm 1.0) \times 10^{-4}$ phot cm^{-2} s^{-1} rad^{-1} and the width of the line $\leq$ 3 keV, consistent with the Doppler broadening expected from differential Galactic rotation. The detection of ^{26}Al was shortly afterwards confirmed with the spectrometer on board SMM (Share *et al.*, 1985). The 1.809 MeV γ-ray emission was interpreted as resulting from the decay of about 3 $M_\odot$ of ^{26}Al. At the epoch of these discoveries, and without spatial distribution information, novae were considered the most plausible origin of ^{26}Al, because supernovae models from that epoch did not produce enough ^{26}Al. However, it was already known that large uncertainties affected some of the nuclear reaction rates responsible for the synthesis of ^{26}Al, such as ^{25}Mg(p,γ)^{26}Al.

A breakthrough in this topic came with the CGRO/COMPTEL observations of the Galactic ^{26}Al, providing the first map of the 1.809 MeV line along the Galaxy (Diehl *et al.*, 1995). It was deduced that the emission originated in rather localized regions, not only concentrated in the inner Galactic disk; smooth intensity distributions as expected from a nova origin could not be reconciled with the observations. It was concluded that massive stars were probably the origin of most of the ^{26}Al detected by COMPTEL (Chen, Gehrels & Diehl, 1995). Extensive research work has been dedicated to disentangling the various possible origins of the Galactic ^{26}Al and its spatial distribution; some particular regions (e.g. Vela, Carina, Cygnus) with enhanced emission have been studied in detail, to understand which of the possible candidates (AGB stars, massive stars either in their WR phase or as supernovae) contributes most (see the review by Prantoz and Diehl (1996)). An important step forward has been the discovery that there is a strong correlation between the 1.809 MeV emission and the 53 GHz microwave free–free radiation, which traces ionized gas in the Galaxy (Knödlseder, 1999). A detailed analysis of that correlation indicated that most of the ^{26}Al nucleosynthesis should occur in the same massive stars responsible for the ionization of the interstellar medium.

An important clue to the origin of ^{26}Al would come from the detection of ^{60}Fe in the Galaxy, another long-lived radioactive isotope ($\tau \sim 2 \times 10^6$ yr) only produced in Type II supernovae and emitting two γ-ray lines at 1.173 and 1.332 MeV. The predicted flux of the ^{60}Fe lines from the inner Galaxy is around 16% of the ^{26}Al flux, provided that Type II supernovae are the only sources of Galactic ^{60}Fe (Timmes *et al.* (1995), based on calculations of SNII nucleosynthesis from Woosley and Weaver (1995)). Subsequently, the RHESSI satellite detected the two ^{60}Fe lines, at a level very close to the predicted one (Smith, 2004); this same instrument detected ^{26}Al at a level similar to the COMPTEL detection (Smith, 2003). Interestingly enough, studies of nucleosynthesis in Type II supernovae produce much higher ^{60}Fe/^{26}Al ratios (Rauscher *et al.*, 2002; Limongi & Chieffi, 2003). If these new calculations and RHESSI detection are accepted, the obvious conclusion is that ^{60}Fe and ^{26}Al are not produced in the same sources; since ^{60}Fe origin in Type II supernovae is broadly accepted, this leads to the conclusion that other sources of ^{26}Al (in addition to Type II supernovae) are needed (Prantzos, 2004). The favorite candidates are winds from Wolf–Rayet stars, but AGB and novae also remain on the list of possibilities. In fact the COMPTEL map did not completely discard any of them. Subsequently, INTEGRAL's SPI (see Section 11.6) has also detected ^{60}Fe (Harris *et al.*, 2005).

An additional property of the ^{26}Al line which deserves particular attention is its shape, since it gives important information about the environment where ^{26}Al is synthesized and propagates. As mentioned above, the observation with HEAO 3, which used high-resolution germanium detectors, provided a width around 3 keV (Mahoney *et al.*, 1984), with a statistical significance of 4.8σ. Before RHESSI and INTEGRAL, only one high-resolution measurement comparable to HEAO 3 was made, with the GRIS (Gamma-ray Imaging Spectrometer) instrument on board a balloon (Naya *et al.*, 1996); the line was found to be broad, with an instrinsic width of $5.4^{+1.4}_{-1.3}$ keV (significance 6.8σ). This Doppler broadening would correspond to velocities around 500 km s^{-1} or equivalent temperatures around 4.5×10^8 K, which posed a problem, since it is not easy to understand how such large temperatures or velocities can be maintained in the interstellar medium gas for a period as long as 10^6 years; various suggestions were made at the time, assuming for instance that ^{26}Al should be concentrated in grains which could maintain their velocities for a longer time than gaseous material. However, that line width has not been confirmed with the more recent observations of the ^{26}Al line with RHESSI (Smith, 2003) and INTEGRAL/SPI (Diehl *et al.*, 2003); both observations reject with high confidence the reported width in Naya *et al.* (1996). RHESSI observations give $2.03^{+0.78}_{-1.21}$ keV and INTEGRAL confirms that the width is smaller than around 3 keV, the exact value depending on the model used for the data analysis. Another interesting result obtained with the INTEGRAL/SPI, thanks to its excellent spectral resolution (thus allowing discrimination between blue-shifted and red-shifted line centers), is that the ^{26}Al source regions corotate with the Galaxy, supporting its Galaxy-wide origin (Diehl *et al.*, 2006).

11.6 Observations with INTEGRAL

The launch of the ESA satellite INTEGRAL on 2002 October 17 has opened new perspectives for the detection of γ-rays from explosive events, with its two major instruments, the spectrometer SPI and the imager IBIS, and the two monitors in the X-ray (Joint European Monitor, JEM X) and optical (Optical Monitoring Camera, OMC) ranges; the four instruments are co-aligned. The orbit of INTEGRAL is very eccentric, in order to minimize the time spent close to the Earth, where the background is most intense. SPI is made of 19 germanium detectors sensitive in the range from 18 keV to 8 MeV and its field of view is around 15°. SPI is the best-suited INTEGRAL instrument for the detection of the γ-ray lines expected from nuclear radioactivity, such as those from novae. Its 3σ sensitivity at 1 MeV, for 10^6 s observation time and narrow lines, is around 2.4×10^{-5} phot cm^{-2} s^{-1}, with 2 keV energy resolution. However, this sensitivity degrades for broad lines, because of the high energy

Table 11.4. *INTEGRAL/SPI 3σ detectability of ^{7}Be (478 keV) and ^{22}Na (1275 keV) lines from classical novae*[a]

Line [E (ΔE) keV]	t_{obs}(ks)	F_{min} (ph cm^{-2} s^{-1})	d(kpc)
478 (8)	2.4×10^3	6.67×10^{-5}	0.17
478 (8)	4.8×10^3	4.71×10^{-5}	0.21
1275 (20)	2.4×10^3	6.07×10^{-5}	0.57
1275 (20)	4.8×10^3	4.29×10^{-5}	0.68

[a] F_{min} are the fluxes that would give a 3σ detection of the lines, with the quoted observation times, which have been computed with the Observation Time Estimator for INTEGRAL OTE (version of 2006 April, AO4 call for proposals). The detectability distances have been computed adopting 2×10^{-6} and 2×10^{-5} ph cm^{-2} s, as model fluxes for the 478 keV and 1275 keV lines, at 1 kpc, for a typical CO and ONe nova, respectively (see Gómez-Gomar *et al.*, 1998; Hernanz *et al.*, 1999).

resolution of the Ge detectors; also, the expected sensitivity some years prior to the launch was about 6×10^{-6} phot cm^{-2} s^{-1}, i.e. four times better. Detecting γ-rays from novae with INTEGRAL will be rather hard, because the detectability distances are very small and, therefore, few novae are expected. This is a result of both the small fluxes expected and the reduced (with respect to pre-launch estimates) inflight measured sensitivities at the relevant energies. We present in Table 11.4 prospects for detectability of novae with SPI, mainly suited for the 478 keV and 1275 keV lines. Very small distances are needed to get a positive detection; nonetheless, novae are important objectives for INTEGRAL and therefore are included in both the Core Programme and in Open Time proposals, for a total observing time between 4 and 5 Ms.

INTEGRAL has in fact already obtained important results concerning the 511 keV line in the Galaxy, with an emission quite concentrated in the bulge and almost absent in the disk (with the data available to date; see Jean *et al.* (2003), Teegarden *et al.* (2005) and references therein for complete information about current and past observations of the annihilation emission, line plus continuum). However, the positrons responsible for that emission are thought to come mainly from supernovae (those from novae contribute less, by some orders of magnitude), but the shape of the emission does not favor supernovae and leaves some room for other origins (Knödlseder *et al.*, 2005).

11.7 Discussion

Gamma-ray astronomy in the MeV range, which is the most interesting for observations of nucleosynthesis products of novae and supernovae, has always faced important challenges from the instrumental point of view. In addition to the general difficulties of γ-ray detection (few signal photons have to be extracted from a very intense background), the MeV range is especially difficult, since it corresponds to the energy range where Compton scattering (with small cross-sections) is dominant; Compton scattering is harder to handle than photoelectric absorption or pair formation, at lower and higher energies respectively. Actually, the present generation of γ-ray instruments in the MeV range makes use of geometrical optics (shadowcasting in modulating aperture systems), or quantum optics (Compton scattering). This kind of instrument is faced with the problem that *bigger does not necessarily mean better*. The reason for this apparent contradiction is that the collection area in traditional γ-ray telescopes should be roughly equal to the detection area. Therefore, the larger the collection area, the larger the detection volume and thus the higher the instrumental background. This means that significant improvements in sensitivity need huge instruments, too expensive for space missions.

An innovative concept for detecting γ-rays in the MeV range, which overcomes this problem and allows for unprecedented sensitivities, consists of focusing the γ-rays from a large collection area onto a small detector. The basic idea under the *γ-ray lens* concept is that γ-rays can interact coherently inside a crystal lattice provided that angles of incidence are small enough. In a crystal diffraction lens, crystals are arranged in concentric rings such that they will diffract the incident radiation of a particular energy on to a common focal spot, where the detector is placed (von Ballmoos *et al.*, 2004). Laue diffraction lenses have demonstrated their potential in laboratory measurements (see von Ballmoos *et al.* (2004) and references therein). The CLAIRE project, developed at CESR in Toulouse (France), was conceived to prove the principle of a Laue diffraction lens for nuclear astrophysics. Its natural continuation is a project for a space mission named MAX (for Max von Laue). The main objective of MAX will be the detection of the 847 keV line from ^{56}Co expected in supernovae, with a dedicated set of Cu rings, encompassing the energy range $\sim$800–940 keV, but an additional set of Ge rings will collect photons in the energy range from $\sim$440 to 540 keV, thus including the ^{7}Be line at 478 keV and the 511 keV line. The latter will not be easily detectable because of the limited field of view inherent to the γ-ray lens concept. In future designs of γ-ray lens, additional crystal rings suited for the detection of higher-energy lines (such as the 1275 keV line of ^{22}Na) will be possible[2].

We would like to comment about a future mission corresponding to the *Beyond Einstein Programme* from NASA: the Energetic X-ray Imaging Survey Telescope (EXIST), which would conduct the first high-sensitivity all-sky imaging survey in the energy range 5–600 keV (Grindlay *et al.*, 2003). With such a mission novae could be detected up to several kiloparsecs, depending on the model, thanks to their annihilation emission (both the 511 keV line and the continuum), provided that further changes in the nuclear reaction rates do not reduce once more the final amount of ^{18}F (or ^{13}N) predicted to be synthesized in novae (see Hernanz, Gómez-Gomar and José (2002), and Hernanz and José (2004), for details).

In spite of the problems it faces, the idea of an Advanced Compton Telescope (ACT) is still being pursued; NASA has funded its feasibility study under its New Visions Programme[3]. Various concepts for the detector are under deep study: Ge, Si, liquid xenon and others (see *New Astronomy Reviews* **48** (2004), for a set of papers about future intrumental developments in the MeV energy range). If a sensitivity of 5×10^{-7} phot cm^{-2} s^{-1} for broad lines such as those in novae, i.e. $\Delta E/E \sim 2\%$ (meaning a sensitivity of $\sim 2 \times 10^{-7}$ phot cm^{-2} s^{-1} for narrow lines) with a large field of view instrument is really attainable, then novae would be detectable through their ^{22}Na line at 1275 keV up to at least 6 kpc (encompassing about 60% of all ONe novae). In addition, all novae would be detectable through their annihilation line

[2] See http://gri.rm.iasf.cnr.it/ for a new concept of γ-ray mission (Gamma Ray Imager-GRI), based partially on the γ-ray lens.

[3] The final report of the Concept Study of an ACT (*'Witness to the fires of Creation'*), presented to NASA at the end of 2005, can be found at http://ssl.berkeley.edu/act/.

emission, leading to a direct knowledge of the nova spatial distribution not possible in other energy ranges because of interstellar extinction.

Thus, if future technology developments are able to build instruments of such sensitivity in the MeV range, novae would be routinely detected in γ-rays and important information would be obtained about the dynamics of the explosion and about the Galactic nova distribution, the latter if the annihilation emission could be detected through a-posteriori analyses of the data collected by large field of view instruments. In the meantime, improvements in the models are needed, especially focused on solving the problem of the initial mixing required to power the explosion and to explain the observed abundances, as well as on reproducing the large ejected masses in some well-observed novae; all these aspects have direct consequences for the predicted γ-ray emission from novae.

Acknowledgements

I sincerely thank my colleague Jordi José for a long-lasting and very fruitful collaboration on nova studies and for a careful reading of this manuscript.

Funds from the Spanish MEC through the project AYA2004-06290-C02-01 are acknowledged.

References

Alibés, A., Labay, J., & Canal, R., 2002, *Ap. J.*, **571**, 326.
Ambwani, K., & Sutherland, R., 1988, *Ap. J.*, **325**, 820.
Arnett, D. W., & Truran, J. W., 1969, *Ap. J.*, **157**, 339.
Arnould, M., & Norgaard, H., 1975, *A&A*, **42**, 55.
Audouze, J., & Reeves, H., 1981, in *Essays in Nuclear Astrophysics*, ed. C. A. Barnes, D. D. Clayton, & D. N. Schramm. Cambridge: Cambridge University Press, p. 355.
Banerjee, D. P. K., Varricatt, W. P., & Ashok, N. M., 2004, *Ap. J.*, **615**, L53.
Bishop, S., Azuma, R. E., Buchmann, L. *et al.*, 2003, *Phys. Rev. Lett.*, **90**, 162501-1.
Bode, M. F., O'Brien, T. J., Osborne, J. P. *et al.*, 2006, *Ap. J.*, **652**, 629.
Boffin, H. M. J., Paulus, G., Arnould, M., & Mowlavi, N., 1993, *A&A*, **279**, 173.
Bussard, R. W., Ramaty, R., & Drachman, R. J., 1979, *Ap. J.*, **228**, 928.
Cameron, A. G. W., 1955, *Ap. J.*, **121**, 144.
Chafa, A., Tatischeff, V., Aguer, P. *et al.*, 2005, *Phys. Rev. Lett.*, **95**, 031101-1.
Chan, K. W., & Lingenfelter, R. E., 1993, *Ap. J.*, **405**, 614.
Chen, W., Gehrels, N., & Diehl, R., 1995, *Ap. J.*, **440**, L57.
Clayton, D. D., 1975, *Nature*, **257**, 36.
Clayton, D. D., 1981, *Ap. J.*, **244**, L97.
Clayton, D. D., & Hoyle, F., 1974, *Ap. J.*, **187**, L101.
Clayton, D. D., & Leising, M. D., 1987, *Phys. Rep.*, **144**, 1.
Coc, A., Mochkovitch, R., Oberto, Y., Thibaud, J.-P., & Vangioni-Flam, E., 1995, *A&A*, **299**, 479.
Coc, A., Hernanz, M., José, J., & Thibaud, J.-P., 2000, *A&A*, **357**, 561.
D'Auria, J. M., Azuma, R. E., Bishop, S. *et al.*, 2004, *Phys. Rev. C*, **69**, 065803-1.
Davids, B., Beijers, J. P., van den Berg, A. M. *et al.*, 2003, *Phys. Rev. C*, **68**, 055805-1.
Della Valle, M., & Livio, M., 1994, *A&A*, **286**, 786.
Della Valle, M., Pasquini, L., Daou, D., & Williams, R. E., 2002, *A&A*, **390**, 155.
De Séréville, N., Coc, A., Angulo, C. *et al.*, 2003, *Phys. Rev. C*, **67**, 052801-1.
Diehl, R., Dupraz, C., Bennett, K. *et al.*, 1995, *A&A*, **298**, 445.
Diehl, R., Knödlseder, J., Lichti, G. G. *et al.*, 2003, *A&A*, **411**, L451.
Diehl, R., Halloin, H., Kretschmer, K. *et al.*, 2006, *Nature*, **439**, 45.
Domínguez, I., Tornambé, A., & Isern, J., 1993, *Ap. J.*, **419**, 268.
Eberhardt, P., 1974, *Earth & Planet. Sci. Lett.*, **24**, 182.
Eberhardt, P., 1979, *Ap. J.*, **234**, L169.
Ferland, G. J., Lambert, D. L., Netzer, H., Hall, D. N. B., & Ridgway, S. T., 1979, *Ap. J.*, **227**, 489.
Fishman, G. J., Wilson, R. B., Meegan, C. A. *et al.*, 1991, in *Gamma-Ray Line Astrophysics*, ed. P. Durouchoux, & N. Prantzos. New York: American Institute of Physics, p. 190.
Fox, C., Iliadis, C., Champagne, A. E. *et al.*, 2004, *Phys. Rev. Lett.*, **93**, 081102-1.
Friedjung, M., 1979, *A&A*, **77**, 357.
García-Berro, E., Ritossa, C., & Iben, I., 1997, *Ap. J.*, **485**, 765.
Gehrels, N., *et al.*, 2006, *Ap. J.*, **611**, 1005.
Gil-Pons, P., García-Berro, E., José, J., Hernanz, M., & Truran, J. W., 2003, *A&A*, **407**, 1021.
Gómez-Gomar, J., Hernanz, M., José, J., & Isern, J., 1998, *MNRAS*, **296**, 913.
Grindlay, J. E., Craig, W. W., Gehrels, N. A., Harrison, F. A., & Hong, J., 2003, *Proc. SPIE*, **4851**, 331.

Harris, M. J., 1996, *A&AS*, **120**, 343.
Harris, M. J., Leising, M. D., & Share, G. H., 1991, *Ap. J.*, **375**, 216.
Harris, M. J., Naya, J. E., Teegarden, B. J. *et al.*, 1999, *Ap. J.*, **522**, 424.
Harris, M. J., Teegarden, B. J., Weidenspointner, G. *et al.*, 2001, *Ap. J.*, **563**, 950.
Harris, M. J., Knödlseder, J., Jean, P. *et al.*, 2005, *A&A*, **433**, L49.
Harrison, T. E., Osborne, H. L., & Howell, S. B., 2004, *AJ*, **127**, 3493.
Hernanz, M., 2002, in *Classical Nova Explosions*, ed. M. Hernanz, & J. José. New York: American Institute of Physics, p. 399.
Hernanz, M., & José, J., 2004, *New Astron. Rev.*, **48**, 35.
Hernanz, M., & José, J., 2006, *New Astron. Rev.*, **50**, 504.
Hernanz, M., Gómez-Gomar, J., & José, J., 2002, *New Astron. Rev.*, **46**, 559.
Hernanz, M., José, J., Coc, A., & Isern, J., 1996, *Ap. J.* **465**, L27.
Hernanz, M., José, J., Coc, A., Gómez-Gomar, J., & Isern, J., 1999, *Ap. J.*, **526**, L97.
Hernanz, M., Smith, D. M., Fishman, J. *et al.*, 2000, in *Fifth COMPTON Symposium*, ed. M. L. McConnell, & J. M. Ryan. New York: American Institute of Physics, p. 82.
Higdon, J. C., & Fowler, W. A., 1987, *Ap. J.*, **317**, 710.
Hillebrandt, W., & Thielemann, F.-K., 1982, *Ap. J.*, **255**, 617.
Iyudin, A. F., Benett, K., Bloemen, H. *et al.*, 1995, *A&A*, **300**, 422.
Iyudin, A. F., Benett, K., Bloemen, H. *et al.*, 1999, in *Proceedings of the 3rd INTEGRAL Workshop*, *Astrophys. Lett. Commun.*, **38**, 371.
Iyudin, A. F., Benett, K., Lichti, G. G., Ryan, J., & Schönfelder, V., 2005, *A&A*, **443**, 477.
Jean, P., Hernanz, M., Gómez-Gomar, J., & José, J., 2000, *MNRAS*, **319**, 350.
Jean, P., Knödlseder, J., von Ballmoos, P. *et al.*, 2001, in *Proceedings of the 4th INTEGRAL Workshop*, ed. A. Gimenez, V. Reglero, & C. Winkler. ESA-SP-459, p. 73.
Jean, P., Knödlseder, J., Lonjou, V. *et al.*, 2003, *A&A*, **407**, L55.
Jenkins, D. G., Lister, C. J., Janssens, R. V. *et al.*, 2004, *Phys. Rev. Lett.*, **92**, 031101-1.
José, J., & Hernanz, M., 1998, *Ap. J.*, **494**, 680.
José, J., Hernanz, M., & Coc, A., 1997, *Ap. J.*, **479**, L55.
José, J., Coc, A., & Hernanz, M., 1999, *Ap. J.*, **520**, 347.
José, J., Hernanz, M., García-Berro, E., & Gil-Pons, P., 2003, *Ap. J.*, **597**, L41.
Knödlseder, J., 1999, *Ap. J.*, **510**, 915.
Knödlseder, J., Jean, P., Lonjou, V. *et al.*, 2005, *A&A*, **441**, 513.
Kolb, U., & Politano, M., 1997, *A&A*, **319**, 909.
Lee, T., Papanastassiou, D. A., & Wasserburg, G. J., 1977, *Ap. J.*, **211**, L107.
Leising, M. D., 1991, in *Gamma-Ray Line Astrophysics*, ed. P. Durouchoux, & N. Prantzos. New York: American Institute of Physics, p. 173.
Leising, M. D., 1993, *A&AS*, **97**, 299.
Leising, M. D., 1997, in *Proceedings of the 4th COMPTON Symposium*, ed. B. L. Dingus, M. H. Salamon, & D. B. Kieda. New York: American Institute of Physics, p. 163.
Leising, M. D., & Clayton, D. D., 1985, *Ap. J.*, **294**, 591.
Leising, M. D., & Clayton, D. D., 1987, *Ap. J.*, **323**, 159.
Leising, M. D., Share, G. H., Chupp, E. L., & Kanbach, G., 1988, *Ap. J.*, **328**, 755.
Leventhal, M., MacCallum, C., & Watts, A., 1977, *Ap. J.*, **216**, 491.
Lewis, R. S., Alaerts, L., Matsuda, J.-I., & Anders, E., 1979, *Ap. J.*, **234**, L165.
Limongi, M., & Chieffi, A., 2003, *Ap. J.*, **592**, 404.
Livio, M., & Truran, J. W., 1994, *Ap. J.*, **425**, 797.
Mahoney, W. A., Ling, J. C., Jacobson, A. S., & Lingenfelter, R. E., 1982, *Ap. J.*, **262**, 742.
Mahoney, W. A., Ling, J. C., Wheaton, W. A., & Jacobson, A. S., 1984, *Ap. J.*, **286**, 578.
Matthews, T. G., Smith, D. M., & Hernanz, M., 2006, *Abstracts for the 9th Meeting of the AAS High Energy Astrophysics Division (HEAD)*, October 4–7, 2006, San Francisco, California.
Naya, J. E., Barthelmy, S. D., Bartlett, L. M. *et al.*, 1996, *Nature*, **384**, 44.
Nofar, I., Shaviv, G., & Starrfield, S., 1991, *Ap. J.*, **369**, 440.
Ore, A., & Powell, J. L., 1949, *Phys. Rev.*, **75**, 1696.
Pozdniakov, L. A., Sobolev, I. M., & Sunyaev, R. A., 1983, *Sov. Sci. Rev. E: Astrophys. Space Phys. Rev.*, **2**, 189.
Politano, M., Starrfield, S., Truran, J. W., Weiss, A., & Sparks, W. M., 1995, *Ap. J.*, **448**, 807.
Prantzos, N., 2004, *A&A*, **420**, 1033.
Prantzos, N., & Diehl, R., 1996, *Phys. Rep.*, **267**, 1.

Prialnik, D., & Shara, M. M., 1995, *AJ*, **109**, 1735.
Prialnik, D., & Kovetz, A., 1997, *Ap. J.*, **477**, 356.
Prialnik, D., Shara, M. M., & Shaviv, G., 1978, *A&A*, **62**, 339.
Prialnik, D., Shara, M. M., & Shaviv, G., 1979, *A&A*, **72**, 192.
Ramaty, R., & Lingenfelter, R. E., 1977, *Ap. J.*, **213**, L5.
Rauscher, T., Heger, A., Hoffman, R. D., & Woosley, S. E., 2002, *Ap. J.*, **576**, 323.
Ritossa, C., García-Berro, E., & Iben, I., 1996, *Ap. J.*, **460**, 489.
Romano, D., Matteucci, F., Molaro, P., & Bonifacio, P., 1999, *A&A*, **352**, 117.
Rowland, C., Iliadis, C., Champagne, A. E. *et al.*, 2004, *Ap. J.*, **615**, L37.
Rudy, R. J., Dimpfl, W. L., Lynch, D. K. *et al.*, 2003, *Ap. J.*, **596**, 1229.
Shafter, A. W., 1997, *Ap. J.*, **487**, 226.
Shafter, A. W., 2002, in *Classical Nova Explosions*, ed. M. Hernanz, & J. José. New York: American Institute of Physics, p. 462.
Shara, M. M., 1994, *AJ*, **107**, 1546.
Shara, M. M., & Prialnik, D., 1994, *AJ*, **107**, 1542.
Share, G. H., Kinzer, R. L., Kurfess, J. D. *et al.*, 1985, *Ap. J.*, **292**, L61.
Shore, S. N., Schwarz, G., Bond, H. E. *et al.*, 2003, *AJ*, **125**, 1507.
Smith, D. M., 2003, *Ap. J.*, **589**, L55.
Smith, D. M., 2004, *New Astron. Rev.*, **48**, 87.
Starrfield, S., Truran, J. W., Sparks, W. M., & Kutter, G. S., 1972, *Ap. J.*, **176**, 169.
Starrfield, S., Truran, J. W., Sparks, W. M., & Arnould, M., 1978a, *Ap. J.*, **222**, 600.
Starrfield, S., Truran, J. W., & Sparks, W. M., 1978b, *Ap. J.*, **226**, 186.
Starrfield, S., Shore, S. N., Sparks, W. M. *et al.*, 1992, *Ap. J.*, **391**, L71.
Starrfield, S., Truran, J. W., Politano, M. *et al.*, 1993, *Phys. Rep.*, **227**, 223.
Starrfield, S., Truran, J. W., Wiescher, M., & Sparks, W. M., 1998, *MNRAS*, **296**, 502.
Teegarden, B. J., Watanabe, K., Jean, P. *et al.*, 2005, *Ap. J.*, **621**, 296.
Timmes, F. X., Woosley, S. E., Hartmann, D. H. *et al.*, 1995, *Ap. J.*, **449**, 204.
Urey, H. C., 1955, *PNAS* **41**, 127.
von Ballmoos, P., Halloin, H., Evrard, J. *et al.*, 2004, *New Astron. Rev.*, **48**, 243.
Ward, R. A., & Fowler, W. A., 1980, *Ap. J.*, **238**, 266.
Weiss, A., & Truran, J. W., 1990, *A&A*, **238**, 178.
Wiescher, M., Görres, J., Thielemann, F.-K., & Ritter, H., 1986, *A&A*, **160**, 56.
Woosley, S. E., & Weaver, T. A., 1995, *Ap. JS*, **101**, 181.

12

Resolved nebular remnants

T. J. O'Brien and M. F. Bode

12.1 Introduction

We may assume that all classical novae eject significant amounts of material ($\sim 10^{-4}\,M_{\odot}$) at relatively high velocities ($\sim 1000\ \mathrm{km\,s^{-1}}$) and hence produce nebulae that ultimately may be spatially resolved (hereafter 'nebular remnants'). The observation and modelling of such nebular remnants are important to our understanding of the classical nova phenomenon from several points of view. For example, combining imaging and spatially resolved spectroscopy allows us to apply the expansion parallax method of distance determination with greater certainty than any other technique (of course without knowledge of the three-dimensional shape and inclination of a remnant this method is still prone to error). On accurate distances hang most other significant physical parameters, including energetics and ejected mass. In addition, direct imaging of the resolved remnants can potentially clarify the role of clumping and chemistry in rapid grain formation earlier in the outburst (see Chapters 6 and 13). Remnant morphology (and potentially the distribution of abundances) can also give vital clues to the orientation and other parameters of the central binary and the progress of the TNR on the white dwarf surface. Finally, a fuller understanding of the nebular remnants of novae has implications for models of the shaping of planetary nebulae and (in at least one case) physical processes in supernova remnants.

At first glance, it was perhaps surprising that at the time of publication of the first edition of this book in 1989, of around 200 known classical novae, only just over 10% had confirmed nebular remnants in the optical (Wade, 1990), 1% in the radio (Seaquist, 1989) and none at any other wavelength. The situation has changed quite distinctly since then. Thanks to further systematic surveys and improvements in technology (e.g. the ubiquity of CCD detectors and the opportunity to observe remnants with HST from early phases of development with low inherent background) around twice as many remnants are now known in the optical, plus several in the infrared. At least one (admittedly unusual) object (GK Per) has been successfully imaged in X-rays. The picture in the radio has also changed markedly.

In this chapter, we concentrate on describing the physical characteristics and evolution of the nebular remnants of classical novae across the electromagnetic spectrum. We also refer as appropriate to the relationship between the central binary and the ejected nebula, particularly in terms of remnant shaping. Evidence for remnant structure in the spectra of unresolved

Classical Novae, 2nd edition, ed. Michael Bode and Aneurin Evans. Published by Cambridge University Press.

novae is briefly described before moving on to discuss resolved nebular remnants, first in the optical and infrared, and then in the radio. Over the past two decades, great strides have also been made in understanding the relationship of remnant shape to the evolution of the outburst and the properties of the central binary. The results of various models are presented. Finally, we briefly discuss the idiosyncratic remnant of GK Per – an object that may provide important clues to the evolution of nova binaries – before concluding with a discussion of outstanding problems and prospects for future work.

12.2 Optical and near-infrared imagery

12.2.1 Optical

Intermediate- to high-resolution spectroscopy of the spatially unresolved remnants of many novae has long been known to show evidence of organized structure (see for example Hutchings (1972) and Chapter 9). This has allowed simple models of remnant geometry to be formulated and the existence of structures such as equatorial and tropical rings and polar caps/blobs to be inferred. However, optical depth effects (for the permitted lines) and variations in ionization/excitation mean such analysis should be treated with caution (again, see Chapter 9). It is only when we spatially resolve the shell that we can (with care) become more confident in our conclusions. A case where structure inferred from early spectroscopy can be compared with imagery of the resolved shell is HR Del, and this is discussed more fully below.

Observations of resolved remnants in the optical are still the most fruitful in terms of determining physical characteristics and gaining insight into shaping mechanisms. Taking a typical nova expansion velocity and distance (Figure 12.1), we would expect to resolve the

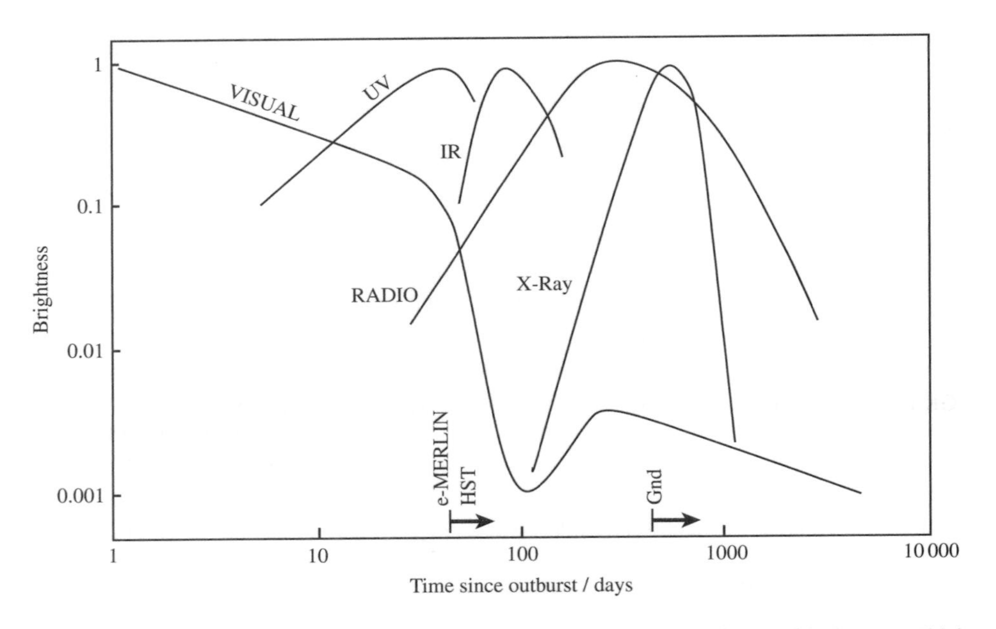

Fig. 12.1. Schematic multi-frequency development of a nova outburst with times at which a remnant with expansion velocity of 1000 km s^{-1} and distance 1 kpc becomes spatially resolved in the radio (e-MERLIN) or optically from space (HST), and on a conventional ground-based optical telescope (Gnd).

Table 12.1. *Classical novae with resolved nebular remnants (ordered by constellation)*

The list includes: the year of outburst; the waveband at which a remnant has been resolved (O – optical, I – infrared, R – radio, X – X-ray); whether an optical image is shown in Figure 12.2; the t_3 time in days; and references to the discovery of the resolved remnant.

Name	Year	Band	Image	t_3/d	Reference
OS And	1986	O	Y	?	This chapter
V500 Aql	1943	O		42	Cohen (1985)
V603 Aql	1918	O	Y	8	Duerbeck (1987b)
V1229 Aql	1970	O		37	Cohen (1985)
V1425 Aql	1995	O	Y	~30	Ringwald *et al.* (1998)
V1494 Aql	1999	R		16	Eyres *et al.* (2005)
T Aur	1891	O	Y	100	Duerbeck (1987b)
QZ Aur	1964	O		23-30	Esenoglu (2002)
V705 Cas	1993	IR		61	Diaz *et al.* (2001), Eyres *et al.* (2000)
V723 Cas	1995	R		173	Heywood *et al.* (2005)
V842 Cen	1986	O	Y	48	Gill & O'Brien (1998)
RR Cha	1953	O	Y	60	Gill & O'Brien (1998)
CP Cru	1996	O	Y	10	Ringwald *et al.* (1998)
V450 Cyg	1942	O	Y	108	Slavin *et al.* (1995)
V476 Cyg	1920	O	Y	16.5	Duerbeck (1987b)
V1500 Cyg	1975	O	Y	3.6	Duerbeck (1987b)
V1819 Cyg	1986	O	Y	89	Downes & Duerbeck (2000)
V1974 Cyg	1992	OIR		42	Paresce *et al.* (1995), Krautter *et al.* (2002), Eyres *et al.* (1996)
HR Del	1967	O	Y	230	Duerbeck (1987b)
DN Gem	1912	O		37	Cohen (1988)
DQ Her	1934	OX	Y	94	Duerbeck (1987b), Mukai *et al.* (2003)
V446 Her	1960	O		16	Cohen (1985)
V533 Her	1963	O	Y	44	Cohen & Rosenthal (1983)
CP Lac	1936	O		10	Adams (1944)
DK Lac	1950	O	Y	32	Cohen (1985)
HY Lup	1993	O		>25	Downes & Duerbeck (2000)
BT Mon	1939	O	Y	?	Duerbeck (1987a)
GK Per	1901	ORX	Y	13	Duerbeck (1987b), Reynolds & Chevalier (1984), Balman & Ögelman (1999)
RR Pic	1925	OX	Y	150	Duerbeck (1987b), Balman (2002)
CP Pup	1942	O	Y	8	Duerbeck (1987b)
DY Pup	1902	O	Y	160	Gill & O'Brien (1998)
HS Pup	1963	O	Y	65	Gill & O'Brien (1998)
V351 Pup	1991	O	Y	26	Ringwald *et al.* (1998)
V3888 Sgr	1974	O		10	Downes & Duerbeck (2000)
V4077 Sgr	1982	O	Y	100	This chapter

Table 12.1. *continued*

Name	Year	Band	Image	t_3/d	Reference
V4121 Sgr	1983	O	Y	?	This chapter
V960 Sco	1985	O	Y	?	This chapter
CT Ser	1948	O		>100?	Downes & Duerbeck (2000)
FH Ser	1970	O	Y	62	Cohen & Rosenthal (1983)
XX Tau	1927	O		42	Cohen (1985)
RW UMi	1956	O	Y	140	Cohen (1985)
LV Vul	1968	O	Y	37	Cohen (1985)
NQ Vul	1976	O	Y	65	Slavin *et al.* (1995)
PW Vul	1984	O	Y	126	Ringwald & Naylor (1996)
QU Vul	1984	OIR	Y	49	Della Valle *et al.* (1997), Krautter *et al.* (2002), Taylor *et al.* (1988)
QV Vul	1987	OI	Y	53	Downes & Duerbeck (2000), Krautter *et al.* (2002)

The unusual objects V605 Aql and CK Vul are not included as they may not be novae (Clayton & de Marco, 1997; Evans *et al.*, 2002).
Where an image is not included in Figure 12.2 it is usually because it was discovered by comparison with a stellar point-spread function or, occasionally, when a reasonable-quality image could not be found in the literature or data archives.

remnant after ~1.5 months with HST and ~1.5 years from the ground (without the aid of adaptive optics).

The first report of a resolved remnant was that for GK Per (in 1916). As noted above, we will return to this rather unusual and important object later in this chapter. Following this, the evolution of the nebular remnants of several other novae, including DQ Her, RR Pic, V603 Aql and T Aur, were also followed in some detail, particularly in the work of Baade and McLaughlin, and as discussed by Mustel and Boyarchuk (1970), for example.

In the early 1980s Judith Cohen, Hilmar Duerbeck and collaborators began the first systematic surveys of novae in an attempt to find, or to recover, their nebular remnants (Cohen & Rosenthal, 1983; Cohen, 1985; Duerbeck, 1987a). A principal aim of the work by Cohen *et al.* was to improve distance determinations via the expansion parallax method and hence to explore the use of the Maximum Magnitude – Rate of Decline (MMRD) relationship which could make classical novae important rungs on the distance ladder (see also Chapters 2 and 14). For most of the objects surveyed, the presence of a nebular remnant was confirmed only by careful fitting and subtraction of point spread functions.

The most extensive recent ground-based imaging surveys have been those of Slavin, O'Brien & Dunlop (1995), conducted from the northern hemisphere, and Gill and O'Brien (1998), for southern objects. These used a combination of modern CCD detectors and large (predominantly 4m-class) telescopes. These two surveys resulted in the discovery of four previously unknown nebular remnants and the detection of previously unobserved features in a total of sixteen others. More recently Downes and Duerbeck (2000) reported on a survey of thirty novae which resulted in the discovery of two new nova shells from ground-based observations and a further three from space-based data (see below). Table 12.1 gives details of all novae with resolved remnants at various wavelengths and Figure 12.2 shows all readily available optical images.

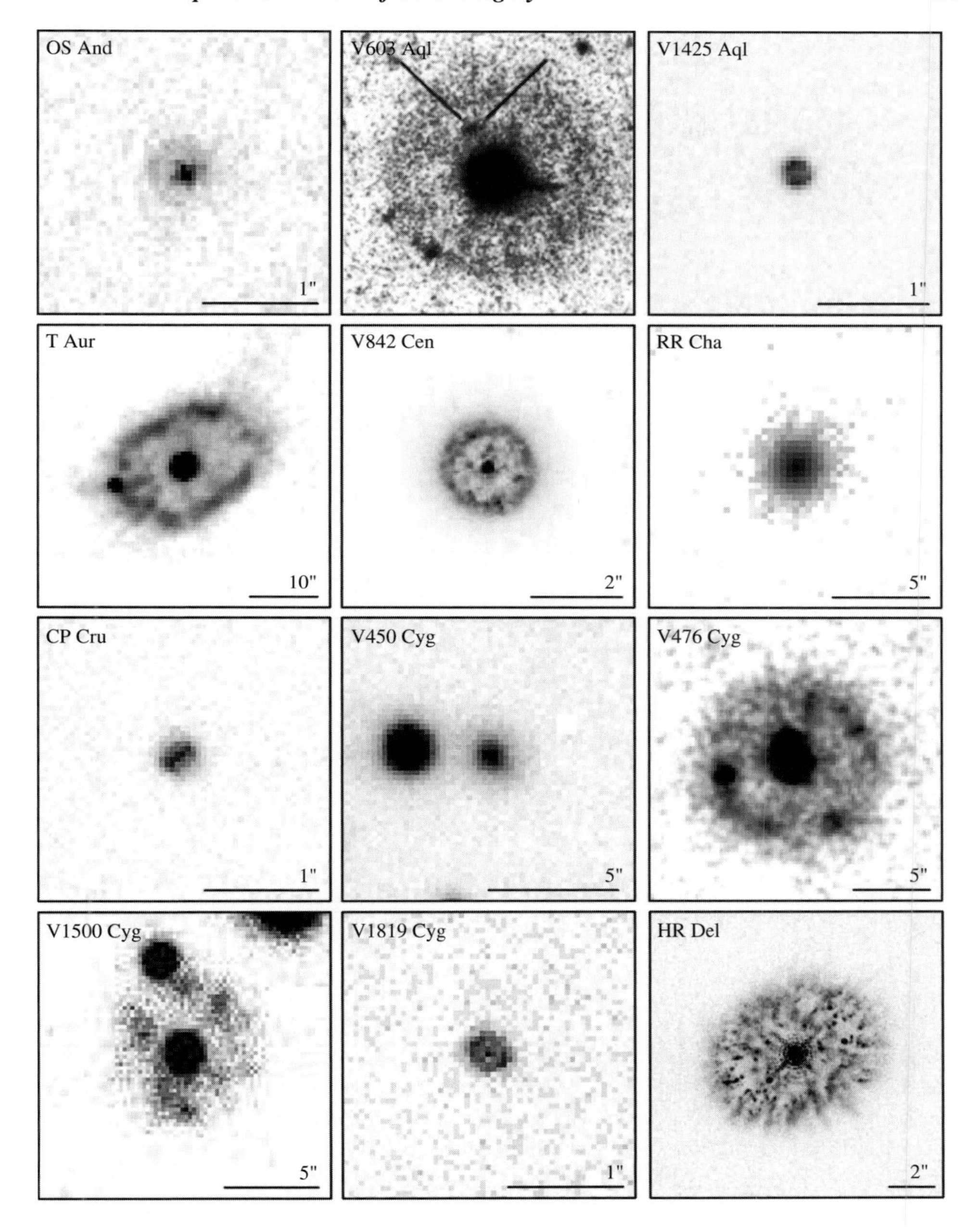

Fig. 12.2. Optical images of nova shells listed in Table 12.1 with scales shown in arcseconds (orientations vary). Wavebands are Hα or Hα/[N II] except for V603 Aql (photographic, band unknown), V1425 Aql ([O III]), V1819 Cyg, V960 Sco and QV Vul (all [N II]). The image of V603 Aql is from Mustel and Boyarchuk (1970), the others are from the authors' own data or from observatories' public archives.

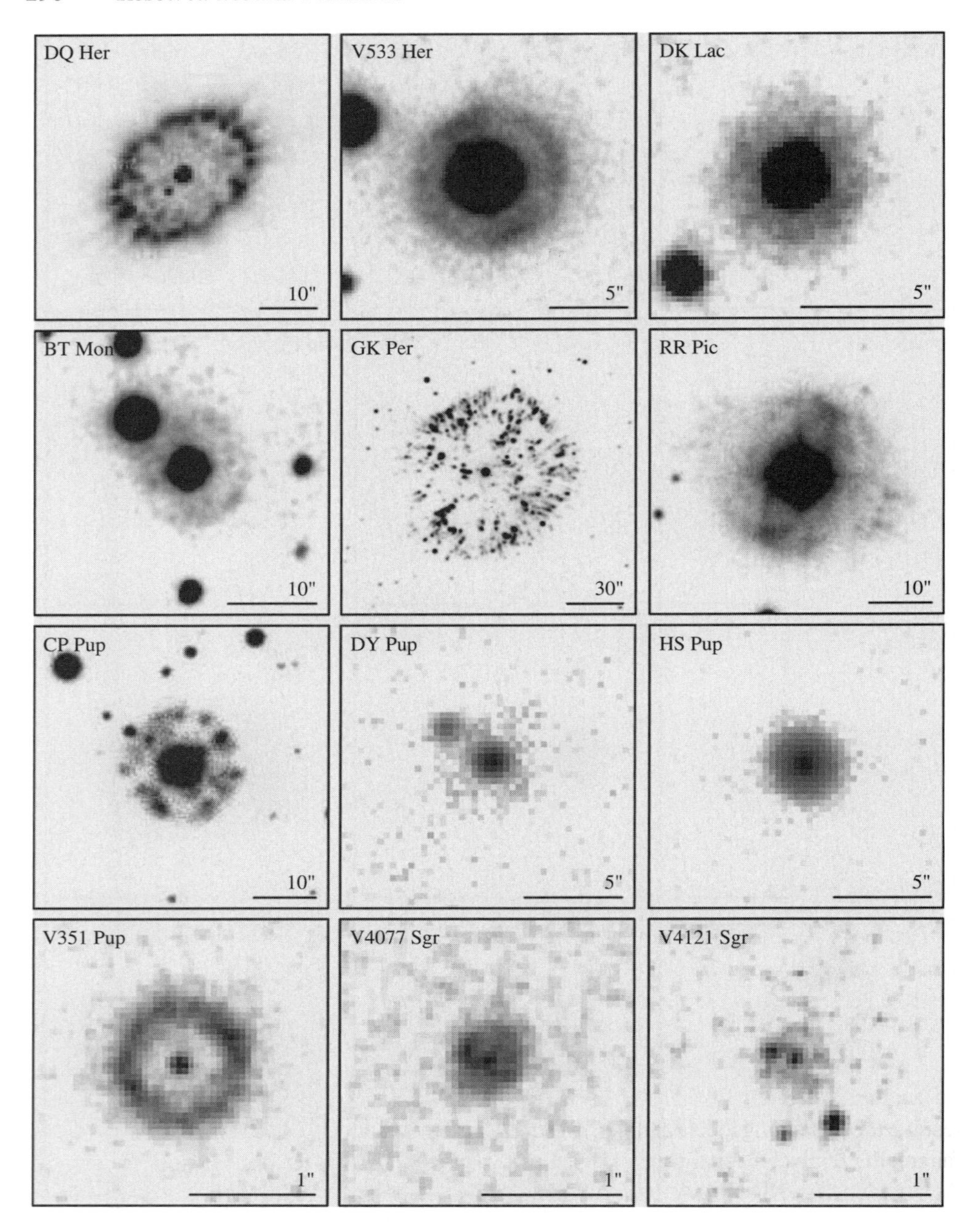

Fig. 12.2. *continued.*

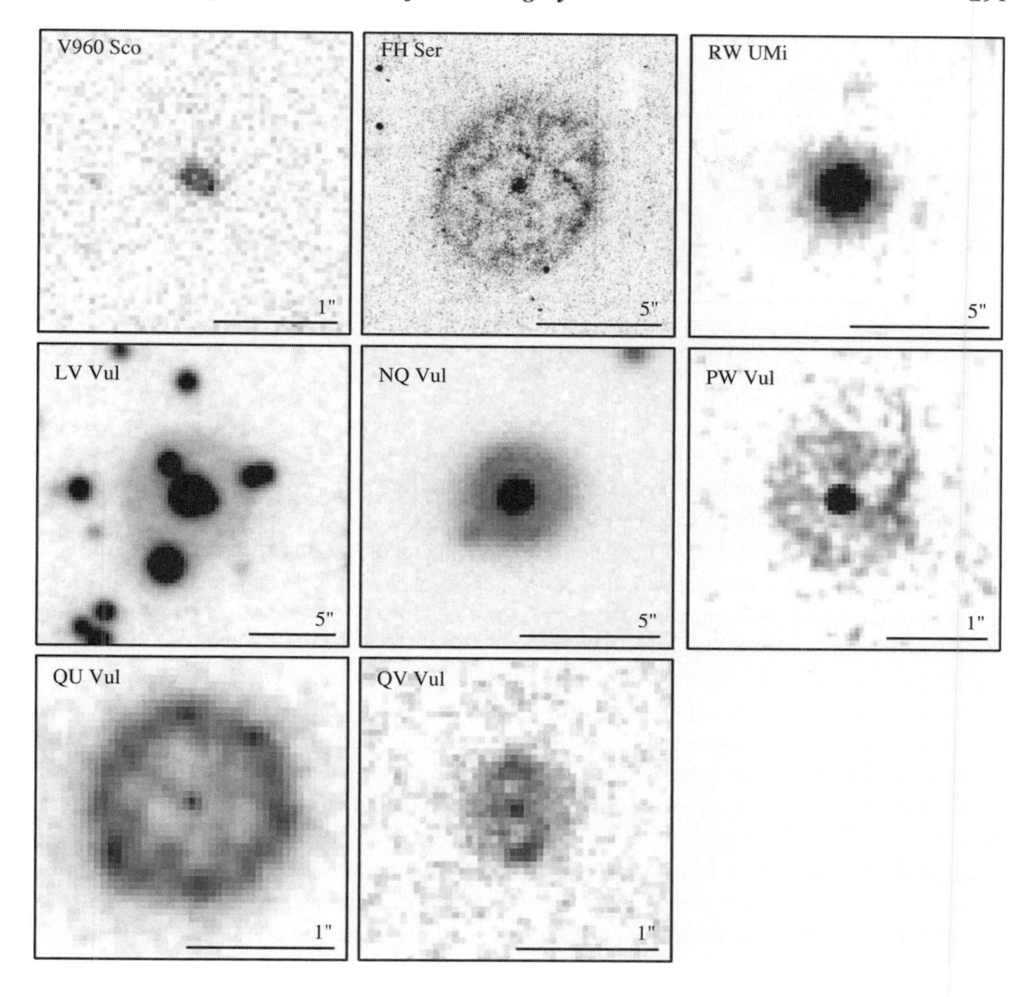

Fig. 12.2. *continued.*

The availability of the HST since 1991 has of course meant that we are not only able to resolve the expanding ejecta at much earlier times, but we can also investigate the structure on far smaller equivalent spatial scales than previously possible in the optical from the ground. Several investigators have subsequently carried out successful HST programmes investigating the structure and evolution of nebular remnants, and we briefly summarize these here. V1974 Cyg was one of the brightest novae of recent times and is probably the nova with the best-studied outburst across the electromagnetic spectrum. HST observations using the Faint Object Camera, reported by Paresce *et al.* (1995), began 467 days after outburst (see Figure 12.3). At that time, the COSTAR image corrector had not been installed and the image quality was relatively poor. The FOC image did, however, clearly show a bright elliptical ring-like structure with condensations surrounding the central nova. Post-COSTAR imaging was continued on days 689, 726 and 818 with the ring best visible in Hα and [O III] lines

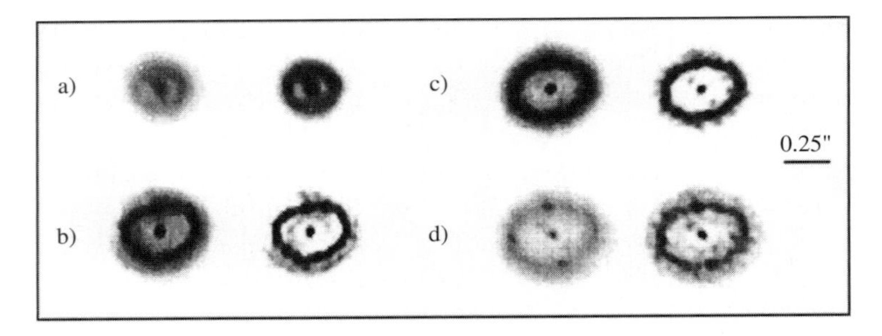

Fig. 12.3. The evolution of the shell of V1974 Cyg. (a) 278 nm filter (F278M) from May 1993: raw HST images are on the left, deconvolved images on the right; narrow-band [O III] F501N images from (b) Jan 1994, (c) Feb 1994, (d) May 1994. From Paresce *et al.* (1995).

and expanding at between approximately 0.2 and 0.3 milliarcsec per day (derived minor and major axis expansion rates respectively).

Using a combination of the HST images, and optical and ultraviolet spectroscopy, Chochol *et al.* (1997) concluded that the nebular remnant comprises two major components: an outer fast-moving, tenuous low-mass envelope, and an inner lower-velocity, high-mass envelope. The HST images showed emission from the inner envelope in the form of an expanding equatorial ring and polar blobs. They also concluded that by day 818, there was evidence of the effect of a strong magnetic field from the white dwarf on the expanding plasma in the form of about 10 meridional arcs resembling a 'water fountain' and interacting with the dense inner shell to form bright spots. Applying a kinematical model of the ejecta, Chocol *et al.* were able to derive an inclination angle for the ellipsoidal shell and hence a distance to the nova of 1.77 ± 0.11 kpc.

Gill and O'Brien (2000) discuss the results of HST observations of the first five of eleven novae imaged in 1997 and 1998 with the Wide Field and Planetary Camera (WFPC2) and the F656N (Hα+[N II]) filter. In this paper, they reported the detection of the shells of FH Ser and V533 Her, with non-detections of extended emission around BT Mon, DK Lac or V476 Cyg. In the second paper of the series, Harman and O'Brien (2003) describe the observations of the bipolar shell of nova HR Del, imaged through filters centred on the lines of Hα, [N II] and [O III]. A primary goal of these observations was to combine them with ground-based optical spectroscopy in order to provide kinematical information (see below).

The other novae observed in this HST programme are discussed in Gill (1999). In the survey by Gill and O'Brien (1998), a bright but only barely resolved (diameter 1–2″) shell was discovered around V842 Cen. The HST image clearly shows a clumpy circular shell with diameter about 2″. There is no obvious structure such as an equatorial ring although there is a suggestion that the shell is thicker in the north-east and south-west. The HST images of RR Pic and T Aur do not improve on earlier ground-based observations showing RR Pic's elliptical shell with a bright equatorial ring and the DQ Her-like elliptical shell of T Aur (e.g. Slavin, O'Brien & Dunlop, 1995). The very faint shell of V1500 Cyg was not detected in the HST observations.

From 1997 to 1999, Fred Ringwald and collaborators had a HST snapshot programme of short (typical exposures of a few minutes) WFPC2 imaging observations of 32 nova remnants. This proved rather successful and has resulted in the discovery of ten new resolved shells, three of which were reported by Ringwald *et al.* (1998), three more by Downes and Duerbeck (2000) and a further four which were discovered by the authors whilst writing this chapter.

Where it can be resolved, the optical images of Figure 12.2 tend to confirm the presence of structure as implied from spectroscopy conducted at earlier phases of outbursts. In particular it is clear that nova shells appear to be extremely clumpy and, in a significant number, for example RR Pic and DQ Her, equatorial (and for DQ Her, tropical) rings are apparent. In addition, in both these objects, 'tails' of emission are evident, streaming away from knots in the shell, suggestive of ablation by faster-moving ejected material. A number of these shells appear to be ellipsoidal and it appears that there is systematically *less* shaping with *increasing* speed class (Slavin, O'Brien & Dunlop, 1995; Downes & Duerbeck, 2000; see below). Indeed, as pointed out in Slavin, O'Brien & Dunlop (1995), the very fast nova GK Per, if placed at the distance of V1500 Cyg, would show a very similar, relatively circular, but otherwise amorphous remnant as seen in ground-based imagery.

12.2.2 Near-infrared

As noted by Krautter *et al.* (2002), infrared observations of nebular remnants are an important addition to those in the optical and radio regimes. In particular, they are not as susceptible to dust extinction as optical imagery and in addition, emission from molecular species may be observed (see also Chapters 8 and 13).

Krautter *et al.* (2002) used the Near-Infrared Camera and Multi-Image Spectrometer (NICMOS) on the HST in 1998 to image novae QV Vul, QU Vul, V1974 Cyg and V723 Cas through six filters from 1.87 to 2.37 μm (see Table 12.1). These targets were chosen to represent a wide variety of individual properties, including a range of speed class and both ONe and CO progenitors. Three of the four novae showed resolved shells, the exception being V723 Cas where the short time since outburst, coupled with the low ejecta expansion velocity of this slow nova ($\sim$300 km s^{-1}) and estimated distance, meant that the shell would be expected to be of the order of the instrumental PSF at this time.

For V1974 Cyg, the Pa-α image (the only band in which the nebular remnant was significantly detected) shows the same overall morphology in the infrared as observed previously in the optical by Paresce *et al.* (1995). Fits to the main elliptical ring of emission, which sits on top of a more extended low-surface-brightness shell, show that the expansion rate has not deviated within the errors from that derived from observations four years previously in both the optical (Paresce *et al.*, 1995) and radio (Hjellming, 1995, see below). The lack of any detectable deceleration of the ejecta over this period placed an upper limit on the density of the pre-outburst medium into which the ejecta are moving of $\leq 5 \times 10^{-22}$ g cm^{-3} for the mass and velocity assumed by Krautter *et al.* for the ejecta. A crude estimate of the inclination of the ring to the plane of the sky gave $i \sim 42°$ compared with $38.7° \pm 2.1°$ from the kinematical model of Chocol *et al.* based on optical imagery. Direct comparison also shows that one or two individual features in the shell have appeared since the last epoch of optical observations.

The remnant of the archetypical neon nova, QU Vul, which condensed a small amount of oxygen-rich dust following outburst, was detected in all three emission-line filters (Pa-α 1.90 μm; Br-γ 2.16 μm; CO band 2.37 μm) as well as in the CO continuum filter (2.22 μm). The emission through the latter filter was stronger than the corresponding line and its true physical origin remains an open question. 'Arcs' of emission, with radius $\sim 0.6''$, are ascribed to limb brightening and both this 'ring' structure and the inner shell are rather inhomogeneous.

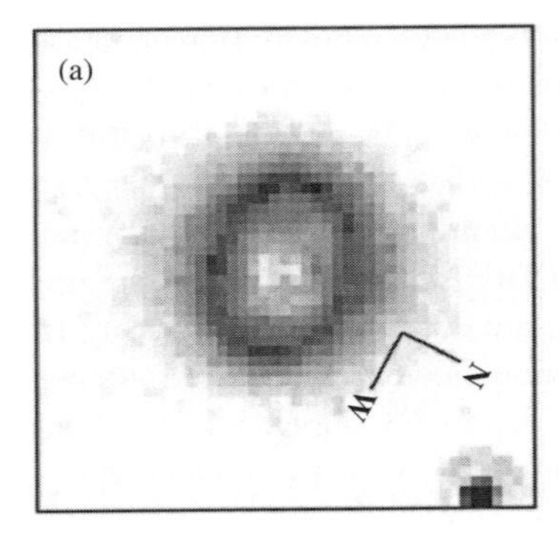

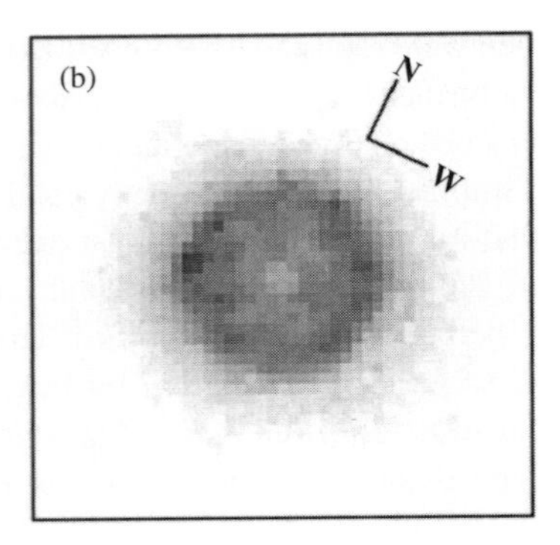

Fig. 12.4. PSF-subtracted images of (a) V1974 Cyg and (b) QU Vul in the Pa-α line. From Krautter *et al.* (2002). Images are $3.56'' \times 3.56''$.

This nova had previously been imaged from the ground by Shin *et al.* (1998), with discrepant results. However, as pointed out by Krautter *et al.*, the NICMOS data are far superior and therefore the conclusions drawn from them more likely to be valid.

The CO nova QV Vul formed an optically thick carbon-rich dust shell. As with V1974 Cyg, the shell was only significantly detected in Pa-α (a 'knot' of emission on the western edge of the shell is ascribed to a serendipitous unrelated stellar source in the line of sight). A compact ellipsoidal remnant is evident, but unlike V1974 Cyg and QU Vul, no ring-like structure appears present. It is suggested that the lack of this feature may be due to the lower ejection velocity of the remnant in QV Vul and insufficient time for the structure to form. There is a discrepancy of a factor of two between the angular expansion rate derived from the NICMOS imagery and calculations based on the black-body angular diameter of the dust shell earlier in the outburst. This may be due to a difference in velocity between the ionized gas envelope seen here and the dust envelope (and presumably therefore the gas from which it formed).

For all three novae, Krautter *et al.* derive expansion parallax distances. They do, however, caution about the uncertainties of the method if the true shell geometry is not known (see below). They also find consistency with the suggested correlation of speed class to remnant shape (again, see below).

12.3 The combination of optical imaging and spectroscopy

The combination of optical (particularly HST) imaging, long-slit optical spectroscopy and simple modelling has been shown to be a very powerful tool in determining the precise geometry of resolved remnants. It is also potentially important in exploring other physical properties, such as possible abundance gradients across the remnants.

12.3.1 Physical conditions

Surprisingly little work seems to have been done using spectroscopy to explore the physical conditions in expanding nebular remnants. Williams *et al.* (1978) derived a remarkably low electron temperature ($T_e \sim 500$ K) in the remnant of DQ Her 42.5 years after outburst. This result was confirmed using IUE spectra by Hartmann and Raymond (1981) and Ferland *et al.* (1984). Subsequently, Williams (1982) found that the nebular remnant of

CP Pup showed similar unusual physical conditions. Full ionization models (Ferland *et al.*, 1984; Martin, 1989a) are able to match the heating and cooling of the nebula at these low temperatures. Martin (1989b) derives a nebular mass of $6 \times 10^{-5}\,M_\odot$ for $T_e = 500$ K and a distance of 485 pc and also notes that, by year 42.5, the level of ionization appears to be higher along the major axis of the nebula relative to the minor. He conjectures that this may be due to differential illumination of the nebula by the central accretion disk, a point that we shall return to below.

The existence of abundance gradients across the ejecta would have significant consequences for our understanding of the progress of the TNR across the white dwarf surface (e.g. Duerbeck, 1987a). Mustel and Boyarchuk (1970) discuss this possibility in the context of DQ Her in particular but favour variations in the excitation conditions as an explanation for the differing strengths of forbidden lines such as [O III] and [N II] across the nebula. Indirect evidence for segregation of elements has been cited in terms of the existence of oxygen-rich and carbon-rich grains in the same nova. For example, Gehrz *et al.* (1992) proposed that the 'fast' and 'slow' ejecta of QV Vul might have very different C/O ratios (see also the discussion in Chapters 6, 8 and 13).

Evans *et al.* (1992) used IUE spectra of the nebular remnants of RR Pic and GK Per to explore variations in abundances. They found that in RR Pic, the C/O ratio is inverted in the southern equatorial ring relative to the western polar blob. Broad emission in the 1400–1700 Å region of the western polar blob is tentatively ascribed to H_2. For GK Per, conclusions are complicated by the likely dominance of shock excitation in some regions rather than photoionization, but line strengths are consistent with non-solar abundances including a substantial over-abundance of nitrogen. It is admitted, however, that there is a need for more rigorous calculation of expected line strengths for models using non-solar abundances.

Enhanced emission in [N II] from equatorial rings (such as that seen in FH Ser, Figure 12.5) may not be due to a simple over-abundance of nitrogen, however. Further discussion on this point is given below, and there is clearly much more work to be done in this regard.

Shore *et al.* (1997) secured spectra of the remnant of V1974 Cyg 1300 days after outburst using the Goddard High Resolution Spectrograph (GHRS) aboard the HST. These observations included those of a knot of emission at the south side of the main ellipsoidal ring with the Small Science Aperture of the GHRS instrument. Comparison of the knot spectrum to that obtained with a larger aperture, effectively covering the entire remnant, indicates that there is an enhancement of the Ne/He ratio in the knot, and possibly a depletion of C compared with He. As the knots are thought to have been created at an early stage of the outburst, such observations can potentially elucidate the role of differential mixing in the accreted layers on the WD prior to the TNR.

12.3.2 Kinematics, remnant shape and distance determination

The nebular remnants of novae provide excellent laboratories for the study of mass-loss from binary systems and the shaping of the resulting circumstellar photo-ionized nebulae. This has relevance not only to novae but also to planetary nebulae where similar processes are thought to take place, albeit on much longer time-scales. In particular, the combination of spatially resolved emission-line spectroscopy and high-resolution imaging allows the construction of detailed spatio-kinematic models for the ejected shells. Good examples are provided by Gill and O'Brien (2000) for FH Ser, and Harman and O'Brien (2003) for HR Del (see also Solf, 1983).

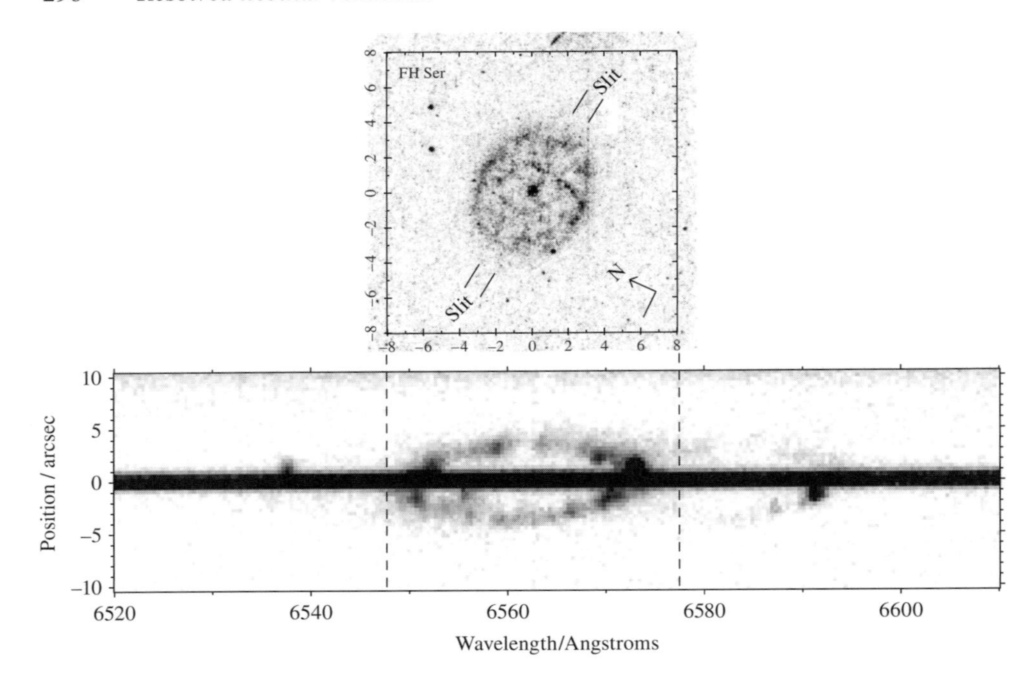

Fig. 12.5. HST image of FH Ser through the Hα/[N II] filter (top) with axes labelled in arcseconds. The lower panel shows a spectrum obtained with the William Herschel Telescope at the slit position shown in the image. The vertical dashed lines in the lower panel indicate the wavelength range of the bandpass of the HST filter, hence only material between these lines will be visible in the image. The spectrum itself shows a clear velocity ellipse due to Hα flanked by fainter corresponding ellipses due to the [N II] doublet. See Gill and O'Brien (2000) for more details.

Gill and O'Brien (2000) obtained ground-based optical spectra of FH Ser following the William Herschel Telescope (WHT) imaging by Slavin, O'Brien & Dunlop (1995) and prior to the HST imaging of its shell. Slit positions were chosen to lie along the major and minor axes of the shell seen in the ground-based imaging. Two wavelength ranges, around Hα/[N II] and Hβ/[O III], were covered with a resolution of 35–50 km s^{-1} – see Figure 12.5.

These models can then be used to derive accurate distances from expansion parallax. Wade, Harlow and Ciardullo (2000) discuss in detail the biases that arise in distance determination via expansion parallax if the prolate nature of nova shells is not taken into account.

In the case of FH Ser, Figure 12.5 shows how spatially resolved spectra can clearly reveal the kinematics of the expanding shell. In this case an equatorial ring is visible in the image and the spectra. In the latter it appears as the bright knots towards the extremes of the [N II] emission below, at the red-shifted end, and above, at the blue-shifted end, the continuum emission from the central system. These data rather neatly demonstrate that the equatorial ring is emitting primarily in [N II] with the image (whose bandpass is indicated by the dashed lines) showing the top (blue-shifted) half of the ring from the brighter 6583 Å line and the lower (red-shifted) half from the fainter 6549 Å line.

Similar evidence for [N II]-enhanced equatorial rings has been seen in DQ Her, V603 Aql and V705 Cas (e.g. Mustel and Boyarchuk, 1970; Petitjean, Boisson & Pequignot, 1990;

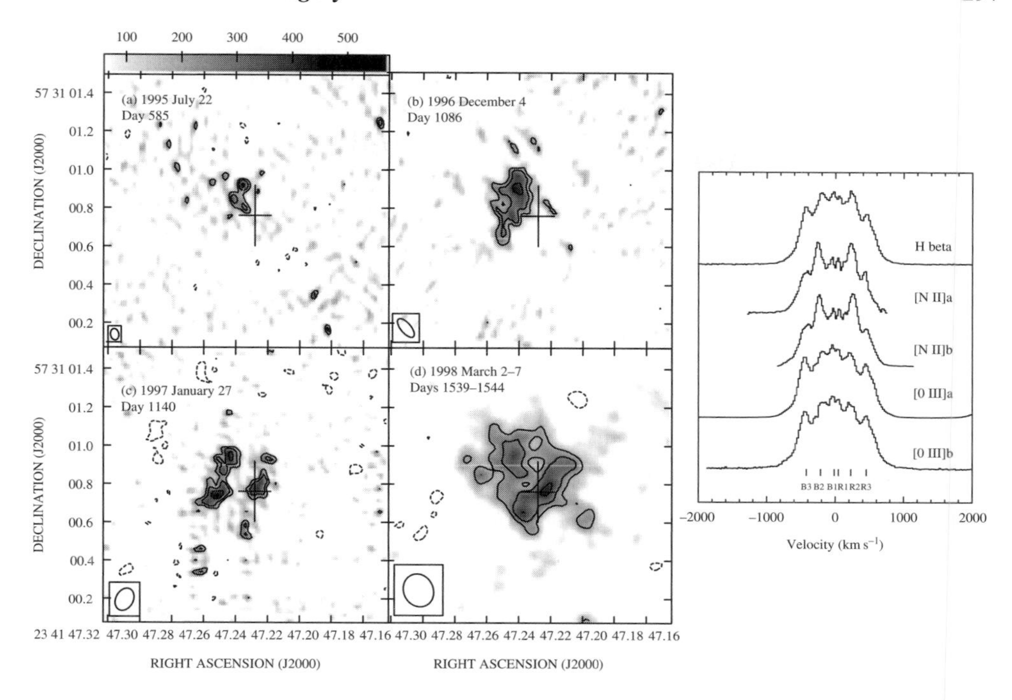

Fig. 12.6. MERLIN maps at 5 GHz (left) showing the radio evolution of the remnant of V705 Cas (each box ~1.4″ on a side) and optical spectra from the William Herschel Telescope (right) taken on day 963. From Eyres *et al.* (2000).

Gill & O'Brien, 1999; Figure 12.6). The origin of this enhancement is still not clear. Gill (1999) carried out some preliminary modelling using *CLOUDY* (Ferland *et al.*, 1998) in which a prolate ellipsoidal shell was illuminated by the central white dwarf plus accretion disc (as suggested by Solf (1983) for the case of HR Del, Martin (1989b) and Petitjean, Boisson and Pequignot (1990) for DQ Her). Allowance was made for the increased illumination towards the poles of the shell when the central system is viewed face-on and shadowing by the accretion disk towards the equatorial regions of the shell. The results suggested that the enhancement in [N II] might be because at higher latitudes nitrogen is ionized beyond the second ionization state. This intriguing suggestion needs to be followed up with more detailed modelling and observational tests of other predicted spectral lines.

The kinematical modelling of the shell of FH Ser enabled an accurate distance, expansion velocity and axial ratio for the shell to be determined. This is by far the best method for determining the distances to novae.

Furthermore it seems likely that, as pointed out by Mustel and Boyarchuk (1970) with regard to DQ Her (an eclipsing binary whose shell is very elliptical) and Solf (1983) for HR Del, the equatorial ring visible in many of these shells corresponds to the plane of the binary orbit. Such modelling then also enables the binary inclination to be estimated.

12.4 Radio imagery

With reference to Figure 12.1, a typical nova can theoretically be resolved by a radio array such as MERLIN whilst still on the optically thick, rising part of the radio light curve.

From a simple model of radio evolution given in Chapter 7, we can derive equations for the peak flux, angular size at peak and time to peak. Thus typically, ~1 yr after outburst, at the time of maximum in the radio light curve, the surface brightness at 5 GHz, $\Sigma_5 \sim 300\,\mu$Jy per beam with MERLIN (cf. ~50 μJy per beam r.m.s. noise) and the expanding remnant should be detectable at least until this time. Thereafter, the flux density $f_\nu \propto t^{-3}$ and $\Sigma_5 \propto t^{-5}$; i.e. unless the ejecta are clumped, or (less likely) T_e is increasing, the remnant will rapidly become undetectable.

At the time of publication of the first edition of this book, Seaquist *et al.* (1989) listed only one nova (QU Vul) with a radio-resolved remnant (we do not include GK Per here for reasons we address below). Table 12.1 shows that in the published literature there are now five novae where the radio emission has been resolved. We suspect, however, that there are more observations of novae in the VLA archive that deserve attention. It should also be noted that the extensive radio observations of V1974 Cyg conducted by the late Bob Hjellming have so far only been published in conference proceedings (e.g. Hjellming, 1995, 1996; see also Chapter 7).

Hjellming (1996) attempted to formulate a unified model of the radio and optical development of V1974 Cyg. This involved a hybrid of the variable wind and Hubble flow models comprising a terminating wind with a linear velocity gradient (see also Chapter 7). In addition, he introduced further free parameters requiring the inner and outer shell boundaries to have different ellipsoidal shapes, and an initial temperature rise was included. From this, he derived reasonable fits to the radio behaviour, but the fits to the spectral line shapes from HST observations were obviously too simplistic. Overall, this was a valiant attempt, and still represents the model which has made the best effort to reconcile the radio light curve, images and optical spectra, but the model was largely phenomenological.

Figure 12.6 shows MERLIN maps of the complex evolution of the remnant of V705 Cas through the optically thick to early optically thin phases (Eyres *et al.*, 2000). Also shown are WHT spectra from day 963 which imply an ordered structure (consistent with expanding equatorial and tropical rings in a remnant with $i = 60°$ – note the bright pair of [N II] lines consistent with an [N II]-enhanced equatorial ring discussed earlier). The optical spectra are difficult to reconcile with the radio observations. Indeed, the radio structure often seems to show dramatic changes which makes straightforward interpretation very difficult.

However, there are several fundamental complications with radio interferometric observations. First of all, larger-scale, lower-surface-brightness emission may be 'resolved out'. More importantly, it is impossible to disentangle the effects of changing optical depth and temperatures without high-spatial-resolution, simultaneous multi-frequency observations of the resolved shell. Currently with MERLIN for example, this is not possible.

These difficulties are further revealed in the observations of V723 Cas (Heywood *et al.*, 2005), a very slow nova which erupted in 1995. MERLIN was used to make nine images of the expanding radio remnant over the following 6 years. The 6 cm light curve was fitted by a Hubble flow model leading to an estimate of ejected mass consistent with other novae (see Chapter 7). However, although the overall scale of the emission seen in the images was consistent with the predictions of this model, the detailed structure was again difficult to reconcile with the model and with features seen in optical spectral line profiles. Heywood (2004) presents a comparison of synthetic images produced by convolving model maps produced by the Hubble flow model with the uv-coverage of the MERLIN array. This analysis suggests that many of the detailed features seen in the observations are likely to be instrumental artefacts resulting from the rather sparse nature of the array combined with the particular scale and brightness of these nova shells. Improvements to MERLIN, specifically the replacement of microwave radio links with optical fibres, will result in a 30-fold increase in sensitivity combined with the potential for multifrequency synthesis and a massive improvement in uv-coverage. This e-MERLIN array will revolutionize our ability to produce high signal-to-noise radio images of resolved nova remnants with faithful reconstruction of detailed structures.

12.5 Remnant shaping

12.5.1 Shaping mechanisms

A number of different mechanisms have been proposed for shaping nova ejecta. These include the common-envelope phase during outburst, the presence of a magnetized white dwarf, the rotation of the white dwarf and a localized thermonuclear runaway. Of these, the most well-developed models have addressed the common-envelope phase and this is discussed in more detail below. There is evidence of a strongly magnetized white dwarf in a number of classical novae (in particular intermediate polars such as DQ Her: see Chapter 2) and clear evidence that at least some of the white dwarfs in nova systems are extremely rapid rotators. The combination of the two may be important in understanding the details of mass-loss from the white dwarf and hence the final structure of the ejected envelope (e.g. Orio, Trussoni & Ögelman, 1992; Livio, Shankar & Truran, 1988). The rotation of the white dwarf has also been considered and, although at first thought to result in oblate remnants in conflict with observation (Fiedler & Jones, 1980), has been shown to help produce a prolate remnant when incorporated into a calculation of the common-envelope phase (Porter, O'Brien & Bode, 1998; see below). Most calculations of the TNR have assumed spherical symmetry (see Chapters 3 and 6) but a number of authors have considered the possibility that a TNR could develop at specific positions at the base of the accreted layer rather than simultaneously across the whole WD (e.g. Orio & Shaviv, 1993; Glasner, Livne & Truran, 1997; Scott, 2000) although as yet with no concensus on whether this could influence the large-scale structure of the resolved nebular remnant.

12.5.2 Speed class and remnant shape

As discussed earlier, analysis of the optical images suggests a relationship between the degree of shaping and speed class in the sense that the slower the nova, the more prolate the nebular remnant. In Figure 12.7 we have combined data from several sources, including corrections for inclination where possible, to demonstrate this relationship. The correlation between speed class and ejection velocity means that the faster the expansion speed the less shaped the remnant becomes.

12.5.3 Hydrodynamical models of shaping

It is easily shown that the ejecta from the white dwarf surface (and indeed the consequent pseudo-photospheric radius) rapidly envelop the secondary star yielding effectively a common-envelope binary. Early results of modelling the effects of the frictional deposition of energy and angular momentum by the secondary on the ejected nova envelope were described by Livio *et al.* (1990). Their 2-D hydrostatic wind models were essentially applicable only to restricted cases of very slow novae.

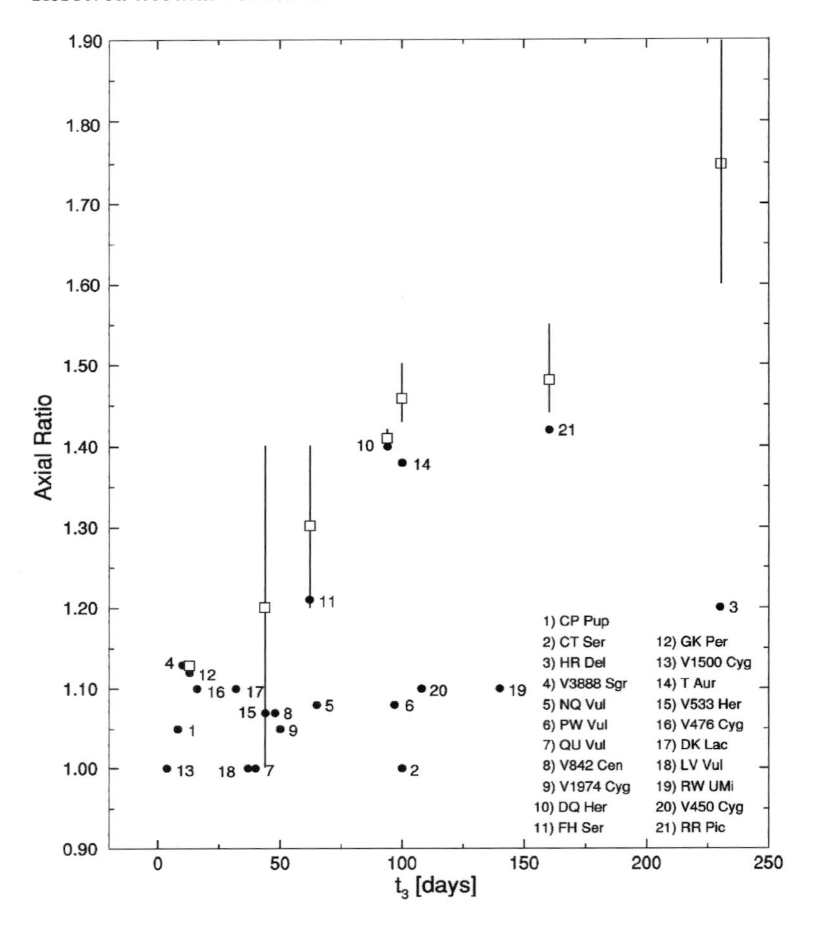

Fig. 12.7. Axial ratio of remnants versus speed class. Filled circles from Downes and Duerbeck (2000), open squares show inclination-corrected results. Data from Slavin, O'Brien and Dunlop (1995), Gill and O'Brien (1998), (2000), Harman and O'Brien (2003).

Subsequently, Lloyd, O'Brien and Bode (1997) used a 2.5-D hydro-code to investigate remnant shaping for a variety of speed classes. This approximation allows rotation about the symmetry axis, although it constrains the flow to be axisymmetric. An important ingredient is the observed relation between ejection velocities and speed class (see Chapter 2). In the case of slower novae, for example, lower ejection velocities would lead to longer effective interaction times between the secondary and the ejecta, and hence more shaping might be expected. The basic model involves ejecta in the form of a wind with secularly increasing velocity and decreasing mass-loss rate. This evolved from a model of the early X-ray emission from V838 Her involving the interaction of ejecta with different expansion velocities early in the outburst (O'Brien, Lloyd & Bode, 1994). The Lloyd, O'Brien and Bode model produces rings, blobs and caps, plus a correlation of speed class to axial ratio in the sense required. It should also be noted that a rather more satisfactory fit to the early radio evolution of V1974

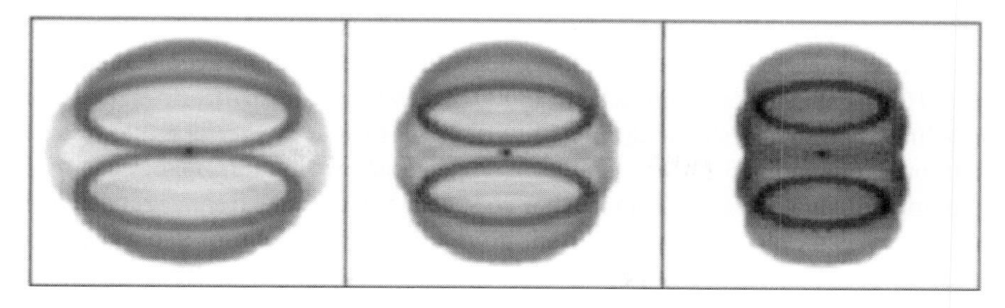

Fig. 12.8. Synthetic images of remnants produced for the parameters of the ejecta and central binary of a moderately fast nova. The case for no rotation of the accreted envelope is shown on the left. The middle and right panels show the cases when the accreted envelope rotates at 0.5 and 0.7 of the Keplerian velocity at the white dwarf surface, respectively. From Porter, O'Brien and Bode (1998).

Cyg may be provided by such a model (Lloyd, O'Brien & Bode, 1996). However, it also produces *oblate* remnants.

Porter, O'Brien and Bode (1998) modified this basic model to include the effects of the rotating accreted envelope on the surface of the white dwarf. From consideration of the effective gravity due to envelope rotation, and its effects on local luminous flux driving mass loss at outburst, a mass-loss rate and terminal velocity of ejecta were derived that are dependent on latitude on the white dwarf. Figure 12.8 shows the comparison of results for a moderately fast nova without and with rotation of the accreted envelope. The Porter, O'Brien and Bode models produce prolate shells as required.

Since these models use a 2.5-D code (in which the ejection is axisymmetric but with angular momentum) the secondary star is modelled as a torus in which energy and angular momentum are added to the wind from the white dwarf. This limits the model to cases where the wind-crossing time is long compared with the orbital time-scale. A more detailed model will require a full 3-D simulation.

12.6 GK Persei

There are several remarkable features associated with GK Per that make it distinct and worthy of individual attention. It may provide a valuable insight not only into the evolution of nova binary systems, but also into the physics of supernova remnants and planetary nebulae.

The nova outburst in February 1901 was accompanied some months later by rapidly expanding nebulosities on arcminute size scales (Ritchey, 1901). Couderc (1939) was the first to formulate detailed models of these nebulosities as reflections off dust clouds lying along the line of sight to the nova. Since then, superluminal motion has of course been observed, and used to great effect, in a wide variety of sources (e.g. the unusual case of V838 Mon: see Bond *et al.* (2003), and other cases discussed therein'. Light echoes have also been searched for in other novae (Schaefer, 1988) and only found in the case of V732 Sgr (Swope, 1940) and V1974 Cyg (Casalegno *et al.*, 2000; see Draine and Tan (2003) for the X-ray halo).

The ejecta proper were first photographed in 1916 by E. E. Barnard and their expansion followed to the present day where they now subtend an angle in excess of an arcminute (see Figure 12.2). Radio observations in the 1980s revealed that GK Per's nebula is a non-thermal radio source, akin to a miniature supernova remnant, and as such very unusual amongst classical novae (see Seaquist *et al.* (1989) and references therein). The radio emission peaks toward the south-west (see Figure 12.9) and is synchrotron emission from relativistic electrons spiralling around enhanced magnetic fields, thought to arise in shocks as the ejecta from the 1901 outburst encounter a pre-existing circumstellar medium. Three-dimensional Fabry–Perot imaging spectroscopy (Lawrence *et al.*, 1995) has confirmed and extended many of the Seaquist *et al.* (1989) conclusions about the morphology of the optical remnant, relating in turn to the radio map. More recently X-ray emission resulting from the same interaction has been imaged by ROSAT and Chandra and is providing further constraints on the development of this 'mini supernova remnant' (Balman, 2005).

IRAS observations were the first to be searched systematically for evidence of the medium into which the ejecta are running. These revealed extended far-infrared emission, symmetrically placed around the nova, and originating from cool dust grains (Bode *et al.*, 1987; Seaquist *et al.*, 1989; Dougherty *et al.*, 1996; $T_\mathrm{d} \sim 25\,\mathrm{K}$). The HIRAS observations of Dougherty *et al.* (1996), coupled with those of CO line emission (Scott, Rawlings & Evans, 1994; see Figures 13.8 and 13.9 in Chapter 13), tended to counter arguments that the far-infrared emission was largely interstellar in origin (Hessman, 1989). Dougherty *et al.*

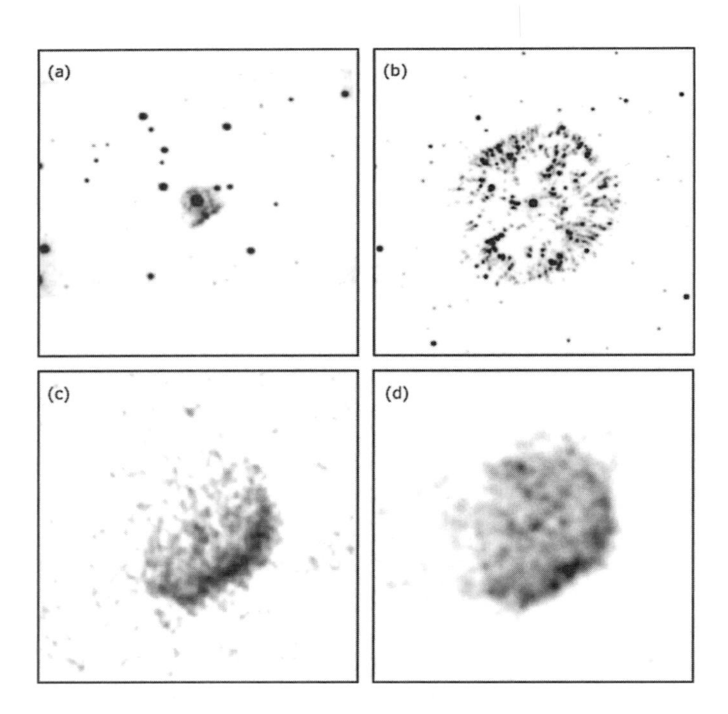

Fig. 12.9. Images of the expanding ejecta of GK Per. (a) Optical image from 1917 November taken by G.W. Ritchey using the Mount Wilson 60″ telescope (Ritchey, 1918); (b) William Herschel Telescope Hα/[N II] image from 1993 September (Slavin, O'Brien & Dunlop, 1995); (c) 5 GHz radio image taken with the Very Large Array in August 1997 showing regions of non-thermal (synchrotron) emission (E. R. Seaquist, private communication) and (d) Chandra observations from February 2000 showing X-ray emission in the 0.4–0.6 keV band (see also Balman, 2002, 2005). All images are 3′ on a side, with north up and east to the left.

proposed that the IRAS nebula was the result of ejection of the outer layers of a 'born again AGB' star phase of what is now the white dwarf in the nova system.

The initial search for the optical counterpart of the large-scale nebulosity was an offshoot of a deep imaging survey of old planetary nebulae conducted by Tweedy and collaborators using the Burrell Schmidt telescope at Kitt Peak. Tweedy (1995) discovered an extensive optical nebulosity in Hα, external to the nova ejecta, and having some features in common with the original light echoes of 1901/02. In 1999, Bode and O'Brien used the Wide Field Camera (WFC) on the 2.5 m Isaac Newton Telescope to image the region through a set of narrow band filters.

As noted in Bode, O'Brien and Simpson (2004) and shown in Figure 12.10, the WFC camera images clearly detect the extended nebulosity, with [O III] emission lying interior to that in the Hα filter, particularly to the south-west. Certain features of this outer nebulosity coincide with those of the superlight nebulae observed immediately following the outburst (including an unexplained jet-like projection to the north-east of the central remnant). Overall, a projected 'hourglass' shape to the outer nebula may be discerned, coincident with the IRAS nebulosity, but 'swept back' at the ends and brighter to the south-west.

Bode, O'Brien and Simpson conjectured that if the outer nebulosity were an ancient planetary nebula, ejected by the central system in a previous evolutionary phase, then the asymmetries we see could be due to motion of the planetary nebula relative to the interstellar medium. Such asymmetries due to interactions with the interstellar medium are well known in ancient planetaries (e.g. Martin, Xilouris & Soker, 2002). If this were so, then the central stellar system would act as a relatively undeflected or decelerated bullet whilst the nebula would be slowed (and distorted) by the interaction. The 1901 nova outburst would then have occurred off-centre with respect to the planetary. Thus, in the 1916 image (Figure 12.9) the 'circular nebulous ring' referred to by Ritchey is consistent with the nova ejecta and the 'straight ray' to the south-west may be where the ejecta are first encountering the inner edge of the material in the ancient planetary nebula. Evidence for the ejecta continuing to plough into the dense medium in the south-west quadrant comes not only from the non-thermal radio emission, but also from X-ray images with the Chandra satellite (see Figure 12.9).

To help to confirm this hypothesis, Bode, O'Brien and Simpson (2004) measured the proper motion of the stellar source in GK Per. They found it is indeed moving toward the south-west, as predicted, at a rate of 0.15 ± 0.002 arcsec yr^{-1} (see Figure 12.10) and has a space velocity of 45 ± 4 km s^{-1}. Using simple dynamical arguments, the main planetary nebula ejection occurred $10^4 - 10^5$ years ago and ceased around 1000 years before the nova outburst. The nova explosion of 1901 would then have been the first of many such events, further emphasizing the importance of GK Per in our understanding of the classical nova phenomenon.

12.7 Concluding remarks

From a situation at the time of the publication of the first edition of this book where only 22 nebular remnants of CNe had been resolved optically (Wade, 1990) and only 2 at any other wavelength (both in the radio), we now have 43 which have been optically resolved, 6 in the radio, 4 in the infrared and 3 in X-rays. These observations have clearly demonstrated that nova shells are often extremely clumpy, sometimes ellipsoidal or bipolar and occasionally encircled by rings of enhanced emission. Analysis of the observed morphologies suggests that the ellipticity is enhanced for slower novae.

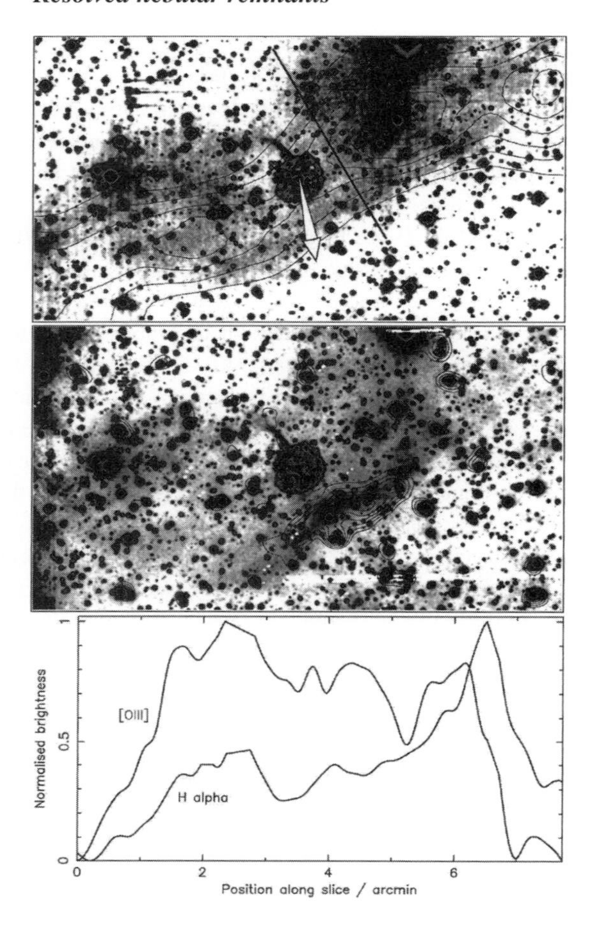

Fig. 12.10. Images of the extended nebulosity of GK Per obtained with the Wide Field Camera of the Isaac Newton Telescope, La Palma, on 1999 November 30 through a Hα filter and 2001 January 18 through an [O III] 5007 Å filter. From Bode, O'Brien and Simpson (2004). (a) [O III] 5007 Å plus contours of 100 μm IRAS emission (from Dougherty *et al.*, 1996), with an arrow indicating the derived direction of proper motion of the central binary and of thickness equivalent to the estimated 1σ error on this direction; (b) an overlay of the contours of the most persistent light echo from 1902 August 13 (Seaquist *et al.*, 1989) on the Hα map; and (c) normalized surface brightnesses in the two filters along the 5$''$ wide slice indicated in (a) crossing the nebula at p.a. = 30° (note that the stars were patched out so that the slice shows only the nebular emission). Compare with Figures 13.8 and 13.9 in Chapter 13. All images are 18$'$ × 10$'$, with north up and west to the right.

It seems certain that the polar axis of the prolate shells is normal to the orbital plane of the central binary system although, despite some advances in modelling the common-envelope phase in particular, the dominant shaping mechanism for the shells is still not fully understood. Further progress will probably require the development of three-dimensional simulations.

Spatio-kinematic modelling of a few nova shells using spatially-resolved spectroscopy and imaging has enabled their structure and distances to be accurately determined. This holds out promise for improving the calibration of the MMRD relation (see Chapters 2 and 14) although, since many bright shells are small, this will probably require very high spatial resolution using either adaptive optics from the ground, or space-based instrumentation. However, since the first edition of this book, little further work has been carried out on the physical conditions and abundances in the spatially resolved remnants. This is crucial to understanding the origin of detailed structures such as the [N II]-enhanced equatorial rings now discovered in several novae. The analysis of any abundance gradients in the resolved remnants would also provide critical constraints on the thermonuclear runaway itself (see Chapters 4 and 6).

New instruments such as Spitzer and Chandra have already begun to provide exciting additional information on some novae. For example, the shock interactions and the evolutionary history of the unique old nova GK Per are now being revealed by combinations of optical, infrared, radio and X-ray imaging. Instruments in development such as e-MERLIN, ALMA and the James Webb Space Telescope will allow new and improved studies of the formation, evolution, structure and content of resolved remnants and further strengthen their crucial role in understanding the classical nova phenomenon.

References

Adams, W. S., 1944, *PASP*, **56**, 218.
Balman, S., 2002, in *Classical Nova Explosions*, ed. M. Hernanz, & J. José. New York: American Institute of Physics, p. 365.
Balman, S., 2005, *Ap. J.*, **627**, 933.
Balman, S., & Ögelman, H. B., 1999, *Ap. J.*, **518**, L111.
Bode, M. F., Seaquist, E. R., Frail, D. A. *et al.*, 1987, *Nature*, **329**, 519.
Bode, M. F., O'Brien, T. J., & Simpson, M., 2004, *Ap. J.*, **600**, L63.
Bond, H. E., Henden, A., Levay, Z. G. *et al.*, 2003, *Nature*, **422**, 405.
Casalegno, R., Orio, M., Mathis, J. *et al.*, 2000, *A&A*, **361**, 725.
Chocol, D., Grygar, J., Pribulla, T. *et al.*, 1997, *A&A*, **318**, 908.
Clayton, G. C., & de Marco, O., 1997, *AJ*, **114**, 2679.
Cohen, J. G., 1985, *Ap. J.*, **292**, 90.
Cohen, J. G., 1988, in *The Extragalactic Distance Scale*, ed. S. van den Bergh, & C. J. Pritchet. San Francisco: Astronomical Society of the Pacific, p. 114.
Cohen, J. G., & Rosenthal, A. J., 1983, *Ap. J.*, **268**, 689.
Couderc, P., 1939, *Ann. Astrophys.*, **2**, 271.
Della Valle, M., Gilmozzi, R., Bianchini, A., & Esenoglu, H., 1997, *A&A*, **325**, 1151.
Diaz, M. P., Costa, R. D. D., & Jatenco-Pereira, V., 2001, *PASP*, **113**, 1554.
Dougherty, S. M., Waters, L. B. F. M., Bode, M. F. *et al.*, 1996, *A&A*, **306**, 547.
Downes, R. A., & Duerbeck, H. W., 2000, *AJ*, **120**, 2007.
Draine, B. T., & Tan, J. C., 2003, *Ap. J.*, **594**, 347.
Duerbeck, H. W., 1987a, *ESO Messenger*, **50**, 9.
Duerbeck, H. W., 1987b, *A Reference Catalogue and Atlas of Galactic Novae*. Dordrecht: Reidel.
Esenoglu, H. H., 2002, in *Classical Nova Explosions*, ed. M. Hernanz, & J. José. New York: American Institute of Physics, p. 548.
Evans, A., Bode, M. F., Duerbeck, H. W., & Seitter, W. C., 1992, *MNRAS*, **258**, 7P.
Evans, A., van Loon, J. Th., Zijlstra, A. A. *et al.*, 2002, *MNRAS*, **332**, L35.
Eyres, S. P. S., Davis, R. J., & Bode, M. F., 1996, *MNRAS*, **279**, 249.
Eyres, S. P. S., Bode, M. F., O'Brien, T. J., Watson, S. K., & Davis, R. J., 2000, *MNRAS*, **318**, 1086.
Eyres, S. P. S., Heywood, I., O'Brien, T. J. *et al.*, 2005, *MNRAS*, **358**, 1019.
Ferland, G. J., Williams, R. E., Lambert, D. L. *et al.*, 1984, *Ap. J.*, **281**, 194.
Ferland, G. J., Korista, K. T., Verner, D. A. *et al.*, 1998, *PASP*, **110**, 761.
Fiedler, R. L., & Jones, T. W., 1980, *Ap. J.*, **239**, 253.
Gehrz, R. D., Jones, T. J., Woodward, C. E. *et al.*, 1992, *Ap. J.*, **400**, 671.
Gill, C. D., 1999, The structure and kinematics of classical nova shells. Unpublished Ph.D. thesis, Liverpool John Moores University.
Gill, C. D., & O'Brien, T. J., 1998, *MNRAS*, **300**, 221.
Gill, C. D., & O'Brien, T. J., 1999, *MNRAS*, **307**, 677.
Gill, C. D., & O'Brien, T. J., 2000, *MNRAS*, **314**, 175.
Glasner, S. A., Livne, E., & Truran, J. W., 1997, *Ap. J.*, **475**, 754.

Hartmann, L., & Raymond, J., 1981, in *The Universe at UV Wavelengths*, ed. R. D. Chapman. NASA Conf. Publ. 2171, p. 495.

Harman, D. J., & O'Brien, T. J., 2003, *MNRAS*, **344**, 1219.

Hessman, F. V., 1989, *MNRAS*, **239**, 759.

Heywood, I., 2004, Radio emission from classical novae. Unpublished Ph.D. thesis, University of Manchester.

Heywood, I., O'Brien, T. J., Eyres, S. P. S., Bode, M. F., & Davis, R. J., 2005, *MNRAS*, **362**, 469.

Hjellming, R., 1995, in *Cataclysmic Variables*, ed. A. Bianchini, M. Della Valle, & M. Orio. Dordrecht: Kluwer, p. 139.

Hjellming, R. M., 1996, in *Radio Emission from the Stars and the Sun*, ed. A. R. Taylor, & J. M. Paredes. San Francisco: Astronomical Society of the Pacific, p. 174.

Hutchings, J. B., 1972, *MNRAS*, **158**, 177.

Krautter, J., Woodward, C. E., Schuster, M. T. *et al.*, 2002, *AJ*, **124**, 2888.

Lawrence, S. S., MacAlpine, G. M., Uomoto, A. *et al.*, 1995, *AJ*, **109**, 2365.

Livio, M., Shankar, A., & Truran, J. W., 1988, *Ap. J.*, **330**, 264.

Livio, M., Shankar, A., Burkert, A., & Truran, J. W., 1990, *Ap. J.*, **356**, 250.

Lloyd, H. M., O'Brien, T. J., & Bode, M. F., 1996, in *Radio Emission from the Stars and the Sun*, ed. A. R. Taylor, & J. M. Paredes. San Francisco: Astronomical Society of the Pacific, p.200.

Lloyd, H. M., O'Brien, T. J., & Bode, M. F., 1997, *MNRAS*, **284**, 137.

Martin, J., Xilouris, K., & Soker, N., 2002, *A&A*, **391**, 689.

Martin, P. G., 1989a, in *Classical Novae*, 1st edn, ed. M. F. Bode, & A. Evans. New York and Chichester: Wiley, p. 121.

Martin, P. G., 1989b, *ibid.*, p. 105.

Mukai, K., Still, M., & Ringwald, F. 2003, *Ap. J.*, **594**, 428.

Mustel, E. R., & Boyarchuk, A. A., 1970, *Ap. &SS*, **6**, 183.

O'Brien, T. J., Lloyd, H. M., & Bode, M. F., 1994, *MNRAS*, **271**, 155.

Orio, M., & Shaviv, G., 1993, *Ap. &SS*, **202**, 273.

Orio, M., Trussoni, E., & Ögelman, H., 2002, *A&A*, **257**, 548.

Paresce, F., Livio, M., Hack, W., & Korista, K., 1995, *A&A*, **299**, 823.

Petitjean, P., Boisson, C., & Pequignot, D., 1990, *A&A*, **240**, 433.

Porter, J. M., O'Brien, T. J., & Bode, M. F., 1998, *MNRAS*, **296**, 943.

Reynolds, S. P., & Chevalier, R. A., 1984, *Ap. J.*, **281**, 33.

Ringwald, F., & Naylor, T., 1996, *MNRAS*, **278**, 808.

Ringwald, F. A., Wade, R. A., Orosz, J. A., & Ciardullo, R. B., 1998, *BAAS*, **30**, 893.

Ritchey, G. W., 1901, *Ap. J.*, **14**, 293.

Ritchey, G. W., 1918, *PASP*, **30**, 163.

Schaefer, B. E., 1988, *Ap. J.*, **327**, 347.

Scott, A. D., 2000, *MNRAS*, **313**, 775.

Scott, A. D., Rawlings, J. M. C., & Evans, A., 1994, *MNRAS*, **269**, 707.

Seaquist, E. R. 1989, in *Classical Novae*, 1st edn, ed. M. F. Bode, & A. Evans. New York and Chichester: Wiley, p. 143.

Seaquist, E. R., Bode, M. F., Frail, D. A. *et al.*, 1989, *Ap. J.*, **344**, 805.

Shin, Y., Gehrz, R. D., Jones, T. J. *et al.*, 1998, *AJ*, **116**, 1966.

Shore, S. N., Starrfield, S., Ake, T. B., & Hauschildt, P. H., 1997, *Ap. J.*, **490**, 393.

Slavin, A. J., O'Brien, T. J., & Dunlop, J. S., 1995, *MNRAS*, **276**, 353.

Solf, J., 1983, *Ap. J.*, **295**, L17.

Swope, H. H., 1940, *Harvard Obs. Bull.*, **913**, 11.

Taylor, A. R., Hjellming, R. M., Seaquist, E. R., & Gehrz, R. D., 1988, *Nature*, **335**, 235.

Tweedy, R. W., 1995, *Ap. J.*, **438**, 917.

Wade, R. A., 1990, in *Physics of Classical Novae*, ed. A. Casstella, & R. Viotti. Berlin: Springer, p. 179.

Wade, R. A., Harlow, J. J. B., & Ciardullo, R., 2000, *PASP*, **112**, 614.

Williams, R. E., 1982, *Ap. J.*, **261**, 170.

Williams, R. E., Woolf, N. J., Hege, E. K., Moore, R. L., & Kopriva, D. A., 1978, *Ap. J.*, **224**, 171.

13

Dust and molecules in nova environments

A. Evans and J. M. C. Rawlings

13.1 Introduction

Dust and molecule formation in the winds of classical novae is now a well-observed, if poorly understood, phenomenon. Although the presence of dust in nova ejecta was not confirmed observationally until the 1970 infrared observations of FH Ser (Geisel, Kleinmann & Low, 1970), the subject has a much longer history. Dust formation in a classical nova was first proposed by McLaughlin (1935) to explain the precipitous deep minimum in the visual light curve of DQ Her; coincidentally, DQ Her was also the first nova to display observational evidence of a molecule (CN; Wilson & Merrill, 1935).

At first sight, the hostile environment of a classical nova is not an obvious place to find molecules and dust. This is because the chemistry that leads to the formation of first diatomic, then polyatomic, molecules, and eventually dust, requires an environment that is well shielded from the increasingly hard radiation field of the stellar remnant.

In this chapter the observational evidence for the presence of dust and molecules in the environments of classical novae is reviewed, together with the current state of our theoretical understanding of chemistry in nova winds, and of the growth and processing of the dust.

13.2 Molecules in nova ejecta

13.2.1 Preamble

Although CN holds the distinction of being the first molecule to be observed in the spectrum of a nova, in the optical spectrum of DQ Her in 1933 (Wilson & Merrill, 1935), molecules are in general most easily detected in the near- and mid-infrared, where common diatomic molecules (such as CO, H_2, SiO) have rotational–vibrational transitions. Attempts to detect purely rotational transitions in diatomic molecules at sub-millimetre wavelengths in both young and old novae have generally proved unsuccessful (Shore & Braine, 1992; Weight *et al.*, 1993; Nielbock and Schmidtobreick, 2003).[1] The exception is the extended CO emission associated with the old (1901) nova GK Per (Scott, Rawlings & Evans, 1994), and which seems to arise in the ancient planetary nebula associated with this object.

[1] The claimed sub-millimetre detection of CO $J = 1 \rightarrow 0$ by Albinson and Evans (1989) was spurious.

Classical Novae, 2nd edition, ed. Michael Bode and Aneurin Evans. Published by Cambridge University Press.

Carbon monoxide is a fundamentally important molecule from the point of view of astrochemistry and in particular, the chemistry that leads to dust nucleation. This is because CO is very stable and chemically relatively inert at the temperatures (~1000–2000 K) at which astrophysical dusts typically form, so that whichever of C and O is under-abundant is locked up in CO and is unavailable for grain formation: the canonical view therefore is that O-rich dusts (e.g. corundum Al_2O_3, various flavours of silicate such as $Mg_xFe_{1-x}SiO_4$) condense in winds having $O > C$ by number, while C-rich dusts (e.g. cohenite Fe_3C, silicon carbide SiC, various forms of carbon) condense otherwise. Indeed, in the environments of evolved stars, this useful rule of thumb applies very well in that the nature of the condensate correlates, in the way expected, with abundances in the star. We shall see, however, that the situation is not so straightforward in nova winds.

As we shall discuss below, despite the very different physical conditions as compared with the interstellar medium (ISM), the formation of molecular hydrogen H_2 also underpins the chemistry in nova winds, and plays a pivotal role in the initial chemistry that eventually leads to other, larger, molecules and dust. Observationally, however, it has proved elusive. Several novae have been observed spectroscopically in the infrared during the early stages when chemistry would have been ongoing, but none has shown any sign of the stronger H_2 features, such as the 1–0 $S(1)$ transition at 2.1218 μm.

The situation is rather better for other (heteronuclear) diatomic molecules, with the first vibrational overtone of CO (with bandhead at 2.29 μm) now routinely detected in infrared spectra (Evans *et al.*, 1996; Rudy *et al.*, 2003).

13.2.2 The observational evidence for molecules in nova winds

Optical

The cyanogen radical (CN, $B\,^2\Sigma^+ - X\,^2\Sigma^+$ bandheads at 359.0 nm, 388.3 nm and 421.6 nm) was present in absorption in the optical spectrum of DQ Her some 2 days after visual maximum (see Payne-Gaposchkin (1957) for a summary of the relevant observations and an illustration of the bands). However, their appearance was extremely transient as they were not present in a spectrum obtained 7 days after maximum. This may be a consequence of molecular destruction, but could also simply result from a change in excitation and/or optical depth effects.

More recently, Bianchini *et al.* (1986) tentatively identified the bandhead of AlO 0–0 $^2\Sigma - X^2\Sigma$ at 4842 Å in the spectrum of GK Per during one of its dwarf nova-like outbursts. The same AlO band has been tentatively identified in RS Car (1895) by Bianchini *et al.* (2001), who have recovered this old nova. This AlO feature is commonly seen in very cool stars (e.g. Evans *et al.*, 2003) and confirmation of its presence in old novae would be valuable. Unlike CO and CN, this feature has not been reported in novae during outburst; prominent features at 1.646 μm and 1.684 μm (*A–X* (2,0)) and at 1.226 μm and 1.242 μm (*A–X* (4,0)) would be worth looking for.

Infrared

Virtually all of the early infrared observations of novae used broadband photometry, and the detection of molecules relied almost entirely on the elevation of the M (5 μm) band flux due to emission in the fundamental $\Delta\upsilon = 1$ transition in CO, which has bandhead at 4.6 μm (Ney & Hatfield, 1978). The equivalent broadband flux in the K (2.2 μm) band

Table 13.1. *Carbon monoxide emission and dust condensation in dusty novae, as of mid-2004. Data based mainly on Gehrz et al. (1998)*

Object	Year	t_3 (days)	t_{CO} (days)	t_{cond} (days)	t_{IRmax}	L_{dust}/L	Type of dust formed
FH Ser	1970	62	<19	60	90	0.5	C
V1229 Aql	1970	37	—	—	—	0.55	C
V1301 Aql	1975	35	—	—	—		C
NQ Vul	1976	65	<19	62	80	1.00	C
V4021 Sgr	1977	70	—	—		≤0.60	C
LW Ser	1978	50	<16	23	75	0.70	C
V1668 Cyg	1978	23	12–23	33	57	0.08	C
V1370 Aql	1982	??	—	≤16	≤37	0.50	C, SiC, SiO_2
PW Vul	1984	97	—	152	≤280	0.003	C
QU Vul[a]	1987	40	—	40–200	240	0.003	SiO_2
OS And[a]	1986	22	—	—	—	—	C?
V842 Cen	1986	48	<25	36	87	1.00	C, SiC, HC
V827 Her[a]	1987	55		43	83	0.10	C
QV Vul	1987	??		56	115	1.00	C, SiO_2, HC, SiC
LMC 1998#1	1988	43	—	59	57	0.06	C?
V838 Her	1991	5		18	25	0.05	C
V705 Cas	1993	84	<6	63	100		C, HC, silicates
V1419 Aql	1993	17–22	?	~90	40		
V1494 Aql	1995	30	—	—	—	0.05	C
V2274 Cyg	2000	33	<18	40			

[a] Nova displaying strong coronal emission

is essentially unaffected by the corresponding emission in the first overtone (bandhead at 2.29 μm), the Einstein coefficient for which ($A_{20} \sim 1\,\mathrm{s}^{-1}$ for $\upsilon = 2 \rightarrow 0$) is much lower than that for the fundamental ($A_{10} \sim 30\,\mathrm{s}^{-1}$ for $\upsilon = 1 \rightarrow 0$). The CO fundamental is therefore expected to be much stronger than the first overtone in nova winds, and is in principle more easily detected if only broadband observations are available: broadband emission in the *K* band tends to see the hot dust (if present), the free–free continuum or the Rayleigh–Jeans tail of the hot stellar remnant rather than CO.

The situation is essentially reversed when spectroscopy is deployed. The atmospheric *K* window is very transparent and first overtone CO emission, if present, is easily observed. Spectroscopy in the *M* window, however, is compromised by the poor transparency and strong emission of the atmosphere around 4.6 μm and, as far as we are aware, there is no spectroscopic observation of the CO fundamental in novae.

In Table 13.1 we summarize the earliest appearance of CO in dusty novae, together with the speed class t_3, the dust condensation time t_{cond}, the time t_{IRmax} of infrared maximum, and the ratio of dust luminosity to the outburst luminosity. Also given is the type of dust formed in the wind of the nova, although it should be appreciated that the data on CO and dust formation for earlier novae are based on broadband photometry over a restricted wavelength range, so that information about the nature of the dust formed is necessarily limited; in particular the identification of the dust as ‘C’ reflects our comment (see below) about the limitations of

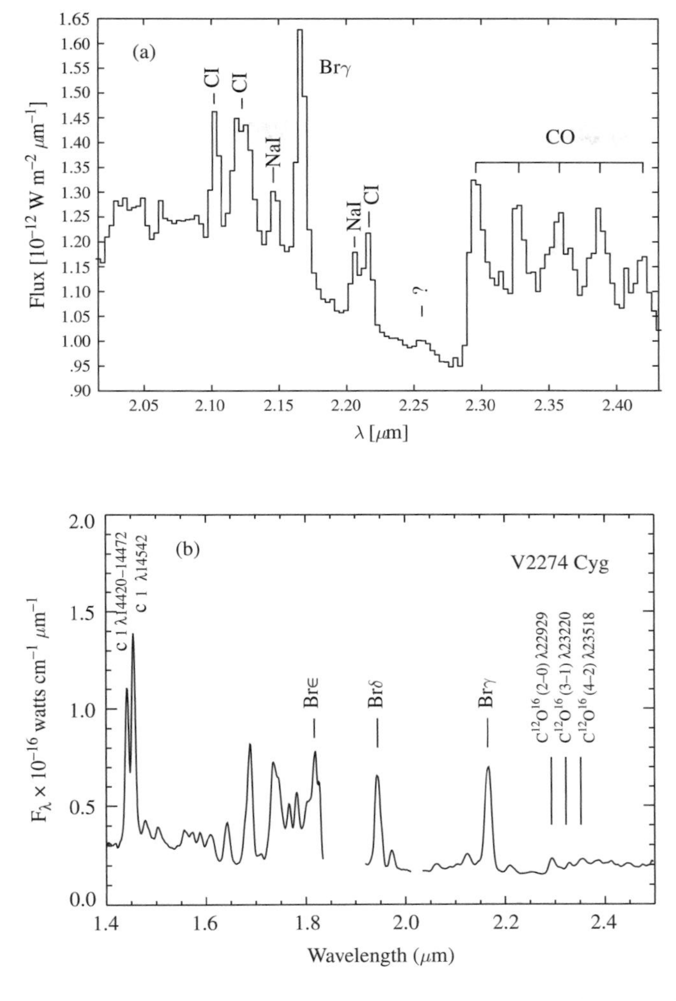

Fig. 13.1. First overtone CO emission in novae. (a) V705 Cas. From Evans *et al.* (1996). (b) V2274 Cyg. From Rudy *et al.* (2003).

broadband photometry. Table 13.1 is based largely on the compilation of Gehrz *et al.* (1998), with additional material from the literature; a null entry in Table 13.1 should be interpreted as inadequate observational coverage rather than absence of CO or of dust. As with the solitary observation of CN in DQ Her, the appearance of CO in novae is transient; the first overtone emission in V705 Cas was present for about 20 days after maximum. Again, this may be a consequence of a decline in the abundance or could just be an excitation effect.

In Figure 13.1 we show first overtone CO emission as seen in two particularly well-observed novae, V705 Cas (Evans *et al.*, 1996) and V2274 Cyg (Rudy *et al.*, 2003). Notice, in both cases, the presence of neutral species with low ionization potential, such as C I (ionization potential 11.26 eV) and even Na I (5.14 eV); this hints at the presence of the neutral region in which the chemistry is proceeding. On the other hand, the weakly identifiable feature marked by '?' in the spectrum of V705 Cas may be CO^+, suggesting that the ionization front is already eating into the neutral zone. However, Dalgarno, Stancil and Lepp (1997) suggest that dissociative recombination

$$CO^+ + e^- \rightarrow C + O$$

is too rapid to allow for a detectable abundance of CO^+ in the ejecta.

In their mid-infrared spectroscopic survey of novae, Smith *et al.* (1995) tentatively identified emission in the fundamental transition $\upsilon = 2 \rightarrow 1$ of SiO in V992 Sco (1992), at an excitation temperature 1500 K, in a spectrum obtained 235 days after maximum. This is plausible, given that this nova displayed a weak silicate feature (Smith *et al.*, 1995) but if confirmed in other silicate-bearing novae, its presence so long after dust condensation is in distinct contrast to the transient nature of CO.

13.3 Observation of dust in nova ejecta

13.3.1 The ultraviolet

Extinction

The presence of dust particles along the line of sight to any astronomical object will result in extinction and reddening. After allowing for the effects of interstellar dust, the presence of any circumstellar dust may be revealed by its extinction and reddening effects. This is not straightforward, as it requires knowledge of the intrinsic flux distribution of the stellar remnant. Shore *et al.* (1994) compared the ultraviolet spectrum of the dusty nova V705 Cas shortly before and shortly after dust condensation, the ratio of the two being a direct measure of the dust optical depth and its wavelength dependence. They concluded that grains grew to dimensions ≥ 0.2 μm very rapidly, seemingly providing independent evidence that nova dust grains are significantly larger than their interstellar counterparts. However, an analysis of the dust emission in V705 Cas by Evans *et al.* (2005) using the *DUSTY* code (Ivezic & Elitzur, 1997) showed that the dust was much smaller by about day 250, indicating that some grain erosion may have taken place.

Dust features

Obervations of the well-known broad 217.5 nm dust-related absorption feature in the ultraviolet spectra of heavily reddened stars was an early coup for ultraviolet observatories such as the International Ultraviolet Explorer (IUE). Although the carrier of this feature remains an enigma, it is widely believed to be some form of carbon. The detection of this feature, and its removal by de-reddening using a standard interstellar extinction law, provides one of the most reliable methods of determining the interstellar reddening and extinction to a nova, essential for modelling the eruption and its aftermath.

In a careful multi-wavelength study of V1370 Aql, Snijders *et al.* (1987) found that, after de-reddening the ultraviolet spectrum for $t = 115$ days (i.e. well after dust condensation), there remained a broad extinction feature peaking at ~250 nm, significantly longward of the interstellar 217.5 nm feature. Moreover, its presence was suspected in ultraviolet spectra at other times, although the data were noisier. Snijders *et al.* (1987) tentatively concluded that the carrier was amorphous carbon although, interestingly, the fact that V1370 Aql sported the 9.7 μm silicate feature (see below) was an early pointer to the 'chemical dichotomy' now known to be common in novae (see below).

Depletion

Ultraviolet spectroscopy has been pivotal in recent years in determining elemental abundances in nova winds, although the following remarks also apply of course to optical and infrared spectroscopy.

As dust condenses, the condensing species (e.g. C in the case of carbon dust, Mg, Si in the case of silicates) are increasingly depleted from the gas phase and this must be reflected in the gas phase abundances. But detecting this effect is fraught with difficulties because not only does the determination of abundances require detailed knowledge of the physical conditions in the ejecta, it often relies on the analysis of emission lines having a restricted range of ionization states. Things are further complicated by deviations from homogeneity, such as the existence of (expanding) clumps within the ejecta.

Nevertheless, evidence for depletion of species from the gas phase as the dust condenses has been presented, for O, Mg, Si, Fe and C (the constituents of the 'silicate' and carbon known to have condensed in the nova), in the case of V1370 Aql (Snijders *et al.*, 1987).

13.3.2 The optical

The visual light curve

The classic symptom of dust formation is of course the minimum, often spectacularly deep, in the visual light curve. As we noted above, the first link between the light curve morphology and dust formation was made for the case of DQ Her (McLaughlin, 1935). However, very few novae manage to produce no dust at all, with novae displaying coronal behaviour being in general poor dust producers; coronal novae are indicated by ^{a} in Table 13.1.

In the case of the most prolific dust producers (such as DQ Her and V705 Cas) the dust visual optical depth at the bottom of the deep minimum is $\sim$6, while in others (e.g. V1668 Cyg) there seems to be little effect on the light curve (implying that the optical depth is $\lesssim 0.1$), the evidence for dust formation being provided by infrared observations. The light curves for two non-dust producers, one prolific producer and one intermediate case are shown in Figure. 13.2. V705 Cas was a prolific dust producer; the optical depth in the visual was $\sim$5.5 at the bottom of the deep minimum evident in Figure 13.2 (Evans *et al.*, 2005).

Optical dust signatures

As well as the well-observed dust-related features in the infrared, there exist a number of features in the optical. The 'diffuse interstellar bands' (Herbig, 1995) are not seen in nova dust, although this is not unexpected: these seem to be weak or absent in circumstellar dust, and may be exclusive to *interstellar* dust.

However, there is a broad emission feature, centred at $\sim$660 nm, which is commonly seen in the optical spectra of objects displaying the so-called Unidentified InfraRed (UIR) features and most spectacularly in the so-called Red Rectangle (HD 44179; Cohen *et al.*, 1975). This feature was seen weakly in the classical nova QV Vul (1987) by Scott, Evans and Rawlings (1994), who attributed it to luminescence in newly hydrogenated amorphous carbon. The significance of this is further explored below.

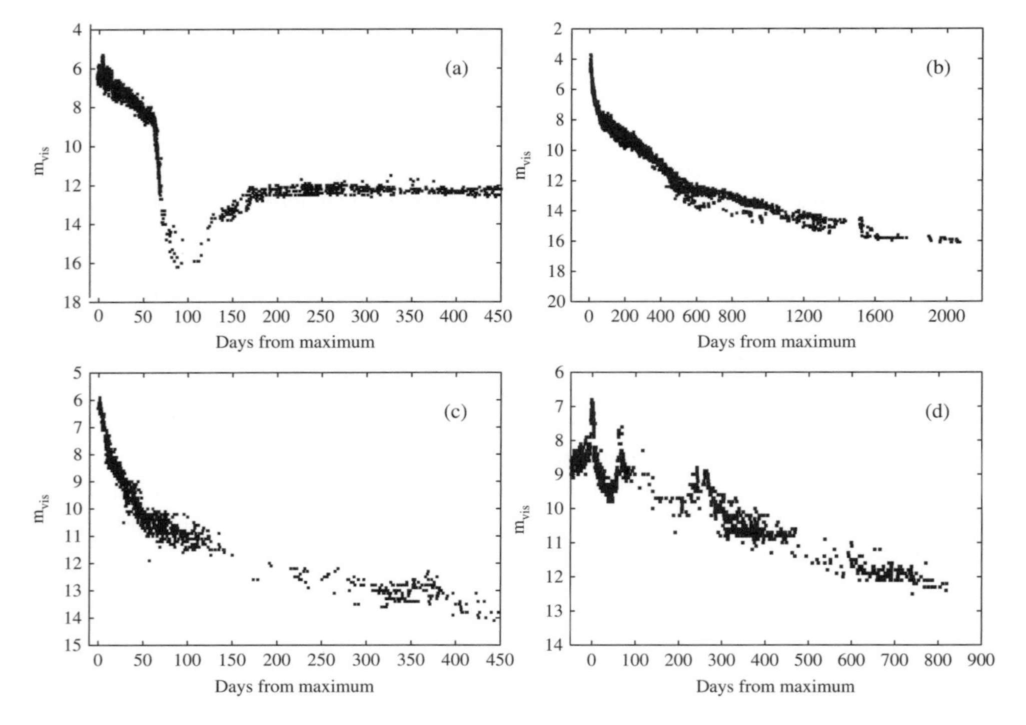

Fig. 13.2. Light curves of dusty and non-dusty novae. (a) V705 Cas, (b) V1974 Cyg, (c) V1668 Cyg, (d) V723 Cas. Data compiled from AFOEV (Association Française des Observateur d'Etoiles Variables).

13.3.3 The infrared

The nature of the data

Unlike broadband optical photometry, which conveys little or no information about the nature of the nova spectrum, broadband infrared photometry has been useful in the investigation of nova dust and molecules.

Early broadband infrared photometry of dusty novae photometry provided very limited information about the nature of nova condensates, other than the temperature (generally determined by fitting a black body to the broadband data) and that the dust emission seemed featureless. On the grounds that CNO are over-abundant in nova ejecta, it was widely believed that the condensate is a simple form of carbon.

However, even the crude broadband photometry was sufficient to demonstrate that novae that do produce dust do so with varying degrees of efficiency. In some cases (such as NQ Vul, FH Ser), the power re-radiated by the dust in the infrared was comparable with the power emitted in the eruption, and the visual light curve displayed a deep minimum indicating that the optical depth in the dust shell is $\gg$1; in these cases it is clear that the dust shell completely covers the nova 'sky'. In other cases $\tau \ll 1$ and the infrared excess is barely discernible.

The now-routine infrared spectroscopy of novae during eruption has revealed a rich variety of mineralogical dust types in nova ejecta, including silicate and hydrocarbons. In addition, other condensates such as SiO_2 and SiC have also been reported (see e.g. the summary in Gehrz *et al.* (1998)).

The dust continuum

Even broadband infrared observations are sufficient to show that dusty novae display a broad ($\sim$2 μm to $\gtrsim$20 μm) continuum, and this is verified by spectroscopy. Given the over-abundance of C relative to solar values in nova winds that results from the thermonuclear runaway (TNR), it seems likely that the continuum is due to some form of carbon, which has relatively featureless emissivity over this wavelength range. What makes this interpretation awkward, for reasons that we shall outline below, is the simultaneous presence of silicate features superimposed on the continuum.

The 9.7 μm silicate feature in novae

The silicate feature at 9.7 μm is well known in the spectra of astronomical objects, and arises from the stretching of the Si−O bond; the weaker 18 μm feature arises from the bending of the O−Si−O bond. For this reason the 9.7 and 18 μm silicate features merely indicate the presence of some form of silicate: they are not by themselves useful diagnostics of the precise nature of the silicate, which requires observations at longer wavelengths. A comprehensive list of features associated with the various crystalline silicates is given by Molster *et al.* (2002).

The 9.7 μm silicate feature was first seen in nova V1370 Aql (1980) by Gehrz, Grasdalen and Hackwell (1985) and Roche, Aitken and Whitmore (1984), and a mini-survey of nova silicate features was given by Smith *et al.* (1995) (see Figure 13.3). Although the observations are relatively sparse, it is clear that a range of silicate mineralogies is present in nova dust. Contrast, for example, the relatively sharp and featureless 9.7 μm feature seen in V705 Cas (which resembles that in the dust-producing O-rich supergiant μ Cep) with the broad and structured feature seen in V838 Her and V1370 Aql. It is tempting to associate the broader 9.7 μm silicate features, such as those seen in V838 Her, with a degree of crystallinity in the silicate (see Figure 13.3), whereas the sharper feature arises in amorphous silicate. However, given the fact that crystalline silicate features appear at wavelengths that are inaccessible from the ground, this can be verified only by observations from space-borne infrared observatories. The corresponding 18 μm feature has been observed in V705 Cas (Evans *et al.*, 1997, 2005), although it was barely discernible above the continuum provided by the (presumably) carbon dust. As discussed below (Section 13.7.2), this indicates that the silicate was freshly condensed.

UIR emission

The presence of broad spectral features centred at 3.29, 3.4, 6.6, 7.7 and 11.3 μm has long been known in a wide range of astronomical sources with strong ultraviolet emission (see Allamandola, Tielens and Barker (1989) for an early, comprehensive, review). Until their identification with emission by polycyclic aromatic hydrocarbon (PAH) molecules, they were variously referred to as the Unidentified InfraRed (UIR) and Overidentified InfraRed (OIR) features; here we shall refer to them as the UIR features. They are generally ascribed to non-equilibrium heating of PAH molecules which occupy the grey area between large

Table 13.2. *UIR features in novae*

λ_{max} (μm)	Identification in novae	Identification in other objects
3.29	Aromatic C–H stretch	Aromatic C–H stretch
3.4	CH_2/CH_3 inclusions in silicate	Aliphatic C–H stretch
6.2	Aromatic C–C stretch	Aromatic C–C stretch
7.7–8.2	$Si-CH_3$ (8.1 μm)	Blend of aromatic C–C stretching bands
8.7		Aromatic C–H in-plane bend
9.2	?	...
11.25	Aromatic C–H out-of-plane bend	Aromatic C–H out-of-plane bend

After Evans *et al.* (2005).

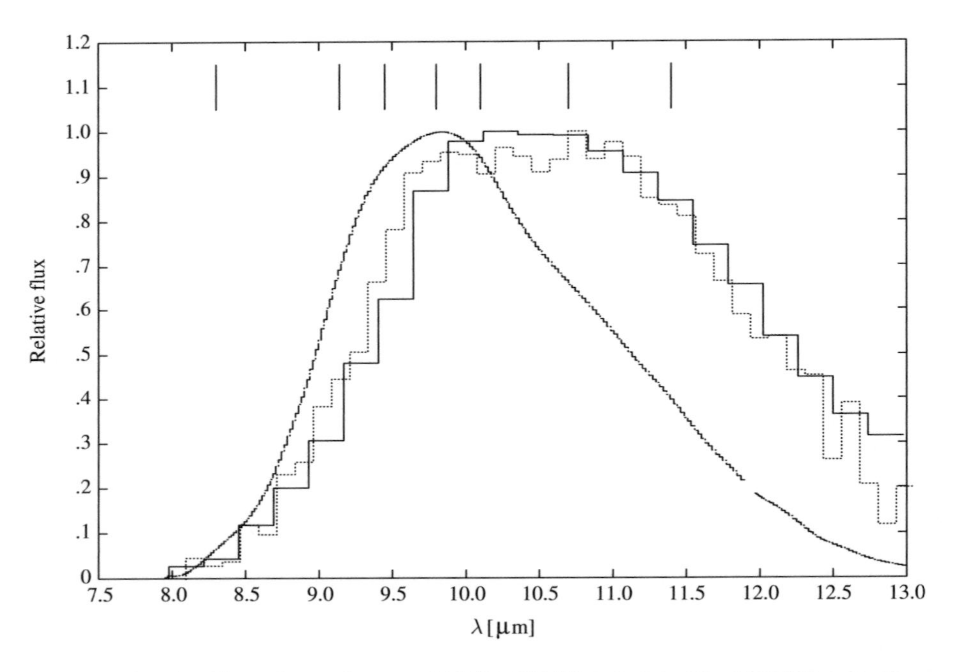

Fig. 13.3. The silicate feature in novae. Full line, V1370 Aql, dotted line V838 Her; data from Smith *et al.* (1995). The dash-dot line is the silicate feature in the O-rich supergiant μ Cep from the ISO archive. The tick marks are the wavelengths of the most prominent crystalline silicate features. After Molster *et al.* (2002).

molecules and small grains (Sellgren, 1984). The most prominent UIR features in novae, their (nearest) equivalents in other astronomical objects and their possible identifications are listed in Table 13.2.

The UIR features can, however, also arise in hydrogenated amorphous carbon (HAC), the corresponding transitions occurring by stretching and bending of C−H bonds on the surfaces of grains having conventional (rather than molecular) dimensions (Duley & Williams, 1981).

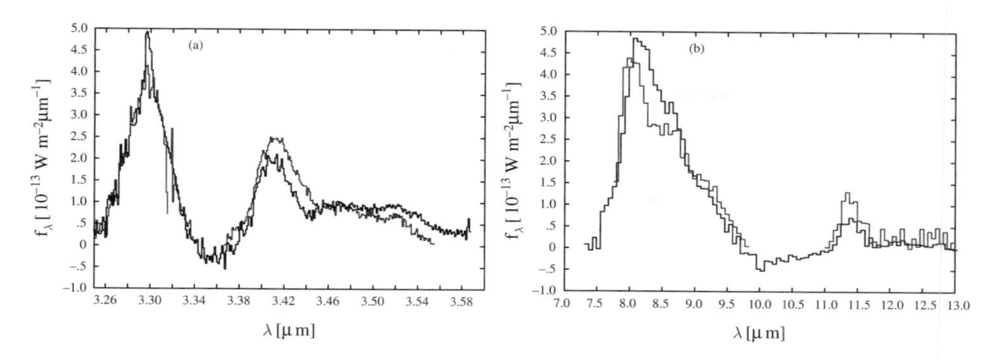

Fig. 13.4. UIR emission in V705 Cas (Evans *et al.*, 1997), (a) in the 3 μm band, (b) in the 7–13 μm band. Note the strength of the '3.4' feature relative to the '3.28', and that the peak wavelength of the '7.7' feature is closer to 8.2 μm.

The UIR features were first seen in a nova (V842 Cen) by Hyland and McGregor (1989), although their presence had been predicted by Mitchell and Evans (1984); they have subsequently been observed in a number of dust-forming novae (Gehrz *et al.*, 1998). It soon became apparent that there are a number of substantial differences between UIR features in novae and those seen in other sources (see Geballe (1997) for a review). In novae the 3.4 μm/3.28 μm flux ratio is higher than it is in objects such as planetary nebulae and H II regions (where the ultraviolet radiation field is also strong), but is similar to that seen in post-AGB stars. Also, the '7.7' feature seemed to be shifted to 8.2 μm in V705 Cas, while the '11.25' feature was at 11.38 μm. The UIR features in V705 Cas are shown in Figure 13.4.

The differences between nova UIR features and those seen in other astrophysical environments suggest that, while the carrier is broadly similar in that it is some form of hydrocarbon, there are significant differences. These differences may arise from the extreme over-abundance of some species (such as N) in the nova wind arising from the TNR. In this context it is of interest that Roche *et al.* (1996) had found that the relative prominence of the UIR 3.4 μm feature is correlated with nitrogen over-abundance. On the other hand, Evans *et al.* (2005) attributed this to CH_2 and CH_3 inclusions in silicate dust.

There has been some discussion as to whether the UIR features in novae arise (as they seem to in other objects in which they are seen) in free-flying PAH molecules, or from HAC. It is likely that both types of particles are present in nova winds, with PAH molecules at the small particle end of a grain size distribution, occupied at the large particle end by HAC particles of dimensions $\gtrsim 0.5$ μm (Evans *et al.*, 1997). However, Evans and Rawlings (1994) have argued that PAH molecules are unlikely to survive the fierce ultraviolet radiation field of novae on the time-scale for which the UIR features are actually observed.

13.3.4 *The millimetre range*

There have been several observations of novae at millimetre and sub-millimetre wavelengths. By and large these observations detect free–free emission from hot gas in the ejecta (Ivison *et al.*, 1993). That the emission is free–free is evidenced by the frequency independence of f_ν once the gas has become optically thin.

Nielbock and Schmidtobreick (2003) concluded, however, that their 1.2 mm observation of V4743 Sgr had detected dust rather than hot gas. This is based on the determination of the so-called β-index for the dust, defined such that the emissivity of the dust is $\propto \nu^{\beta}$ at very long (e.g. sub-millimetre) wavelengths (see e.g. Mennella *et al.* (1998)). Since dust emission is invariably optically thin at these wavelengths at late times, it is easily seen that $f_\nu \propto \nu^{2+\beta}$, so that β is determined if there are observations at two or more wavelengths; indeed, this is a common method of determining the β-index for dust in mass-losing stars. In the case of a nova the value of the β-index can, in principle, discriminate between dust and free–free emission. Nielbock and Schmidtobreick (2003) came to the conclusion that not only did the 1.2 mm emission have its origin in dust, but that the dust was old, very likely pre-dating the 2002 eruption.

13.4 Chemistry in nova winds

The existence of molecules and the (often complex) chemical pathways to their formation are very well known in the context of the ISM. However, compared with the ISM, the ejecta of novae are very much hotter (typically ~500–5000 K) and are irradiated by a harsh, photodissociating and photoionizing ultraviolet radiation field that is many orders of magnitude stronger than the interstellar radiation field.

At first sight, these conditions may not seem at all auspicious for the formation of any form of molecular species. However, from a purely phenomenological viewpoint we know that in a major subset of all novae, efficient molecule formation *must* be occurring. This is because, in these novae, efficient dust formation is seen. The transition from a partially ionized, atomic gas to macroscopic dust grains clearly must involve molecules as intermediates. Whether those molecules are detectable and/or the chemical pathways identifiable is open to debate.

As discussed above, there is direct spectral evidence for the formation of simple diatomic molecules, such as CN and, most notably, CO. These simple molecules are usually detected in the early stages of a nova's evolution – certainly prior to the dust-formation epoch (see Table 13.1), although simple diatomic molecules have been detected in the resolved remnants of 'old' novae (see Section 13.8), long after dust formation.

Whether or not these simple pre-dust-nucleation molecules are important intermediates to the formation of larger molecules and dust grains, or are products of chemical processes that are unrelated to the dust-formation process, is not immediately obvious, but has been the subject of considerable scrutiny, as described below.

In addition to the simple molecules, there is also evidence for the existence of more complex hydrocarbons (e.g. the UIR features discussed above). It would seem reasonable to consider such molecules as the obvious molecular intermediates between atomic carbon and solid state carbon grains. However, their presence in the pre-dust epochs has so far been very elusive. The detections are made well after dust formation indicating that if what we are really seeing are free-flying hydrocarbon/PAH molecules then they are most probably the 'left-overs' of grain formation, or else decomposition products of dust grains.

13.4.1 Formation of simple molecules

Although, as outlined above, nascent molecules are subjected to a harsh environment and will have only a short lifetime, there are a variety of factors which are conducive to rapid and efficient molecule formation:

(1) First, and most obviously, the very high densities in novae ejecta ($\sim 10^{10} - 10^{13}\,\mathrm{cm}^{-3}$, as compared with $\sim 10^{4} - 10^{6}\,\mathrm{cm}^{-3}$ in dark clouds in the ISM) mean that the reaction

rates (which scale as the square of the density for bimolecular reactions) are significantly faster than in interstellar clouds. Indeed, a variety of 'non-standard' reaction types, such as three-body reactions, are found to be significant in novae.

(2) Second, the very high temperatures can also result in enhanced reaction rates for many (but not all) reactions, and certain additional reaction pathways, which are prohibited in cold environments because of the presence of chemical activation energy barriers, become available.

(3) Third, the very high metallicities of nova winds encourage the formation of molecules. Apart from the obvious sense that enhanced abundances of carbon and oxygen lead to enhanced abundances of carbon- and oxygen-bearing molecules etc., there is an important indirect effect to be considered, namely the effective partial blanketing of the harsh radiation field by ionization continua and bound–bound absorption features.

Putting these various factors together, we find that the typical time-scales for the formation (and destruction) of molecules in young novae are of the order of seconds to hours. This should be compared with the time-scales of 10^4-10^6 years in interstellar clouds. The implication is that the nova chemistry is in a quasi-equilibrium state, with the abundances of the various molecular species determined by the balance of the formation and destruction mechanisms whose rates are defined by the (time-dependent) physical constraints.

The earliest attempts to constrain the efficiency of CO formation indicated that in the test case of nova NQ Vul, some $\sim 1-10\%$ of the carbon was locked up in the form of CO, depending on what one assumes for the ejecta configuration (Rawlings, 1988). This result was interesting in that whilst it implied that efficient CO formation must be occurring in the ejecta, the net formation efficiency is not sufficient for CO saturation to occur. This immediately suggests that the usual paradigm for carbon/silicate dust formation in carbon/oxygen-rich outflows (as described in Section 13.2.1 and Section 13.6 below) needs to be re-assessed.

The initial approach to modelling the formation of molecules (Rawlings, 1988) was to assume that the ejecta are spherically symmetric and are illuminated by a radiation field characterized by a single, time-dependent, black-body temperature. As that temperature rises (owing to photospheric contraction) and the ejecta disperse, the expanding wind is overtaken by a succession of ionization fronts.

In fact the photophysics is quite complex. At high optical depths, the photodissociation absorption bands of H_2 become a quasi-continuum which overlaps most of the carbon ionization continuum. Therefore, in an extreme example of what have become known as photon-dominated regions, the neutral carbon region has an H_2:H ratio which is essentially unlimited by photolysis reactions, whereas in the C^+ ionization zone, the hydrogen is essentially all atomic. Moreover, at high optical depths the CO and H_2 photodissociation bands overlap so that CO is also protected against photodissociation in a C I region.

Although the conditions are very different from what is found in the ISM, some aspects of the chemistry are similar. Most notably, in the absence of H_2, the initiating reactions of chemical networks are radiative association reactions, e.g.

$$C^+ + H \rightarrow CH^+ + h\nu.$$

However, these are invariably slow and, bearing in mind the rapidity of the reverse reaction in unshielded regions, inefficient processes. Thus, as in the ISM, H_2 is found to be an essential pre-requisite to a vigorous chemistry taking place. Reactions such as

$$C^+ + H_2 \rightarrow CH^+ + H$$

are much faster than radiative association reactions (especially at the high temperatures appropriate to nova ejecta). So, we find that molecule formation can only occur efficiently in an essentially neutral (C I) region. In the context of the simple physical model of the nova wind/ionization, the existence of such a region necessarily implies the presence of a dense shell bounding the outflow.

However, unlike the ISM, gas-phase chemical reactions must be relied upon for the formation of H_2. At the highest densities (i.e. very shortly after outburst), three-body formation of H_2 is efficient:

$$H + H + H \rightarrow H_2 + H,$$

but thereafter the high temperatures and ionization levels drive the alternative dominant formation mechanism via the H^- ion:

$$H + e^- \rightarrow H^- + h\nu$$

$$H^- + H \rightarrow H_2 + e^-.$$

The H^- ion is susceptible to photodetachment:

$$H^- + h\nu \rightarrow H + e^-$$

but this mechanism is more efficient than alternative H_2 formation routes. In hot, dense regions where the H_2 is shielded against photodissociation, the main loss channel for H_2 is collisional dissociation:

$$H_2 + H \rightarrow H + H + H.$$

Reactions of this type result from collisional excitation of the vibrational ladder, leading to excitation into the vibrational continuum. As a result they are highly sensitive to *both* the density and the temperature.

A severely limiting factor in the calculations is our extremely poor comprehension of the photoreaction cross-sections/reaction rates. Also, unlike the 'cold' assumption made in interstellar chemical calculations, the issue is further complicated by the fact that the molecules are likely to exist in highly excited ro-vibrational, if not Rydberg, states. In fact, more recent studies suggest that the radiation field may be less of a controlling factor in the chemistry than was previously thought to be the case. For example, over the period 1992–1997, Hauschildt *et al.* (see Chapter 5) have developed spherically symmetric, non-LTE, line-blanketed models for the photospheres of novae in expanding atmospheres. The 'optically thick wind' phase has a very complicated equation of state, with the ultraviolet spectrum becoming dominated by extreme blanketing by a forest of Fe II absorption lines. This is the epoch (of limited duration) that can support regions of relatively low ionization and is the most conducive to molecule (and dust) formation. Pontefract and Rawlings (2004) have re-considered the models of the early phase chemistry in the light of these recent developments in our understanding of the radiative transfer and also on the basis of improved TNR modelling (as descibed in Chapter 4). With these, very significant, modifications the rather suprising result is that the chemistry is found to be essentially not photon-dominated, but dominated by simple neutral–neutral reactions and, to a lesser extent, by ion–molecule reactions.

Generally, the dominant reaction types are: neutral–neutral reactions (including direct radiative ionizations), collisional dissociations, negative ion reactions, three-body reactions and photodissociation/ionizations.

The direct radiative association reaction

$$C + O \rightarrow CO + h\nu,$$

which is important in the hydrogen-poor environment of SN1987A (Dalgarno, Du & You, 1990; Rawlings & Williams, 1990) is not important in typical nova outflows. Instead, in the early stages, CO is formed by

$$C + H_2 \rightarrow CH + H$$

$$CH + O \rightarrow CO + H$$

and

$$C^+ + H_2 \rightarrow CH^+ + H$$

$$CH^+ + O \rightarrow CO^+ + H$$

$$CO^+ + H \rightarrow CO + H^+.$$

The molecules C_2 and CN also have simple formation channels, such as

$$N + OH \rightarrow NO + H$$

$$C + NO \rightarrow CN + O$$

and

$$C + CO \rightarrow C_2 + O$$

$$C + CH \rightarrow C_2 + H.$$

Typical results from the model are shown in Figure 13.5. The models confirm that H_2 is a vital prerequisite for the chemistry and that the molecule/dust-forming regions must be carbon-neutral (and hence confined to dense shells at the outer periphery of the wind). Substantial abundances of H_2, CO and CN can only be produced if the density is high. However, if the ejecta are too hot, then collisional dissociation – the dominant loss process for many key species – suppresses the abundances of most molecular species. Importantly, in all model calculations it was found that the CO abundance only reaches saturation levels for a short period of time. The abundance decline at late times occurs as a result of a variety of ion–molecule and neutral–neutral reactions. This result is strongly suggestive of the fact that carbon dust forms in an environment in which free oxygen is present.

13.4.2 Formation of nucleation sites

Novae only contribute a very small fraction ($\sim$0.1%) of the total budget of Galactic dust that originates from stellar outflows but may be major sources of astrophysically significant isotopes, such as ^{22}Na and ^{26}Al. The dust shells have a wide range of optical depths and compositions. However, both observational evidence and the TNR models (see Chapter 4) suggest that $C < O$ in the ejecta of most novae (although theory suggests $C > O$ for the highest white dwarf masses) and that the chemical and physical characteristics of novae

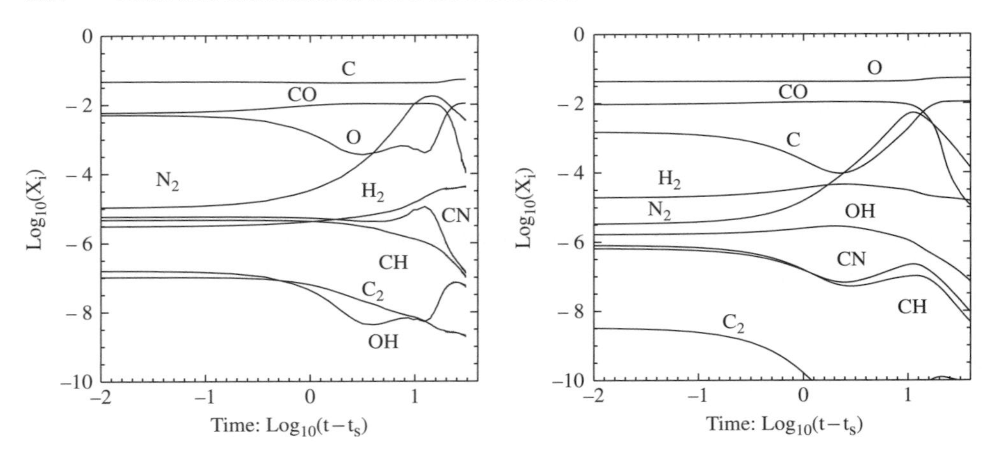

Fig. 13.5. Results from a model of the time-dependent chemistry in the early stages of a nova wind. Left, for a carbon-rich (C > O) chemistry; right, for an oxygen-rich (O > C) chemistry. Results are given as the log of the fractional abundances (relative to hydrogen) vs. the log of the time, measured in units of days. $t_s = 3$ days is the start time, post visual maximum. From Pontefract and Rawlings (2004).

are not too diverse. How such a wide variety of dust properties can be produced from these objects is something of a mystery and suggests that no one single dust-formation mechanism is applicable to all dust-forming novae.

Whilst the photophysics of small molecules is poorly constrained, the photodissociation properties (and indeed, even the structures) of larger molecules enter the realm of pure conjecture. For this reason, there have been very few attempts to study the chemistry leading up to the dust grain nucleation epoch. Rawlings and Williams (1989) were able to deduce some qualitative conclusions, but owing to the huge uncertainties were able to make only very rough quantitative predictions.

At the outset, it must be made clear that the chemical pathways for the formation of dust nucleation sites must be very different from laboratory-determined mechanisms. Detailed carbon combustion chemistries have been developed which invoke reaction mechanisms between large molecules (aromatic, PAH-like). Such channels, although they are appropriate to the dense chemically stable conditions within flames, are quite inappropriate to the nova environment, where large molecules are expected to have only a very short lifetime against photodissociation. Indeed, even the smaller molecules, such as acetylene (C_2H_2), which are often invoked as carbon-insertion monomers, are highly unstable.

Apart from the more obvious thermodynamic constraints, usually stated in the form of a nominal condensation temperature for any particular condensate (silicate, carbon etc.), dust nucleation can occur only if the chemical conditions are conducive to the formation of nucleation sites. It is important to realise that nova ejecta are very far away from being in local thermodynamic equilibrium. As established above, dust nucleation occurs in a C I region. Simple radiative associations and homogeneous nucleation mechanisms such as

$$C_n + C \rightarrow C_{n+1} + h\nu$$

are very slow and inefficient.

In Rawlings and Williams (1989), a model of the hydrocarbon chemistry up to and including C_7- and C_8-bearing species was developed. These species were identified as representative of the kinetic bottleneck in the chemistry: larger species are typically more stable against photodissociation and other destruction pathways, primarily as a result of ring closure. The models indicated that a sufficient abundance of nominal nucleation sites could be produced, but only if a significant pre-existent H_2 abundance was present. The gas is not sufficiently hot or ionized to support efficient H_2 formation in this epoch. Consequently, the formation of higher hydrocarbons was found to take place in a 'burst' of molecule formation, typically lasting a few days, after which time the supply of H_2 becomes exhausted. The key species in the growth pathways are limited by the photochemistry to be atomic and very small molecular ions: C, C^+, C_2, C_2^+, CH, CH^+ and C_2H^+. A fractional ionization between 10^{-5} and 10^{-3}, and a pre-existent H_2 fractional abundance of 10^{-3}, are required for effective nucleation site formation. This latter constraint is itself determined by the thermal and chemical history of the ejecta.

A more rigorous attempt at modelling the formation of nucleation sites has been performed by Pontefract and Rawlings (2007). Two approaches were investigated. The first was an updated form of the hydrocarbon model of Rawlings and Williams (1989, 1990), incorporating oxygen chemistry, using a more realistic determination of the photorates (following the developments in the model atmosphere calculations) and replacing many of the hypothetical reactions and rates by empirically confirmed channels. The results indicated that, unlike the previous models, rapid and efficient formation of large hydrocarbon species is not problematic. The growth mechanisms are simple, largely involving C atoms and H_2 molecules. The presence of oxygen inhibits hydrocarbon formation, via neutral exchange reactions, e.g.

$$CH + O \rightarrow CO + H$$

but, importantly, it was still possible to produce viable abundances of large hydrocarbons, even with significant quantities of free oxygen atoms present. An example of the results is shown in Figure 13.6 which shows the time-dependent evolution of the C_n-type species for several different combinations of ejecta density, temperature and fractional abundance of atomic oxygen. Note the high abundances and extended lifetimes of the large molecular species.

A second approach used a modified version of the speculative hydrocarbon chemistry described above, coupled with an empirically determined PAH chemistry. The latter was based on the models of Cherchneff, Barker & Tielens (1992) which was developed for the purpose of studying hydrocarbon formation in cool carbon-rich stellar envelopes, up to and including cyclopenta[cd]pyrene ($C_{18}H_{12}$). Such models are appropriate to regions of lower temperature and ultraviolet flux strengths than are found in novae. They are largely based on acetylene (C_2H_2) flame chemistries (e.g. Frenklach *et al.*, 1984), which rely on condensation reactions involving C_2H_2 and the vinyl radical (C_2H_3) to form phenyl (C_6H_5) and benzene (C_6H_6). More complicated species are then formed by successive additions of C_2H_2 to create multiple ring (PAH) molecules. In the context of nova winds, the models do not include photoreactions and are overly dependent on the C_2H_2 growth monomer which, as pointed out above, is not stable in the nova wind environment. However, by careful inclusion of the 'missing' chemical reaction pathways it was possible to generate a model which is, again, very successful at producing putative nucleation sites with the required abundances. The models were capable of producing *stable* abundances of large hydrocarbons. As with

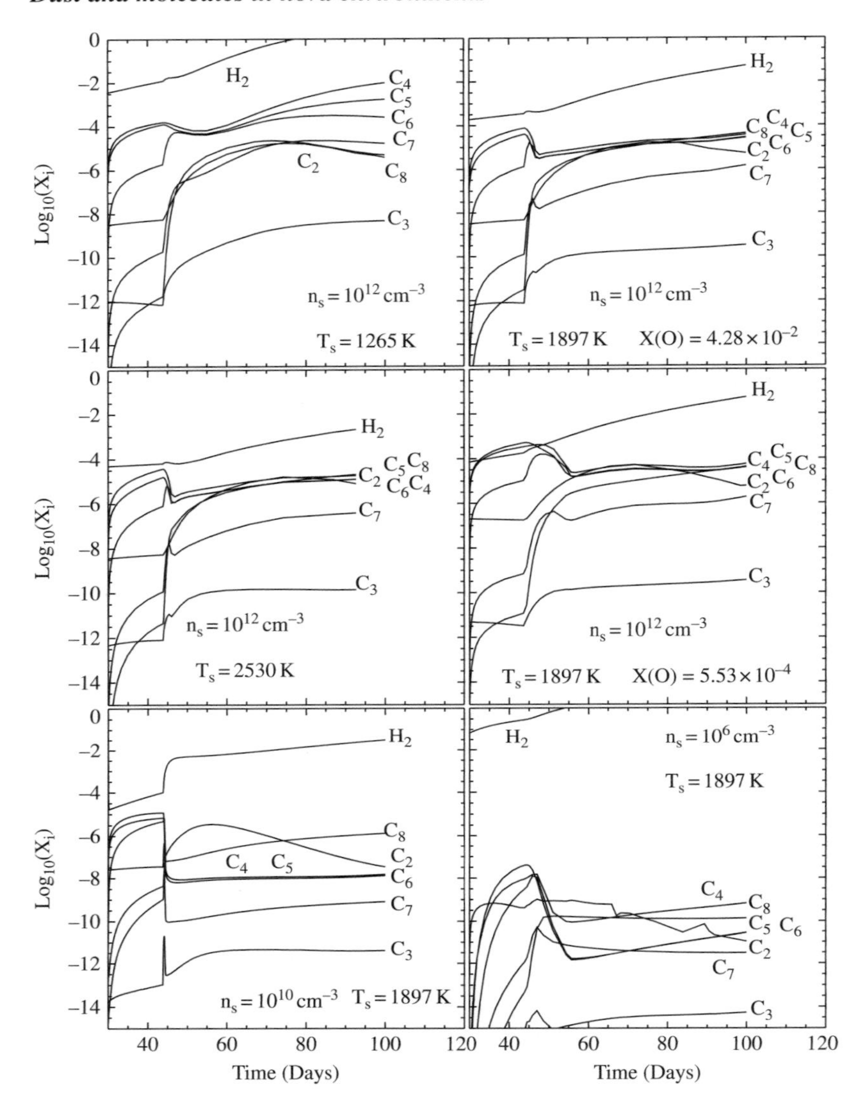

Fig. 13.6. Examples of results from a model of the time-dependent chemistry leading to the formation of hydrocarbon-based dust nucleation sites. Results are presented as the log of fractional abundance (relative to hydrogen) as functions of time since outburst, for several values of the ejecta density, temperature and fractional abundance of oxygen atoms. From Pontefract and Rawlings (2007).

the simpler models it was found that the formation efficiency is very temperature-sensitive and the presence of oxygen does not seriously inhibit the hydrocarbon chemistry. In a sense these models are *too* successful in producing nucleation sites in that they do not clearly discriminate between those novae that are able to produce dust and those that do not. The various models are still highly speculative in many respects, but seem to converge on several key conclusions:

(1) Provided the conditions are conducive – that is to say, the ejecta are sufficiently cool, neutral and dense – rapid and efficient formation of large hydrocarbon species, which can be thought of as dust grain nucleation sites, can occur.
(2) The chemical pathways are dominated by reactions with small molecular species and photodissociation. This is most unlike the situation in, for example, cool stellar atmospheres.
(3) The presence of free oxygen atoms, whilst depressing the hydrocarbon chemistry, does not inhibit it completely.

13.5 Dust formation and growth

The temperature T_d of a dust grain situated a distance $r = V_\mathrm{ej}t$ from a star of bolometric luminosity L is given from elementary black-body physics by the balance between absorption and emission of radiation by the grain:

$$T_\mathrm{d} = \left[\frac{L}{16\pi r^2\sigma}\frac{\langle Q_\mathrm{a}\rangle}{\langle Q_\mathrm{e}\rangle}\right]^{1/4} = \left[\frac{L}{16\pi V_\mathrm{ej}^2 t^2\sigma}\frac{\langle Q_\mathrm{a}\rangle}{\langle Q_\mathrm{e}\rangle}\right]^{1/4} \tag{13.1}$$

where $\langle Q_\mathrm{a}\rangle$ and $\langle Q_\mathrm{e}\rangle$ are respectively the Planck mean absorptivity and emissivity of the grain material. At its very simplest, therefore, dust formation can be outlined as follows. The nova radiates at constant bolometric luminosity L and the ejecta move with velocity V_ej away from the site of the explosion. If the grain material has a 'condensation temperature' T_cond, then

$$t_\mathrm{cond} = \left[\frac{L}{16\pi V_\mathrm{ej}^2\sigma T_\mathrm{cond}^4}\frac{\langle Q_\mathrm{a}\rangle}{\langle Q_\mathrm{e}\rangle}\right]^{1/2}$$

for the condensation time t_cond. Taking graphitic carbon as a simple example, $T_\mathrm{cond} \simeq 1200$ K, $\langle Q_\mathrm{a}\rangle \simeq 1$ and $\langle Q_\mathrm{e}\rangle \simeq 0.01aT^2$, where a is grain radius in centimetres and T grain temperature in Kelvin,[2] we have

$$t_\mathrm{cond} \sim 0.078\left(\frac{L}{\mathrm{L}_\odot}\right)^{1/2}\left(\frac{V_\mathrm{ej}}{1000\,\mathrm{km\,s^{-1}}}\right)^{-1} \text{ days.}$$

With typical values of L and V_ej for dust-forming novae, we get $t_\mathrm{cond} \sim 30$ days, as observed (see Table 13.1). Indeed, given the speed-class dependence of both L and V_ej (see Chapter 2), a dependence of t_cond on speed class might be expected. However, despite efforts to find one (e.g. Gallagher, 1977; Bode & Evans, 1982), there seems to be no general correlation between the speed class of a nova and its ability to produce dust. We shall argue below that the dust-forming ability of a nova is in fact determined by several parameters.

The situation is of course far more complex than this. Shore and Gehrz (2004) and Rawlings and Evans (in preparation) have noted the following necessary conditions for grain formation:

(1) There must exist nucleation sites on which dust grains can form. This has posed the most severe challenge, as the chemical routes to the formation of nucleation sites require an environment that is well shielded from the hard radiation field of the nova (see Section 13.4).

[2] More generally $\langle Q_\mathrm{e}\rangle \propto aT^\beta$, where the β-index for the dust is defined in Section 13.3.4.

(2) Nucleation can only occur in regions where the gas density is much higher than the average in nova ejecta. This requires the existence of significant density enhancements.
(3) The temperature of the forming grains must be less than the Debye temperature for the grain material.
(4) The formation of dust must be thermodynamically feasible; condensation can only occur at a 'condensation radius' determined by the ejecta density, the nova radiation field and the nature of the condensate.
(5) Carbon dust forms preferentially in an environment in which C > O, and vice versa for silicate (or any O-rich grain).

Condition (4) essentially requires that the region in the ejecta in which dust formation occurs must occupy the appropriate region of the phase diagram for the condensing material. In general, taking into account the mass-loss in the wind and the rising effective temperature of the stellar remnant, the *average* density in the ejecta is *too low* for dust nucleation and condensation to occur: condensation requires the presence of a region of the ejecta that is *denser* than average; this points to the outer edge of the ejecta, as noted in Section 13.4.

In addition, the discussion in the previous section indicates that dust nucleation and growth must occur in a region that is carbon-neutral (C I). Thus the density-enhanced region in which dust formation is potentially possible must reach the nominal condensation radius before it is overtaken by the outward-moving carbon ionization front. This requirement is shown, schematically, in Figure 13.7. In the upper half of the figure the dense shell is overtaken by the C II front before it reaches the condensation radius (r_{cond}) and therefore will not produce dust. In the lower half of the figure the physical conditions are such that (at least some part of) the dense shell reaches r_{cond} before it is ionized and is therefore a suitable site for dust formation and growth.

Once nucleation sites are available dust is expected to condense – indeed 'precipitate' – very quickly. Standard thermodynamic arguments give the condensation temperatures T_{cond} but in a circumstellar gas, T_{cond} depends not only on the gas pressure but also on the C:O ratio, nitrogen abundance and overall metallicity (Lodders & Fegley, 1999). In view of the wide range of C:O ratios, nitrogen abundance and metallicity encountered in novae, it is not therefore possible to list condensation temperatures for dusts in a 'typical' nova wind. Instead,

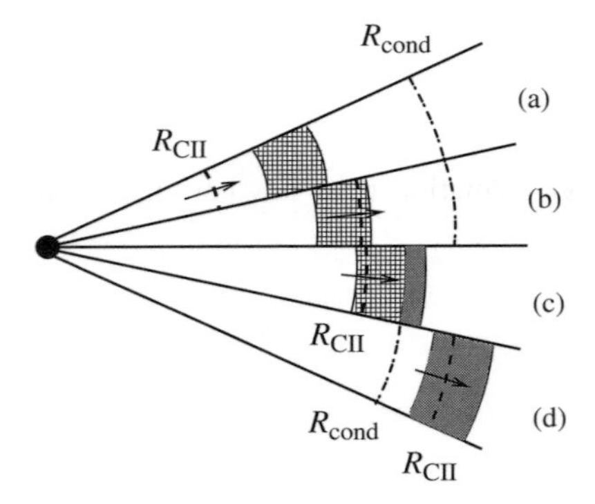

Fig. 13.7. A schematic depiction of the progress of a potential dust-forming ejecta shell (hatched area) and the carbon ionization front (R_{CII}) in a dustless nova (a and b) and a dust-forming nova (c and d). From Rawlings and Evans (in preparation).

Table 13.3. *Condensation temperatures for actual and potential condensates in nova winds, for C:O ratio indicated*

Condensate	Chemical formula	T_{cond} (K) for C:O ratio		
		0.85	1.00	1.15
Olivine	Mg_2SiO_4	1390	1190	1150
Ortho-pyroxene	$MgSiO_3$	1260	—	—
Graphitic C	C	—	1150	1690
Silicon carbide	SiC	—	1630	1720
Corundum	Al_2O_3	1690	—	1240
Cohenite	Fe_3C	—	1160	1470

After Ebel (2000)

in Table 13.3, we give the condensation temperatures for dusts commonly seen in nova winds for a gas pressure $P = 10^3$ dyne cm^{-2} and the C:O ratios indicated, the gas having solar composition for other species. The values are estimated from Ebel (2000) and a null entry indicates that condensation is unlikely for that combination of parameters. Also given, in view of the over-abundances of the constituent atoms, are the condensation temperatures for possible nova condensates.

The higher condensation temperature for carbon compared with any form of silicate for $C > O$ is consistent with the early appearance of amorphous C, but no silicate, in V705 Cas.

13.6 The 'chemical dichotomy' in nova dust

As discussed above, the conventional view is that an O-rich condensate will form if $O > C$ by number, otherwise a C-rich condensate forms. However, it is not unusual to see both O-rich (silicate) and C-rich condensates simultaneously in nova winds. This 'chemical dichotomy' is commonly seen in evolved stars, and is understood in terms of two separate mass-loss episodes, one when the star is O-rich, and a second – following a thermal pulse – when the star is C-rich (Zijlstra *et al.*, 2001).

In view of the belief that the C:O ratio was the crucial factor in determining the nature of the condensate, the dichotomy in nova dusts led to speculation about steep abundance gradients in the ejecta, non-uniform TNR on the white dwarf surface, or rapid depletion of C (or O) resulting in the ejecta flipping from C-rich to O-rich (or vice versa). However, a ready explanation is found in the fact that CO formation goes nowhere near saturation in nova winds, rendering the C:O ratio irrelevant in determining the nature of the condensate.

13.7 Processing of dust

13.7.1 Grain size

It is clear (Evans *et al.*, 1997) that nova grains are considerably larger (~ 0.5 μm) than those in the interstellar medium ($\lesssim 0.2$ μm), and that they attain these dimensions relatively quickly (Shore *et al.*, 1994; Evans *et al.*, 1997). Some circumstantial evidence for grain growth in the ejecta of V705 Cas was presented by Evans *et al.* (1997). In this case grains eventually grew to 0.7 μm, and the deduced initial mass-loss rate is $\sim 10^{24}$ g s^{-1}.

A more detailed analysis (Evans *et al.*, 2005), however, pointed to a mix of amorphous carbon and silicate grains, and to the fact that the grains might have been eroded since the early IUE observations of Shore *et al.* (1994).

13.7.2 *The silicate*

Nuth and Hecht (1990) have shown, on the basis of laboratory experiments, that the relative strength of the 9.7 μm and 18 μm silicate feature is governed by the 'freshness' of the silicate, in that newly formed silicate has a very low '18/9.7' ratio. As the silicate 'ages' by annealing in the circumstellar environment the ratio increases and the 18 μm feature becomes relatively more prominent. This is certainly borne out by the infrared spectroscopic observations of V705 Cas (Evans *et al.*, 1997, 2005), which displayed a barely discernible 18 μm feature and a strong, sharp 9.7 μm feature.

13.7.3 *The carbon*

From Equation (13.1) we expect that $T_d \propto t^{-2/(\beta+4)}$, so the dust temperature should decline monotonically with time as the grains move away from the site of the explosion. However, for cases where the dust optical depth is high (with consequent deep minimum in the visual light curve) the dust temperature may remain independent of time, or may even rise. This behaviour is (confusingly) known as the 'isothermal phase' of the dust shell.

One way of accounting for this is that there occurs a period of grain destruction (so that $da/dt < 0$ and $dT_d/dt \geq 0$). As neither grain evaporation (at temperatures $\lesssim 1000$ K) nor sputtering (in a gas at $\sim 10^4$ K, considerably below the threshold) were believed capable of delivering the necessary grain erosion, Mitchell and co-workers (Mitchell & Evans, 1984; Mitchell, Evans & Albinson, 1986) have suggested chemical sputtering of carbon grains by gas-phase H, leading to surface CH groups and UIR emission. More recently, however, both physi- and chemi-sputtering by H^+ are now known to be efficient at all relevant temperatures (Roth & García-Rosales, 1996).

The isothermal effect may, however, occur without recourse to grain destruction, as the physical nature of carbon grains changes (e.g. by annealing) thus changing the β-index, and as the H ionization front engulfs the grains, thus exposing them to the (hitherto unseen) Lyman continuum (Evans & Rawlings, 1994).

13.8 Nova remnants

The resolved remnants of old novae have not, in general, provided fertile ground for hunters of molecules and dust. A search for extended CO emission in relatively young nova remnants using the NICMOS instrument on the HST (Krautter *et al.*, 2002) was not successful, although Evans (1991) claimed a detection of extended H_2 emission in the resolved nebula associated with DQ Her.

The only nova which *has* shown evidence of molecules and dust in an extended 'remnant' is GK Per, although the dust and molecules in this extended envelope are unconnected with the 1901 nova eruption. From data in the Infrared Astronomy Satellite (IRAS) survey, this nova was found by Bode *et al.* (1987) to be at the centre of extended far-infrared emission, which is identified with the fossil planetary nebula ejected by the system in its pre-nova incarnation. Further studies of this system by Dougherty *et al.* (1996), using maximum entropy reconstruction of the IRAS images, revealed that the emission is resolved into discrete 'blobs', symmetrically disposed about the central star (see Figure. 13.8). The extent of the

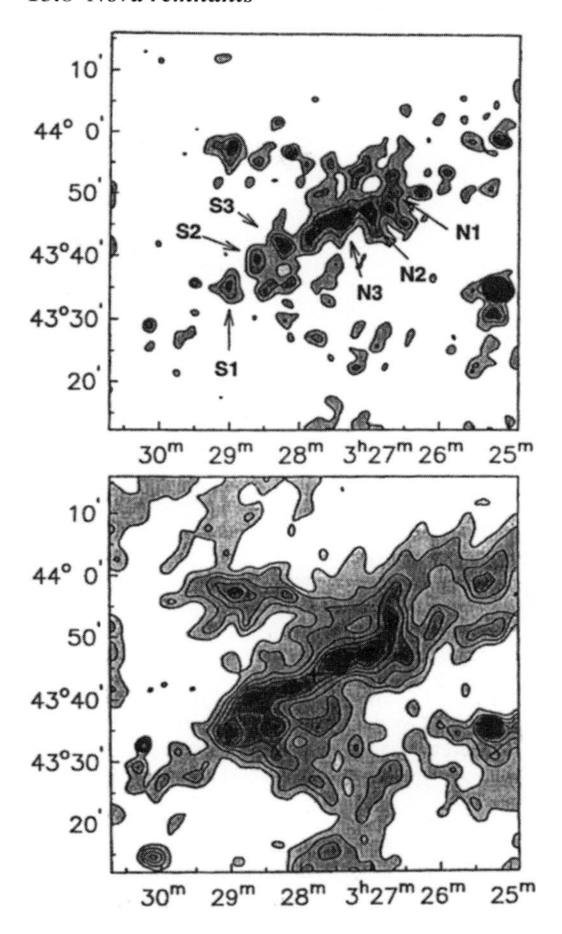

Fig. 13.8. The extended far-infrared emission associated with GK Per; (a) 60 μm map, (b) 100 μm. The 'blobs' labelled S1, N1 etc. may have resulted from separate mass ejection events. From Dougherty *et al.* (1996). Compare with Figures 12.9 and 12.10 in Chapter 12.

IRAS emission (Dougherty *et al.*, 1996), coupled with a reasonable value for the wind velocity, is consistent with the fact that the envelope material pre-dates not only the 1901 eruption but also the classical nova phase of the GK Per system. The dust mass and temperature, both of which are somewhat model-dependent, are $\sim 0.03\,M_\odot$ and ~ 25 K respectively; the total mass in the envelope depends on the gas-to-dust ratio assumed, but is typically $\sim 3\,M_\odot$ (Dougherty *et al.*, 1996).

The IRAS emission is accompanied by extended bipolar emission in the $J = 2 \rightarrow 1$ rotational transition of ^{12}CO (see Figure. 13.9; Scott, Rawlings & Evans, 1994); this is also symmetrically disposed with respect to the central nova, lending strong support to the 'fossil planetary nebula' hypothesis of Bode *et al.* (1987). Dougherty *et al.* (1996) argued that the envelope had been ejected by the white dwarf in the GK Per system following Roche lobe overflow from the (evolved) secondary onto the white dwarf, re-igniting the latter in a thermal pulse.

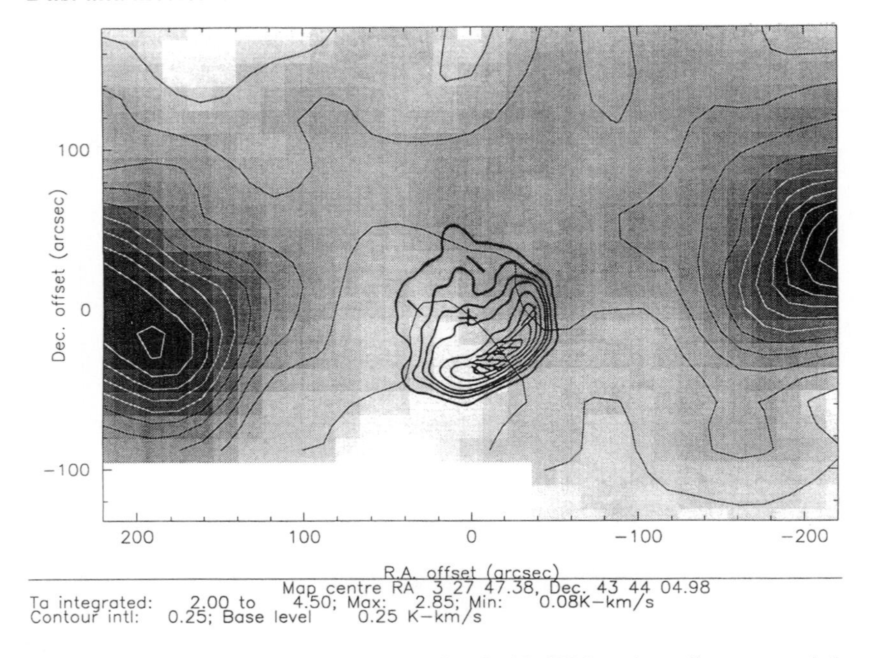

Fig. 13.9. The extended CO remnant associated with GK Per; the stellar remnant is located at (0,0). The CO emission is represented by the grey-scale and the thin-lined contours. The non-thermal radio remnant is represented by the thick-lined contours. From Scott, Rawlings and Evans, 1994. Compare with Figures 12.9 and 12.10 in Chapter 12.

13.9 Isotopic ratios

José *et al.* (2004) have emphasized the importance of determining isotopic abundances in nova ejecta to test TNR models to their limits: while relative elemental abundances are useful, it is the determination of *isotopic ratios* that really provides the most severe tests. José *et al.* (2004) have presented the results of a series of computations of elemental and isotopic ratios for TNR occurring on CO and ONe white dwarfs; they have further determined the possible condensates in the ejecta of a 1.35 $M_\odot$ ONe white dwarf.

In principle, the easiest isotopic ratio to determine in nova winds during eruption is $^{12}C/^{13}C$, using the fundamental and first overtone CO emission now commonly seen in the earlier phases of nova evolution. This is easily done with spectroscopy at modest resolution, as the ^{12}CO and ^{13}CO bandheads (at 2.29 μm and 2.5 μm respectively) are easily separated (see Figure 13.10). Some results are summarized in Table 13.4; the limit for DQ Her is from an analysis of CN (see Section 13.2.2) in the optical spectrum by Sneden and Lambert (1975).

In principle, the UIR features also furnish a means of determining the $^{12}C/^{13}C$ ratio, as laboratory spectroscopy shows that their peak wavelength depends in a simple way on the $^{12}C/^{13}C$ ratio for the carrier (Wada *et al.*, 2003). However, their value in this respect may be compromised by the idiosyncrasies of the nova UIR features.

In the context of nova dust, one of the most exciting developments of recent years has been the likely identification of nova dust in meteorites, although this is still a matter of controversy (see Chapter 6). The key to identification is the determination of the isotopic

Table 13.4. *The $^{12}C/^{13}C$ ratios in nova winds*

Nova	t_3 (days)	$^{12}C/^{13}C$	Ref.
V705 Cas	84	>5	Evans *et al.* (2005)
V2274 Cyg	33[a]	0.83 ± 0.3	Rudy *et al.* (2003)
V842 Cen	48	2.9 ± 0.4	Wichmann *et al.* (1990)
NQ Vul	65	>3	Ferland *et al.* (1979)
DQ Her	94	>1.5	Sneden & Lambert (1975)

[a] Based on t_2 in Rudy *et al.* (2003).

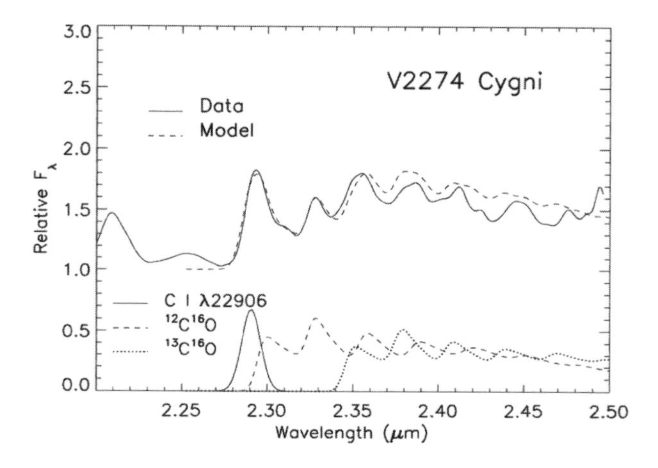

Fig. 13.10. The ^{12}CO and ^{13}CO emission in V2274 Cyg. From Rudy *et al.* (2003).

ratios for specific species, such as $^{14}N/^{15}N$, $^{12}C/^{13}C$, $^{26}Al/^{27}Al$. José *et al.* (2004) have drawn attention to the considerable potential and relevance of meteoritics to the study of nova dust.

13.10 Concluding remarks

Dust in nova winds is no doubt regarded as an inconvenience (and even an irrelevance) by some but as in the ISM, its existence and an understanding of its properties will almost certainly throw light on several facets of the nova phenomenon.

A great success of the TNR theory was the prediction that the nova eruption takes place at constant bolometric luminosity. At the time this was difficult to verify but in some ways it was a theory that surfaced at just the right time, with the advent of ultraviolet and infrared observations to complement the optical. It is for good reason the nova FH Ser is referred to as the 'Rosetta Stone' of nova studies: for the first time the energy budget of a nova could be integrated over the ultraviolet (Gallagher & Code, 1970), optical and infrared (Geisel, Kleinmann & Low, 1970), demonstrating conclusively that, as the flux in the optical declined, the flux at longer and shorter wavelengths rose to maintain constant bolometric luminosity: the optical fading was merely part of a much broader picture.

The detection of molecules (CN in the optical, CO in the infrared), and particularly the optical (deep minimum) and infrared evidence for dust, demonstrates that Nature can achieve with consummate ease that which our current understanding of the astrophysics and astrochemistry of dust finds excruciatingly difficult.

In particular, understanding the chemistry that drives the formation of the diatomic molecules and the nucleation sites has pushed the application (and hence our understanding) of astrochemical reaction networks to the limit, which will no doubt have a positive impact on the problems for which they were orginally devised, namely astrochemistry in molecular clouds.

References

Albinson, J. S., & Evans, A., 1989, *MNRAS*, **240**, P47.
Allamandola, L. J., Tielens, G. G. M., & Barker, J. R., 1989, *Ap. JS*, **71**, 733.
Bianchini, A., Sabbadin, F., Facero, G. C., & Dalmeri, I., 1986, *A&A*, **321**, 625.
Bianchini, A., Alford, B., Canterna, R., & Della Valle, M., 2001, *MNRAS*, **321**, 625.
Bode, M. F., & Evans, A., 1982, *MNRAS*, **200**, 175.
Bode, M. F., Seaquist, E. R., Frail, D. A. *et al.*, 1987, *Nature*, **329**, 519.
Cherchneff, I. C., Barker, J. R., & Tielens, A. G. G. M., 1992, *Ap. J.*, **401**, 269.
Cohen, M., Anderson, C. M., Cowley, A. *et al.*, 1975, *Ap. J.*, **196**, 179.
Dalgarno, A., Du, M. L., & You, J. H., 1990, *Ap. J.*, **349**, 675.
Dalgarno, A., Stancil, P. C., & Lepp, S., 1997, *Ap. &SS*, **251**, 375.
Diaz, M. P., Costa, R. D. D., & Jatenco-Pereira, V., 2001, *PASP*, **113**, 1554.
Dougherty, S. M., Waters, L. B. F. M., Bode, M. F. *et al.*, 1996, *A&A*, **306**, 547.
Duley, W. W., & Williams, D. A., 1981, *MNRAS*, **196**, 269.
Ebel, D. S., 2000, *J. Geophys. Res.*, **105**, 10 363.
Evans, A., 1991, *MNRAS*, **251**, 54P.
Evans, A., & Rawlings, J. M. C., 1994, *MNRAS*, **269**, 427.
Evans, A., Geballe, T. R., Rawlings, J. M. C., & Scott, A. D., 1996, *MNRAS*, **282**, 1049.
Evans, A., Geballe, T. R., Rawlings, J. M. C., Eyres, S. P. S., & Davies, J. K., 1997, *MNRAS*, **292**, 192.
Evans, A., Geballe, T. R., Rushton, M. T. *et al.*, 2003, *MNRAS*, **343**, 1054.
Evans, A., Tyne, V. H., Smith, O. *et al.*, 2005, *MNRAS*, **360**, 1483.
Ferland, G. J., Lambert, D. L., Netzer, H., Hall, D. N. B., & Ridgway, S. T., 1979, *Ap. J.*, **227**, 489.
Frenklach, M., Clary, D. W., Gardiner, W. C., & Stein, S. E., 1984, in *Twentieth International Symposium on Combustion*. Pittsburgh: The Combustion Institute, p. 887.
Gallagher, J. S., & Code, A. D., 1970, *Ap. J.*, **189**, 303.
Gallagher, J. S., 1977, *AJ*, **82**, 209.
Geballe, T. R., 1997, in *From Stardust to Planetesimals*, ed. Y. J. Pendleton, & A. G. G. M. Tielens, San Francisco: Astronomical Society of the Pacific, p. 119.
Gehrz, R. D., Hackwell, J. A., Grasdalen, G. L. *et al.*, 1980, *Ap. J.*, **239**, 570.
Gehrz, R. D., Grasdalen, G. L., & Hackwell, J. A., 1985, *Ap. J.*, **298**, L47.
Gehrz, R. D., Truran, J. W., Williams, R. E., & Starrfield, S., 1998, *PASP*, **110**, 3.
Geisel, S. L., Kleinmann, D. E., & Low, F. J., 1970, *Ap. J.*, **161**, L101.
Herbig, G. H., 1995, *ARA&A*, **33**, 19.
Hyland, A. R., & McGregor, P., 1989, in *Interstellar Dust*, IAU Symposium 135, Contributed papers, ed. L. J. Allamandola, & A. G. G. M. Tielens. Springfield: NASA Conference Publication, p. 101.
Ivezic, Z., & Elitzur, M., 1997, *MNRAS*, **287**, 799.
Ivison, R. J., Hughes, D. H., Lloyd, H. M., Bang, M. K., & Bode, M. F., 1993, *MNRAS*, **263**, L43.
José, J., Hernanz, M., Amari, S., Lodders, K., & Zinner, E., 2004, *Ap. J.*, **612**, 414.
Krautter, J., Woodward, C. E., Schuster, M. T. *et al.*, 2002, *AJ*, **124**, 2888.
Lodders, K., & Fegley, B., 1999, in *Asymptotic Giant Branch Stars*, *IAU Symposium 191*, ed. T. Le Bertre, A. Lébre, & C. Waelkens. San Francisco: Astronomical Society of the Pacific, p. 279.
McLaughlin, D. B., 1935, *Publ. AAS*, **8**, 145.
Mennella, V., Brucato, J. R., Colangeli, L. *et al.*, 1998, *Ap. J.*, **496**, 1058.

Mitchell, R. M., & Evans, A., 1984, *MNRAS*, **209**, 945.
Mitchell, R. M., Evans, A., & Albinson, J. S., 1986, *MNRAS*, **221**, 663.
Molster, F. J., Waters, L. B. F. M., Tielens, A. G. G. M., Koike, C., & Chihara, H., 2002, *A&A*, **382**, 241.
Nielbock, M., & Schmidtobreick, L., 2003, *A&A*, **400**, L5.
Ney, E. P., & Hatfield, B. F., 1978, *Ap. J.*, **219**, L111.
Nuth, J. A., & Hecht, J. H., 1990, *Ap. &SS*, **163**, 79.
Payne-Gaposchkin, C., 1957, *The Galactic Novae*. New York: Dover.
Pontefract, M., & Rawlings, J. M. C., 2004, *MNRAS*, **347**, 1294.
Rawlings, J. M. C., 1988, *MNRAS*, **232**, 507.
Rawlings, J. M. C., & Williams, D. A., 1989, *MNRAS*, **240**, 729.
Rawlings, J. M. C., & Williams, D. A., 1990, *MNRAS*, **246**, 208.
Roche, P. F., Aitken, D. K., & Whitmore, B. W., 1984, *MNRAS*, **211**, 535.
Roche, P. F., Lucas, P. W., Hoare, M. G., Aitken, D. K., & Smith, C. H., 1996, *MNRAS*, **280**, 924.
Roth, J., & García-Rosales, C., 1996, *Nucl. Fusion*, **36**, 1647.
Rudy, R. J., Dimpfl, W. L., Lynch, D. K. *et al.*, 2003, *Ap. J.*, **596**, 1229.
Scott, A. D., Evans, A., & Rawlings, J. M. C., 1994, *MNRAS*, **269**, L21.
Scott, A. D., Rawlings, J. M. C., & Evans, A., 1994, *MNRAS*, **269**, 707.
Sellgren, K., 1984, *Ap. J.*, **277**, 623.
Shore, S. N., & Braine, J., 1992, *Ap. J.*, **392**, L59.
Shore, S. N., & Gehrz, R. D., 2004, *A&A*, **417**, 695.
Shore, S. N., Starrfield, S., Gonzalez-Riestra, R., Hauschildt, P. H., & Sonneborn, G., 1994, *Nature*, **369**, 539.
Smith, C. H., Aitken, D. K., Roche, P. F., & Wright, C. M., 1995, *MNRAS*, **277**, 259.
Sneden, C., & Lambert, D. L., 1975, *MNRAS*, **170**, 533.
Snijders, M. A. J., Batt, T. J., Roche, P. F. *et al.*, 1987, *MNRAS*, **228**, 329.
Wada, S., Onaka, T., Yamamura, I., Murata, Y., & Tokunaga, A. T., 2003, *A&A*, **407**, 551.
Warner, B., 1995, *Cataclysmic Variable Stars*. Cambridge: Cambridge University Press.
Weight, A., Evans, A., Albinson, J. S., & Krautter, J., 1993, *A&A*, **268**, 294.
Wichmann, R., Krautter, J., Kawara, K., & Williams, R. E., 1990, in *The Infrared Spectral Regions of Stars*, ed. C. Jaschek, & Y. Andrillat. Cambridge: Cambridge University Press.
Wilson, O. C., & Merrill, P. W., 1935, *PASP*, **47**, 53.
Zijlstra, A. A., Chapman, J. M., te Linkel Hekkert, P. *et al.*, 2001, *MNRAS*, **322**, 280.

14

Extragalactic novae

Allen W. Shafter

14.1 Introductory remarks

Observations of extragalactic novae date back to the early twentieth century, and were influential in the debate concerning the nature of the spiral nebulae (see van den Bergh (1988) for a review of the early history). Initially, the identification of extragalactic novae was fraught with confusion, as the distinction between classical novae, with typical absolute magnitudes ranging from $M_{pg} \sim -7$ to $M_{pg} \sim -9$, and supernovae, which are of order ten thousand times more luminous, was not yet appreciated. The best-known example of this confusion concerns the report by Hartwig (1885) of a 'nova' in the great nebula in Andromeda. This object, S And, is now recognized as the first and only supernova to be observed in M31. Just a decade later another bright star was discovered very near the spiral nebula NGC 5253 by Fleming during her examination of Draper Memorial photographs (Pickering & Fleming, 1896). This object, Z Cen, is also now recognized as a supernova (SN 1895B). No additional nova candidates were associated with spiral nebulae until the discovery on 19 July 1917 by Ritchey (1917a) of a 14th-magnitude transient star in the outer portion of NGC 6946. This discovery set off a systematic search of archival plates from the Mt Wilson 1.5 m reflector dating back to 1908. By the end of 1917, a total of 11 'novae', or 'temporary stars', as they were sometimes called, had been identified in various spiral nebulae (Shapley, 1917). Of these objects, one turned out to be a Galactic variable star not associated with an extragalactic nebula, and seven, including Ritchey's 1917 nova, Z Cen, and S And, are now recognized as supernovae. The remaining three were in fact bona fide classical novae in M31, and represent the first true discoveries of classical novae outside the Milky Way. The credit for the initial discovery goes to Ritchey (1917b) for his re-examination of the first Mt Wilson 1.5 m plates of M31 taken back in August and September of 1909. Remarkably, two novae were identified, both of which were recorded at maximum light on 16 September 1909. During the period between 1917 and 1922 an additional 19 novae were discovered during sporadic monitoring of M31.

The potential role of novae as extragalactic distance indicators was recognized early on. Shapley (1917), in his 'Notes on the Magnitudes of Novae in Spiral Nebulae', was already contemplating the potential ramifications of placing the Andromeda nebula at a distance

Classical Novae, 2nd edition, ed. Michael Bode and Aneurin Evans. Published by Cambridge University Press.

of ~50 times that of the average Galactic nova, as suggested by a comparison of the apparent magnitudes of Galactic and M31 novae (with the one exception of S And). The minimum luminosity implied for S And, $M \sim -15$, was considered fantastical at the time. In the end, the difficulty in reconciling the enormous luminosity implied for S And, coupled with the perceived difficulty in reconciling van Maanen's (erroneous) measurement of internal proper motion in M101 (van Maanen 1916) with its extragalactic nature, led Shapely to seriously question the 'island universe' interpretation of the spiral nebulae.

The role of novae in measuring extragalactic distances was further developed by Lundmark (1922) in his work to calibrate the absolute magnitudes of Galactic novae. Based on rather uncertain parallax measurements of ten Galactic novae (only seven of which are now recognized as classical novae), Lundmark estimated a mean absolute magnitude of $M_{\rm pg} = -6.2$. In a later paper Lundmark (1923) summarized much of what was known about Galactic novae up to that time, which included an interesting comparison between the apparent magnitudes at maximum of Galactic novae with those in the field of the Andromeda nebula.[1] Specifically, by comparing the apparent magnitudes of novae in the Sagittarius region of the Milky Way (which displayed a relatively small dispersion) and those novae (excluding S And) discovered in the Andromeda nebula, he estimated that M31 was approximately 9 mag fainter (a factor of ~60 more distant) than the Sagittarius novae. Thus, taking a modern estimate of 8 kpc to the Galactic center (Gwinn, Moran & Reid, 1992) yields a distance to M31 of ~500 kpc, which is in closer agreement with the modern value of 765 kpc (Freedman *et al.*, 2001) than Hubble's 1929 Cepheid distance of 275 kpc (Hubble, 1929). Indeed, if not for the confusion between novae and supernovae, the distance to M31 would probably have been determined first (and more accurately) through the use of novae.

Starting in the autumn of 1923, Hubble began an annual monitoring program to study the statistical properties of novae in M31. This program, which by 1927 had identified an additional 63 objects in M31, represented the first systematic study of extragalactic novae. Hubble's early work established several properties of M31 novae that are still accepted today. Specifically, novae exhibited a frequency distribution at maximum light characterized by $\langle m_{\rm pg} \rangle \simeq 16.5$, and a spatial distribution that generally follows the nebular light. Remarkably, Hubble deduced an overall nova rate for the galaxy of $\sim 30\,{\rm yr}^{-1}$, which is in excellent agreement with virtually all subsequent studies (Arp, 1956; Capaccioli *et al.*, 1989; Shafter & Irby, 2001). Although Hubble's inaccurate Cepheid-based distance to M31 caused him to underestimate the luminosities of M31 novae by ~2 mag, his observations clearly established their similarity to Galactic novae, while clearly distinguishing normal novae from anomalously bright objects such as S And. Indeed, when discussing the 86 M31 novae in his classic paper on the Andromeda nebula, Hubble (1929) remarked 'The nova of 1885 is clearly an exceptional case, and the eighty-five photographic novae must be considered as normal.' It would be another five years before the term 'supernova' was coined by Baade and Zwicky (1934) to describe the brighter class of objects.

Although our current understanding of classical novae as arising from a thermonuclear runaway on the surface of an accreting white dwarf in a semi-detached binary system would have to await the now classic work in the 1960s and 1970s by Kraft (1964a,b), Warner and Nather (1971), and Starrfield *et al.* (1972) (see Chapter 4), an impressive collection of

[1] Curiously, he also noted that the amplitudes, $A\ (= m_{\rm min} - m_{\rm max})$, of nova outbursts are anticorrelated with $m_{\rm max}$ as shown in his Figure 3, which is not surprising, and must be true for any distance-limited sample of novae.

observational data on both Galactic and extragalactic novae continued to accumulate in the intervening years. The next major extragalactic nova survey following Hubble's pioneering work was a survey of M31 conducted by Arp (1956) using the 1.5 m telescope on Mt Wilson. This study was remarkable in terms of the intensity of temporal coverage. Almost 1000 plates were taken on 290 nights between June 1953 and January 1955. A total of 30 novae were identified. As a result of the dense coverage, only five novae were believed to have their maxima missed by more than a day. During the 2-year M31 survey, Arp photographed five other local group members: M32, NGC205, M33, NGC147, and NGC 185. The first two were included on the M31 plates, and thus received essentially the same coverage. The latter three galaxies were monitored once or twice a week. No novae were detected in any of these relatively low-mass systems.

Broadly speaking, the results of Arp's M31 survey confirmed Hubble's earlier conclusions. Arp found a global nova rate of 26±4 per year, a nova spatial distribution intermediate between the flattened disk of M31 and its nearly spherical bulge component, and light curve properties similar to those of Galactic novae. A particularly noteworthy finding was that the frequency distribution of nova magnitudes at maximum light was apparently bimodal with peaks near $m_{pg} = 16.0$ and $m_{pg} = 17.5$ (see Figure 14.1). The bimodal nature of the distribution has been evoked in recent years in support of the intriguing possibility that there may be two populations of novae: 'bulge' novae and 'disk' novae, with the former thought to be generally less luminous and 'slower' than their 'disk' counterparts. We now turn to a discussion of the stellar population of novae.

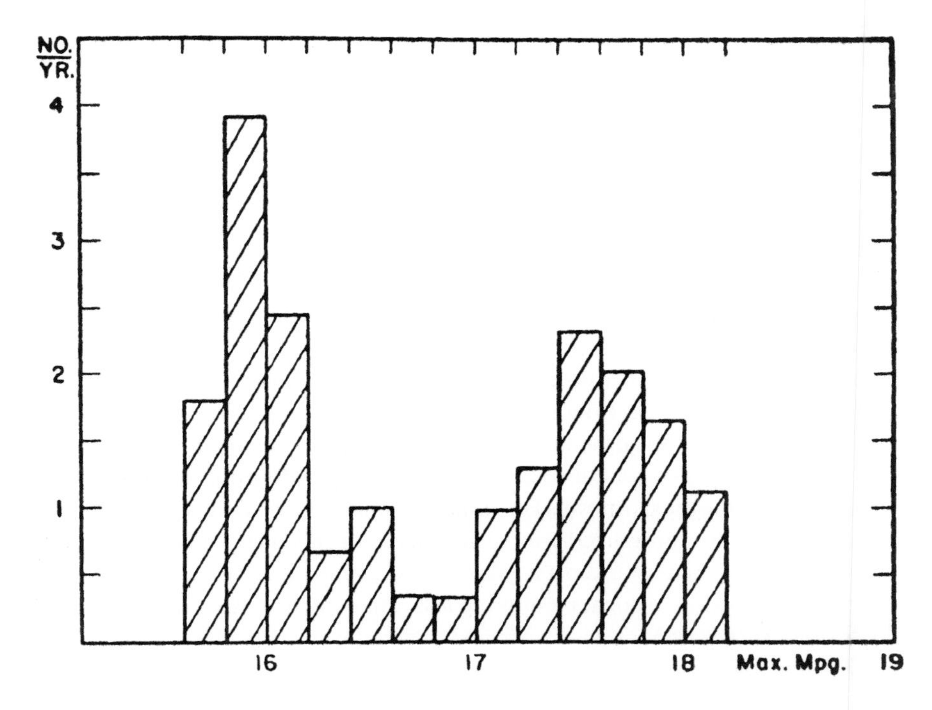

Fig. 14.1. The distribution of magnitudes at maximum light for the sample of 30 M31 novae. Note the bimodal nature of the distribution with peaks near $m_{pg} = 16.0$ and $m_{pg} = 17.5$. From Arp (1956).

14.2 Nova populations

The stellar population of novae has been a subject of discussion since Baade introduced the concept of stellar populations more than half a century ago (Baade, 1944). It was recognized early on that any attempt to elucidate the population of novae from their apparent Galactic space distribution will be confounded by the effects of patchy interstellar extinction. Nevertheless, several attempts were made to assign novae to a particular stellar population with no real success. Perhaps the situation was best summed up by Plaut (1965), who noted that given the apparent concentration of novae both towards the Galactic plane and the Galactic center, 'Classification according to the simple population I and II scheme is therefore somewhat ambiguous'. Duerbeck (1984, Table 1) gives a convenient summary of the early attempts to determine the Galactic distribution of novae up to that time. In view of the difficulties associated with Galactic observations, it is not surprising that observation of the spatial distribution of novae in external galaxies soon became the focal point in the study of the stellar population of novae.

After Arp's classic study concluded in 1955, novae continued to be discovered in M31 as part of ongoing surveys over roughly the next 30 years, primarily in Italy, Crimea and Latvia (Rosino, 1964, 1973; Rosino *et al.*, 1989; Sharov & Alksnis, 1991). The most extensive survey was conducted at the Asiago Observatory where Rosino and collaborators discovered a total of 142 novae, some with sufficient data for light curves to be characterized. In agreement with the findings of Hubble and Arp, Rosino (1964) concluded that 'In general, the light curves of novae found at Asiago show the same variety of forms of the Galactic novae, to which they are strikingly similar.' One change in the Asiago survey concerned observations in the latter years of the survey, which were obtained through a UG1 ultraviolet filter to provide better contrast against the bright nuclear region. Despite the attempt to detect novae in the nuclear region, the nova distribution fell off significantly near the nucleus as first noted by Arp. Rosino (1973) concluded that 'the region close to the nucleus seems really devoid of novae'. Another surprising early result of the Rosino study concerned the distribution of nova magnitudes at maximum light. When the Hubble, Arp, and Asiago samples were combined, the bimodality seen in the Arp data was no longer apparent (Rosino (1973), his Figure 14).[2]

Little additional progress was made in our understanding of the M31 nova populations for the next decade. Then, in the Fall of 1981, the foundations were laid for a new approach in the study of nova populations, when Ciardullo *et al.* (1983) discovered four anomalously bright $H\alpha$ sources in on-band, off-band images taken during a search in M31's bulge for $H\alpha$-bright planetary nebulae, and other potentially interesting emission line sources, such as the then recently discovered SS433 (Margon *et al.*, 1979). Follow-up spectroscopic observations revealed the sources to be novae during their decline from maximum light (see Figure 14.2). Although it had been known for some time that novae developed strong $H\alpha$ emission shortly after maximum, the use of a nova's $H\alpha$ emission as an aid to discovery, and as a potential standard candle, was not appreciated prior to this time. Not only does a nova's strong $H\alpha$ emission provide increased contrast against a galaxy's background light, the fade rate is considerably slower in $H\alpha$, making frequent monitoring less important than with broadband observations. The first $H\alpha$ survey of M31 was conducted during the period 1982–1986 by Ciardullo *et al.* (1987) using telescopes at Kitt Peak National Observatory, McDonald Observatory, and the

[2] Capaccioli *et al.* (1989) have shown that the bimodality is preserved when the analysis is restricted to the highest quality data.

NOVAE IN M31

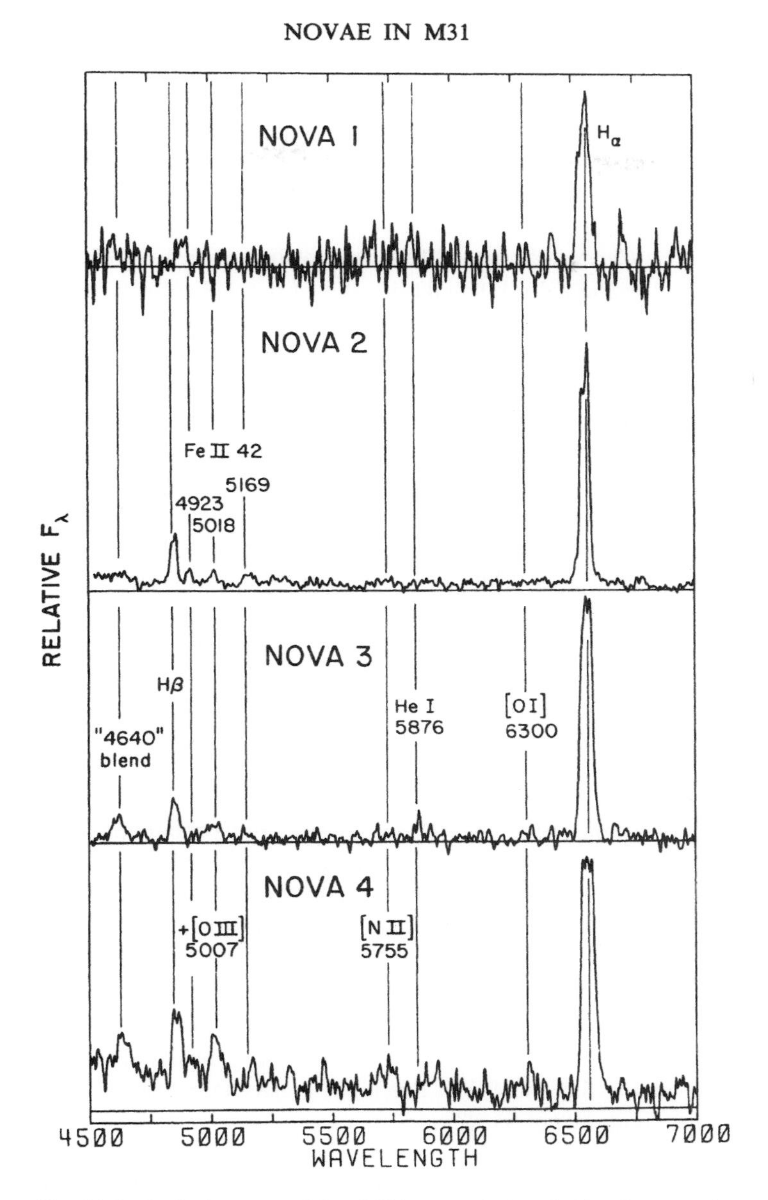

Fig. 14.2. Optical spectra of four M31 novae during decline from maximum light. Note the prominent Hα emission lines. From Ciardullo *et al.* (1983).

Wise Observatory in Israel. The results of this study yielded an improved nova rate for the bulge of M31, and clearly established, contrary to the conclusions of Arp and Rosino, that the nova distribution follows the light all the way to the center of the galaxy.

Perhaps the most surprising, and controversial, result of the study concerned the extended radial distribution of novae. When the Arp (1956) novae (which extended to larger galactocentric radii than the Hα observations) were included in the Hα sample, a comparison of the radial nova distribution with model bulge and disk luminosity profiles revealed that the nova rate per unit B light in the bulge of M31 was at least an order of magnitude greater than the corresponding rate in the disk, and was formally consistent with *all* the novae arising from M31's bulge population. This result was unexpected considering that Galactic observations, although hampered by extinction, appeared to suggest that novae belonged to an old disk population (e.g. Patterson, 1984; cf. Wenzel & Meinunger, 1978). The association of novae primarily with M31's bulge was corroborated in a comprehensive analysis of available M31 nova data undertaken by Capaccioli *et al.* (1989) who estimated that $\sim$85% of the novae arise in the galaxy's bulge and halo. It should be pointed out, however, that since M31 is observed at a relatively high inclination to the line of sight, it is difficult to determine the true position of a nova unambiguously from the position projected onto the sky. The difficulty in interpreting the spatial distribution of both Galactic and M31 novae has been illustrated quite effectively by the Monte Carlo simulations of Hatano *et al.* (1997a,b).

The possibility that the spatial distribution of novae in M31 has been biased towards bulge novae because of extinction in M31's disk is not easy to rule out. Although the use of Hα imaging provides a modest improvement over earlier studies, the Ciardullo *et al.* (1987) Hα bulge survey relied on B-band observations from Arp's study (1956) to extend their spatial coverage. In an attempt to explore further the effect extinction in M31's disk may have on the spatial distribution of novae, Shafter and Irby (2001) extended Hα observations further out along M31's major axis and well into the disk. To assess the degree to which extinction may be biasing the novae results, they also analyzed the spatial distribution of M31's planetary nebulae (PNe), which should be at least as affected by extinction as the novae since the former were detected via their [O III] 5007 Å emission. Since the stellar death rate of a system of stars (and presumably the rate of PNe formation) is not expected to be sensitive to the age, metallicity, or initial mass function of the underlying stellar population (Renzini & Buzzoni, 1986), the radial PNe surface density profile should provide a fiducial by which to compare the radial nova density distribution. As shown in Figure 14.3, unlike the nova distribution, the planetary nebula distribution drops off more slowly with distance from the nucleus than does the bulge light, but is in good agreement with the radial distribution of the overall galaxy's background B light. Thus, regardless of the effects of extinction, the nova distribution is clearly more centrally concentrated than the PN distribution, and is consistent with an association with M31's bulge. A similar conclusion was reached by Darnley *et al.* (2004, 2006) in their analysis of POINT-AGAPE microlensing data. These authors found a bulge nova rate per unit r' flux more than five times greater than that in the disk, and a surprisingly high global nova rate of $65^{+16}_{-15}\,\mathrm{yr}^{-1}$.

In addition to M31, the spatial distribution of novae has also been studied in another nearby spiral galaxy, M81, with conflicting results. Shara *et al.* (1999) analyzed 23 novae in M81 discovered on 5 m Palomar plates taken in the early 1950s by a number of observers, including Humason, Sandage, Baade, Baum, Hubble, and Minkowski, and found evidence for an appreciable outer disk/spiral arm nova population. The overall spatial distribution of novae was found to be considerably more extended than the background galactic light. In a more recent and exhaustive Hα study, Neill and Shara (2004) conclude, in agreement with the M31 results, that the spatial distribution of novae in M81 follows the bulge light much better than the disk or total light. As was the case with the earlier broadband B surveys of M31 by Arp (1956) and Rosino (1964, 1973), it is likely that the spatial distribution derived from the early Palomar data was biased by the difficulty in finding novae in the bright central regions of M81.

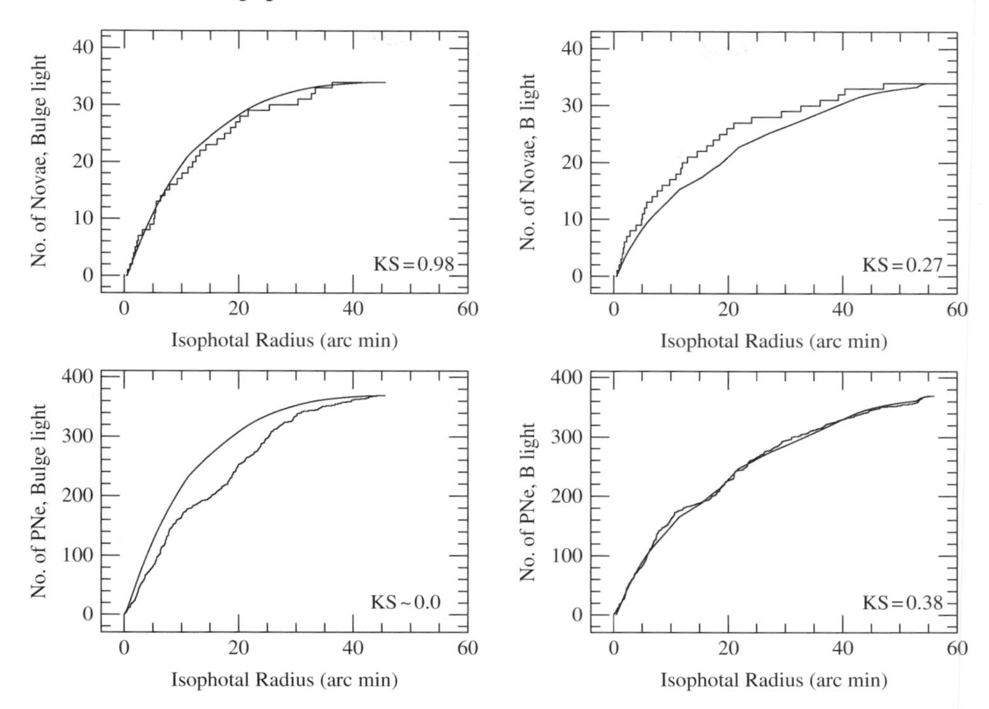

Fig. 14.3. The cumulative distributions of M31 novae and PNe are compared with the background *B* light and with the bulge light. The nova distribution matches the bulge light better than the background light, while the opposite is true for the PNe. Since the PNe are discovered from their [O III] 5007 Å emission, and the novae from their Hα emission, the difference cannot be due to extinction.

The surprising result that M31's nova population may be bulge-dominated led Ciardullo *et al.* (1987) to speculate that the bulge nova rate may be enhanced by nova binaries that were spawned in M31's globular cluster system, and subsequently injected into the bulge through three-body encounters in clusters (e.g. McMillan, 1986), or by tidal disruption of entire clusters, or both. It has long been recognized that the number of X-ray sources per unit mass is of order a hundred to a thousand times higher in globular clusters compared with the rest of the Galaxy (Clark, 1975; Katz, 1975). A similar enhancement of X-ray sources is seen in M31's globular cluster population (e.g. Crampton *et al.*, 1984; Di Stefano *et al.*, 2002). The realization that these X-ray sources are the result of captures between neutron stars and low-mass main-sequence companions has led to the expectation that globular clusters should produce an even greater number of close red dwarf/white dwarf binaries, including classical novae (e.g. Hertz & Grindlay, 1983; Rappaport & Di Stefano, 1994). To date at least one and probably two classical novae have been observed in the cores of Galactic globular clusters: T Sco in M80 (Luther, 1860; Pogson, 1860; Sawyer, 1938), and an anonymous nova near the core of M14 in 1938 (Hogg & Wehlau, 1964). After initially disappointing searches, recent observations with the Hubble Space Telescope (HST) and Chandra have started to reveal increasing numbers of cataclysmic variables in Galactic globular clusters (Knigge *et al.*, 2002; Pooley *et al.*, 2002; Heinke *et al.*, 2003; Edmonds *et al.*, 2003).

Attempts to detect directly novae in M31's globular cluster system have been undertaken by Ciardullo, Tamblyn and Phillips (1990), and by Tomaney (1992). Both studies make use of the fact that at maximum light, the luminosity of an average nova is comparable to the integrated luminosity of a typical globular cluster. Thus, careful photometry of M31 clusters can detect a nova eruption. In the first study, using a Fourier point-spread-function matching technique, Ciardullo, Tamblyn and Phillips (1990) measured the brightnesses of 54 M31 globular clusters that fell in the fields covered by the Ciardullo *et al.* (1987) nova survey. Over an effective survey time of ~2 yr, no cluster showed a brightness increase indicative of a nova outburst, whereas Ciardullo, Tamblyn and Phillips (1990) claimed as many as three would be expected if the enhancement of novae in globular clusters were comparable to that of the X-ray sources found by Crampton *et al.* (1984). In a different approach, Tomaney (1992) used a multi-fiber spectrograph on the McDonald Observatory 2.7 m reflector to search for novae through their expected Hα emission. A total of over 200 globular clusters were observed over an effective survey time of ~1 yr. The failure to detect enhanced Hα emission from any cluster led Tomaney to conclude that the enhancement of novae in M31's globular cluster system was unlikely to be as high as that of low-mass X-ray binaries.

Not all searches for novae in extragalactic globular cluster systems have been fruitless. In a recent development, Shara and Zurek (2002) have discovered a nova coincident with a globular cluster of M87. Taken at face value, the detection of one nova in the 1057 globular clusters of M87 over an effective survey time of 90 days suggests a globular cluster nova rate of $\sim 0.004\,\mathrm{yr}^{-1}$ $\mathrm{cluster}^{-1}$, which is ~100 times more frequent than one would expect if novae are not enhanced in the clusters. Of course, it is possible (though unlikely) that the nova is a chance superposition of a field nova, or that they were just very lucky to have found this one example. Clearly, additional monitoring of extragalactic globular cluster systems, and the improved statistics that will come as a result, will be necessary to reach any definitive conclusions regarding cluster nova rates.

14.2.1 Two populations of novae?

The idea that novae from differing stellar populations may have distinct outburst characteristics finds support in theoretical studies of nova outbursts (e.g. Shara, Prialnik, & Shaviv, 1980; Shara, 1981; Prialnik *et al.*, 1982; Livio, 1992; Prialnik & Kovetz, 1995), which have shown that the character of the outburst (e.g. peak luminosity and decline rate) depends on properties such as the white dwarf's mass, luminosity, and accretion rate, some or all of which may vary systematically with the underlying stellar population. The strength of the nova outburst is most sensitive to the mass of the accreting white dwarf. The increased surface gravity of a more massive white dwarf results in a higher pressure at the base of the accreted envelope at the time of thermonuclear runaway, resulting in a more violent outburst. In addition, since a smaller mass of accreted material is required to achieve the critical temperature and density necessary for a runaway, nova outbursts produced on massive white dwarfs are expected to have shorter recurrence times and faster light-curve evolution. Population synthesis studies have shown that the mean white dwarf mass in a nova system is expected to decrease as a function of the time elapsed since the formation of the progenitor binary (de Kool, 1992; Tutukov & Yungelson, 1995; Politano, 1996). Thus, the proportion of fast and bright novae, which are associated with massive white dwarfs (Prialnik & Kovetz, 1995; Livio, 1992), might be expected to be higher in a younger stellar population. In addition, as discussed by Yungelson, Livio & Tutukov (1997), the later Hubble type galaxies, and in particular, low-mass late-type galaxies such as M33 and the Magellanic Clouds, with their younger stellar populations, should be more prolific nova producers than are their earlier Hubble type counterparts.

By 1990 the possibility that there may be two distinct populations of novae with differing outburst evolution was beginning to gain significant observational support. In a study of the spatial distribution of Galactic novae, Duerbeck (1990) found that the observed number counts of novae showed an inflection point near $m = 6$ beyond which the number counts increased as expected for the contribution of a separate and more distant population. Based both on a tendency for novae in the Galactic bulge to be 'slower' in outburst development when compared with nearby disk novae, and on similar differences in speed class noted for novae in M31 and those in the LMC, Duerbeck became the first to postulate formally the existence of two populations of novae: a relatively young population that he called 'disk novae', which were found in the solar neighborhood and in the LMC, and 'bulge novae', which were concentrated towards the Galactic center and found in the bulge of M31, and were characterized by generally slower outburst development. The argument in favor of two populations of novae was further developed by Della Valle *et al.* (1992), who showed that the average scale height above the Galactic plane for 'fast' novae ($t_2 < 13$ days)[3] is smaller than for novae with slower rates of decline. At about the same time, Williams (1992) was proposing that classical novae could be divided into two classes based on their spectral properties: specifically, the strengths of Fe II relative to He and N emission lines. Novae with prominent Fe II lines ('Fe II novae') usually show P Cygni absorption profiles, and tend to evolve more slowly, have lower expansion velocities, and have a lower level of ionization, than novae that exhibit strong lines of He and N (the 'He/N novae'). In addition, the latter novae display very strong neon lines, but not the forbidden lines that are often seen in the Fe II novae. Following up on their earlier work, Della Valle and Livio (1998) noted that Galactic novae with well-determined distances that were classified as He/N were concentrated near the Galactic plane, and tended to be faster, and more luminous compared with their Fe II counterparts.

The available evidence in support of two nova populations from extragalactic data is mixed. Supporting evidence has been described in a series of papers by Della Valle and collaborators (Della Valle *et al.*, 1994; Della Valle, 1995, 2002), who point out that novae in 'disk dominated' galaxies, such as the LMC, appear to be on average faster and brighter than novae arising from the bulge of M31. Available spectroscopic evidence, although limited, appears consistent with this picture. In spectroscopic observations of a total of 13 novae in the inner region of M31 (presumably from the bulge), both Ciardullo *et al.* (1983) and Tomaney and Shafter (1992) find no examples of the violent eruptions and high ejection velocities that are commonly observed in Galactic 'disk' novae. Only one nova observed by Tomaney and Shafter (McD89 No.1) appears consistent with classification as a He/N nova.

Not all extragalactic data, however, support the two-population scenario. Despite the bimodal maximum magnitude distribution seen in the M31 data (Arp, 1956; Capaccioli *et al.*, 1989), there does not appear to be any correlation between rate of decline of M31 novae and their spatial position within the galaxy (Sharov, 1993; Shafter, 2002). Such a correlation should be expected in a two-population scenario given the relationship between maximum magnitude and rate of decline (see the MMRD relation in Section 14.4). Finally, a challenge

[3] See Chapter 2 for definition of t_2.

to the idea of distinct nova populations has come from a recent and pioneering effort by Ferrarese, Côté and Jordan (2003), who conducted the first HST survey specifically designed to discover novae in another galaxy. In a 55 day observing campaign targeting the Virgo elliptical NGC 4472 (M49), Ferrarese *et al.* discovered a total of nine novae. Perhaps the most interesting result from this study does not concern the nova rate in M49, which will be discussed in the next section, but rather the properties of the M49 nova light curves themselves. Specifically, Ferrarese *et al.* found that M49 appeared to lack a significant population of slow, faint novae compared with the Milky Way and M31. Instead, they found that the decline rates were remarkably similar to the faster novae in the LMC (see Figure 14.4).

Despite the intriguing nature of these findings, there are a few caveats to be considered. Since only a relatively small number of nova light curves were available in the study, the light-curve properties may not be representative of the global M49 nova population. Another potential concern is that the stellar population of M49 and other radio-loud ellipticals may be contaminated through mergers with late-type galaxies. In particular, Della Valle and Panagia (2003) have shown that radio-loud elliptical galaxies exhibit an overproduction of Type Ia supernovae, which they speculate may result from contamination by a ~1 Gyr stellar population. Since some fraction of Type Ia supernova progenitors may be related to the cataclysmic variables, and to recurrent novae in particular (e.g. Livio, 2000), it is possible that the nova rate in radio-loud ellipticals may also be enhanced by the mergers (see Chapter 3 for a discussion of this point). If so, it would not be surprising if properties of the nova population in these systems showed similarities to those of late-type galaxies.

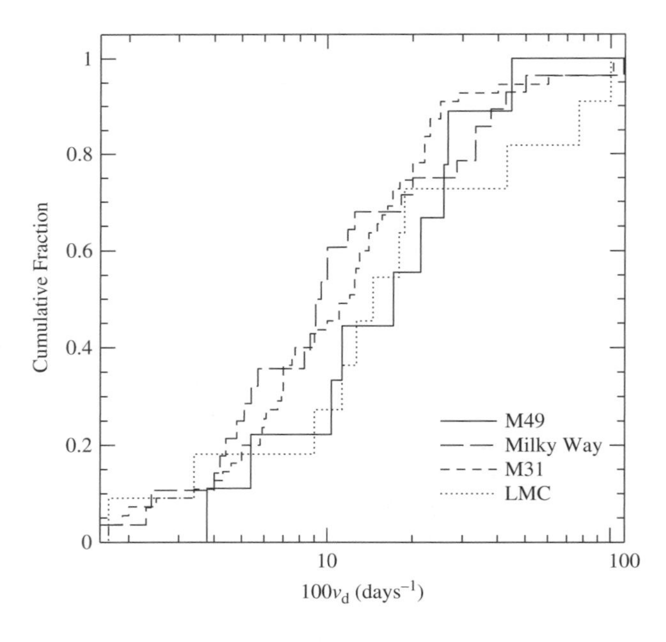

Fig. 14.4. The cumulative distribution of novae as a function of fade rate for several galaxies. Note that the fade rate distribution of the M49 novae matches that of the LMC better than those of M31 and the Galaxy. From Ferrarese *et al.* (2003).

14.3 Extragalactic nova rates

By the early 1990s novae had been detected in a total of eight extragalactic systems, in six with sufficient numbers for an estimate of their nova rates. The first study comparing nova rates in a broad range of galaxies was that of Ciardullo *et al.* (1990a) who compared the nova rate in NGC 5128 with those in the SMC, LMC, M31, M33, and a sample of Virgo ellipticals. In order to compare nova rates in different galaxies, it is necessary to normalize the rates by the stellar mass surveyed. Because novae arise from an evolved stellar population, Ciardullo *et al.* chose to normalize the nova rates by the galaxy's infrared K magnitude, which they adopted as a convenient proxy for the mass in evolved stars. The resulting normalized nova rates are referred to as Luminosity-Specific Nova Rates (LSNRs). When comparing LSNRs, Ciardullo *et al.* found no evidence for a systematic variation with the Hubble type of the galaxy (see Figure 14.5), although the error bars for the individual galaxies were quite large.

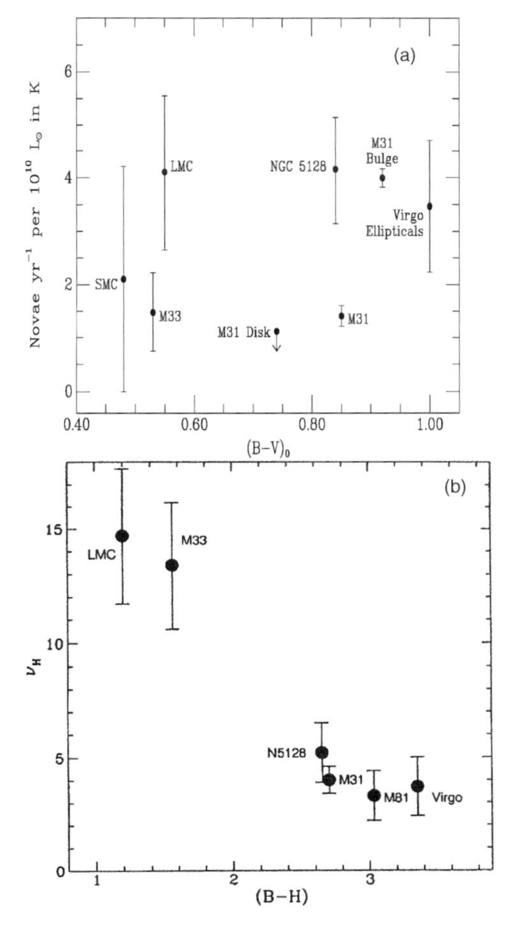

Fig. 14.5. LSNRs plotted as a function of galaxy color. Ciardullo *et al.* (1990a) find no dependence on galaxy type (a), while Della Valle *et al.* (1994) find that early-type galaxies are more prolific nova producers (b).

A few years after the publication of the Ciardullo *et al.* (1990a) analysis, a similar study was published by Della Valle *et al.* (1994), who came to very different conclusions regarding the variation of the LSNR with the Hubble type of the galaxy. Despite the considerable overlap in the galaxies studied (Della Valle *et al.* included a recent estimate of the nova rate in M81, while excluding the poorly known nova rate estimate for the SMC), the latter authors found that early-type galaxies, specifically the LMC and M33, were more prolific nova producers, with LSNRs that were roughly a factor of three greater than their earlier-type counterparts (see Figure 14.5). The population synthesis models of Yungelson, Livio and Tutukov (1997), published a few years later, which showed that nova rates should be enhanced in galaxies with active star formation, appeared to provide a theoretical foundation to the observational results of the Della Valle *et al.* study.

14.3.1 Determining the LSNR

The determination of the LSNR of a galaxy involves several steps, and represents a challenging observational task. Broadly speaking, the procedure can be divided into two principal tasks: estimating the nova rate, usually in a limited region of the galaxy, and estimating the infrared luminosity in this surveyed region. Regardless of galaxy type, uncertainty in the nova rate arises because of difficulty in characterizing the extent of completeness in the nova surveys. As a result of practical restrictions on telescope availability, most surveys have been synoptic in nature and have relied on estimates of the length of time the average nova remains detectable to compute nova rates. Since the absolute magnitude of novae at maximum light and their rate of decline are variable, and are likely to depend on stellar population, a determination of nova detectability is problematic. Generally, two approaches have been used: a mean nova lifetime approach, and a Monte Carlo approach. In both cases a representative sample of nova light curves is needed, and the requisite mean nova lifetime relations have been calibrated from an assumed absolute nova rate in the bulge of M31 (e.g. Capaccioli *et al.* (1989) for broadband B data; Ciardullo *et al.* (1990b) for Hα data).

The mean lifetime approach, used by both Ciardullo *et al.* (1990b) and Della Valle *et al.* (1994), is based on a procedure first used by Zwicky (1942) to study the frequency of supernovae. For a total of $N(M < M_{\text{lim}})$ novae observed brighter than a survey's limiting absolute magnitude M_{lim}, the nova rate is given by

$$\mathcal{R} = \frac{N(M < M_{\text{lim}})}{T(M < M_{\text{lim}})}, \tag{14.1}$$

where $T(M < M_{\text{lim}})$ – the total number of days the survey is able to detect an average nova – is known as the effective survey time. If, following Ciardullo *et al.* (1990b), τ_{lim} is defined to be the mean nova lifetime (the length of time in days an average nova remains brighter than the limiting magnitude of the survey), then for multi-epoch observations, the effective survey time is given by

$$T(M < M_{\text{lim}}) = \tau_{\text{lim}} + \sum_{i=2}^{n} \min\,(t_i - t_{i-1}, \tau_{\text{lim}}), \tag{14.2}$$

where t_i is the time of the ith epoch of observation. If $\tau_{\text{lim}}(M_{\text{lim}})$ is known, $\mathcal{R}$ can be calculated directly from the nova observations and a knowledge of the galaxy's distance. Based on an annual nova rate of $23.2 \pm 4\,\text{yr}^{-1}$ in M31's bulge (Capaccioli *et al.*, 1989), and Hα light-curve data for 40 M31 bulge novae observed over a 7 year period, Ciardullo *et al.*

(1990b) estimated log $\tau_{\rm lim} \simeq 5.6 + 0.48 M_{\rm lim}$ over the typical range of nova luminosities. If additional Hα light-curve data from the M31 study of Shafter and Irby (2001) are included, the second-order relationship

$$\log \tau_{\rm lim} \simeq -4.78 - 2.10 M_{\rm lim} - 0.162 M_{\rm lim}^2, \tag{14.3}$$

shown in Figure 14.6, provides a slightly better fit to the data.

The mean nova lifetime approach followed by Della Valle *et al.* (1994) is similar, and relies on the B-band light curves of the nova sample given in Capaccioli *et al.* (1989). In either case, the reliability of the mean nova lifetime method relies both on an accurate knowledge of M31's bulge nova rate, and on the assumption that the light-curve properties of the M31 nova sample is representative of novae in the galaxy being studied. Since M31's bulge rate may be in error, and the properties of the bulge novae may not be characteristic of novae from disk populations, nova rates computed using the mean nova lifetime approach should be viewed with appropriate caution.

A better approach for estimating extragalactic nova rates involves the use of numerical simulations, which do not depend on a knowledge of the absolute nova rate in M31's bulge. Here, for a given assumed global nova rate, $\mathcal{R}$, a model population of novae at various stages in their outburst evolution is produced using a sample of known light-curve properties (e.g. from the M31 bulge sample) and compared with number of novae observed, $N_{\rm nova}$. The number of novae detectable will depend on the frequency of observation and the limiting absolute magnitude of the survey. The completeness as a function of magnitude over the

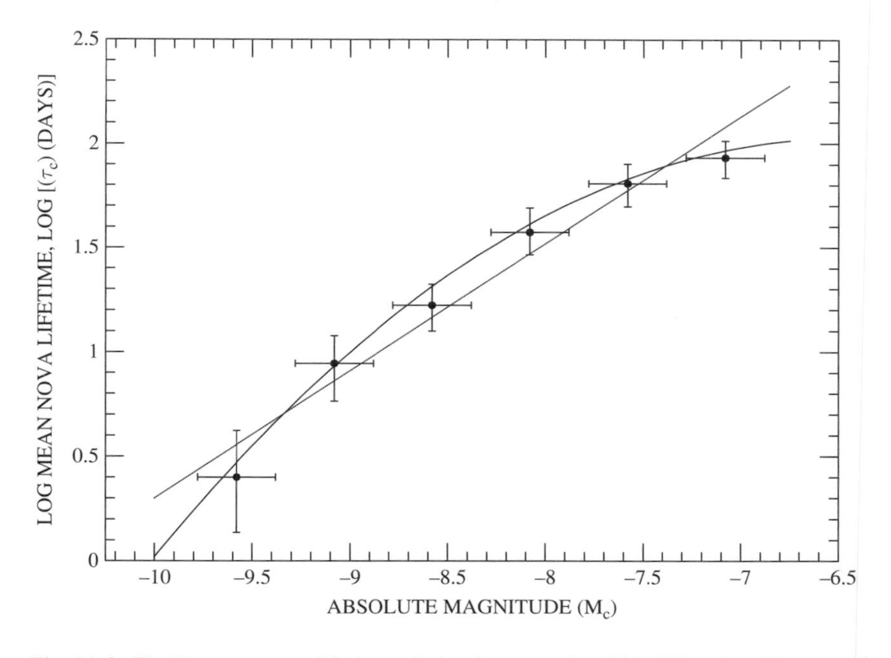

Fig. 14.6. The Hα mean nova lifetime relation for a sample of 64 M31 novae. The mean nova lifetime is an estimate of the average number of days a nova remains brighter than a specified absolute magnitude in Hα. The solid curve is the best second-order fit to the data, as given by Equation (14.3). From Shafter and Irby (2001).

surveyed region, $C(m)$, can be estimated from artificial star tests and then convolved with the model nova luminosity function, $N(m, \mathcal{R})$, to predict the number of novae detected in the survey (e.g. Williams & Shafter, 2004):

$$N_{\rm obs}(\mathcal{R}) = \int C(m) * N(m, \mathcal{R})\, {\rm d}m. \tag{14.4}$$

The nova rate is given by the value of $\mathcal{R}$ that produces the best match between $N_{\rm obs}(\mathcal{R})$ and $N_{\rm nova}$.

Both the mean nova lifetime and Monte Carlo approaches have a principal drawback: the currently available light-curve data, which are derived mainly from novae in the bulge of M31, may not accurately reflect the distribution of decline rates in different galaxies. If, for example, the novae in the target galaxy are generally faster than those from M31's bulge, then the nova lifetime will be overestimated, and the rate underestimated. To guard against this possibility, Neill and Shara (2004) in their recent study of novae in M81 have stressed the need for near-continuous coverage in extragalactic surveys to make sure that the fastest novae are not overlooked. Clearly, continuous coverage will reduce the uncertainty in the nova rate calculations, and should be used when feasible. The only concern is that the limiting magnitude of the survey goes deep enough to assure that the slower (and less luminous) novae are not missed. If only the brightest novae are observed, even a continuous coverage survey will depend on the distribution of speed class within the target galaxy.

Over the past decade, LSNRs have been determined for several additional galaxies. In an attempt to better constrain the LSNRs in late-type galaxies, Shafter, Ciardullo and Pritchet (2000) initiated a nova survey of the spiral galaxies NGC 5194/5 (M51) and NGC 5457 (M101). For comparison, the early-type giant elliptical galaxy, NGC 4486 (M87), was also monitored. Although the results of this program found that the LSNRs in the spiral galaxies were somewhat lower than that of M87, within the errors of measurement the differences were not significant. Overall, the conclusions were consistent with those of Ciardullo *et al.* (1990a), namely there was no compelling evidence that the LSNR varied systematically with the Hubble type of the galaxy. In their HST study of novae in the Virgo elliptical M49, Ferrarese, Côté and Jordan (2003) found an LSNR virtually identical with that found by Shafter *et al.* for M87. Upon considering the full sample of galaxies with measured nova rates, Ferrarese *et al.* conclude that the LSNR in M49 is fully consistent with that measured in all other galaxies for which data are available, with the possible exception of the LMC.

A comparison of LSNRs among galaxies is given in Williams and Shafter (2004), who have completed a multi-epoch survey of M33 in an attempt to reconcile the large discrepancy between published nova rates for this galaxy. Their revised M33 nova rate of 2.5 $\rm yr^{-1}$ yields a LSNR that is consistent with those of most other galaxies. Specifically, they find that the LSNR is constant across a wide range of Hubble types at a value of $\sim 2\times 10^{-10}\,{\rm yr}^{-1}\,L_{K,\odot}^{-1}$, with the exceptions of the Magellanic Clouds, for which the authors conclude that the LSNRs appear to be roughly a factor of three higher (see Figure 14.7). The SMC now joins the LMC as a high LSNR galaxy as a result of an upward revision of its nova rate from the hitherto poorly determined value of $0.3 \pm 0.2\,{\rm yr}^{-1}$ (Graham, 1979) to a value of $0.7 \pm 0.2\,{\rm yr}^{-1}$ (Della Valle, 2002). The higher SMC rate results from the inclusion of recent nova discoveries made possible through microlensing surveys of the Magellanic Clouds.

Two other studies have found relatively high LSNRs in early type galaxies. In a preliminary study of HST archival images, Shara and Zurek (2002) have reported the discovery of over

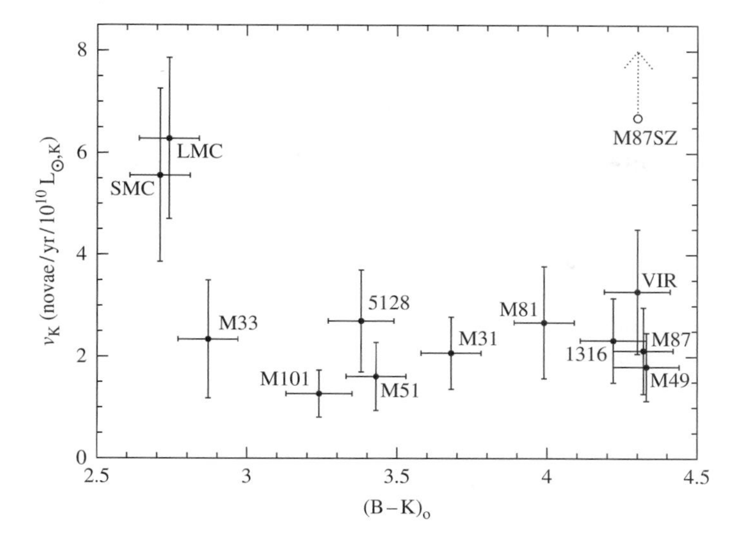

Fig. 14.7. The LSNR plotted as a function of galaxy color. All galaxies appear to have similar LSNRs with the exceptions of the Magellanic Clouds and possibly M87. The M87SZ point is based on the Shara and Zurek (2002) nova rate. From Williams and Shafter (2004).

400 classical nova candidates in M87, leading to an estimated annual nova rate of at least 300 yr^{-1}. Subsequently, Neill and Shara (2005) have made preliminary estimates of the LSNRs in the local group dwarf ellipticals M32 and NGC 205 that are the highest measured for any galaxy. If these results are confirmed, it will necessitate a radical reassessment of the current ideas regarding nova rates in differing stellar populations (e.g. Yungelson, Livio & Tutukov, 1997; see also Chapter 3). The galactic chemical evolution models of Matteucci *et al.* (2003) may provide a step in that direction, as they predict nova rates should scale closely with galaxy mass, with little dependence of the LSNR on stellar population. Thus, giant elliptical galaxies such as M87 are predicted to have particularly high nova rates. Despite the difference between the predictions of Yungelson *et al.* and Matteucci *et al.*, it is clear that all attempts to model Galactic nova rates must depend on input parameters such as the fraction of white dwarfs that end up in nova binaries, the time delay from zero-age main sequence to nova production, and the recurrence time between nova outbursts (see Chapter 3 for detailed discussion), all of which are uncertain and are likely to vary with galaxy morphology.

14.3.2 Uncertainties in the LSNRs

A major source of uncertainty in the global nova rates concerns the role of extinction in shielding an unknown number of novae from detection and in hampering an accurate estimate of galaxy luminosity. Clearly, this uncertainty is much more of a concern in the dusty environments of irregular galaxies and the disks of spirals, where the optical depth may be higher than usually assumed (Disney, Davies & Phillipps, 1989). Although the use of Hα imaging in these late-type galaxies can minimize the impact of extinction, we are unlikely to understand the full extent of the problem until nova surveys in the near-infrared have been conducted. Advances in large-format infrared imaging detectors have now made such surveys feasible. In the near-infrared, a possible alternative to Hα imaging would be to image in O I 8446 Å, which has been observed to be strongly in emission in erupting novae (e.g. Strittmatter *et al.*, 1977).

In cases where the entire galaxy is not surveyed, uncertainty in the global nova rate is also introduced during the process of extrapolating the measured nova rate to the entire galaxy from a comparison of the light sampled in the survey with that of the galaxy as a whole. Not only is an estimate of the luminosity sampled in spiral systems complicated by patchy extinction and numerous H II regions, but the extrapolations implicitly assume that the LSNR is constant throughout the galaxy, which, given the radial population gradient, it may well not be. Generally, determining the LSNR in elliptical galaxies, which contain little dust and a nearly homogeneous stellar population, is relatively straightforward compared with dusty, multi-population spiral systems where identification of the stellar population of the nova progenitors is a problem.

Finally, the computation of a LSNR requires a measurement of the infrared luminosity of the galaxy. Until recently only B magnitudes were available for many galaxies and estimates of $(B - K)$ color were necessary to convert available photometry to the infrared magnitudes required to compute the LSNRs. Often the $(B - K)$ color of a galaxy was either poorly known, or was measured from aperture photometry for the inner region of the galaxy only. That situation has changed recently with the publication of the 2MASS catalog, which contains K-band data for nearby galaxies of large angular size. Unfortunately, as noted by Williams and Shafter (2004) in their study of novae in M33, the published 2MASS K-band magnitudes of nearby, large angular diameter are subject to large errors because of difficulty in background subtraction. Future infrared studies should not only resolve this discrepancy, but should aid in the detection of novae in the dusty regions of late type galaxies.

14.4 Novae as distance indicators

Given both their high outburst luminosities ($-7 > M_V > -9$), which make them among the most luminous objects in the cosmos, and their frequent outbursts ($30\,\mathrm{yr}^{-1}$ in a typical spiral galaxy like the Milky Way; Shafter, 2002), novae are of obvious interest as extragalactic distance indicators (de Vaucouleurs, 1978; van den Bergh, 1981; van den Bergh & Pritchet, 1986; Jacoby *et al.*, 1992; Livio, 1992; Della Valle & Livio, 1995; Livio, 1997; Gilmozzi & Della Valle, 2003). In addition to being up to ~2 mag brighter on average than the longest-period Cepheid variables, novae are found in both Population I and Population II environments. Thus, they are found in elliptical galaxies, and in uncrowded, dust-free regions of spiral galaxies where Cepheids, for example, do not exist (van den Bergh, 1981).

As discussed previously, the principal obstacle in the early years to the use of novae as distance indicators concerned their large dispersion in absolute magnitude, which was particularly acute until the distinction between novae and supernovae was finally appreciated in the 1930s. However, as the observations of novae in M31 by Hubble (1929) had shown, there was still a considerable spread of around 3 magnitudes in luminosity, even with the exclusion of supernovae. Although Hubble had pointed out that bright novae in M31 faded faster than faint novae, the real breakthrough came when McLaughlin (1945) calibrated a relationship between a nova's luminosity and its rate of decline, which he referred to as the 'life–luminosity relation'.[4] This relation, which became widely known in subsequent years as the *maximum magnitude–rate of decline*, or MMRD, relation is usually cast in a form with the absolute magnitude given as a linear function of $\log t_n$, where, as usual, t_n is the time in days a nova takes to decline n magnitudes from maximum light (usually $n = 2$ or 3). The relation has been calibrated for Galactic novae numerous times over the years (Kopylov, 1952, 1955; Schmidt-Kaler, 1957; Pfau, 1976; de Vaucouleurs, 1978; Duerbeck, 1981; Cohen, 1985; Downes & Duerbeck, 2000); further discussion of these relationships may be found in Chapter 2.

Much of the scatter in the Galactic MMRD relation, which is caused by difficulty in measuring accurate distances to novae, can be minimized by observing a nearly equidistant sample of novae in nearby galaxies such as M31 and the Magellanic Clouds. Although the linear MMRD relations provide acceptable fits for Galactic novae, more complicated empirical relations have been proposed in recent years to improve the fit when extragalactic data are included. One such calibration for novae in M31 and the LMC is given by Della Valle and Livio (1995):

$$M_V = -7.92 - 0.81 \arctan \frac{1.32 - \log t_2}{0.23}, \tag{14.5}$$

and is shown graphically in Figure 14.8. The flattening at the bright end of the relation is expected from theoretical modeling of the nova eruption (Livio, 1992), and is the result of the white dwarf mass approaching the Chandrasekhar limit. The reality of the flattening at the faint end has been questioned by Warner (1995), and may be due to observational bias. For comparison, data for Galactic novae (Downes & Duerbeck, 2000) are shown in Figure 14.8 and compared with a similar relation. As expected, the scatter in the Galactic data is somewhat larger, and the data can be fitted just as well by a linear relation (not shown):

$$M_V = -(11.32 \pm 0.44) + (2.55 \pm 0.32) \log t_2. \tag{14.6}$$

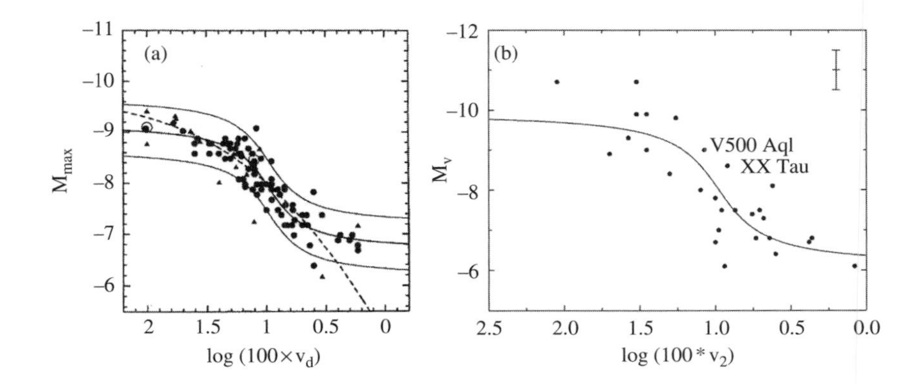

Fig. 14.8. MMRD relationship for novae in M31 (left panel, from Della Valle and Livio (1995)) and for Galactic novae (right panel, from Downes and Duerbeck (2000)). The 'S'-shaped curves are empirical fits of the form given in Equation 14.5. The dashed line represents a theoretical relation from Livio (1992).

[4] Zwicky (1936) is often given credit for being the first to discover the 'life–luminosity' relation for novae; however, Zwicky's analysis, which failed to distinguish between novae and supernovae, led to the misimpression that bright novae faded more slowly than their fainter counterparts.

In addition to the MMRD relation, there have been numerous techniques developed that allow novae to be used in the determination of extragalactic distances. As discussed in Chapter 2 (cf. Table 2.4), another calibration of nova magnitudes was proposed by Buscombe and de Vaucouleurs (1955), who reasoned that if fast novae faded more rapidly than slow novae, their light curves must intersect. Based on a sample of well-observed Galactic nova light curves, they found that at approximately 15 days after maximum light the absolute visual magnitude of novae was given by $\langle M_{15} \rangle = -5.2 \pm 0.1$ (probable error) irrespective of speed class. Like the MMRD relation, there have been various attempts to improve the calibration of this quantity (e.g. Capaccioli *et al.*, 1989; Ferrarese *et al.*, 2003; Downes & Duerbeck, 2000), but it is generally not considered as reliable as the MMRD relation.

The potential of using novae for measuring extragalactic distances was further developed by van den Bergh and Pritchet (1986), who pointed out that quantities derived from the nova luminosity function could also be used in lieu of the MMRD relation. Based on the Arp (1956) and Rosino (1964, 1973) M31 nova samples, van den Bergh and Pritchet proposed extrapolating the linear portion of the broadband blue integrated luminosity function of the magnitudes at maximum light to obtain the quantity '$m^*_{\rm pg}$', which they proposed as a 'standard candle' (see Figure 14.9). Unfortunately, this technique will not work for Hα data. As shown by the M31 observations of Ciardullo *et al.* (1990b), the Hα luminosity function of novae is a power law, and cannot be used to measure distances. Furthermore, the fade rate of a nova's Hα emission is not well correlated with its Hα flux at maximum light; thus, there appears to be no useful Hα MMRD relation either.

In another technique, van den Bergh and Pritchet argued that the mean period of nova visibility, as calibrated by Arp's M31 sample, could also be used as a distance indicator. In particular, they found that the M31 data were well represented by the relation

$$\log \langle t \rangle = 0.67\, m_{\rm pg}(\rm lim) - 11.0, \tag{14.7}$$

which shows that the period that a nova remains visible depends critically on the limiting magnitude of the survey (see Figure 14.9). Although promising, both of these have significant

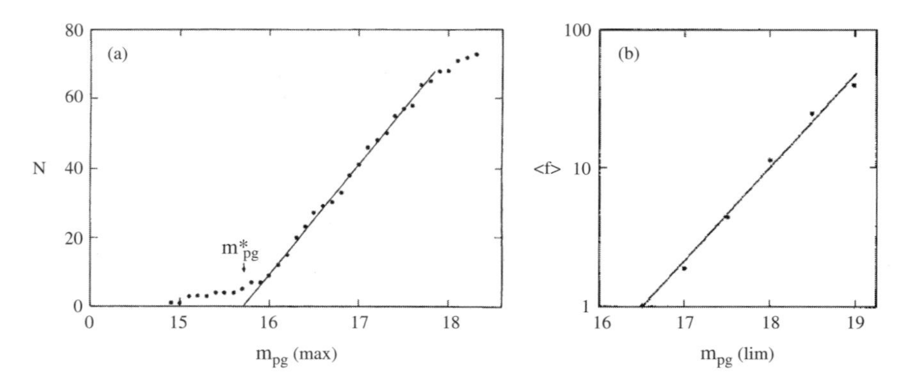

Fig. 14.9. (a) The cumulative broadband blue luminosity distribution of M31 novae at maximum light. From van den Bergh and Pritchet (1986), who proposed that extrapolation of the linear portion of the distribution, '$m^*_{\rm pg}$', could be used as a standard candle. (b) The mean period of visibility of M31 novae is plotted as a function of limiting magnitude. Also from van den Bergh and Pritchet (1986). The linear regression line is given by Equation (14.7).

drawbacks that have curtailed their use. First, both techniques implicitly assume that the relative frequency of novae with differing rates of decline is constant from galaxy to galaxy, which may not be the case. Second, a practical difficulty arises since both techniques rely on properties of the nova luminosity function, because a large sample of complete nova light curves must be obtained. The MMRD relation is more robust in that a large sample of nova light curves (although desirable) is not required, and in that it is insensitive to the relative proportion of fast and slow novae. All that is required is that the relation itself be universal.

14.4.1 Recent developments

Despite their virtues, the use of novae as distance indicators has not become particularly widespread. As just discussed, a major impediment has been the severe impositions on telescope time required to characterize nova light curves and determine fade rates used to calibrate the luminosity. Since the magnitude at maximum light must be observed, densely spaced monitoring programs must be in place to discover and follow novae over many consecutive weeks. In addition to these practical hurdles, the nagging questions about the putative effect of stellar population on nova properties, and thus on the universality of the MMRD relation, have dampened interest in using novae for distance measurements.

Although local group galaxies such as the LMC, M31, and M33 have been primarily used to calibrate the MMRD relation as discussed earlier, distances to these galaxies have also been determined through the use of MMRD relations calibrated in the Galaxy (Capaccioli *et al.*, 1989, 1990). Practical challenges notwithstanding, a few notable efforts to derive extragalactic distances have been undertaken. Perhaps the most ambitious was that carried out by Pritchet and van den Bergh (1987) in their study of novae in three Virgo cluster elliptical galaxies, NGC 4365, NGC 4472 (M49), and NGC 4649 (M60). During observations that spanned 15 nights over a roughly one-month period, Pritchet and van den Bergh were able to identify a total of nine novae (eight of which were in M49). The coverage was sufficient to estimate light-curve properties (maximum magnitude and rate of decline) for six of these novae. A comparison with the MMRD data from the Arp (1956) and Rosino (1964, 1973) sample of novae, shown in Figure 14.10, yields a relative distance modulus between M31 and M49, $\Delta(m - M)_B = 6.8 \pm 0.43$.[5]

As a check, Pritchet and van den Bergh used the 'nova visibility' technique described earlier, noting that the mean period during which seven of their reasonably well-observed novae remained brighter than $m_B = 25.0$ was 17.1 days. Substituting this value into Equation (14.7) above yields $\Delta(m - M)_B = 6.74$ mag, in good agreement with the value derived from the MMRD relation. Adopting the currently accepted value for the M31 distance modulus of 24.48 ± 0.05 (Freedman *et al.*, 2001) yields a Virgo distance modulus $(m - M)_{\mathrm{Virgo}} = 31.28 \pm 0.40$. Although this value is $\sim$0.2 magnitude fainter than the currently accepted M49 distance of $(m - M)_{\mathrm{M49}} = 31.07 \pm 0.08$ (Ferrarese, Côté & Jordan, 2003), the agreement is noteworthy given the relatively small number of nova light curves available in the Pritchet and van den Bergh study.

Other than the Virgo study, attempts to use the MMRD relation or the nova visibility technique to determine distances beyond the local group have been relatively few, and when employed have been based on small samples of novae. Della Valle (1988) compared the

[5] This value includes the 0.2 ± 0.1 correction found by the authors in their numerical simulations of sampling biases.

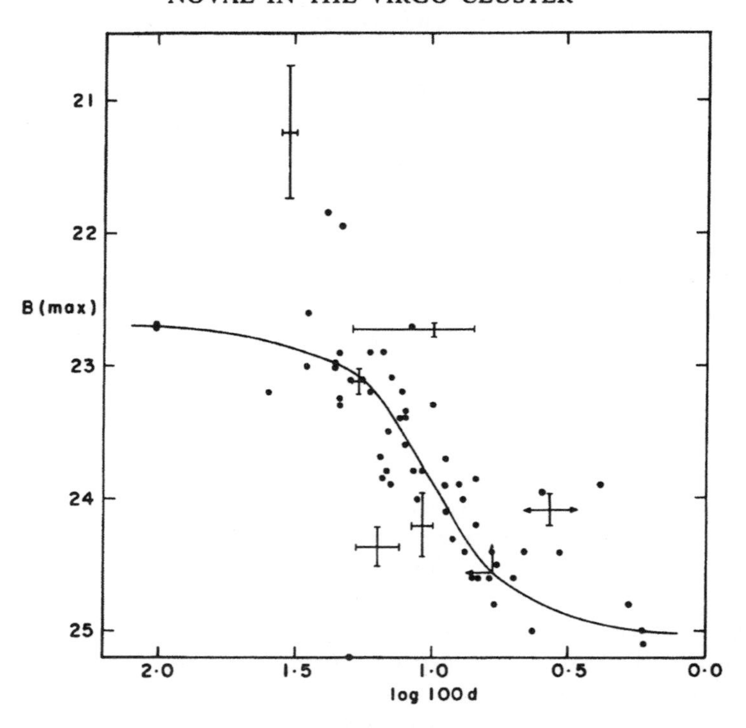

Fig. 14.10. The MMRD relation for six Virgo novae compared with the combined Arp (1956) and Rosino (1964, 1973) samples of M31 novae. From Pritchet and van den Bergh (1987).

period of nova visibility in M31 with that of five novae in M33 to estimate that the latter galaxy is ~0.3 mag more distant than the former. A distance modulus of $(m\text{–}M)_0 = 31.0\pm0.3$ to M100 was obtained by Ferrarese *et al.* (1996) through observations of the light curve of a single nova, and has the distinction of being the first nova-based distance derived from HST data. More recently, Shara, Sandage and Zurek (1999) identified one nova out of the 23 discovered in the Palomar 5 m survey of M81 described earlier that had sufficient coverage for its maximum magnitude and rate of decline to be well determined. Analysis of this single nova using a (theoretically calibrated) MMRD relation from Shara (1981) yielded a distance modulus $(m - M)_0 = 27.75$ mag for M81. An independent distance estimate was obtained using all 23 novae through a simple comparison of their apparent magnitude at discovery distribution with that for M31 novae from the survey of Ciardullo *et al.* (1987). The result was a differential modulus, $\Delta(m - M) = 3.4 \pm 0.3$ mag, which coupled with the Freedman *et al.* (2001) M31 modulus (24.48 ± 0.05) and an estimated extinction of 0.1 mag (Sandage & Tammann, 1987) yields $(m - M)_0 = 27.8 \pm 0.3$. This value is in excellent agreement not only with the MMRD value given above, but with the recent Cepheid-based distance to M81, $(m - M)_0 = 27.8 \pm 0.08$, reported in Freedman *et al.* (2001).

Although the observations from the Ferrarese *et al.* (2003) HST nova study were not used by the authors to derive an independent distance to M49, they did compare the light-curve properties of their M49 novae with both existing Galactic and M31 MMRD relations, finding only 'marginal agreement'. In addition, they used the M49 light-curve data to recalibrate the Buscombe–de Vaucouleurs parameter, finding a value of $M_{V,15} = -6.36 \pm 0.19$, which is substantially brighter than that found from most earlier Galactic nova calibrations. In particular, it is inconsistent with the van den Bergh and Younger (1987) determination of $M_{V,15} = -5.23 \pm 0.16$, marginally consistent with both the Cohen (1985) value of $M_{V,15} = -5.60 \pm 0.45$ and the Capaccioli *et al.* (1989) value of $M_{V,15} = -5.59 \pm 0.42$, but in reasonable agreement with the determination of $M_{V,15} = -6.05 \pm 0.44$ found by Downes and Duerbeck (2000). Ferrarese *et al.* conclude their M49 study on a pessimistic note vis-a-vis the use of novae as extragalactic distance indicators. They argue, given the observed scatter in the MMRD and M_{15} calibrations, and the practical difficulties associated with acquiring sufficient numbers of high-quality light curves, that the surface brightness fluctuation method (Tonry *et al.*, 2001) and potentially the globular cluster luminosity function both offer more effective methods for determining distances to Population II galaxies.

A more optimistic view is provided by the work of Della Valle and collaborators, who have touted the use of 10 m class telescopes to significantly improve the efficiency of nova detections in galaxies beyond the Local Group. In a pilot program designed to test the improvements in nova detection efficiency made possible with the latest generation of large optical telescopes, Della Valle and Gilmozzi (2002) used the Very Large Telescope (VLT) to detect four novae in NGC 1316 (Fornax A) during nine nights between December 1999 and January 2000. Although the VLT observations were not dense enough to permit a distance determination through the MMRD relation, Della Valle and Gilmozzi were able to use the M_{15} method to constrain the distance modulus of the galaxy to $31.3 \pm 0.25 < (m - M)_0 < 31.5 \pm 0.25$. NGC 1316 now holds the record as the most remote galaxy whose distance has been estimated through observations of classical novae.

As is the case with many other approaches, a potential concern when using novae to measure extragalactic distances involves the Malmquist (1922) bias. In a magnitude-limited sample of novae in a distant galaxy, only the most luminous novae will be observed. This bias appears to pose a particular problem for novae, as the existence of a class of *super-bright* novae, which are not well characterized by the MMRD relation, has been noted by several authors (van den Bergh & Pritchet, 1986; Della Valle, 1991; Shara & Zurek, 2002). The existence of this bright nova population is readily revealed as a bright tail in the integrated nova luminosity function of M31 (see Figure 14.9). It is clear that any attempt to use novae for measuring extragalactic distances should make sure that the observations extend sufficiently far down the luminosity function to characterize the MMRD relation properly.

14.5 Directions for future studies

Despite the promise (or lack thereof) of using novae as extragalactic distance indicators, recent years have seen the emphasis of extragalactic novae studies shift from the use of novae as distance indicators to the prospect of using them to probe the evolution of interacting close binary stars in differing stellar populations. In this context, it is the study of the aggregate properties of novae from galaxies with differing stellar populations that has taken center stage. Indeed, the central question that must be answered before meaningful progress can be made in our understanding of extragalactic novae is whether the nova rate and the distribution of speed classes depend on stellar population. An understanding of the speed class distribution is perhaps more fundamental since the computation of nova rates typically depends on a knowledge of the mean light-curve properties in the galaxy under study. Current understanding of variations in speed class between different stellar populations, which comes primarily from the samples of light curves from the LMC and the bulge of M31, needs to be improved. Not only would a larger sample of light curves improve the situation, an extensive sample of light curves from a purely Population II environment would be helpful, since the sample of novae attributed to the bulge of M31 is contaminated by an unknown number of disk novae projected onto M31's bulge (Hatano *et al.*, 1997a,b). An ideal target would be one of the massive elliptical galaxies in Virgo such as M49 or M87, which have high absolute nova rates. In view of the potential effect of galaxy mergers discussed earlier (Della Valle & Panagia, 2003), and the possible role of M87's jet in enhancing the nova rate (Livio, Riess & Sparks, 2002), it would useful to include other radio-quiet Virgo ellipticals as well. A larger sample of Population I light curves should become available as by-products of various microlensing surveys toward the Magellanic Clouds in the MACHO and EROS2 surveys (Alcock *et al.*, 2000; Lasserre, 2002), and toward M31 in the POINT-AGAPE survey (Darnley, 2004, 2006).

Once variations in the speed class distribution between galaxies are better understood, not only will measurements of absolute nova rates become more reliable, but it will be possible to explore observationally the effect the metallicity of the accreted material has on the nova outburst, and to test the universality of the MMRD relation. Until recently the consensus has been that, for a given white dwarf mass, the properties of the eruption were expected to be relatively insensitive to the metallicity of the accreted material (e.g. Livio & Truran, 1994). However, calculations by Starrfield *et al.* (2000; see also Chapter 4) have shown that increasing the opacity of the accreted material will reduce the amount of material accreted prior to the thermonuclear runaway, presumably leading to a 'faster' nova. At this juncture there are reasons to expect that the MMRD relation will be shown to be universal. As pointed out by Gilmozzi and Della Valle (2003), although an older stellar population (characterized by nova binaries having a lower mean white dwarf mass) may produce generally fainter novae that erupt less frequently compared with a young stellar population, the zero-point of the MMRD relation should not be affected. The only difference between nova properties should be the relative proportion of fast (bright) and slow (faint) novae.

Finally, the question of whether the LSNR depends on the underlying stellar population will require improved confidence in the completeness of the surveys. Clearly, if past surveys have missed a significant fraction of novae either through problems with extinction in the spiral and irregular systems, or through overlooking a population of fast novae in surveys with poor temporal sampling, or both, then any conclusions regarding extragalactic nova rates, and how they may or may not depend on stellar population, will be suspect. Continuous-coverage surveys, like that carried out for M81 by Neill and Shara (2004), will ensure that unusually fast novae are not missed. However, the problem of extinction in the dusty environments encountered in spiral and irregular systems is likely to remain a problem in perhaps all but the deepest surveys. Future infrared imaging surveys offer a promising step forward in this regard. At present, however, the question of whether the rate of nova production varies with stellar population, and hence the Hubble type of the galaxy, remains unclear.

Acknowledgements

I am grateful to R. Ciardullo and M. Della Valle for a thorough reading of the manuscript, and for their insightful comments that helped improve this review.

References

Alcock, C., Allsman, R. A., Alves, D. *et al.*, 2000, *Ap. J.*, **542**, 281.
Arp, H. C., 1956, *AJ*, **61**, 15.
Baade, W., 1944, *Ap. J.*, **100**, 137.
Baade, W., & Zwicky, F., 1934, *Phys. Rev.*, **45**, 138.
Buscombe, W., & de Vaucouleurs, G., 1955, *Observatory*, **75**, 170.
Capaccioli, M., Della Valle, M., Rosino, L., & D'Onofrio, M., 1989, *AJ*, **97**, 1622.
Capaccioli, M., Della Valle, M., D'Onofrio, M., & Rosino, L., 1990, *Ap. J.*, **360**, 63.
Ciardullo, R., Shafter, A. W., Ford, H. C. *et al.*, 1983, *Ap. J.*, **356**, 472.
Ciardullo, R., Ford, H. C., Neill, J. D., Jacoby, G. H., & Shafter, A. W. 1987. *Ap. J.*, **318**, 520.
Ciardullo, R., Tamblyn, P., Jacoby, G. H., Ford, H. C., & Williams, R. E., 1990a, *AJ*, **99**, 1079.
Ciardullo, R., Shafter, A. W., Ford, H. C. *et al.*, 1990b, *Ap. J.*, **356**, 472.
Ciardullo, R., Tamblyn, P., & Phillips, A. C., 1990, *PASP*, **102**, 1113.
Clark, G. W., 1975, *Ap. J.*, **199**, L143.
Cohen, J. G., 1985, *Ap. J.*, **292**, 90.
Crampton, D., Hutchings, J. B., Cowley, A. P., Schade, D. J., & van Speybroeck, L. P., 1984, *Ap. J.*, **284**, 663.
Darnley, M. J., Bode, M. F., Kerins, E. J. *et al.*, 2004, *MNRAS*, **353**, 571.
Darnley, M. J., Bode, M. F., Kerins, E. J. *et al.*, 2006, *MNRAS*, **369**, 257.
Della Valle, M., 1988, in *The Extragalactic Distance Scale*, ed. S. van den Bergh, & C. J. Pritchet. San Francisco: ASP Conference Series, p. 73.
Della Valle, M., 1991, *A&A*, **252**, L9.
Della Valle, M., 1995, in *Cataclysmic Variables*, ed. A. Bianchini, M. Della Valle, & M. Orio. Dordrecht: Kluwer, p. 503.
Della Valle, M., 2002, in *Classical Nova Explosions*, ed. M. Hernanz, & J. José. Melville: American Institute of Physics, p. 443.
Della Valle, M., & Livio, M., 1995, *Ap. J.*, **452**, 704.
Della Valle, M., & Livio, M., 1998, *Ap. J.*, **506**, 818.
Della Valle, M., & Gilmozzi R., 2002, *Science*, **296**, 1275.
Della Valle, M., & Panagia, N., 2003, *Ap. J.*, **587**, L71.
Della Valle, M., Bianchini, A., Livio, M., & Orio, M., 1992, *A&A*, **266**, 232.
Della Valle, M., Rosino, L., Bianchini, A., & Livio, M., 1994, *A&A*, **287**, 403.
de Kool, M., 1992, *A&A*, **261**, 188.
de Vaucouleurs, G., 1978, *Ap. J.*, **223**, 351.
Disney, M., Davies, J., & Phillipps, S., 1989, *MNRAS*, **239**, 939.
Di Stefano, R., Kong, A. K. H., Garcia, M. R. *et al.* 2002, *Ap. J.*, **570**, 618.
Downes, R. A., & Duerbeck, H. W., 2000, *AJ*, **120**, 2007.
Duerbeck, H. W., 1981, *PASP*, **93**, 165.
Duerbeck, H. W., 1984, *Ap. &SS*, **99**, 93.
Duerbeck, H. W., 1990, in *Physics of Classical Novae*, ed. A. Cassatella, & R. Viotti. Berlin: Springer, p. 34.
Edmonds, P. D., Gilliland, R. L., Heinke, C. O., & Grindlay, J. E., 2003, *Ap. J.*, **596**, 1177.
Ferrarese, L., Côté, P., & Jordán, A., 2003, *Ap. J.*, **599**, 1302.
Ferrarese, L., Livio, M., Freedman, W. *et al.*, 1996, *Ap. J.*, **468**, 95.
Ferrarese, L., Mould, J. R., Kennicutt, R. C. *et al.*, 2000, *Ap. J.*, **529**, 745.

Freedman, W. L., Madore, B. F., Gibson, B. K., *et al.* 2001, *Ap. J.*, **553**, 47.
Gilmozzi, R., & Della Valle, M., 2003, in *Stellar Candles for the Extragalactic Distance Scale*, ed. D. Alloin, & W. Gieren. Berlin: Springer, p. 229.
Graham, J. A., 1979, in *Changing Trends in Variable Star Research*, ed. F. M. Bateson, J. Smak, & I. H. Urch. Hamilton NZ: University of Waikato, p. 96.
Gwinn, C. R., Moran, J. M., & Reid, M. J., 1992. *Ap. J.*, **393**, 149.
Hartwig, E., 1885, *AN*, **112**, 355.
Hatano, K., Branch, D., Fisher, A., & Starrfield, S., 1997a, *MNRAS*, **290**, 113.
Hatano, K., Branch, D., Fisher, A., & Starrfield, S., 1997b, *Ap. J.*, **487**, L45.
Heinke, C. O., Grindlay, J. E., Edmonds, P. D. *et al.*, 2003, *Ap. J.*, **598**, 516.
Hertz, P. & Grindlay, J. E., 1983, *Ap. J.*, **267**, 83.
Hogg, H. S., & Wehlau, A., 1964, *AJ*, **69**, 141.
Hubble, E. P., 1929, *Ap. J.*, **69**, 103.
Jacoby, G. H., Branch, D., Ciardullo, R. *et al.*, 1992, *PASP*, **104**, 599.
Katz, J. I., 1975, *Nature*, **253**, 698.
Knigge, C., Zurek, D. R., Shara, M. M., & Long, K. S., 2002, *Ap. J.*, **579**, 752.
Kopylov, I. M., 1952, *Izu. Krym. Astrofiz. Obs.*, **9**, 116.
Kopylov, I. M., 1955, *A. Zh.*, **32**, 48.
Kraft, R. P., 1964a, *Ap. J.*, **139**, 457.
Kraft, R. P., 1964b, *Astron. Soc. Paci. Leaflets*, **9**, 137.
Lasserre, T., 2002, *Prog. Particle Nucl. Phys.*, **48**, 289.
Livio, M., 1992, *Ap. J.*, **393**, 516.
Livio, M., 1997, in *The Extragalactic Distance Scale*, ed. M. Livio. Cambridge: Cambridge University Press, p. 186.
Livio, M., & Truran, J. W., 1994, *Ap. J.*, **425**, 797.
Livio, M., 2000, in *Type Ia Supernovae, Theory and Cosmology*, ed. J. C. Niemeyer, & J. W. Truran. Cambridge: Cambridge University Press, p. 33.
Livio, M., Riess, A., & Sparks, W., 2002, *Ap. J.*, **571**, 99L.
Lundmark, K., 1922, *PASP*, **34**, 225.
Lundmark, K., 1923, *PASP*, **35**, 95.
Luther, R., 1860, *AN*, **53**, 293.
Malmquist, K. G., 1922, *Ark. Math. Astr. Fys.*, 16, No. 23, *Medd. Lund. Obs.* (I), No. 100.
Margon, B., Grandi, S. A., Stone, R. P. S., & Ford, H. C., 1979, *Ap. J.*, **233**, L63.
Matteucci, F., Renda, A., Pipino, A., & Della Valle, M., 2003, *A&A*, **405**, 23.
McLaughlin, D. B., 1945, *PASP*, **57**, 69.
McMillan, S. L. W., 1986, *Ap. J.*, **307**, 126.
Neill, J. D., & Shara, M. M., 2004, *AJ*, **127**, 816.
Neill, J. D., & Shara, M. M., 2005, *AJ*, **129**, 1873.
Patterson, J., 1984, *Ap. JS*, **54**, 443.
Pfau, W., 1976, *A&A*, **50**, 113.
Pickering, E. C., & Fleming, W. P., 1896, *Ap. J.*, **3**, 162.
Plaut, L., 1965, in *Galactic Structure*, ed. A. Blaauw, & M. Schmidt. Chicago: University of Chicago Press, p. 311.
Pogson, N., 1860, *MNRAS*, **21**, 32.
Politano, M., 1996, *Ap. J.*, **456**, 338.
Pooley, D., Lewin, W. H. G., Homer, L. *et al.*, 2002, *Ap. J.*, **569**, 405.
Prialnik, D., & Kovetz, A., 1995, *Ap. J.*, **445**, 789.
Prialnik, D., Livio, M., Shaviv, G., & Kovetz, A., 1982, *Ap. J.*, **257**, 312.
Pritchet, C. J., & van den Bergh, S., 1987, *Ap. J.*, **318**, 507.
Rappaport, S., & Di Stefano, R., 1994, *Ap. J.*, **437**, 733.
Renzini, A., & Buzzoni, A., 1986, in *Spectral Evolution of Galaxies*, ed. C. Chiosi, & A. Renzini. Dordrecht: Reidel, p. 195.
Ritchey, G. W., 1917a, *PASP*, **29**, 210.
Ritchey, G. W., 1917b, *PASP*, **29**, 257.
Rosino, L., 1964, *Ann. Astrophys.*, **27**, 498.
Rosino, L., 1973, *A&AS*, **9**, 347.
Rosino, L., Capaccioli, M., D'Onofrio, M., & Della Valle, M., 1973, *AJ*, **97**, 83.

Sandage, A., & Tammann, G. A., 1987, *A Revised Shapley-Ames Catalog of Bright Galaxies (Publ. 635)*. Washington DC: Carnegie Institution.
Sawyer, H. B., 1938, *JRASC*, **32**, 69.
Schmidt-Kaler, Th., 1957, *Z. Ap.*, **41**, 182.
Shafter, A. W., 2002, in *Classical Nova Explosions*, ed. M. Hernanz, & J. José. Melville: American Institute of Physics, p. 462.
Shafter, A. W., & Irby, B. K., 2001, *Ap. J.*, **563**, 749.
Shafter, A. W., Ciardullo, R., & Pritchet, C. J., 2000, *Ap. J.*, **530**, 193.
Shapley, H., 1917, *PASP*, **29**, 213.
Shara, M. M., 1981, *Ap. J.*, **243**, 926.
Shara, M. M., & Zurek, D. R., 2002, in *Classical Nova Explosions*, ed. M. Hernanz, & J. José. Melville: American Institute of Physics, p. 457.
Shara, M. M., Prialnik, D., & Shaviv, G., 1980, *Ap. J.*, **239**, 586.
Shara, M. M., Sandage, A., & Zurek, D. R., 1999, *PASP*, **111**, 1367.
Sharov, A. S., 1993, *AstL*, **19**, 230.
Sharov, A. S., & Alksnis, A., 1991, *Ap. &SS*, **180**, 273.
Starrfield, S., Truran, J. W., Sparks, W. M., & Kutter, G. S., 1972, *Ap. J.*, **176**, 169.
Starrfield, S., Sparks, W. M., Truran, J. W., & Wiescher, M. C., 2000, *Ap. JS*, **127**, 485.
Strittmatter, P. A., Woolf, N. J., Thompson, R. I., *et al.*, 1977, *Ap. J.*, **216**, 23.
Tomaney, A. B., 1992, Unpublished Ph.D. thesis, University of Texas, Austin.
Tomaney, A. B., & Shafter, A. W., 1992, *Ap. JS*, **81**, 683.
Tonry, J. L., Dressler, A., Blakeslee, J. P., *et al.*, 2001, *Ap. J.*, **546**, 681.
Tutukov, A., & Yungelson, L., 1995, in *Cataclysmic Variables*, ed. A. Bianchini, M. Della Valle, & M. Orio. Dordrecht: Kluwer, p. 495.
van den Bergh, S., 1981, *JRASC*, **75**, 169.
van den Bergh, S., 1988, *PASP*, **100**, 8.
van den Bergh, S., & Pritchet, C. J., 1986, *PASP*, **98**, 110.
van den Bergh, S., & Younger, P. F., 1987, *A&AS*, **70**, 125.
Van Maanen, A., 1916, *Ap J*, **44**, 210.
Warner, B., 1995, *Cataclysmic Variable Stars*. Cambridge: Cambridge University Press.
Warner, B., & Nather, R. E., 1971, *MNRAS*, **152**, 219.
Wenzel, W., & Meinunger, I., 1978, *AN*, **299**, 239.
Williams, R. E., 1992, *AJ*, **104**, 725.
Williams, S. J., & Shafter, A. W., 2004, *Ap. J.*, **612**, 867.
Yungelson, L., Livio, M., & Tutukov, A., 1997, *Ap. J.*, **481**, 127.
Zwicky, 1936, *PASP*, **48**, 191.
Zwicky, 1942, *Ap. J.*, **96**, 28.

Object index

Page numbers in bold font refer to entries in a table or figure.

Miscellaneous nomenclature

Subject index

Page numbers in bold font refer to entries in a table or figure.